ANALYTICAL CHEMISTRY SYMPOSIA SERIES - volume 9

affinity chromatography and related techniques

theoretical aspects/industrial and biomedical applications

Proceedings of the 4th International Symposium, Veldhoven, The Netherlands, June 22–26, 1981

ANALYTICAL CHEMISTRY SYMPOSIA SERIES

Volume 1 Recent Developments in Chromatography and Electrophoresis. Proceedings of the 9th International Symposium on Chromatography and Electrophoresis, Riva del Garda, May 15–17, 1978
edited by A. Frigerio and L. Renoz

Volume 2 Electroanalysis in Hygiene, Environmental, Clinical and Pharmaceutical Chemistry, Proceedings of a Conference, organised by the Electroanalytical Group of the Chemical Society, London, held at Chelsea College, University of London, April 17–20, 1979
edited by W.F. Smyth

Volume 3 Recent Developments in Chromatography and Electrophoresis, 10. Proceedings of the 10th International Symposium on Chromatography and Electrophoresis, Venice, June 19–20, 1979
edited by A. Frigerio and M. McCamish

Volume 4 Recent Developments in Mass Spectrometry in Biochemistry and Medicine, 6. Proceedings of the 6th International Symposium on Mass Spectrometry in Biochemistry and Medicine, Venice, June 21–22, 1979
edited by A. Frigerio and M. McCamish

Volume 5 Biochemical and Biological Applications of Isotachophoresis. Proceedings of the First International Symposium, Baconfoy, May 4–5, 1979
edited by A. Adam and C. Schots

Volume 6 Analtyical Isotachophoresis. Proceedings of the 2nd International Symposium on Isotachophoresis, Eindhoven, September 9–11, 1980
edited by F.M. Everaerts

Volume 7 Recent Developments in Mass Spectrometry in Biochemistry, Medicine and Environmental Research, 7. Proceedings of the 7th International Symposium on Mass Spectrometry in Biochemistry, Medicine and Environmental Research, Milan, June 16–18, 1980
edited by A. Frigerio

Volume 8 Ion-selective Electrodes. Proceedings of the Third Symposium, Mátrafüred, October 13–15, 1980
edited by E. Pungor

Volume 9 Affinity Chromatography and Related Techniques. Theoretical Aspects/ Industrial and Biomedical Applications. Proceedings of the 4th International Symposium, Veldhoven, The Netherlands, June 22–26, 1981
edited by T.C.J. Gribnau, J. Visser and R.J.F. Nivard

ANALYTICAL CHEMISTRY SYMPOSIA SERIES - volume 9

affinity chromatography and related techniques

theoretical aspects/industrial and biomedical applications

Proceedings of the 4th International Symposium, Veldhoven, The Netherlands, June 22–26, 1981

edited by

T.C.J. Gribnau
Scientific Development Group, Organon International B.V., Oss, The Netherlands

J. Visser
Department of Genetics, Agricultural University, Wageningen, The Netherlands

and

R.J.F. Nivard
Department of Organic Chemistry, Katholieke Universiteit, Nijmegen, The Netherlands

ELSEVIER SCIENTIFIC PUBLISHING COMPANY
Amsterdam – Oxford – New York 1982

ELSEVIER SCIENTIFIC PUBLISHING COMPANY
Molenwerf 1
P.O. Box 211, 1000 AE Amsterdam, The Netherlands

Distributors for the United States and Canada:

ELSEVIER/NORTH-HOLLAND INC.
52, Vanderbilt Avenue
New York, N.Y. 10164

Library of Congress Cataloging in Publication Data
Main entry under title:

Affinity chromatography and related techniques.

(Analytical chemistry symposia series ; v. 9)
Includes index.
1. Affinity chromatography--Congresses. 2. Affinity chromatography--Industrial applications--Congresses. 3. Biological chemistry--Technique--Congresses. 4. Chemistry, Clinical--Technique--Congresses. I. Gribnau, T. C. J. II. Visser, J., Ir. III. Nivard, R. J. F. (Rutger Jan Frans) IV. Series. [DNLM: 1. Chromatography, Affinity--Congresses. W3 AN54 v.9 / QU 25 A255 1981]
QP519.9.A35M33 574.19'285 81-15255
ISBN 0-444-42031-2 AACR2

ISBN 0-444-42031-0 (Vol. 9)
ISBN 0-444-41786-9 (Series)

Printed in The Netherlands

CONTENTS

CONTENTS

* manuscript not available

* manuscript not available

ORGANIZING COMMITTEE

Prof. Dr. E.J. Ariëns
Department of Pharmacology
Faculty of Medicine
Katholieke Universiteit
Nijmegen - The Netherlands

Prof. Dr. H. Bloemendal
Department of Biochemistry
Faculty of Science
Katholieke Universiteit
Nijmegen - The Netherlands

Dr. T.C.J. Gribnau (secretary)
Scientific Development Group
Organon International B.V.
Oss - The Netherlands

Prof. Dr. R.J.F. Nivard
Department of Organic Chemistry
Faculty of Science
Katholieke Universiteit
Nijmegen - The Netherlands

Mr. P.J.M. Toll (treasurer)
Faculty of Science
Katholieke Universiteit
Nijmegen - The Netherlands

Dr. J. Visser
Department of Genetics
Landbouwhogeschool
Wageningen - The Netherlands

ACKNOWLEDGEMENTS

The Organizing Committee gratefully acknowledges the assistance of the Secretarial and Technical Staff of Organon International B.V. (Scientific Development Group - Biochemical Research and Development Laboratories) and of the Katholieke Universiteit Nijmegen (Faculty of Science). Special thanks are due to Mrs. J.H.M. Romme.

Thanks are also expressed to the following Organizations for their financial support of the Symposium:

Akzo N.V.
Albic B.V.
Amstelstad B.V.
Amicon B.V.
Applied Science Europe B.V.
Boehringer Mannheim B.V.
Boehringer Mannheim GmbH
Burroughs Wellcome Co.
Chrompack Nederland B.V.
Diosynth B.V.
Hoffmann-La Roche Inc.
IBM Nederland N.V.
KabiVitrum AB
Katholieke Universiteit Nijmegen
Koninklijke Luchtvaart Maatschappij N.V.
LKB-Produkten B.V.
E. Merck Nederland B.V.
Miles Nederland
Ministerie van Onderwijs en Wetenschappen
Nederlands Congres Bureau
Organon International B.V.
Pharmacia Fine Chemicals AB
Pierce Eurochemie B.V.
Réactifs IBF-Pharmindustrie
Röhm Pharma GmbH
Société Chimique Pointet Girard
Unilever N.V.
Waters Associates B.V.

Vignette design by Ernst van Cleef (Agricultural University, Wageningen, The Netherlands).

PREFACE

In this book we have collected the lectures and the titles of the posters presented at the 4th International Symposium on Affinity Chromatography and Related Techniques which was organized in the Conference Centre "De Koningshof" in Veldhoven, The Netherlands, from June 22 to 26, 1981.

The meeting was attended by about 300 participants representing 24 countries inside and outside of Europe. Approximately 40% of them were affiliated with chemical, biochemical or pharmaceutical industries. Since one of the main goals of this Symposium was to stimulate the interaction between industrial and academic research, we feel from this figure that the conditions have been favourable for such a process.

The past has shown that the development of affinity chromatographic techniques has been largely induced by empirical studies. The consideration of theoretical aspects was emphasized therefore at this meeting, and represented the first topic of the scientific programme. As a second subject, the preparation and properties of polymeric matrices and methods for ligand immobilization were reviewed from different points of view, and new developments in this field were reported. Attention was also focused on industrial and biomedical applications, the third major theme of this meeting. Almost a third of all lectures was presented by scientists from industry, which indicates that affinity techniques have gradually found their way outside the university laboratories.

In our opinion it has been beneficial that several speakers, although themselves being no experts in affinity chromatography, were invited to cover some of the border areas between organic chemistry, physical chemistry and biochemistry, in relation to this separation technique. Examples can be found in contributions about the thermodynamical and mechanical forces involved in molecular interactions, about the chemical reactivity of natural and synthetic polymers and about the structure of natural polysaccharides.

This Symposium has furthermore shown the feasibility of various large scale applications mainly in isolating and purifying blood plasma proteins and vaccines. The development of new carrier materials, both synthetic and natural, continues to be a topic of interest. Progress, crucial for new industrial and analytical applications, could be observed.

Solving problems in science takes time: almost 15 years elapsed between the publication of the CNBr activation procedure and the final unravelling of the underlying chemistry. The chemistry of reactive dyes, and their application, is well known already from the textile industry. These dyes do not require an additional activation of the matrix. They serve as readily available and cheap ligands, and will certainly have a tremendous impact on new developments in affinity chromatography. Lectures and posters presented numerous examples of analytical and preparative (column chromatography/affinity partition) applications, which involved both nucleic acids and proteins.

Materials and methods based on affinity principles are also widely applied in diagnostic research. The use of immobilized reagents in immunoassays was reviewed and progress with respect to the development of new labels was reported. Extracorpor blood treatment with affinity adsorbents or immobilized enzymes, and the synthesis a application of selective, polymer-bound drugs, were discussed as two other promisin but still rather tricky, biomedical applications.

The development of High Performance Liquid Affinity Chromatography (HPLAC) ju began to emerge when the first preparations for this conference were made. From data presented at this meeting it seems to become a fertile hybrid of two existing techniqu promising new horizons for affinity chromatography.

T.C.J. Gribnau

J. Visser

R.J.F. Nivard

OPENING ADDRESS BY THE NETHERLANDS MINISTER FOR SCIENCE POLICY, Dr. Ir. A.A.Th.M. VAN TRIER, AT THE FOURTH INTERNATIONAL SYMPOSIUM ON AFFINITY CHROMATOGRAPHY AND RELATED TECHNIQUES AT VELDHOVEN ON JUNE 22, 1981

Mr. Chairman, Ladies and Gentlemen,

Speaking on behalf of the Dutch government, I should like first of all to join Dr. Gribnau in welcoming you here. Like your Dutch fellow researchers, we are very honoured that you have chosen the Netherlands as host country for the fourth international symposium on affinity chromatography and related techniques, since conferences of this nature are not only important to the people directly involved, but tend generally to exert a healthy influence on the scientific climate of the country.

I must confess that the specialist subject which is to occupy you for the rest of the week is pretty much a closed book to me. However, from what little I have been told about it, I understand that affinity chromatography is a fairly recent but very promising technique which could lead to significant advances in various areas of the life sciences. This fourth symposium offers again an opportunity for a review of the present state of affairs in your special area from which to draw inspiration for further development of the technique and its applications. Your findings will also benefit research and development in related fields. Symposia like this one are important in another respect, too, in that they come about through close collaboration between the academic world and industry. It is not so long ago that relations between part of our academic community and the private sector left much to be desired, but in the past few years there has been a definite improvement. People at our universities are becoming increasingly interested in industry and commerce, and increasingly open to them, as evidenced again by this symposium. It is an encouraging development and one which is also strongly supported by the government. In your own field it is particularly encouraging because up to about five years ago biochemistry, unlike other branches of chemistry, offered little in the way of industrial applications. Now we see a very different picture and the changes have come faster than many people predicted. Improvements in affinity chromatography have made their contributions and will continue to do so.

Until recently affinity chromatography was regarded more or less as an isolated purification technique, the biggest advantages of which included a high degree of precision, sensitivity and reliability, although one major drawback was that it could only be used on a small scale. This now seems to be changing: the opportunities fo use on a larger scale are becoming increasingly real, thanks partly to progress in ot specialist fields within the life sciences, in which this technique is increasingly becoming involved. I could name for example, recombinant DNA technology, or hyb doma technology which offers the possibility of producing large quantities of pure ar bodies of a desired type.

There can be no doubt that affinity chromatography will benefit from the progress made in these and related areas. A direct example can be found in the fact that pure and specific proteins - notably antibodies - and other natural substances are now wi available.

An indirect benefit accrues from the fact that these technologies give us better insight into the relationship between the structure and the effect of interactions betw one macromolecule and another and between macromolecules and other substances. reverse is equally true, as I have already pointed out: certain areas of the life scien can benefit greatly from continuing progress in affinity chromatography. As a separa and detection technique it can be a great aid to diagnosis and to methods of biocher analysis in general. Other valuable applications which you will be discussing lie in medical field for example extracorporal blood treatment. Medicine is in fact the fir field in which, thanks to the progress made in affinity chromatography, a start can made with gradually working at larger scales. For these types of valuable applicatic the present high cost of scaling up operations is justified by the enormous benefits.

One other very important field of applications is of course biotechnology. In this area we can benefit from the fact that by working with this separation technique knowledge can be obtained about the immobilisation of substances on solid support and about the way such immobilisation affects the physical and chemical behaviou of the substance.

In other fields than biotechnology, this knowledge has enabled us in recent year to make considerable progress in obtaining increasingly specific and accurate and e smaller quantities of a particular material. It now looks as if affinity chromatograp has reached such a stage, that it can also contribute to the development of method for large-scale or even industrial applications. The specialised subjects in the life

sciences which I have referred to, each of which represents a field of application for affinity chromatography, are also receiving a great deal of government attention. I can illustrate this with a single example from the field of biotechnology.

Just under a month ago, the Dutch government, after extensive consultations and preparations, set up a committee to be responsible for the final elaboration and supervision of an "innovation-oriented research programme" for biotechnology. The committee is composed of representatives of the research institutions, industry and government. The innovation programme for biotechnology is the first in a series of such programmes for various fields of science which are expected to make a strong contribution to technological innovation in the Netherlands. The aim of the programmes is to key Government-financed research in such fields more clearly to the needs of society and industry. They are medium to long-term programmes, in which cooperation between companies, or groups of companies and universities or other (semi-)governmental research institutions is an essential element.

Universities will be involved in the innovation programme for biotechnology in two ways. The plan is initially to appoint three of them to stimulate interdisciplinary biotechnological research, with a view to achieving the best possible integration between a number of disciplines in the life sciences and process technology.

In addition, fundamental sciences like microbiology, molecular genetics and biochemistry which are all important to biotechnology will continue to receive stimulation from the second flow of funds chanelled through the Netherlands Organisation for the Advancement of Pure Research, ZWO.

In addition to programmes for research in agriculture and environmental protection, a campaign is also in preparation for industry-oriented research. The Netherlands Organisation for Applied Scientific Research (TNO), a state supported institution, which with 5,000 employees is the largest research establishment in the Netherlands, will play an important part in the campaign. It has already long been involved in biotechnological projects, and will in this instance be concerned mainly with biomolecular engineering and process technology.

The innovation-oriented research programme for biotechnology, which by the way is not the only government initiative in this area, will in my opinion meet a strongly felt need for coordination in this branch of scientific research. I also expect that it will help orientate public research more towards the needs of industry. Industry will have the job of using the opportunities offered to good effect.

I hope, ladies and gentlemen, that this symposium will in its own way fulfil a cor parable integrating function and that it will, if you will forgive the expression, promo affinities among you. From your different disciplines and branches of research it can bring you closer to your common goal, namely to obtain a better understanding of affinity chromatography and to extend its applications.

I wish you all a pleasant stay in the Netherlands and above all a fruitful exchange of ideas.

Thank you.

Dr. Ir. A.A.Th.M. van Trier

INTRODUCTION

INTERNATIONAL SYMPOSIUM
ON AFFINITY CHROMATOGRAPHY
AND RELATED TECHNIQUES

T.C.J. Gribnau, J. Visser and R.J.F. Nivard (Editors),
Affinity Chromatography and Related Techniques

AFFINITY CHROMATOGRAPHY-HISTORICAL SURVEY- PRESENT STATUS - FUTURE ASPECTS*

Jerker PORATH**
Groupe de Neurobiochimie Cellulaire et Moléculaire, U.E.R. de Biochimie, Université Pierre et Marie Curie, 96 boulevard Raspail, 75006 PARIS (France)

The term "affinity chromatography" was originally proposed for chromatography based on biological recognition (1). This is unfortunate and not satisfactory since the word "affinity" has a much broader meaning and was used for more than hundred years.Not surprisingly and quite logically the concept of affinity chromatography has now been broadened to include, besides biospecific affinity,hydrophobic affinity, metal ion affinity and other types of interactions. At this Symposium, we are covering affinity chromatography in a broad sense.

Biological recognition phenomena were observed and studied more than 90 years ago by H. Stillmark and P. Ehrlich. The discoveries were made as a consequence of separation of biological matter and they lead logically to a principle for fractionation. It is also evident that precipitation was the first bioaffinity separation method to be applied. In retrospect it may seem strange that bioaffinity dependent adsorption-desorption methods were not elaborated to a high degree of perfection already at the beginning of this century.

However, attempts were made at that time starting in 1910 when Starkenstein purified amylase on starch. Around 1950 Campbell and Lerman prepared the first cellulose based affinity adsorbents. The development of the charge-free highly permeable molecular sieves, Sephadex (2) and agarose (3) and their conversion to adsorbents and immobilized enzymes (4,5) opened the way to modern affinity chromatography (6). The discovery of the cyanogen bromide coupling was also a major contributing factor (4). Dr M. Wilcheck,one of the originators of modern affinity chromatography, and R. Schwyzer's group including Dr Gribnau have made important contributions to the understanding of the limitations in the practical use of the cyanogen bromide reaction with hydroxylic polymers. Other coupling methods developed before and after

---

*This contribution is dedicated to Professor Georg Manecke, Freie Universität, Berlin, on his 65 years'birthday.

**Permanent adress : Institute of Biochemistry, Uppsala University Biomedical Center, Box 576, Uppsala, Sweden.

the breakthrough in the field are necessary complements especially where the CNBr-coupling is not applicable or fails to give satisfactory products (7;8).

I do not intend to give a detailed historical review but will instead refer to an article by L. Sundberg and myself (9). To save time and space I will also refer to the preceeding three international Symposia and to some of the excellent books (10-16) which cover much of the contemporary state of the art. The following will contain some personal views on the topic.

The modern development of affinity chromatography is closely connected with the elaboration of methods for preparing and studying immobilized enzymes. The work tha my collaborators and I made in the field was, at least on my part, greatly influenced by the important contributions by Ephraim Katchalski-Katzir and Georg Manecke. By the development of hydrophobic type immobilized enzymes Georg Manecke was ahead of his time. His contributions will be more clearly recognized when organic solvents become more frequently exploited in biotechnology. During this period, I became soon convinced of the need for improved hydrophilic supports to cope with our problems namely protein fractionation and enzyme immobilization in aqueous medi There was also a need for more efficient methods for attaching ligands to the supports.

Agarose,including improved gels such as CL-Sepharose, Ultrogels, Magnogels and Biogels while approaching the ideal matrix, does not meet the demand for low compressibility and elasticity required for a material to be used in high pressure chromatography. Silica gels and porous glass can withstand high pressure but suffe from other shortcomings. For example, they are not resistant to alkali and undesired protein adsorption cannot be totally avoided. However, the introduction of "glucophase" bonded silica by F.E. Regnier and collaborators is an important improvement (17). Other supports such as Spheron and Toyopearl can be used in high pressure chromatography. However, the risk for uncontrolled protein adsorption to these supports has not yet been eliminated.

Efforts to improve chromatographic performance are presently directed to submicrogram analysis of biogenic materials. To save time, high pressure techniques are being rapidly developed. However, there is also a need to extend affinity chromatography at the other end of the scale upwards for possible replacement of old-fashened less efficient purification procedures in laboratories and industrial plants. This interesting branch of affinity chromatography will be treated in detail at this Symposium.

Bioaffinity methods are particularly attractive for large scale operations since they require comparatively small quantities of a contacting insoluble or immiscible phase containing "affinity ligands" (an affinity adsorbent or a soluble affinity polymer). Such batch-wise operations may be better suited for handling extracts of tons of raw materials. Large scale affinity procedures should make many biochemicals less expensive (or at least counteract the present inflation-caused persistant rise in expenses for biochemicals which impedes progress in biosciences). They should also for the same reason open new areas of biotechnology. In addition, new substances present in living matter in extreme dilutions may be discovered.

At the "Separation Center Unit" in Uppsala my collaborators, Drs B. Ersson and K.J. Brink, are isolating 10-100 grams of lectins in one-step operations from 10-1000 liters of extract. Frequently, but not always, the extract can be contacted with an affinity adsorbent without prefractionation. Sometimes, a simple ammonium sulfate precipitation improves chromatography or prevents damage of the gel. The lectins are rapidly concentrated and purified in single operations from hundreds of liters of extract using a volume of the gel which may be of the order of one percent or less than that of the sample. Inexpensive adsorbents for purification are prepared from agarose by activating the gel with divinylsulfone followed by coupling of the desired sugar or aminosugar. The isolated lectins are coupled to Sepharose and used by Dr T. Kristiansen and collaborators to produce "split vaccines" and other immunogenic viral coat components. In this work, selective affinity desorption is frequently used to increase the efficiency of purification and to avoid harsh treatment. With proper handling the adsorbents may be used repeatedly over periods of months or years.

Group specific adsorbents are usually less expensive than those containing highly selective ligands. In general, therefore, it may be more practical to use a stepwise purification process starting with group fractionation to obtain subfractions of smaller volumes making it easier to subject the extracts to an affinity adsorbent of higher specificity. Selective desorption is also easier on a small scale, especially if expensive affinity displacers are to be used, or if the chromatogram is to be developed by pH or salt gradient elution.

Sofisticated group affinity methods are likely to be more frequently used in the future. They are based on some particular interaction parameter such as hydrophobicity or charge-charge interaction. Affinity chromatography employing general ligands is based on the chemical nature of the ligands rather than on a specified type of physico-chemical interaction with the solute(s) to be purified. The same is true for the lectin-based adsorption methods.

At an early stage in the development of modern affinity chromatography I tried to interest a chemical company to produce series of agarose gels to which different amino acids should have been coupled. Any kind of such amino-acid agarose should specifically adsorb certain groups of proteins :enzymes and carrier proteins. Only lysine-Sepharose has been extensively exploited so far.

Another way of approaching the problem of synthesizing group adsorbents is to select ligands so that a certain characteristic parameter will maximally operate in the adsorption process while others are suppressed. For example, for the sake of simplicity and making fractionation strategy easier, ionic and hydrophobic adsorption should be kept apart. In other words ion exchangers should not contain hydrophobic groups and hydrophobic adsorbents should not contain ionizable ligands. This was clearly stated when we introduced the latter type of amphiphilic adsorbents in 1973. Of course in certain cases, a mixed type adsorbent may be preferable but as a rule it is not true. When the ligand structure complexity increases it becomes mo difficult to predict its properties as an interaction center in an affinity adsorbe By chance, Cibachron Blue was found very useful. Certainly, every dye coupled to a gel matrix may be useful for some purpose but elaboration of such adsorbents should be made on a rational basis if belonging to the realm of separation science. Such work should at least be planned with the intention to discover structure-adsorption relationships of advantage to further improve chromatographic methods even if the fundamental principles behind them are not clearly understood.

Dye ligand interactions comprise simultaneously operating ionic, hydrophobic, charge transfer interactions and sometimes perhaps hydrogen bond formation making i terpretation exceedingly difficult. We will learn more about that at this Symposiur by Drs Inman, Atkinson and Müller. In my opinion charge-transfer interactions may be very important and are perhaps the most characteristic factor influencing dye-type affinity adsorption.

Charge-transfer interactions have been exploited rather early in gas chromatogra phy and thin layer chromatography by using adsorbents prepared by using various ki of coating procedures. We were the first to try to explore charge-transfer affinity chromatography in aqueous solutions with the electron donor or acceptor molecules covalently attached to an adsorption-inert matrix such as Sephadex or agarose.

An efficient donor or acceptor ligand should presumably possess a delocalized electron donor system. This means that the ligand must have either an unsaturated aliphatic structure or an aromatic character and in neither case can hydrophobic interaction be avoided. If amino groups have been used for ligand attachment ionic forces may be involved in the adsorption as was the case with the first hydrophobi affinity adsorbents designed for protein fractionation.

Among simple ligands with π-electrons we find groups such as allyl, phenyl and naphtyl. To increase the tendency for electronic charge-transfer, introduction of electron-withdrawing or accepting groups into the aromatic ring system is necessary. Substances with easily displacable lone electron pairs may also form weak complexes by interaction with electron accepting groups in the solutes to be purified. Such interactions, including hydrogen bond formation, may be looked upon as reactions between an insoluble Lewis base and an acid solute or vice versa. It see as if hydroxylic hydrogen in the matrix or the vicinal water interacts with π-electron rich compounds and thus cause or add to the well-known "aromatic adsorption" that we first observed on cross-linked dextran even before the latter was exploited as a molecular sieve (Sephadex).

So far we have not found the ideal simple electron donor substance for matrix attachment. Instead Dr J.M. Egly and I have started model studies on acriflavin gels. They behave not only as hydrophobic adsorbents. The yellow colored gels turn brown when loaded with large amounts of nucleotides. This indicates that charge-transfer adsorption complexes are formed.

After or at the complex formation, the partner molecules will orient themselves in space to seek an energy minimum. Therefore, it may perhaps be possible to tailor ligand components which preferentially bind to molecules or regions of macromolecules that fit the sterical requirements for maximal orbital overlapping. Cibachron Blue, reported to be sterically similar to nicotine-adenine dinucleotide, and gels to which this dye has been bonded may therefore act with sterical specificity.

Metal (ion) affinity chromatography (14) as applied to fractionation of biopolymers may be considered as a special kind of charge-transfer chromatography either in a narrow sense as when, for example, the π-electrons in unsaturated compounds are forming complexes with gel-bonded Ag^{+} or Hg^{2+}, or in a broader sense as when lone electron pairs in amino nitrogen are coordinating with immobilized metal ions. One difficult problem to solve before high separation efficiency can be achieved concerns band spreading due to the heterogenous nature of the adsorption sites. The heterogeneity is a consequence of the presence of many competing ligands in the liquid as well as in the solid matrix. However, by proper selection of metal ions, ligand structure, pH and salt composition a fairly high degree of specificity has been reached already at the present stage of development. In my laboratory Drs M.A. Vijayalakshmi and P. Hubert have studied the simple model substances in the hope to better understand the behaviour of proteins and nucleic acids.

Peptides can be separated on Cu^{2+}, Zn^{2+} gels according to their histidine content (unpublished) and oligonucleotides according to their content of purinic bases. Metal affinity chromatography is already a useful method for group fractionation of protein mixtures. Considerable improvements are anticipated when the operational parameters are under better control. Theory may shed light upon these complicated ligand-ligate interactions.

Hydrogen bond and covalent bond formation are two principles used for affinity chromatography. The former (rather undeveloped as yet) is likely to be more frequently used for separation of oligo- and polynucleotides and the latter perhaps as an aid in synthesis of artificial polymers on solid supports.

Electrophoresis, gel filtration and density perturbation employing the affinity concept are techniques introduced recently for the study of complex formation and enzyme mechanisms. Particularly exiting is affinity chromatography in hydroorganic solvents at subzero temperatures. P. Douzou and his collaborators have started the developmental work in this direction (18).

Related to the topic of this Symposium is counter-current affinity extraction with Albertsson's polymer phase systems (14). Such extraction methods will become valuable alternatives or complements to chromatography especially for fractionation of particles in suspensions or emulsions that are not well suited for processes involving percolation through beds of granular adsorbents.

Theoretical penetration of the sorption-desorption processes is of paramount importance to provide the guidelines for further development and to create a sound physical background to the art of affinity chromatography.

Progress in affinity chromatography will influence further development of other techniques based on affinity principles such as immunodiagnostics, the therapeutic use of cell,tissue or organ specific drugs, certainly industrial applications of immobilized enzymes and solid phase synthesis of biopolymers. The molecules of life and their artificial equivalents, synzymes for example, caught in networks of supportpolymers may, we hope, in the future serve mankind in many areas : for improving health and environment and last but not the least for further progress in the basic biosciences.

REFERENCES

1 P. Cuatrecasas, M. Wilchek and C.B. Anfinsen, Proc. Natl. Acad. Sci. U.S.A. 61(1968)636.
2 J. Porath and P. Flodin, Nature 183(1959)1657.
3 S. Hjertén, Arch. Biochem. Biophys. 99(1962)446.
4 R. Axén, J. Porath and S. Ernback, Nature 214(1967)1302.
5 J. Porath, R. Axén and S. Ernback, Nature 215(1967)1491.
6 J. Porath, Nature 218(1968)834.
7 J. Porath and R. Axén, in K. Mosbach (Ed.), Methods Enzymol. 44, Immobilized Enzymes, Academic Press, New York, 1976.
8 T.C.J.Gribnau,Coupling of Effector-Molecules to Solid Supports. Thesis. Drukkerij van Mameren B.V., Nijmegen, The Netherlands 1977.
9 J. Porath and L. Sundberg in M.L. Hair (Ed.), The Chemistry of Biosurfaces, Marcel Dekker, New York, 1972, pp.633-661.
10 W.B. Jakoby and M. Wilckek (Eds.), Methods in Enzymology, Vol. 22, Enzyme Purification - Part B, Affinity Techniques, Academic Press, New York, 1974.
11 C.R. Lowe and P.D.G. Dean, Affinity Chromatography, John Wiley and Sons, London 1974.
12 E. Ruoslahti (Ed.), Immunoadsorbents in Protein Purification, University Park Press, Baltimore 1976.
13 J. Turkowa, Journal of Chromatography Library, Vol. 12, Affinity Chromatography, Elsevier, Amsterdam, 1978.
14 M. Lederer (Ed.), Chromatographic Reviews, Vol. 22, No.1, First Tiselius Symposium on Modern Biochemical Separation Techniques, Uppsala June 13-17, 1977, Elsevier, Amsterdam, 1978.
15 R. Epton (Ed.), Chromatography of Synthetic and Biological Polymers, Vols. 1 and 2 Published for the Chemical Macromolecular Group, Ellis Horwood, Chichester, 1978.
16 C.R. Lowe, An Introduction to Affinity Chromatography in (T.S. and E. Work Eds.) Laboratory Techniques in Biochemistry and Molecular Biology Series, North-Holland Amsterdam, 1979.
17. F.E. Regnier and K.M. Gooding, Anal. Biochem. 103(1980)1.
18. P. Douzou and C. Balny, Advances in Protein Chemistry, Vol. 32(1978)pp.77-189.

CHAPTER I

THEORETICAL ASPECTS

4th INTERNATIONAL SYMPOSIUM
ON AFFINITY CHROMATOGRAPHY
AND RELATED TECHNIQUES

T.C.J. Gribnau, J. Visser and R.J.F. Nivard (Editors),
Affinity Chromatography and Related Techniques

MOLECULAR INTERACTIONS IN AFFINITY CHROMATOGRAPHY

J. LYKLEMA
Laboratory for Physical and Colloid Chemistry, Agricultural University, De Dreijen 6, 6703 BC Wageningen, The Netherlands.

ABSTRACT

The purpose of this paper is to analyse the various physico-chemical unit steps and their integration in the binding of biopolymers to substrates. The various types of forces are classified into mechanical and thermodynamical ones. Coulomb forces, three types of van der Waals forces, H-bridging and hydrophobic bonding are discussed in some detail. The way in which these forces add up in complex molecules, containing several types of groups, is described and the analysis is illustrated by the adsorption of human plasma albumin on negative polystyrene latex. In this example, and probably also in many other cases, entropic contributions due to structural rearrangements in the biomolecule play an important and sometimes decisive role. Depending on conditions and the nature of the system, increasing the ionic strength can either increase or reduce the amount bound. Co-adsorption of simple ions of the electrolyte is probably a more general feature, at any rate it deserves wider attention.

INTRODUCTION

Over the past decades, considerable progress has been made in the development and application of affinity chromatography. A wide variety of macromolecular substances of biological origin can now be separated and/or purified. However, the understanding of the physics and chemistry of the binding between biopolymer and ligand has by no means kept abreast of these developments. For instance, a recent book on affinity chromatography (ref. 1) contains virtually no information on the analysis of the forces responsible for the binding of the biopolymer. The reason for this lack of progress is primarily that this binding is as a rule the compounded result of a number of simultaneously occurring molecular processes. Unravelling these processes is extremely cumbersome and any attempt to solve this problem for a particular case is psychologically thwarted by the often remarkable specificity of the binding: if the mechanism has been solved for a particular case, it is questionable whether the same solution applies also to other systems.

Because of these intrinsic difficulties, a number of "escape routes" have been developed that have proven to work well in practice. Examples are the development of special ligands (and/or spacers) on the basis of general experience on the interaction in solution between these ligands and the biopolymer to be bound and on the influence thereupon of important variables like ionic strength, pH and temperature. It is likely that such methods will remain useful during the oncoming decade.

The present paper chooses a more deductive approach, starting from a description of elementary binding forces between isolated molecules. Next, the additivity of these forces in complex molecules will be discussed. It is hoped that this procedure will be helpful in recognizing various contributions to the binding of biopolymers to surfaces and/or ligands and hence stimulate the development of new affinity chromatography systems on a more *a priori* basis. The discussion will be followed by a brief analysis of the adsorption of proteins on solid surfaces to illustrate the developed picture.

GENERAL CONSIDERATIONS

In literature, several procedures can be found to measure the strength or affin of a given bond.

Perhaps the most straightforward way is to give directly the force f(r) as a function of the distance of separation. This force is expressed in Newtons (N). This way of representation is for instance popular in (molecular) mechanics, especially if the interaction concerns simple, isolated molecules or solid bodies.

Another way is to translate the binding in terms of equilibrium constants K, which is possible if the participating molecules can all be identified. This procedure is common among (bio)chemists, and is particularly appropriate for the description of chemical equilibria. The dimension of K depends on the number of reactants and products.

Perhaps the most general description is in terms of standard Gibbs energies ΔG^o of the binding process, expressed in J mol^{-1}. It is the isothermal reversible work of bringing the system from some specified standard state to the bound situation, and ΔG^o is related both to f(r) and to K if the interaction is indeed reversible.

Reversibility in the attachment of biopolymers to surfaces or other matrices is matter that deserves attention. In affinity chromatography this reversibility is presupposed because it is inherent to the technique that a biopolymer, once bound can be liberated in its original form. Admittedly, to that end some changes in conditions are needed (such as a change of solvent, ionic strength, pH etc.), but if the liberated product is entirely identical to the injected molecule this does not detract from the reversibility: if the molecule temporarily alters its confirmation, say, upon binding, ΔG^o contains a contribution due to this structural change.

At this stage it is expedient to introduce the distinction between *mechanical* and *thermodynamical* forces.

In statistical thermodynamics, the distinctions between mechanical and thermodynamical quantities in general is familiar (ref. 2). Mechanical properties can be defined without introducing the notion temperature. Examples are: volume, pressure, energy, number of molecules, etc.. Thermodynamic quantities require the temperature for their definition. Examples are: the chemical potential, entropy, and free energy. In principle, thermodynamic variables are obtained by statistical averaging of mechanical variables.

The force f(r) between two isolated molecules, separated by a distance r, is often (but not always) a mechanical quantity. This is, for instance, the case with a chemical bond (which can be analysed with the Schrödinger equation, which does not contain T) and with the Coulomb force between two charges (q_1 and q_2, sign included) which vectorially can be written as

$$\vec{f}_c(r) = \frac{q_1 q_2 \vec{r}}{4\pi\varepsilon\varepsilon_o r^3} \tag{1}$$

where ε is the dielectric permittivity of the medium and $\varepsilon_o = 8.854 \times 10^{-12}\ C^2N^{-1}m^{-2}$. If the vectorial character needs not be indicated, the quotient $\vec{r}/r^3$ can be replaced by r^{-2}. According to (1), f(r) is counted negative for attraction and positive for repulsion, a convention that will be adhered to in this paper. Other examples of mechanical forces are H-bridges and the gravitational attraction between undeformable bodies.

A typical example of a thermodynamic force is that due to a hydrophobic bond. This force is due to the fact that the statistical averaging of water dipoles around two isolated hydrocarbon chains differs from the same when the two chains are in contact. Briefly stated, this force is of an *entropic* nature: the bonding can therefore also occur when the chains repel each other enthalpically, that is: when enthalpically molecule-solvent attraction prevails.

Paying now attention to the binding of and between biopolymers, it is readily realized that such interactions will, as a rule, be of a thermodynamical rather than of a mechanical character. It is likely that upon binding the positions and orientations of several parts of the biopolymer molecule will change with respect to each other; the extent to which this happens will be dependent on T. It follows immediately that such interactions must therefore preferably be described in terms of free energies or free enthalpies rather than energies or enthalpies, respectively. This is why interpretation in terms of ΔG^o values is the most general procedure.

This consideration brings us back to the three ways to formulate binding (through f(r), K or ΔG^o) and the relations between them.

For purely mechanical systems, energy and force are related as follows:

$$\vec{f}(r) = -\text{grad } U = -\nabla U \tag{2}$$

with

$$\text{grad } U \equiv \vec{i}\,\frac{\partial U}{\partial x} + \vec{j}\,\frac{\partial U}{\partial y} + \vec{k}\,\frac{\partial U}{\partial z} \tag{3}$$

where $\vec{i}$, $\vec{j}$ and $\vec{k}$ are unit vectors. For a two-body system at distance of separation r, one can simply write

$$f(r) = -\frac{dU(r)}{dr} \tag{4}$$

or, upon integration,

$$U(r) = -\int_{\infty}^{r} f(r)dr \tag{5}$$

This expression is nothing else than the classical expression for mechanical work. It applies also for electrical work: if (1) is substituted in (5), the Coulombic energy of two charges q_1 and q_2 at distance r is obtained:

$$U_c(r) = \frac{q_1 q_2}{4\pi\varepsilon\varepsilon_o r} \tag{6}$$

The minus sign in (5) stems from the fact that if the two molecules attract each other (both f(r) and U(r) negative), the energy of the pair decreases if their dis tance is reduced from infinity to r.

If the interaction is not purely mechanical, the energy in these expressions ha to be replaced by the Gibbs or Helmholtz (free) energy. If, as is usually the case in affinity chromatography, the pressure p is constant, the Gibbs (free) energy is needed:

$$f(r) = -\frac{dG(r)}{dr} = -\frac{dH(r)}{dr} + T\,\frac{dS(r)}{dr} \tag{7}$$

which may be considered a generalization of (4). The Gibbs energy is the reversibl isothermal work to bring the interacting particles from infinite distance (or from an otherwise specified standatd state) to distance r. Only if the entropy does not depend on r (i.e. if it does not vary if the two interacting units are brought together) is the force of a purely energetic (or enthalpic) nature.

Eq.(7) is general. In the case of formation of a chemical bond, the function G(will exhibit a minimum at given r, which is the equilibrium situation. Then f(r) = Under such conditions the appropriate way to describe the final situation is in te of equilibrium constants. The relationship with the standard Gibbs energy ΔG^o of t bonding process is

$$\frac{\Delta G^o}{RT} = pK = -\ln K \tag{8}$$

where the concentrations of the participating components in K must be expressed as mole fractions, so that K is dimensionless (otherwise it is impossible to take a l rithm of it).

THE BUILDING BRICKS: INTERACTIONS BETWEEN ISOLATED MOLECULES

Six types of attractive forces between individual molecules can possibly contribute to interactions in biopolymeric systems: Coulomb, Keesom - van der Waals, Debye-van der Waals, London-van der Waals, hydrogen bridges and hydrophobic bonding. In equilibrium, the attractive force is balanced by the repulsive force due to the overlap of the outer electron shells, according to the Pauli exclusion principle. This is the familiar Born repulsion.

Of these, the *Coulomb interaction* has already been given in (1) and (6). It depends on the charges on the molecules but not on their specific natures (size, shape). Coulomb energies decay with r^{-1} (6) and therefore have a long range. The medium affects the interaction through ε, which may be interpreted as a measure of the screening power. Water ($\varepsilon \sim 78$ at room temperature) screens much stronger than apolar solvents ($\varepsilon \lesssim 8$). Coulomb forces are mechanical forces.

The three types of *van der Waals interactions* have in common that they are attractive, that they act between uncharged molecules and that the energy decays with r^{-6} (provided r is not very high in which case London-van der Waals forces are retarded). Generally [1],

$$u_w = -\frac{\beta}{r^6} \qquad (9)$$

Van der Waals forces between isolated molecules have therefore a short range, their influence is not felt beyond distances of a few molecular diameters. As the additivity of the three types of van der Waals forces in assemblies of molecules is different, it is necessary to discuss them in some more detail.

The Keesom-van der Waals equation has been derived for the case that the two interacting molecules are ideal dipoles, each with dipole moment $\vec{\mu}$(C m). We write μ_1 and μ_2 if the two molecules are dissimilar. A dipole produces an electrical field $\vec{E}$ that is proportional to $\vec{\mu}$ and depends on the distance towards the dipole and the angle between the direction of the dipole and the line connecting the dipole and the location in space ($\vec{r}$) where $\vec{E}$ is measured (16). If a second dipole is positioned at $\vec{r}$, it experiences an energy $-(\vec{E}\cdot\vec{\mu}) = -E\mu\cos\alpha$, where α is the angle between the directions of $\vec{E}$ and the second dipole. For two spatially fixed dipoles, u_w would therefore be proportional to μ^2 (first order effect). However, the dipoles are not fixed but free. If they could orient themselves independently, the chances of having an attractive and a repulsive position would be equal, so that u_w would be zero. As, however, attractive positions are energetically more favourable than repulsive ones, the former prevail and statistical averaging leads to an overall attraction. This is a second-order effect, and it turns out that $u_w \sim \mu^4$.

---

1) In the following discussion we shall use small symbols for energies or enthalpies per molecule or per group, and capitals for corresponding molar quantities, or for the interactions between assemblies of many molecules.

The Keesom equation reads (ref. 3):

$$u_{wK} = - \frac{2\mu^4}{3(4\pi\varepsilon\varepsilon_o)^2 kTr^6} \quad \text{(J)} \qquad (10)$$

The factor kT in the denominator is typical. It enters the equation due to the st tistical averaging and is ultimately due to the fact that thermal motion countera the mutual orientation of the two dipoles. For this reason Keesom-van der Waals f ces are thermodynamic forces.

At this instance a note must be made on the ε in the denominator. It enters th equation because the attraction is of a Coulomb nature. Keesom himself derived (1 for molecules in the gas phase ($\varepsilon = 1$), but in the systems that are now at issue t is always a medium, mostly water. For water $\varepsilon \sim 78$ at room temperature, but it is questionable if in actual cases this high value is needed. The point is that Kees type forces become operative only at distances of a few molecular diameters, so that between the dipoles there can be only a few molecular diameters, so that bet the dipoles there can be only a few solvent (water) molecules. The high value $\varepsilon =$ for water applies to *bulk* water, where all H_2O molecules are free. However close surfaces and polar molecules, due to saturation, ε might well be much lower and rather resemble the effective relative dielectric constant in so-called Stern lay (i.e. parts of double layers in immediate contact with the surface). This value i usually of order 8-15. Similar reasoning applies to ε in the Debye equation.

Debye-van der Waals forces operate between a permanent dipole and an apolar mo cule. Apolar molecules, if brought in an electric field, can be polarized. The in duced dipole moment μ_i is proportional to the field strength and has the same dir tion:

$$\vec{\mu}_i = \alpha\vec{E} \quad \text{(C m)} \qquad (11)$$

Here, α is the polarizability of the molecule. In the S.I. system the dimension o is $C^2m^2J^{-1}$, which means that $\alpha/4\pi\varepsilon_o$ has dimension m^3; it appears that this quotie is almost identical to the molecular volume. In the Debye picture, $\vec{E}$ is the field the dipole, and it is this field that induces a dipole in the apolar molecule; th permanent and induced dipole attract each other. Again, some averaging over all orientations is needed; Debye forces are also second-order interactions. The fina result is (ref. 4):

$$u_{wD} = - \frac{2\alpha\mu^2}{(4\pi\varepsilon\varepsilon_o)^2 r^6} \quad \text{(J)} \qquad (12)$$

Debye forces have a mechanical character, the adjustment of the induced dipole be momentary. It is important to note that they increase very strongly with increasi radius of the apolar molecule. Anticipating the coming discussion, we can already state that if biopolymers adsorb on a polar surface there is always an intrinsic tendency to turn the biggest groups of the molecule towards the surface.

London-van der Waals forces operate between apolar molecules and are therefore the most ubiquitous. The attraction is a quantummechanical phenomenon (the final equation contains the Planck constant h), but physically the picture may be understood in the following simplified fashion: consider the interacting molecules to behave like very rapidly oscillating dipoles (electrons, especially the outer ones, circulating around a nucleus), then these oscillations will influence each other when the intermolecular distance is very short. Just like in the previous types, the attractive orientations are energetically favoured over the repulsive ones, so that overall attraction ensues. For two similar molecules, the London equation reads as follows (ref. 5):

$$u_{wL} = -\frac{3}{4}\frac{\hbar\omega\alpha^2}{(4\pi\varepsilon'\varepsilon_o)^2 r^6} \quad \text{(J)} \qquad (13)$$

Here $\hbar = h/2\pi$ and ω is the angular frequency of the outer electron in the ground state. London-van der Waals forces therefore persist at $T \rightarrow 0$. They may be considered as mechanical forces as long as higher electronic levels are not excited. In the denominator we have put ε' instead of ε because London-type interactions are of an electromagnetic origin. Therefore, not the static dielectric permittivity is needed but the permittivity "seen" by electromagnetic waves. Strictly speaking, a range of permittivities at various frequencies $\varepsilon(\omega)$ are needed. In the so-called macroscopic approach this is made explicit. Note that again u_{wL} decreases with the 6th power of distance between participating molecules. At larger distances, u_{wL} decays more strongly with r due to electromagnetic retardation, eventually u_{wL} becomes $\sim r^{-7}$. However, the distances at which retardation becomes noticeable are so large that u_{wL} is then already unmeasurably small.

For more extensive discussions on van der Waals forces see refs.(6 - 10), and on the electrostatic foundations see ref.(11).

For practical purposes it may be useful to assess the orders of magnitude of u_w. As stated before, they decay so strongly with r that u_w becomes negligible as compared with kT if r is beyond one or a few molecular diameters. However, u_w does become appreciable if the molecules come at distance of closest approach (that is the value of r at which the van der Waals attraction just cancels against the Born repulsion) and then they do contribute substantially to the "fine structure" that is representative of biological specificity in protein binding. Because of the great variety of values of α and μ, the uncertainty with respect to ε and the extreme sensitivity to packing it is difficult to give general rules. We may, however, assess the relative order of the three types of forces. From (10, (12) and (13),

$$u_{wK} : u_{wD} : u_{wL} = \frac{2\mu^4}{3\varepsilon^2 kT} : \frac{2\alpha\mu^2}{\varepsilon^2} : \frac{3\hbar\omega\alpha^2}{4(\varepsilon')^2} \qquad (14a)$$

Substituting $\varepsilon \simeq 10$, $\varepsilon' = n^2 = 1.77$, $kT = 4 \times 10^{-21}$ J, $\hbar\omega \sim$ ionization energy $= e^2/8\pi\varepsilon_o a$ (e = elementary charge and a = radius), $e = 1.6 \times 10^{-19}$ C, $\alpha = 4\pi\varepsilon_o a^3$, with $4\pi\varepsilon_o$ =

$= 1.11 \times 10^{-10}\ C^2N^{-1}m^{-2}$, and expressing μ in Debyes ($1D = 3.3 \times 10^{-30}$ Cm) and a in Ångstrøm units (10^{-10} m) one obtains the ratio

$$u_{wK} : u_{wD} : u_{wL} = 1.97\ \mu^4 : 0.24\ \mu^2 a^3 : 34\ a^5 \qquad (14b)$$

from which it is inferred that London forces can compete very well with Keesom forces (for instance for water, $a = 2.7$ Å and $\mu = 1.85$ D), which is mainly due tc the much lower dielectric permittivity in the denominator: if ε' would be identica to ε, as is the case in vacuum, the last term would have been $1.07\ a^5$. The differe with the last term of the RHS in (14b) underlines the importance of the medium.

Hydrogen bonds or -bridges play an important role in biopolymer-ligand binding, the more so since also the internal conformational stability of the former is to a large extent determined by it. H-bonds occur between a proton donor group AH, wher A is an electronegative atom (O, N, S, Cl, Br, I, ...) and an acceptor group B, wh is a lone electron pair or the π-electron orbital of an unsaturated bond. The bonc is strongly directional, although variation in the angle AHB can occur. The strenc of the bond is as a rule greater than that of a van der Waals binding and in conne tion with this the distance AB tends to be smaller than the sum of the two van der Waals radii ($r_A + r_B$). Bond energies vary from one system to the other, but often t are in the range of 12-40 kJ mole^{-1}. By comparison, for U_{wK} with $\varepsilon = 1$, $r = 2a$ and t substitutions discussed above, we obtain -2.6×10^{-20} J per bond for $a = 0.15$ nm anc $\mu = 1.86$ D, corresponding with 14.4 kJ mole^{-1}, i.e. the strongest van der Waals for are in the range of the weakest H-bonds. If, however, the distance r is slightly increased, the H-bond breaks, whereas U_w remains effective, although it diminishes rapidly.

For bibliographical references on hydrogen bonds, see refs. (12-14).

Hydrophobic bonding (Hϕ) is a typical example of a thermodynamic force. It occu only in a medium, i.e. it is not encountered in the interaction between two molec in the gas phase. If of two molecules 1, dissolved at large distance from each oth in medium 2, one is fixed and the other is brought to a position in close proximi of the fixed one, one (or an equivalent number of) molecule(s) of the solvent has be transported the other way round. Upon this exchange, energies u_{11} and u_{22} are gained but u_{12} is lost twice. Expressed per pair, the energy change is therefore $u = u_{12} - \frac{1}{2}u_{11} - \frac{1}{2}u_{22}$. It is customary to write this in terms of the Flory-Huggins parameter:

$$kT\chi = z(h_{12} - \tfrac{1}{2}h_{11} - \tfrac{1}{2}h_{22}) \qquad (15)$$

where z is the coordination number and the (molecular) energies have been replace by enthalpies, the pressure having negligible effect under usual conditions.

Eq.(15) is general; it applies irrespective of the natures of the interactions that give rise to the three enthalpies. They could be any of the five types of forces discussed above, and hence the RHS could be either negative or positive, i.e. the interaction can be either attractive or repulsive.

It is typical for Hϕ that the bonding takes place by factors *in addition* to the enthalpic ones of (15), specifically: by entropic forces. In terms of (7), the last term of the RHS plays a decisive role. The most common type of Hϕ occurs with water as the solvent. Water is a strongly associated liquid. If a hydrocarbon chain is dissolved in it, a number of H-bridges has to be severed. The overall process is energetically unfavourable, because the Debye-type of interaction energy between hydrocarbon chain segments and water dipoles (12) is usually not enough to compensate this loss of energy due to H-bond disruption. As a result, the H_2O molecules adjoining a hydrocarbon chain tend to make as many mutual H-bridges as possible, but this leads automatically to a lower mobility and, hence, to a lower entropy. Sometimes such adjoining water is therefore called "icelike", although it is difficult to stress the structural similarity between such "water of lower mobility" and ice. Dissolution of hydrocarbons in water is therefore an entropically unfavourable process. If now two hydrocarbon (or, for that matter apolar) molecules associate with each other, part of this "frozen" water is liberated and the ensuing increase of entropy is the driving force behind the process, irrespective of the sign of (15).

Hydrophobic bonding plays a very important role in the attachment of biopolymers to surfaces or to ligands. So it does in the establishment of the tertiary structure of proteins. In particular the possibility must be considered that hydrophobic parts inside a protein have an affinity to a hydrophobic patch on some substrate, that is high enough to compete with other forces and hence can induce the molecule to undergo some conformational alteration. Hydrophobic bonding is also predominantly responsible for the formation of micelles and vesicles from amphipolar substances such as surfactants and lipids, respectively.

For further reading see e.g. refs.(15 - 18).

INTEGRATION OF FORCES: INTERACTIONS BETWEEN ASSEMBLIES OF MOLECULES

In affinity chromatography, binding does not take place according to just one of the interaction types described above, but to a combination of them, in addition to chemical bond formation, so that as a next step a discussion must be given of the additivity of these interactions. However, for a full description this is not by a long way enough, because the fine structure in the arrangement of the various surface groups on biopolymers, suitable ligands and -surfaces can lead to the highly specific interactions that are so conducive to selective chromatographic separations. It is obviously impractical to discuss all feasible types of interactions, taking these structural details in consideration: the present contribution can therefore achieve not more than reviewing some important features.

The first problem to be discussed is that of the rigidity of the biopolymer molecule. If the mutual spatial disposition of all groups in it does not undergo any alteration upon binding or adsorption, the molecule is rigid. In that case, interaction is mechanical and relatively easy to formulate. Especially in biochemical liter ture, rigidity is often tacitly taken for granted. However, the following semiquanti tative argument shows that complete rigidity is not always likely.

Gibbs energies of binding ΔG of biopolymers depend on the nature of the bond, but let us for the sake of argument assume that they are of order 100 kJ mole^{-1}. If the internal strength of the molecule is characterized by an enthalpy of comparable orde of magnitude, the molecule can partly unfold and for purely enthalpic reasons loose some of its structure to enable the binding to be formed. However, any loss of struc ture is also accompanied by an increase of entropy and not very much structure break down is needed to obtain TΔS values comparable with ΔG. Severing one internal bond inside the biopolymer molecule will lead to an increase of several degrees of freedo each contributing to ΔS by about R ln 2 (depending on the nature and the extent of freedom) which value corresponds to TΔS = 1.72 kJ mole^{-1}. It appears that only a few tens of internal bonds need to be broken to attain an entropic contribution that is of a magnitude comparable to that of ΔG. Given the fact that the molecular masses o most biopolymers correspond with several hundreds of monomers per macromolecule, we arrive at the conclusion that only a small fraction of these monomers needs to be in volved. Such relatively minor structural alterations are difficult to detect. If they do occur, the interaction assumes thermodynamic character. Below we shall brie ly discuss an example where there is evidence for substantial entropic contributions involving changes in the structure of adsorbing proteins.

If the binding of complex biomacromolecules to a surface or ligand (briefly: to a substrate) takes place, many types of forces will as a rule be operative simultaneously. The question is: are these forces additive? Let us, therefore, consider the interplay of the six different types of interactions.

In answering this question, one has first to consider the ranges of action of these various forces. If this range is very short and if the molecule does not chan its conformation, molecules at not too close proximity do not feel each other's presence, so that the total effect is the sum of a number of individual interactions. This is obviously the case with chemical bonds, hydrogen bridges and Hϕ.

For Keesom- and Debye-van der Waals forces a distinction must be made between media consisting entirely of the same molecules (such as water) and media in which only a few lone dipolar groups occur (as in many proteins). Although water is stron ly associated because of the polarity and tetrahedral structural of the H_2O-molecul (that is: because of dipolar, quadrupolar and H-bridge contributions), liquid water as a whole will not attract proteins or surfaces by Keesom- or Debye forces, becaus the effects of the many dipoles cancel due to internal compensation. In other words bulk water has no net moment , although some preferential orientation may occur at phase boundaries. If only a few isolated (i.e. non-interacting) dipoles are present in a complex molecule, Keesom and Debye forces remain operative, provided dipole-dipole or dipole-apolar contacts can be made with the substrate. Dipoles buried in the interior of the biomolecule will be ineffective because the attraction decays rapidly with distance, so that only surface groups need to be considered, groups that become exposed upon binding included. The precise value of the interaction energy depends on the rotational mobility of the dipole involved. In (10) and (12) complete free mobility was presupposed, but if the dipole is bound to a bigger molecule the mobility is of course substantially inhibited. For Keesom forces it then depends on the mutual average orientations of the two dipoles how strong this force is; with Debye forces it is the angle between the dipole vector and the line connecting the centres of the dipole and the apolar molecule that counts. The following equations may be useful. The average electric field at vectorial distance $\vec{r}$ from the centre of an ideal dipole $\vec{\mu}$ is (ref. 11):

$$\overline{\vec{E}(\vec{r})} = \frac{3\overline{(\vec{\mu}\cdot\vec{r})}\vec{r}}{4\pi\varepsilon_o \varepsilon r^5} - \frac{\vec{\mu}}{4\pi\varepsilon\varepsilon_o r^3} \qquad (16)$$

The first term in the RHS is the radial component of $\vec{E}$ and the second is the component parallel to the direction of $\vec{\mu}$. $\overline{(\vec{\mu}\cdot\vec{r})}$ stands for the averaged scalar product of $\vec{\mu}$ and $\vec{r}$ and is given by

$$\overline{(\vec{\mu}\cdot\vec{r})} = \mu r \overline{\cos\alpha} \qquad (17)$$

where α is the angle between the vectors $\vec{\mu}$ and $\vec{r}$. If at $\vec{r}$ there is another dipole, it experiences an energy

$$u = -\overline{(\vec{\mu}\cdot\vec{E})} = -\mu\overline{E}\,\overline{\cos\beta} \qquad (18)$$

if β is the angle between the second dipole and the connecting line. If from other considerations the probabilities of finding the two dipoles in certain orientations can be estimated, $\overline{\cos\alpha}$ and $\overline{\cos\beta}$ can be evaluated and hence u. For completely freely rotatory dipoles $u = u_{wK}$ according to (10). If at $\vec{r}$ there is an apolar molecule, a dipole moment

$$\overline{\vec{\mu}_i} = \alpha\overline{\vec{E}} \qquad (19)$$

is induced and the ensuing energy is

$$u = -\tfrac{1}{2}\mu_i\overline{E} = -\tfrac{1}{2}\alpha\overline{E}^2 \qquad (20)$$

Carrying out the averaging leads to (12) if the dipole can rotate freely.

For London-van der Waals interactions the situation is different, because these forces are to a large extent additive even if the involved molecules are very close by. This is related to the extent of "tuning" of the molecular frequencies in the field of action of more than one neighbouring molecule. Sparnaaij (ref. 19) estimated that the pairwise additivity is realized for 70 - 90 %. In practice, this means the following. If the London-van der Waals interaction is sought between, say, a ligand and a big molecule at distance d, u_{wL} according to (13) between the ligand a[nd] all the atoms inside the big molecule can simply be added; the answer is then corre[ct] within 10 - 30 %. It is clear that the atoms farther away from the ligand contribute progressively less, because of the r^{-6} in the denominator, but on the other hand th[e] number of participating atoms increases rapidly with r. The upshot is that the over-all interaction between a small ligand and a big molecule (that by comparison may b[e] considered as semi-infinite) decays with d^{-3} and that the interaction per unit area between two semi-infinite media decays with d^{-2}. Because of this much slower decrea[se] and their ubiquity, London-van der Waals forces acquire a long range and therefore play an important role in colloid stability, adhesion, protein adsorption and affin[i]ty chromatography.

For the more detailed computation two alleys are open, currently known as the *microscopic* and the *macroscopic* approach. The first is easier but more approximativ[e] but as the results usually do not differ by more than some tens of a percent, most authors prefer to use it. This theory has been developed by Hamaker and De Boer (re[fs.] 20 - 22). Briefly, the principle is that full additivity is assumed and that screeni[ng] is neglected, i.e. in (13) $\varepsilon' = 1$. Material properties are accounted for by so-calle[d] *Hamaker constants* A. For the interaction between two homogeneous substances 1 in vacuum it reads

$$A_{11} = \pi\beta_1\rho_{n1}^2 \qquad (21)$$

where β is the constant in the London eq. (9) and ρ_{n1} is the number density of atoms in 1. For nonhomogeneous substances, A_1 is more complex. For the interaction betwee[n] two dissimilar homogeneous substances 1 and 2, by good approximation

$$A_{12} \cong \sqrt{A_{11}A_{22}} \qquad (22)$$

Eq. (22) is a consequence of the Berthelot principle. If the interaction between 1 and 2 takes place in a medium 3, also the properties of the medium enter the expression and the Hamaker constant becomes

$$A_{12(3)} = A_{12} - A_{13} - A_{23} + A_{33} \cong (\sqrt{A_{11}} - \sqrt{A_{33}})(\sqrt{A_{22}} - \sqrt{A_{33}}) \qquad (23)$$

This is for instance the situation with a protein (1) interacting with a substrate (2) in water (3). If 1 and 2 are identical, (23) simplifies to:

$$A_{11(3)} = A_{11} - 2A_{13} + A_{33} \cong (\sqrt{A_{11}} - \sqrt{A_{33}})^2 \qquad (24)$$

of which the middle expression resembles the RHS of (15) and of which the square indicates that the force is always attractive.

The relation between U and A depends on the geometry of the system. For our purpose two of them are relevant. For two semi-infinite plates at distance d (say, a flat protein parallel to a flat surface), the energy per unit area is

$$U_A = - \frac{A_{12(3)}}{12\pi d^2} \qquad (25)$$

and for a sphere of radius a and a flat semi-infinite plate (say, a ligand and a flat surface) with distance d between the surfaces,

$$U_A = - \frac{A_{12(3)}}{6} \left[\frac{a}{d} + \frac{a}{d+2a} + \ln \frac{a}{a+2d} \right] \qquad (26)$$

A collection of Hamaker constants can be found in literature (ref.23 - 24). For various polymers in water they are of order $kT = 4 \times 10^{-21}$ J. In a sphere-plate system at short distance of separation ((26) with $d \ll a$) $U_A \sim -Aa/6d$, therefore the attraction starts to exceed the thermal energy if $d < a/6$.

The macroscopic approach follows an entirely different path. The interacting bodies are considered macroscopic entities. Inside these bodies there are spontaneous random electromagnetic fluctuations. If two such bodies approach each other closely, the fluctuations in them are no longer random but influence each other mutually, and the attractive modes prevail over the repulsive ones just as in the case of two approaching free dipoles. This picture has been elaborated by Lifshits and coworkers (ref. 25 - 26). The final equations are scientifically satisfactory because they are rigorous and contain all the features relevant for practice, such as the influences of particle geometry, retardation and extent of additivity. For routine computations, especially with biological materials, they are perhaps too unwieldy because one needs to know the dielectric dispersion $\varepsilon(\omega)$ of all materials over a wide range of frequencies, and such information is not usually available, although interpolation procedures have now been developed for cases where only part of the dispersion spectrum is known (ref. 27). We shall not pursue this theory here further, except for repeating that in those cases where a mutual check has been possible the microscopic and macroscopic approach agree within some tens of a percent. A number of special cases (including sphere-plate and plate-plate interaction) has been treated in ref.(10).

Coulombic energies have a long range and play almost always a role. All water-compatible materials dispersed in aqueous solution bear a surface charge, except if special precautions are taken, such as choosing the pH so as to bring the surface in its point of zero charge. Because of their screening action, electrolytes can considerably reduce this range of action.

In the binding of biopolymers to substrates two kinds of Coulombic interaction ca be distinguished. First, there can be a Coulombic contribution to the template-like specific bonding. Second, there is the overall electrostatic interaction between bic polymer and substrate. This can be attractive or repulsive, depending on the signs c the two surface charges. The contributions to specific site binding must be individua ly counted, but for the calculation of the overall interaction it is often warrantec to treat the charges as smeared-out. This simplifies the computation considerably.

What is needed is the electrical contribution to the Gibbs energy of interaction $\Delta_{el}G$. When two charged particles, each surrounded by an electrical double layer, ap- proach each other, the distribution of ions in this double layer changes. Hence the is an important entropical contribution and the interaction is therefore of a therm dynamical nature. If upon overlap also the structure of the biopolymer changes, the change of $\Delta_{el}G$ due to this must also be taken into account.

As with the previous types of interaction, the precise value of $\Delta_{el}G$ depends on the geometry of the system, on the electrolyte, the surface charge, etc., so that n specific elaborations will be given. Quite generally the procedure to obtain $\Delta_{el}G$ i through a charging process. The starting point is the system in its uncharged state Then the charge is put on it step by step, at each moment during the process the di tribution of ions is left to come to equilibrium. Proceeding this way, the isotherm reversible work of charging is obtained, that is the Gibbs energy. In formula (ref.2

$$\Delta_{el}G = \int_0^{\sigma_o} \psi_o' d\sigma_o' \qquad (J\ m^{-2}) \qquad (27)$$

where σ_o is the surface charge and ψ_o' and σ_o' stand for the surface potential and th surface charge during the charging process. If for a given system the relation betw ψ_o' and σ_o' is known, the integration can be carried out. For protein-substrate inter action, integration (27) has to be carried out three times: (i) for the isolated pr tein, (ii) for the isolated substrate, and (iii) for the protein bound to the sub- strate; the electrical Gibbs energy of binding is then obtained as $\Delta G(iii) - \Delta G(ii) -$ $\Delta G(i)$.

It may be good to point out that the attachment of proteins to substrates is oft accompanied by co-adsorption of protons and other ions (ref. 28) to help balancing charge contrasts. For a complete description this co-adsorption must be taken into account (ref. 29); its contribution to the overall ΔG can be substantial. Because o this incorporation of ions, $\Delta_{el}G$ is not very sensitive to the charges on protein an substrate.

AN ILLUSTRATION: ADSORPTION OF HUMAN PLASMA ALBUMIN ON LATICES

A few years ago, Norde *et al.* published a series of papers on the adsorption of human plasma albumin (HPA) on various monodisperse polystyrene latices (ref. 29 - 31 From these studies we shall select a few features that may serve as an illustration some aspects of the above discussion.

The structure of HPA molecules is susceptible to pH. The conformational stability is a masimum at the isoelectric point (pH $\sim$ 4.7) and decreases with increasing distance to this point, both to the positive and to the negative side (ref. 32). This is reflected in the HPA adsorption on negative latices: as a function of pH, the plateau adsorption Γ_m passes through a maximum at the i.e.p.. On purely electrostatic grounds this feature is difficult to explain because electrostatics would rather predict an adsorption continually decreasing with increasing pH (pH > i.e.p.: electrostatic attraction, pH < i.e.p.: repulsion). The suggestion forces itself upon us that at the i.e.p. the adsorbed layer is thicker than outside this point, because the molecule progressively unfolds and flattens the more the pH differs from the i.e.p.. In this way, relatively simple adsorption measurements point already to structural, and hence to entropical, contributions. Experiments with the counterpart RNase which has a much more rigid molecule confirm this: for this molecule Γ_m is almost independent of pH.

That entropic factors do contribute to the binding could be inferred from two independent observations. (a) Careful studies on the initial parts of the isotherms showed a tendency for Γ to *in*crease with temperature, except at the i.e.p.. As adsorption always involves loss of translational degrees of freedom, the majority of all adsorption processes are enthalpically driven; for those cases $d\Gamma/dt < 0$. For HPA on negative latices, however, an increase of S must be the driving force. (b) Under some specified conditions of pH, T and ionic strength, the adsorption is spontaneous ($\Delta G < 0$) although the microcalorimetrically determined enthalpy of adsorption is positive ($\Delta H > 0$), i.e., although the process is endothermal. This is possible only if upon adsorption $T\Delta S > 0$.

Further analysis of the factors responsible for the binding led to the unravelling of ΔG into various enthalpic and entropic contributions. This analysis followed in broad lines the suggestions of the previous sections and was based on a number of additional experimental observations. Two striking features deserve special mentioning. The first is that the carboxyl groups of the protein, notwithstanding their negative charge, tend to accumulate near the (also negative) surface of the latex. The evidence stems from proton titrations of the adsorbed protein. Two reasons could be forwarded to explain this unexpected observation: the carboxyl groups are, because of their size, highly polarizable and therefore their London attraction to the surface of the latex (also polarizable because of the π-electrons) is relatively strong. The second reason, and at the same time the second striking feature, is that upon adsorption cations are co-adsorbed, apparently to avoid the development of too negative potentials in the contact region where the negative carboxyl groups and the negative surface groups accumulate. Direct proof of this co-adsorption stems from electrophoresis measurements (the total electrokinetic charge of the protein + latex system is more positive than the sum of the electrokinetic charges on protein and latex before adsorption). Later the uptake could be confirmed by direct measurements, using labelled cations (ref. 32).

The analysis of the adsorption enthalpy led to the following conclusions. The var der Waals contribution was small because the Hamaker constants for water and proteir are not too different. Hϕ does play a role, but this role is predominantly entropica ly determined: its enthalpic contribution is small. The more hydrophobic the sur face, the more important Hϕ. Very important contributions are those related to the c adsorption of cations, the electrical and the structural term. From all of this it m be concluded that protein adsorption studies should preferably be accompanied by ior uptake experiments. In passing, the influence of the ionic strength ω on the binding may be mentioned. The primary effect is the following: if adsorption proceeds *becaus of* electrostatic attraction, increase of ω will lead to a reduction of Γ because the attraction is screened. However, if adsorption takes place due to other factors (e.g entropical) *against* the electric interaction, $d\Gamma/d\omega > 0$. The latter is the case with HPA on negative latices. There are, however, also secondary electrolyte effects that may obscure or reinforce the primary trend, depending on conditions. Of these, the effect on the conformational stability (and hence on $T\Delta S$) deserves special mentionir All told, salt addition can thus work either way, a fact that is also known from exp rience in affinity chromatography.

Having established that the increase in entropy is the main driving force for th adsorption of HPA on negative latices, the source of this entropy rise must be detec ted. In the system under consideration three such contributions could be identified: the entropy change due to Hϕ, that due to structural changes of the HPA molecule, ar a fraction stemming from the increase in $\Delta_{el}S$ upon adsorption.

It is likely (and in a number of cases even proven) that the trends observed with HPA on negative latices apply also to other systems, although there will be differer ces of a quantitative nature.

In conclusion it may be stated that protein binding to substrates is a very complex and intrigueing process, for which simple solutions are not likely to apply. Th molecular building bricks of the analysis are, however, available and systematic exp riments with well-defined systems remain desirable and interesting.

REFERENCES

1 W.H. Scouten, Affinity Chromatography, Bioselective Adsorption on Inert Matrices, in P.J. Elving and J.D. Winefordner (Eds.), A Series of Monographs on Analytica Chemistry and its Applications, Vol. 59, J. Wiley, New York etc., 1981.
2 T.L. Hill, An Introduction to Statistical Thermodynamics, 2nd edn., Addison-Wesle Reading (USA), London, 1962, Sec.1-2.
3 W.H. Keesom, Proc. Koninkl. Nederl. Akad. Wetenschap., 18 (1915) 636; ibid., 23 (1920) 939; Physik. Z., 22 (1921) 129, 643.
4 P. Debye, Physik. Z., 21 (1920) 178; ibid., 22 (1921) 302.
5 F. London. Z. Physik, 63 (1930) 245.
6 J.O. Hirschfelder, C.F. Curtiss and R.B. Bird, Molecular Theory of Gases and Liquids, J. Wiley and Chapman & Hall, London, 1954, Ch. III.

7 K.S. Pitzer, Inter- and Intramolecular Forces and Molecular Polarizability, in I. Progogine (Ed.), Advan. Chem. Phys., Vol. 2, Interscience, New York etc., 1959, 59.
8 J.O. Hirschfelder (Ed.), Intermolecular Forces, in Advan. Chem. Phys. Series, Vol. 12, Interscience, New York etc., 1967.
9 H. Margenau and N.R. Kestner, Theory of Intermolecular Forces, 2nd edn., Pergamon Press, Oxford etc., 1971.
10 J.N. Israelachvili and D. Tabor, Van der Waals Forces, Theory and Experiment, in J.F. Danielli, M.D. Rosenberg and D.A. Cadenhead (Eds.), Progress in Surface and Membrane Science, Vol. 7, Academic Press, New York etc., 1973, p. 2.
11 C.J.F. Böttcher, Theory of Dielectric Polarization, 2nd edn., revised by O.C. van Belle, P. Bordewijk and A. Rip, Vol. I, Elsevier, Amsterdam etc., 1973.
12 G.C. Pimentel and A.L. McClellan, The Hydrogen Bond, W.H. Freeman Publ., San Francisco, 1960.
13 S.N. Vinogradov and R.H. Linnell, Hydrogen Bonding, Van Nostrand Reinhold Cy., New York etc., 1971.
14 M.D. Joesten, L.J. Schaad, Hydrogen Bonding, Marcel Dekker, New York, 1974.
15 C. Tanford, The Hydrophobic Effect. Formation of Micelles and Biological Membranes, J. Wiley-Interscience, New York etc., 1973.
16 F. Franks, Aqueous Solution Interaction of Low Molecular Weight Species. The Application of Model Studies in Biochemical Thermodynamics, in M.N. Jones (ed.), Biochemical Thermodynamics, Vol. 1, Elsevier, Amsterdam etc., 1979, Ch. 2.
17 F. Franks, The Hydrophobic Interaction, in F.Franks (Ed.), Water, a Comprehensive Treatise, Vol. 4, Plenum Press, New York etc., 1975, Ch. 1.
18 A. Ben-Naim, Water and Aqueous Solutions: Introduction to a Molecular Theory, Plenum Press, New York, 1974.
19 M.J. Sparnaaij, Physica, 25 (1959) 217.
20 H.C. Hamaker, Rec. Trav. Chim., 55 (1936) 1015; ibid., 56 (1937) 3, 727.
21 H.C. Hamaker, Physica, 4 (1937) 1058.
22 J.H. de Boer, Trans. Faraday Soc., 32 (1936) 11, 21.
23 J. Visser, Advan.Colloid Interface Sci., 3 (1972) 331.
24 S. Nir, Van der Waals Interactions between Surfaces of Biological Interest, Progr. Surface Sci., 8 (1977) No. 1.
25 E.M. Lifshits, Zhur.Eksp. i. Teor. Fiz., 29 (1955) 94 (transl. Sovjet Physics JETP, 2 (1956) 23).
26 I.E. Dzyaloshinskii, E.M. Lifshits and L.P. Pitaevski, Zhur. Eksp. i. Teor. Fiz., 37 (1959) 229 (transl. Sovjet Physics JETP, 10 (1960¾ 161).
27 J. Mahanty and B.W. Ninham, Dispersion Forces, Acad. Press, New York etc., 1976.
28 E.J.W. Verwey and J.Th.G. Overbeek, Theory of the Stability of Lyophobic Colloids, Elsevier, Amsterdam, 1948.
29 W. Norde and J. Lyklema, J. Colloid Interface Sci., 66 (1978) 277.
30 W. Norde and J. Lyklema, J. Colloid Interface Sci., 71 (1979) 350.
31 W. Norde and J. Lyklema, J. Colloid Interface Sci., 66 (1978) 257, 266, 285, 295.
32 J.F. Foster, in V.M. Rosenoer, M. Oratz and M.A. Rothschild (Eds.), Albumin Structure and Uses, Pergamon Press, Oxford etc., 1973, 53.
33 P. van Dulm, W. Norde and J. Lyklema, J. Colloid Interface Sci., in press (1981).

T.C.J. Gribnau, J. Visser and R.J.F. Nivard (Editors),
Affinity Chromatography and Related Techniques

ROLE OF ATTRACTIVE AND REPULSIVE VAN DER WAALS FORCES IN AFFINITY AND HYDROPHOBIC CHROMATOGRAPHY

C.J. VAN OSS[I], D.R. ABSOLOM[II,III] AND A.W. NEUMANN[III]

[I]Dept. of Microbiology, State University of New York at Buffalo, Buffalo, NY 14214, U.S.A.; [II]The Hospital for Sick Children, Toronto, Ontario M5G 1X8, Canada; [III]Dept. of Mechanical Engineering, University of Toronto, Toronto, Ontario M5S 1A4, Canada

ABSTRACT

In affinity as well as in hydrophobic chromatography the interactions between ligand and substrate are, in varying proportions, of a Coulombic and/or a van der Waals nature. The mechanisms that make Coulombic interactions attractive or repulsive are well known. We demonstrate here that van der Waals interactions between different materials, immersed in liquid media, also can be made attractive or repulsive at will. General procedures are given for making Coulombic as well as van der Waals interactions attractive or repulsive, to favor respectively the attachment and the elution step in affinity and hydrophobic chromatography.

INTRODUCTION

In many enzyme-substrate and in most antigen antibody interactions, van der Waals (or hydrophobic) forces play an important role, frequently (but not always) in combination with electrostatic forces. By means of changes in pH and/or ionic strength, or through the addition of plurivalent cations, or on the contrary, by admixture of complexing agents, it is easy to modify the electrostatic components of the attractive or repulsive forces in the desired direction. However, it is less generally realized that van der Waals interactions between different components, immersed in liquids, also easily can be made either attractive or repulsive, by modifying the van der Waals constant (or Hamaker coefficient) of the liquid. This is done by modulating the surface tension γ_{LV} of the liquid medium (where L stands for liquid and V for vapor) (ref. 1). These arguments essentially also apply to hydrophobic chromatography, where however van der Waals interactions play the preponderant role (ref. 2). Thus in affinity as well as in hydrophobic chromatography the attachment or the elution step can be favored, by controlling the sign of the electrostatic at the same time as that of the van der Waals interaction.

ANTIGEN ANTIBODY INTERACTIONS

Antigens (Ag) and antibodies (Ab) bind to one another via their antigenic determinants and antibody-active sites, by non-covalent "weak" bonds, consisting of Coulombic (or electrostatic) or van der Waals interactions, or a combination of both (refs. 3, 4, 5).

Coulombic attractions can be solely responsible for Ag-Ab interactions in exceptional cases where the Ag is strongly charged and where the corresponding Ab then has a considerable charge of the opposite sign, whilst the van der Waals interaction in aqueous media is negligible, due to the strong hydrophilicity of both Ag and Ab, resulting from their pronounced degree of ionization. An example of a virtually purely electrostatic Ag-Ab interaction is the DNA-anti-DNA system (ref. 6).

Van der Waals interactions can represent the sole Ag-Ab binding mechanism in some cases, where electrostatic charges are absent from the Ag determinant or the Ab-active site, e.g., with 3 azopyridine as the Ag determinant (refs. 7, 8), or with dextran-anti-dextran (refs. 9, 10). Other weak interactions that sometimes are separately invoked as playing a role in Ag-Ab interactions, such as hydrogen bonds (which, however only rarely play a role, ref. 11) and "hydrophobic bonds" (refs. 3, 4) also should be considered as van der Waals interactions (ref. 12).

The majority of Ag-Ab interactions and practically all Ag-Ab interactions in which the Ag is a protein (the Ab, being an immunoglobulin, of course always is a protein) comprise both Coulombic and van der Waals bonds in various proportions (ref. 5). The surface area of the Ag-Ab binding site usually is of the order of 200-400 $Å^2$ (refs. 5, 9). The free energy of binding, ΔF, which is generally obtained from equilibrium constants, varies from -4 to -12 kcal/mole (ref. 5), or from -3 to -17 ergs/cm^2 (= -3 to -17 mJ/m^2) (ref. 5).

In aquous media, with $\gamma_{LV} \approx 72.5$ ergs/cm^2 and at physiological pH ($\approx$ 7.4), the reaction

$$Ag + Ab \rightleftharpoons AgAb$$

virtually always is favored in the direction of AgAb. Thus the binding step in affinity chromatography usually takes place spontaneously, with Ag as well as with Ab ligands. For the elution step however both the Coulombic and the van der Waals interactions generally have to be reversed, which in each case requires careful study of the appropriate changes to be made in the physicochemical properties of the liquid medium.

HYDROPHOBIC CHROMATOGRAPHY

Contrary to the implication of the name that has been given to the method, this process most often applies to the (reversible) binding of fairly hydrophilic molecules (e.g., protiens) to hydrophobic ligands, in aqueous media (refs. 2, 12). The principal binding force in this case also generally is a van der Waals interaction (ref. 12), which can be made attractive or repulsive.

ATTRACTIVE AND REPULSIVE VAN DER WAALS INTERACTIONS

Theory

Hamaker's classical paper on London-van der Waals interactions in liquids (ref. 13) suggested the possibility that, depending on the properties of both the liquid medium and of different suspended (or dissolved) particles (or macromolecules), the van der Waals interaction between different particles (or molecules) in liquids, can be either attractive or repulsive, or even zero (refs. 1, 14, 15, 16). If one may postulate, for simplicity's sake, that the interaction between two different suspended materials 1 and 2 in a liquid medium 3 be expressed as occurring between two parallel semi-infinite slabs, the Hamaker equation for its free energy of interaction (ref. 13) is:

$$\Delta F_{132} = -A_{132}/12\,\pi d_o^2 \qquad (1)$$

where A_{132} is the effective Hamaker coefficient between materials 1 and 2 immersed in liquid medium 3 and d_o their equilibrium separation distance. The effective Hamaker coefficient can be calculated from eq. (1), once the free energy of interaction ΔF_{132} is known, which may be equated with the free energy of adhesion (refs. 14 - 16):

$$\Delta F^{adh}_{132} = \gamma_{12} - \gamma_{13} - \gamma_{23} \qquad (2)$$

where γ_{ij} is the interfacial tension between materials i and j. The various values for γ_{ij}, γ_{iV} and γ_{jV} (V standing for vapor) can be derived from contact angle data and an equation of state (refs. 16 - 18). With the help of a computer program, values for γ_{iV} (ref. 17) and for γ_{i3} (ref. 19) can be derived for different values of the contact angle θ obtained with a liquid of given γ_{3V}. For this purpose recently compiled conversion tables of contact angles to surface tensions now also can be used (ref. 20). The surface tensions γ_{3V} of liquids are readily determined with any method for measuring liquid surface tensions, such as the Wilhelmy plate method (ref. 21).

The effective Hamaker coefficient can also be obtained via Hamaker's combining rule (refs. 13, 22):

$$A_{132} = A_{12} + A_{33} - A_{13} - A_{23} \qquad (3)$$

where

$$A_{ij} = 12\pi d_o^2\,\Delta F_{ij} \qquad (4)$$

in which

$$\Delta F_{ij} = \gamma_{ij} - \gamma_{iV} - \gamma_{jV} \tag{5}$$

and

$$\Delta F_{ii} = -2\ \gamma_{iV} \tag{6}$$

where γ_{ij} and γ_{iV} may be determined as indicated above (refs. 16, 21). As long as the d_o values in eqs. (1) and (4) are identical, the validity of eq. (3) follows directly from eqs. (1), (2), (4), (5), (6); (refs. 16, 23). According to our own findings (refs. 14, 16) and in good agreement with Israelachvili's data (refs. 24, 25) the best value for d_o to be used in the calculation of A_{132} (eq. (4)) or of A_{ij} in general (eq. (4)), is very close to 1.8 A. As eq. (3) yields a value for the effective Hamaker coefficient A_{132} as the (often rather small) difference between two (fairly large) numbers, small experimental errors in values for A_{ij} may well give rise to significant errors in A_{132} (refs. 16, 23). It thus is preferable to determine A_{132} directly via eqs. (1) and (2).

It has been shown (refs. 1, 14, 16, 22) that the value for the effective Hamaker coefficient A_{132} (see eq. (3)) is negative when:

$$A_{11} > A_{33} > A_{22} \tag{7}$$

or when:

$$A_{11} < A_{33} < A_{22} \tag{8}$$

Under all other conditions A_{132} is positive (or zero, when $A_{11} = A_{22} = A_{33}$). From eqs. (4), (6), (7), and (8) it follows that A_{132} is negative when:

$$\gamma_{1V} < \gamma_{3V} < \gamma_{2V} \tag{9}$$

or when:

$$\gamma_{1V} > \gamma_{3V} > \gamma_{2V} \tag{10}$$

Thus the effective Hamaker coefficient is negative and the net van der Waals interaction between two different materials immersed in a liquid repulsive, when the surface tension of the liquid medium has a value intermediate between the values of the surface tensions of the two different materials (refs. 1, 2, 10, 15, 16).

Practice

We confirmed the validity of the above rule (eqs. (9), (10)) in a number of experimental situations, summarized in Table 1.

TABLE 1

Experimental verification of the effect of negative Hamaker coefficients

Experimental system	Result	References
Rejection or engulfment of particles by solidifying melts.	Rejection at negative A_{132}; engulfment at positive A_{132}.	1, 14, 16, 27
Phase separation or compatibility of polymer mixtures in solution.	Phase separation at negative A_{132}; polymer compatibility at positive A_{132}.	1, 15, 16, 27
Desorption of protein from latex particles at lowered liquid surface tension.	Irreversible adsorption of γ globulin on polystyrene particles at positive A_{132} and desorption at negative A_{132}.	26

Table 1 illustrates that repulsive van der Waals forces can elucidate a number of phenomena that were hitherto difficult to explain.

Table 2 gives a number of examples in which net repulsive van der Waals forces are used to dissociate Ag-Ab complexes, or to elute adsorbed proteins or cells from hydrophobic ligands. The lowering of γ_{3V} from ≈ 72.5 ergs/cm^2

TABLE 2

Applications of net repulsive van der Waals forces in Ag Ab dissociation and in hydrophobic chromatography

Experimental system	Results	References
Dissociation of Ag-Ab complexes at γ_{3V} ≈ 50 ergs/cm^2 and pH 4.0 or 9.5.	Complete dissociation of Ag-Ab complexes at negative A_{132}.	1, 8, 27
Elution of blood group Abs from human red cells at γ_{3V} ≈ 50.5 ergs/cm^2 and pH 9.5	Complete desorption of A, Kell and Rh blood group Abs from red cells at negative A_{132}.	28
Hydrophobic chromatography of serum proteins; elution at γ_{3V} ≈ 70.6 to 66.8 ergs/cm^2.	Adsorption of proteins onto hydrophobic ligands at positive A_{132} and elution at negative A_{132}.	1, 2, 27
Elution of human granulocytes from nylon fibers at γ_{3V} ≈ 69 ergs/cm^2, at pH 7.2 and in the presence of 0.1% Na_2EDTA.	Adhesion of granulocytes to nylon fibers at positive A_{132} and complete detachment of functionally and morphologically intact granulocytes at negative A_{132}.	1, 29

to the appropriate values for the various dissociations and separations was done through the admixture of ethylene glycol, dimethyl sulfoxide, dimethyl acetamide, propanol or ethanol, to the initial aqueous medium. Through the concept of net repulsive van der Waals interactions, the empirically established usefulness of ethylene glycol as an eluant in hydrophobic chromatography (refs. 30, 31) could now be explained (refs. 1, 2, 27). Repulsive van der Waals interactions also form the basis for novel separations (see Table 2) (refs. 1, 8, 27, 28, 29), of direct import to affinity chromatography.

Recent results with an exclusively van der Waals system (dextran-anti-dextran) indicate that anti-dextran Ab, specifically bound to a column packed with cross-linked dextran beads (Sephadex G-10), can be eluted by lowering γ_{3V} to $\approx$ 47 ergs/cm^2, at pH 7.2 (ref. 32).

HYDROPHOBIC INTERACTIONS

The preferential attraction between hydrophobic determinants immersed in an aqueous medium, frequently alluded to as the "hydrophobic effect", also can be entirely ascribed to van der Waals interactions, as recently has been shown thermodynamically (ref. 33). In addition, as emerges rather strikingly from the results obtained with hydrophobic chromatography (see Table 2 and refs. 1, 2, 27) of a number of serum proteins, which recently have been shown to be very hydrophilic (ref. 34), there also is a marked attraction between a hydrophobic and a hydrophilic determinant, in water (ref. 33). Only the attraction between two hydrophilic determinants in water is small enough to be easily overwhelmed by the electrostatic repulsion that usually prevails between such entities (ref. 33).

Figure 1 illustrates the magnitude of the free energies of attraction between a typical hydrophobic ligand, phenyl sepharose, (material 1), and human serum albumin (material 2), in water (material 3). Taking the value for the surface tension of phenyl sepharose as γ_{1V} = 41 ergs/cm^2 (ref. 1) and for albumin γ_{2V} = 70 ergs/cm^2 (ref. 34) whilst the surface tension of water γ_{3V} = 72.5 ergs/cm^2, we can, via eq. (2) and the appropriate computer programs (refs. 17, 19) or tables (ref. 20) compute the various free energies of interaction. Figure 1 also shows that contrary to popular belief (ref. 35) there is no repulsion between the molecules of the aqueous medium and the hydrophobic ligand (ΔF_{13}); on the contrary, that attraction is quite strong. The attraction between the hydropholic protein (albumin) and water (ΔF_{23}) also is very strong.

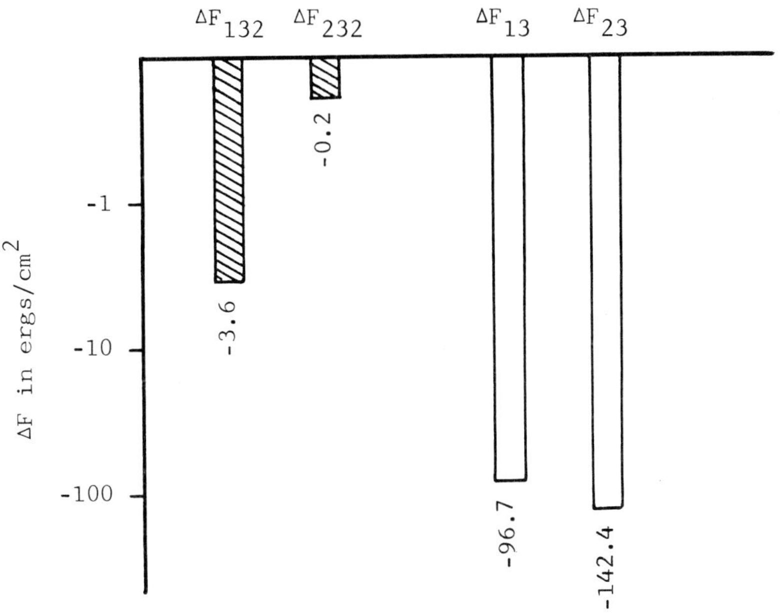

Fig. 1. Free energies of attraction in ergs/cm^2 (mJm^{-2}) between a hydrophobic ligand (phenyl sepharose) and albumin in water: ΔF_{132}, and between two albumin molecules in water: ΔF_{232} (shaded bars). For comparison, the free energies of attraction between water molecules and the hydrophobic ligand: ΔF_{13}, and between albumin and water molecules: ΔF_{23}, are also shown (open bars).

COULOMBIC INTERACTIONS

Whilst the Coulombic interactions that often occur between Ag and Ab generally are accompanied by van der Waals interactions, enhanced by the small equilibrium separation distance d_o between Ag and Ab, under exceptional circumstances they can occur alone, e.g., when the A_{132} of the Ag-Ab interaction is virtually zero, as is the case with DNA anti DNA complexes. In such cases only the Coulombic bonds need to be dissociated (ref. 6).

Contrary to the hitherto prevailing (but usually only partly successful) practice of attempting to dissociate Ag-Ab bonds at low pH, we have found that it is generally more effective and much less denaturing to both Ag and Ab, to use a fairly high pH combined with and appropriately lowered liquid surface tension γ_{3V}. In cases of relatively low Ag-Ab affinity, Coulombic bonds also may be dissociated at high ionic strength (ref. 6). However, the Coulombic bonds of Ag-Ab complexes of high affinity only can be broken under extreme conditions of pH (ref. 6).

In hydrophobic chromatography Coulombic interactions also can occur (refs. 2, 27,36). Usually however the hydrophobic ligands as well as the macromolecules that are to be separated are (fairly weakly) negatively charged (ref. 27), a circumstance which rarely impedes the attachment step, but somewhat facilitates the elution.

In the presence of multivalent cations (e.g., Ca^{++}) however, bridging between two negatively charged components is not uncommon (refs. 37-39). It thus does not always suffice to make both ligand and substrate negatively charged to achieve electrostatic repulsion but, in the presence of multivalent cations it may be essential to add a complexing agent (e.g., Na_2EDTA) to the eluting liquid (ref. 29).

CONCLUSIONS

Attractive van der Waals interactions prevail when the liquid surface tension γ_{3V} is larger or smaller than both the surface tensions γ_{1V} and γ_{2V} of ligand and substrate.

Ag-Ab bonds that are exclusively of the van der Waals type (e.g., 3-azopyridine-anti-3-azopyridine) can be completely dissociated by lowering γ_{3V} of the suspending liquid to ≈50 ergs/cm^2, through the sole admixture of ≈60% ethylene glycol, ≈ 50% dimethyl sulfoxide, or ≈ 3% propanol, whilst typical protein-anti-protein complexes can be dissociated in the same manner by also simultaneously neutralizing the electrostatic attraction through an increase in pH to 9.5, or a decrease in pH to 4.0 (ref. 8). Blood group Abs can be completely eluted from red cells by the admixture of ≈ 47% dimethyl sulfoxide, at pH 9.5 (ref. 28).

In another exclusively van der Waals system, anti-dextran Abs can be eluted from columns packed with Sephadex beads (consisting of a cross-linked dextran gel) with ≈55% dimethyl sulfoxide, at pH 7.2 (ref. 32).

In hydrophobic chromatography the attachment step of proteins to hydrophobic ligands is favored by raising the liquid surface tension γ_{3V} by increasing the salt concentration. Elution of the proteins is effected by lowering γ_{3V} (at low salt concentrations) through the addition of, e.g., ≈10 to 40% ethylene glycol (ref. 2). The elution of functional and undamaged polymorphonuclear leukocytes, that have been made to adhere to nylon fibers, can be achieved with ≈ 10% dimethyl sulfoxide or dimethyl acetamide, or ≈2.5% ethanol, in the presence of 0.1% Na_2EDTA (ref. 29).

REFERENCES

1. C.J. van Oss, D.R. Absolom and A.W. Neumann, Colloids & Surfaces, 1(1980)45-56.
2. C.J. van Oss, D.R. Absolom and A.W. Neumann, Separ. Sci. & Technol., 14(1979) 305-317.

3. E.A. Kabat, Structural Concepts in Immunology and Immunochemistry, 2nd edn., Holt, Rinehart & Winston, New York, 1976, Ch. 1, p. 1.
4. M.W. Steward, in L.E. Glynn and M.W. Steward (Eds.), Wiley, Chichester, New York, 1977, Ch. 7, p. 233.
5. C.J. van Oss and A.L. Grossberg, in N.R. Rose, F. Milgrom and C.J. van Oss (eds.), 2nd edn., Macmillan, New York, 1979, Ch. 5, p. 65.
6. E.R. de Groot, M.C. Lamers, L.A. Aarden, R.J.T, Smeenk and C.J. van Oss, Immunol. Commun. 9(1980)515-528.
7. A.L. Grossberg and D. Pressman, Biochem., 7(1968)272-279.
8. C.J. van Oss, D.R. Absolom, A.L. Grossberg and A.W. Neumann, Immunol. Commun., 8(1979)11-29.
9. E.A. Kabat, Structural Concepts in Immunology and Immunochemistry, 2nd edn., Holt, Rinehart & Winston, New York, 1976, Ch. 6, p 119.
10. C.J. van Oss and A.W. Neuman, Immunol. Commun., 6(1977)341-354.
11. D. Pressman and A.L. Grossberg, The Structural Basis of Antibody Specificity, Benjamin, Reading, MA. 1973, p. 55.
12. C.J. van Oss, D.R. Absolom and A.W. Neumann, Colloid & Polymer Sci., 258 (1980)424-427.
13. H.C. Hamaker, Physica, 4(1937)1058-1072.
14. A.W. Neumann, S.N. Omenyi and C.J. van Oss, Colloid & Polymer Sci., 257 (1979)413-419.
15. C.J. van Oss, S.N. Omenyi and A.W. Neumann, Colloid & Polymer Sci., 257 (1979)737-744.
16. S.N. Omenyi, Attraction and Repulsion of Particles by Solidifying Melts, Ph.D. Dissertation, University of Toronto, 1978.
17. A.W. Neumann, R.J. Good, C.J. Hope and M. Sejpal, J. Colloid & Interface Sci., 49(1974)291-304.
18. C.J. van Oss, C.F. Gillman and A.W. Neumann, Phagocytic Engulfment and Cell Adhesiveness, Marcel Dekker, New York, 1975.
19. A.W. Neumann, O.S. Hum, D.W. Francis, W. Zingg and C.J. van Oss, J. Biomed. Mater. Res., 14(1980)499-509.
20. A.W. Neumann, D.R. Absolom, D.R. Francis and C.J. van Oss, Separ. & Purif. Meth., 9(1980)69-163.
21. A.W. Neumann and R.J. Good, Surface & Colloid Sci., 11(1979)31-91.
22. J. Visser, Adv. Colloid & Surf. Sci., 3(1972)331-363.
23. A.W. Neumann, S.N. Omenyi and C.J. van Oss, in preparation.
24. J.N. Israelachvili, J. Chem. Soc., Faraday Tran., 2, 69(1973)1729-1738.
25. J.N. Israelachvili, Quart. Rev. Biophys. 6(1974)341-387.
26. H.G. de Bruin, C.J. van Oss and D.R. Absolom, J. Colloid & Interface Sci., 76(1980)254-255.
27. C.J. van Oss, A.W. Neumann, S.N. Omenyi and D.R. Absolom, Separ. & Purif. Meth., 7(1978)245-271.
28. C.J. van Oss, D. Beckers, C.P. Engelfriet, D.R. Absolom and A.W. Neumann, Vox Sang., in the press.
29. D.R. Absolom, C.J. van Oss and A.W. Neumann, Transfusion, in the press.
30. B.H.J. Hofstee, Analyt. Biochem., 52(1973)430-448.
31. J. Rosengren, S. Pahlman, M.Glad and S. Hjertén, Biochim. Biophys. Acta, 412(1975)51-61.
32. D.R. Absolom, C.J. van Oss and A.W. Neumann, in preparation.
33. C.J. van Oss, D.R. Absolom and A.W. Neumann, Colloid & Polymer Sci., 258 (1980)424-427.
34. C.J. van Oss, D.R. Absolom, A.W. Neumann and W. Zingg, Abstr. Second Chem. Congr. North Amer. Continent, Las Vegas (Aug. 1980)No. 108.
35. C. Tanford, Science 200(1978)1012-1017.
36. R. Srinivasan and E. Ruckenstein, Separ. & Purif. Meth., 9(1980)267-370.
37. A.S.G. Curtis, Biol. Rev., 37(1962)82-129.
38. A.W. Neumann, D.R. Absolom, C.J. van Oss and W. Zingg, Cell Biophys., 1(1979) 79-92.
39. D.R. Absolom, C.J. van Oss, R.J. Genco, D.W. Francis and A.W. Neumann, Cell Biophys., 2(1980)113-126.

T.C.J. Gribnau, J. Visser and R.J.F. Nivard (Editors),
Affinity Chromatography and Related Techniques

STUDIES ON THE MECHANISM OF PROTEIN ADSORPTION ON HYDROPHOBIC AGAROSES

H.P. JENNISSEN*, A. DEMIROGLOU* and E. LOGEMANN§

*Institut für Physiologische Chemie der Ruhr-Universität Bochum and
§Institut für Rechtsmedizin der Universität Freiburg, West Germany

ABSTRACT

Although phosphorylase b is strongly adsorbed to butyl agarose at high ionic strength attempts to detect a significant binding of ^{14}C-butylamine and ^{14}C-hexylamine in solution were unsuccessful. We conclude that there are no accessible, high-affinity alkyl binding sites or pockets on the phosphorylase b molecule at high ionic strength. However a newly discovered temperature dependent sol-gel transition of the enzyme indicates the presence of a number of low-affinity hydrophobic binding areas on the surface. Kinetic studies of the adsorption of phosphorylase b to immobilized butyl residues support a biphasic binding mechanism. Conformational changes may be associated with the adsorption of the enzyme. This is indicated by the finding that freeze-inactivation of phosphorylase b is prevented by the adsorption to butyl agarose.

INTRODUCTION

Previous studies have shown that phosphorylase b is cooperatively adsorbed to butyl agarose (1) and exhibits adsorption hysteresis (2). A multivalent binding mechanism has been proposed (1,3). Desorption kinetics have supported the model of negative cooperative protein adsorption (4). Recently (5) a rationale for the synthesis of butyl agaroses with finite distribution coefficients has been described. However a number of questions have remained unanswered, e.g.: How many hydrophobic binding sites exist on phosphorylase b in solution? Can they be detected by binding studies with soluble alkylamines? What information can be gained from an analysis of the adsorption kinetics? And finally what role do conformational changes of the enzyme play? Experiments devised to answer some of these questions will be described in this paper.

MATERIALS AND METHODS

Activation of Sepharose 4B with CNBr and the coupling of n-^{14}C-al amines was either performed according to ref. 6 (freeze-inactivation experiments) or according to the modified procedure, ref. 7 (all oth experiments). ^{14}C-Methylamine and 1-^{14}C-ethylamine were obtained fro New England Nuclear, n-1-^{14}C-hexylamine was obtained from Amersham. n-1-^{14}C-butylamine was synthesized from n-1-^{14}C-butanol as described

The degree of substitution has been determined in our laboratory by the addition of radioactive alkylamine tracers to the 2M solution of the amine to be coupled. Since at that time n-1-^{14}C-butylamine wa not available the degree of substitution of butyl agaroses was deter mined with the tracer 1-^{14}C-ethylamine (6). However the question re- mained if different ^{14}C-alkylamines added as tracer to the n-butylam solution yield identical results. Therefore the degree of substituti of agarose with butylamine was determined with the three tracers ^{14}C methylamine, 1-^{14}C-ethylamine and n-1-^{14}C-butylamine (Table 1). If t

TABLE 1

Differential labelling of butyl Sepharose with tracers of homologous ^{14}C-alkylamines

BrCN mg/ml	IMMOBILIZED RESIDUE CONCENTRATION µMOL/ml PACKED GEL				
	^{14}C-METHYLAMINE	^{14}C-ETHYLAMINE	^{14}C-BUTYLAMINE	$^{14}C_1$ / $^{14}C_2$	$^{14}C_2$ / $^{14}C_4$
8	82.3 ± 0.3 (3)	18.3 ± 1.2 (3)		4.6	
15	130.9 ± 6.0 (3)	30.9 ± 0.4 (3)		4.2	
30	180.0 ± 8.2 (3)	38.8 ± 2.0 (3)		4.6	
8		22.4 ± 0.5 (3)	18.3 ± 0.3 (3)		1.2
15		30.9 ± 1.2 (2)	25.5 ± 0.2 (2)		1.2
30		42.9 ± 1.4 (3)	35.6 ± 0.3 (3)		1.2

Activation and coupling were performed according to ref. 7. The anal sis of the gels is described in ref. 6. $^{14}C_1$, $^{14}C_2$ and $^{14}C_4$ denote t tracers of methylamine, ethylamine and butylamine respectively. The ber of analyses is given in parentheses. For further details see te

ethylamine tracer is employed the apparent degree of substitution is ca. 1.2-fold higher than the value obtained with the butylamine tracer. The significance of this difference is being examined with higher members of the homologous series. However addition of the methylamine tracer leads to ca. 5.5-fold higher results than expected. The calculated ratios (see Table 1) are independent of the CNBr concentration employed for activation. The basis for the differential reactivity of the alkylamines is still unclear since the basicity of the amine does not correlate with the enhanced labelling effect (8). It may be that steric or apolar effects play an important role.

All methods pertaining to the preparation and determination of activity of phosphorylase b (ca. 80 U/mg) and the radioactive labelling of the enzyme (^{3}H-phosphorylase b_r, ca. 45 U/mg) have been extensively described (1, 2). The high ionic strength buffer A employed contains 10 mM tris(hydroxymethyl) aminomethane/maleate, 5 mM dithioerythritol, 1.1 M ammonium sulfate, 20% sucrose, pH 7.0.

Binding measurements of alkylamines to phosphorylase b were permed in a flow dialysis apparatus of Feldmann (9) according to the method of Colowik and Womak (10). The upper chamber of the cell was filled with 0.5 ml buffer A with or without the enzyme (80-100 mg/ml) to which the ^{14}C-alkylamine was added to a final concentration of 1-2 mM and ca. 10^6 cpm/ml at 5°C. The lower chamber (ca. 0.03 ml) was perfused with buffer A (18 ml/hr) and 0.3 ml fractions were collected every minute. For radioactivity measurements 0.2 ml aliquots were taken from the fractions and counted in 2 ml scintillation fluid (Quickscint 212, Zinsser, see ref. 2).

The methods for measuring sorption kinetics have been previously described (4).

Phosphorylase b is irreversibly freeze-inactivated (11) under standard conditions by freezing the enzyme (0.3-3 mg/ml) in buffer B (20 mM sodium ß-glycerophosphate, 1 mM EDTA, 20 mM mercaptoethanol, pH 7.0) for 3 hours at -18 to -20° in a freezer and thawing the enzyme for 5 min at 30°. Slow thawing at 5° leads to the same result. For the preparation of solute free phosphorylase b the native enzyme (10 mg/ml) was dialyzed for 24 hrs against four changes of the 600-fold volume of double distilled water in a nitrogen atmosphere. In spite of the precautions the specific activity of the enzyme decreased to ca. 50 U/mg. The freeze-inactivation experiment with adsorbed phosphorylase b was performed on two columns in the following way: A sample of 1 ml phosphorylase b (3.5 mg/ml) was applied to 1 ml packed butyl Sepharose (activated with 20 mg/ml CNBr and coupled according to ref. 6) on a small column (0.8 cm i.d. x 12 cm) in buffer B. After washing t gel with 5 ml buffer B the column was allowed to run dry by gravity. The control column remained at 5°C. The gel of the second column was removed, frozen at - 20°C for 3 hours and then thawed at 5° after ad ding 1 ml buffer B. After filling the thawed gel back into the colum both columns were eluted with buffer B in which the pH had been lowered to pH 5.6 with HCl (6). The eluted enzyme was then analyzed.

RESULTS AND DISCUSSION

Binding Studies

Binding of immobilized alkyl residues to protein. One of the most important parameters governing the adsorption of a protein to immobilized alkyl residues is the surface concentration of residues (5,6). the protein contains more than one available binding site for the im mobilized residue a sigmoidal binding curve (adsorbed protein vs. im mobilized residue concentration) will be obtained (for review see re 4,5). The adsorption of phosphorylase *b* on butyl Sepharose is a good example for the cooperative binding of immobilized alkyl residues to a protein .

Binding of soluble alkyl residues to protein. The qualitatively different aspect in the interaction of a protein with immobilized or soluble residues lies in the fact that in the former case a two-dim sional concentration (mol/m^2) and in the latter case a three-dimensio (mol/l) concentration of residues is effective. This difference beco fully evident in the case of phosphorylase *b*. Fig. 1 shows a flow di alysis experiment in which 1-2 mM n-1-^{14}C-hexylamine in buffer A is cubated with phosphorylase *b* (0.85 mM monomer units). A similar expe ment was conducted with n-1-^{14}C-butylamine (not shown). In contrast the experiments with immobilized butyl residues (1,2) no significant binding of soluble butyl- or hexylamine to phosphorylase *b* can be de ted (see Fig. 1). Thus the binding constant of the protein-hexylami interaction must be so low that a mixture of mM concentrations of t binding partners does not lead to a significant saturation of putat hydrophobic areas or pockets on phosphorylase *b*. In addition no bind was observed at higher temperatures. We therefore conclude that the are no high-affinity alkyl binding sites on phosphorylase *b*. Specif high-affinity binding sites for small alkanes have however been fou on a variety of other proteins e.g. hemoglobin, myoglobin and ß-lac globulin (12). For the binding of pentane the binding constant (K_1) o the specific,high-affinity binding site was between 1-8 x 10^3 M^{-1} f

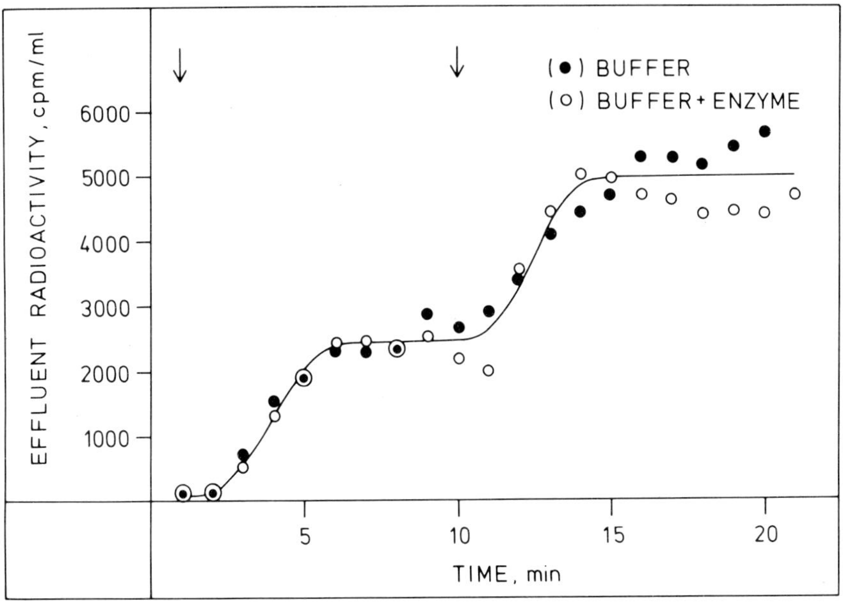

Fig. 1. Binding analysis of n-1-^{14}C-hexylamine to phosphorylase b at high ionic strength in a flow-dialysis cell.

The arrows indicate the addition of 10 µl 0.05 M n-1-^{14}C-hexylamine (5×10^7 cpm/ml) to the upper chamber containing 0.5 ml solution (●),(○). Samples were taken at the indicated times from effluent fractions of the lower chamber (1 fraction/min). The radioactivity of the amine in the fractions is given on the ordinate. For further details see ref. 9, 10 and Materials and Methods.

the three mentioned proteins (12) whereas the low affinity binding sites showed constants (K_2) in the range of $1\text{-}6 \times 10^2$ M^{-1}. From the apparent association constants of half-maximal saturation ($K_{0.5}$) for the binding of phosphorylase b to butyl agaroses (20-30 µmol/ml packed gel) of $9\text{-}16 \times 10^4$ M^{-1} (1,2) and the estimated minimum number (i.e. 3-4) of butyl residues interacting with the enzyme during nucleation (2,3) apparent association constants for a monovalent interaction of the enzyme with one butyl residue can be calculated according to $(9\text{-}16 \times 10^4)^{1/3-4}$ to be of the magnitude 17-54 M^{-1}. Binding constants in this range are below our detection limit (see Fig. 1) and also demonstrate that the binding of phosphorylase b to butyl Sepharose requires a multivalent type of mechanism.

Detection of low-affinity hydrophobic binding sites. For soluble, globular proteins the existence of hydrophobic areas on the accessible surface has been suggested by studies on protein structure (13,14). Such information is as yet unavailable for phosphorylase b. We could

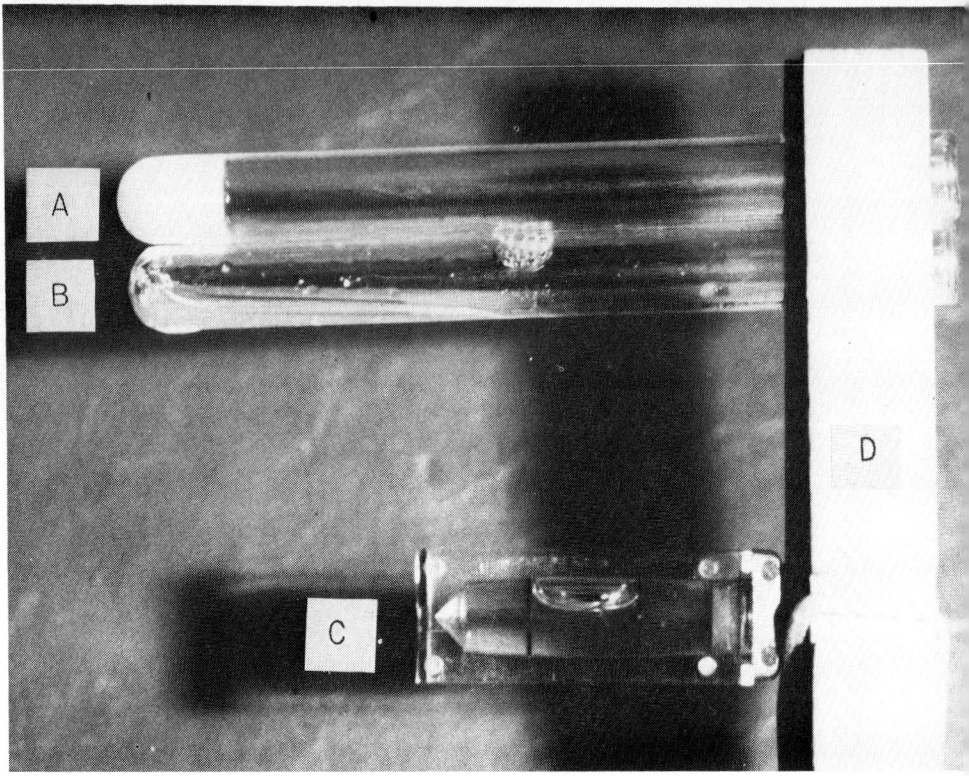

Fig. 2. Temperature induced sol-gel transition of phosphorylase b.
A. Gel-state of phosphorylase b aliquot after warming to 30° C.
B. Sol-state of second aliquot of enzyme kept at 0° C.
C. Spirit-level. D. Clamp.

however obtain strong evidence for the existence of such hydrophobic surface sites on phosphorylase b through the discovery of a fully reversible, temperature dependent sol-gel transition of the enzyme at hi ionic strength (see Fig. 2):

$$\text{phosphorylase } \underline{b}\text{-sol} \underset{0^{\circ}}{\overset{30^{\circ}}{\rightleftharpoons}} \text{phosphorylase } \underline{b}\text{-gel} \qquad (1$$

Since the gel developes after increasing the temperature from 0° to 1 endothermic, hydrophobic interactions apparently underly gel formatio This indicates the presence of numerous low-affinity hydrophobic surf sites on the enzyme which allow polymerization under these conditions

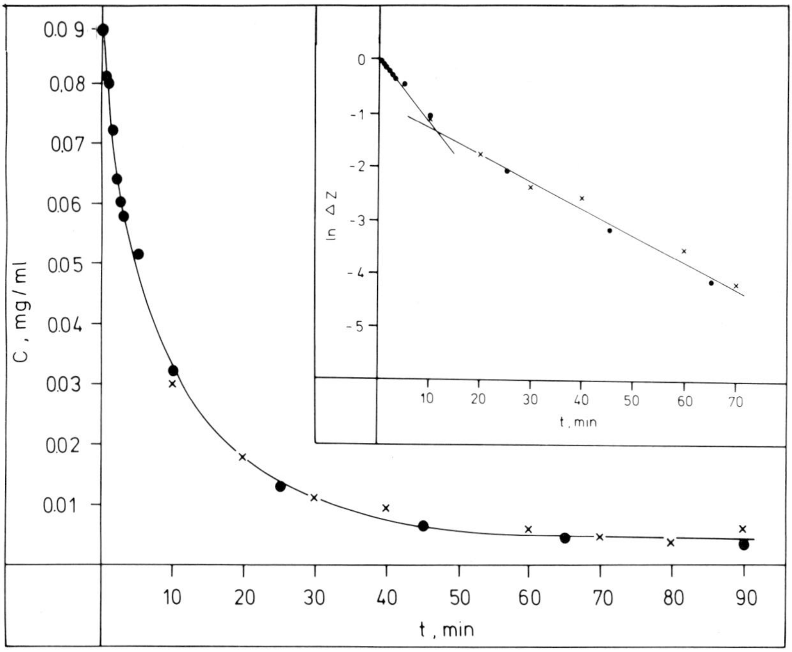

Fig. 3. Time dependence of the adsorption of ^{3}H-phosphorylase $\underline{b}_r$ on butyl Sepharose in buffer A.

In an incubation mixture of 60 ml (2,4) the reaction was started by addition of 1.5 ml packed butyl Sepharose (ca. 20 μmol/ml packed gel) corresponding to a final concentration of binding units of ca. 10 nM. The initial concentration (C) of enzyme was ca. 0.09 mg/ml (0.9 nM). At the indicated times samples were taken and counted (2). Insert: Semilogarithmic plot of the data. For definition of ln ΔZ see ref. 15. For further details see ref. 2,4 and the text. (●) and (X) denote two separate experiments under identical conditions.

Kinetic Studies

Adsorption kinetics. Fig. 3 shows the adsorption of ^{3}H-phosphorylase $\underline{b}_r$ on butyl Sepharose under pseudo-first order conditions i.e. $[Aga] \geq 10\,[E]$, where Aga denotes binding units (3,4) and E the enzyme ligand. The semilogarithmic plot in the insert shows that adsorption occurs in a biphasic manner: a fast phase ($t_{1/2} \simeq 5$ min) and a slow phase ($t_{1/2} \simeq 15$ min). Previously only the first phase was considered (4). From these phases the relaxation times (15) were calculated at increasing concentrations of butyl agarose (see Fig. 4). The reciprocal relaxation times of the fast phase appear to depend linearly on the concentration of butyl Sepharose, however those of the slow phase are practically independent of the concentration of excess component. The concentration independence of the latter relaxation times is strong evidence against a

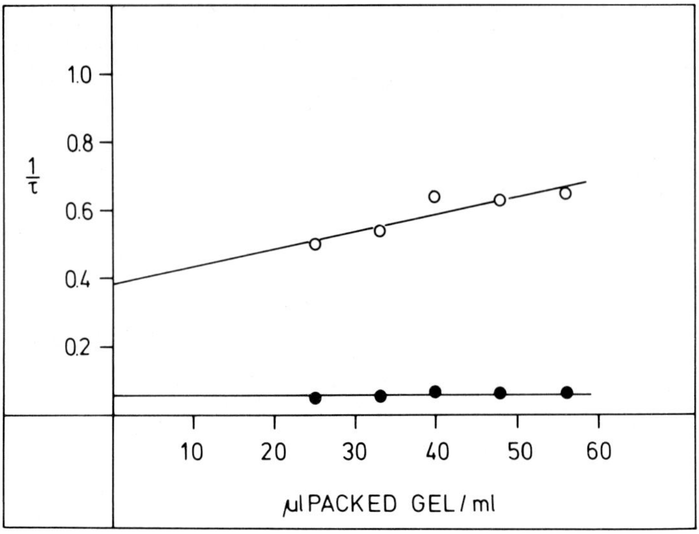

Fig. 4. Dependence of the reciprocal relaxation times (1/τ) of the fa (O) and slow (●) binding reactions on the concentration of butyl agar in the incubation mixture.

The relaxation times were derived from data analyzed as shown in Fig. For further details see legend to Fig. 3 and ref. 4 and 15.

mechanism based on two concurrent reactions due e.g. to a heterogenei of ligand or binding units. Therefore not a concurrent but a consecut pathway is proposed for initial binding:

$$E + Aga \underset{k_{-1}}{\overset{k_{+1}}{\rightleftharpoons}} E\text{-}Aga \underset{k_{-2}}{\overset{k_{+2}}{\rightleftharpoons}} E^{*}\text{-}Aga \qquad (2$$

(E^{*}) denotes a different species of bound ^{3}H-phosphorylase $\underline{b}_{r}$ due to increase in the number of butyl residues interacting with the enzyme (2) or possibly to an altered conformation. Preliminary values for th magnitude of the rate constants derived from Fig. 4 for the fast binding reaction of ^{3}H-phosphorylase $\underline{b}_{r}$ are k_{+1} = 13.5 mM^{-1} min^{-1} and k_{-} 0.332 min^{-1}. The respective constants for the second slow phase were calculated to be k_{+2} = 0.068 min^{-1} and k_{-2} = 0.038 min^{-1}. Further pro pagation steps (2) have as yet not been amenable to study.

In other experiments it has been found that the initial adsorptio rate is a hyperbolic function of the free enzyme concentration (4,16 This is especially interesting since the binding curves of phosphorylase b show negative cooperativity (1). On the other hand the result is in agreement with a saturable rate-limiting step of adsorption (see eq. 2.

Role of Enzyme Conformation

In the studies of alkane binding to proteins (17) it was found that conformational changes of the protein can exert a strong influence on the binding of hydrocarbons. Similarly conformational changes may be associated with the adsorption of proteins on alkyl agaroses and possibly with the phenomenon of adsorption hysteresis. Freeze-inactivation of phosphorylase b also appears to be related to conformational changes.

Freeze-inactivation of phosphorylase b. It has long been known that phosphorylase b can be reversibly inactivated by lowering the temperature to ca. 0° (18). On freezing however the enzyme is irreversibly denatured and precipitated (11). This freeze-inactivation can be prevented if certain cryoprotective solutes e.g. glucose, betain or hydroxprolin are added to buffer B in moderate concentrations (50-100 mM) (11). The mechanism of the inactivation and cryoprotection however remained obscure until it was found that the buffer constituents (i.e. buffer salts and mercaptoethanol) are responsible for the inactivation. Freeze-inactivation is almost entirely eliminated after removal of all buffer components from the enzyme solution as is shown in Table 2.

Table 2

Influence of freezing on phosphorylase b in the absence of buffer salt and mercaptoethanol.

	Betain mM	Phosphorylase b % activity
Unfrozen control	0	100
Frozen samples		
1	0	84
2	0.1	88
3	1	85
4	10	91

The concentration of solute free enzyme prepared as described in Materials and Methods was 0.3 mg/ml in buffer B. Details of the freeze-inactivation procedure are given in the same section.

After freezing there is only a small ca. 15% loss in activity and therefore no significant cryoprotective effect of betain. The highest concentration of betain was only 10 mM since the cryoprotective effect inc ses when the concentration of buffer B components is reduced. Since has been shown (18) that phosphorylase b can bind cystein in stoichi metric amounts and that the bound cystein can be exchanged by mercar ethanol, freeze-inactivation may be due to a direct interaction of m captoethanol and possibly buffer salts with the enzyme. This may lea to conformational changes of the enzyme. The following reaction sche is proposed for the freeze inactivation of phosphorylase b:

$$\text{Phosphorylase } \underline{b}_{(\text{cryostable})} \rightleftharpoons \text{Phosphorylase } \underline{b}_{(\text{cryolabile})} \qquad ($$

Thus the equilibrium of eq. 3 may be shifted from the cryostable for to the cryolabile form through the buffer constituents. Alternativel this might result at the high local solute concentrations occurring during freezing. This shift can be reversed by addition of cryoprote tants or by the removal of the inactivating buffer solutes (Table 2)

Further evidence for the importance of enzyme conformation in fre inactivation is the finding that phosphorylase b can be protected by adsorption to butyl Sepharose (Table 3).

Table 3

Protection of phosphorylase b against freeze inactivation by adsorpt on butyl Sepharose.

	Eluted enzyme U/mg	Yield %
Control gel (5°)	84	83
Frozen gel (-20°)	69	84

The amount of enzyme adsorbed was ca. 3 mg/ml packed gel which was taken as 100% in calculating the yield. The eluted fractions were pooled and analyzed. For further details see Materials and Methods.

Table 3 demonstrates that the enzyme adsorbed on butyl Sepharose in buffer B retains 82% of its initial specific activity, which it woul have fully lost by freezing in solution alone. We therefore conclude that the binding of immobilized butyl residues to phosphorylase b stabilizes the enzyme in a conformation not liable to freeze-inactiv tion.

ACKNOWLEDGMENTS

We thank Mrs. G. Botzet and Mrs. I. Bichbäumer for excellent technical assistence. This work was supported by Grant Je 84/6-5 from the Deutsche Forschungsgemeinschaft and by the Fonds der Chemie.

REFERENCES

1. H.P. Jennissen, Biochemistry, 15 (1976) 5683-5692
2. H.P. Jennissen and G. Botzet, Int. J. Biolog. Macromolecules, 1 (1979) 171-179
3. H.P. Jennissen, J. Chromatogr., 159 (1978) 71-83
4. H.P. Jennissen, in G. Weber (Ed.), Advances in Enzyme Regulation Vol. 19, Pergamon Press, New York, 1981 in press
5. H.P. Jennissen, J. Chromatogr., (1981) in press
6. H.P. Jennissen and L.M.G. Heilmeyer, Jr., Biochemistry, 14 (1975) 754-760
7. H.P. Jennissen, Protides Biol. Fluids Proc. Colloq., 23 (1976) 675-679
8. E. Logemann and H.P. Jennissen, Hoppe-Seyler's Z. Physiol. Chem., 361, (1980) 295-296
9. K. Feldmann, Anal. Biochem., 88 (1978) 225-235
10. S.P. Colowik and F.C. Womak, J. Biol. Chem., 244 (1969) 774-777
11. B. Schobert and H.P. Jennissen, Hoppe-Seyler's Z. Physiol. Chem., 361 (1980) 329-330
12. A. Wishnia, Biochemistry, 8 (1969) 5064-5070
13. I.M. Klotz, Arch. Biochem. Biophys., 138 (1970) 704-706
14. C. Chotia, J. Mol. Biol., 105 (1976) 1-14
15. C.F. Bernasconi, Relaxation Kinetics, Academic Press, New York, 1976, pp. 21-29, 141-147
16. H.P. Jennissen, J. Solid-Phase Biochem., 4 (1979) 151-165
17. A. Wishnia and T.W. Pinder, Biochemistry, 3 (1964) 1377-1384
18. S. Shaltiel, J.L. Hedrick and E.H. Fischer, Biochemistry, 8 (1969) 2429-2436

T.C.J. Gribnau, J. Visser and R.J.F. Nivard (Editors),
Affinity Chromatography and Related Techniques

THE CONCEPT OF GENERAL MULTISPECIFICITY - APPLICATION TO AFFINITY AND DYE-LIGAND CHROMATOGRAPHY

J.K. INMAN

National Institute of Allergy and Infectious Diseases, National Institutes of Health, Bethesda, Maryland 20205, USA

ABSTRACT

Affinity chromatography is optimally employed whenever it is feasible to carry out both adsorption and desorption biospecifically. Where the natural or recognized ligands are not available in amount and purity needed for these applications, it may be possible to search for and find alternative, specific ligands from synthetic and natural sources. The hypothesis of general multi-specificity of ligand-binding proteins predicts that, in any large collection of diverse, mutually unrelated structures (screened as potential ligands), there are a few members that bind specifically, in the same sense as does the "natural" ligand, and exhibit a binding strength in the range that would be useful in an affinity separation. The chances of finding such disparate ligands may be estimated from a statistical model which makes extrapolations from some existing information on monoclonal antibodies. The collection of potential ligands can be biased by *a priori* considerations to increase the frequency for finding suitable ligands. It is proposed that dye-ligand chromatography may be just such an endeavor. A direct experimental approach to validating the concept of general multispecificity is described which involves affinity chromatography for screening trials and for quantitative assessment of equilibrium association constants.

INTRODUCTION

In many instances where affinity chromatography has been applied, the substance to be isolated is biospecifically adsorbed from a crude mixture and then eluted by nonspecific means as, for example, with a change of pH and/or ionic strength or by introducing a denaturing agent (urea, chaotropic ions, etc.). Since some impurities are nonspecifically adsorbed at the same time, and the desired substance is often a minor component, nonspecific elution will remove a relatively large background of impurities from the support. The procedure still may be a useful purification step, but the potential of an affinity separation for giving very high separation factors has been sacrificed by not employing a specific component to effect desorption. Unfortunately, the use of naturally occurring, biospecific components, bound on supports and as soluble eluting species, is often prevented or severely restricted by one or more of the following circumstances: The adequately purified component (ligand or ligand-binding protein) is (1) not available or very expensive, (2) too insoluble, (3) too unstable, (4) gives too high a binding affinity, or (5) has a chemical structure unsuited for covalent attachment. In certain cases, analogues of an unavailable or unsuitable ligand have been synthesized and used successfully for affinity separations. In situations where neither this solution nor the attachment of a natural component is feasible, investigators have relied on less specific methods such as hydrophobic or hydrophobic/ion-exchange chromatography.

I wish to discuss a new, general approach to affinity chromatography that, if validated, could serve to overcome all of the above problems and allow additional opportunities to be explored. This approach is based on the hypothesis that all ligand-binding proteins (enzymes, antibodies, transport proteins, cell surface receptors, etc.) are generally multispecific; that is, they can bind occasional substances that have no recognizable connection with or structural resemblance to the natural ligand or to one another. Further, such disparate-structure bindings that have sufficiently high affinities to be useful in specific separations can be discovered with reasonable effort in a systematic search. The new approach starts, therefore, with the screening of a suitable set of diverse, synthetic ligands for observable binding. I would also like to propose that dye-ligand chromatography, now being practiced with increasing frequency, is, indeed, a special area of this approach which, in its general reach, could be termed, multispecific affinity chromatography (MAC).

THE CONCEPT OF GENERAL MULTISPECIFICITY

Traditional concepts of specificity (meaning selectivity) of antibodies and, to some extent, other ligand-binding proteins, have been derived largely from experience with immunological and serological systems. The observed phenomena are somewhat more subtle and complex than their readouts suggest, yet much of what we have believed about the specificity of antibodies has resulted from observing the selectivity of precipitation, agglutination and lytic assays, for example. Considering all that has been said and done since the turn of this century, surprisingly little data exists that bears directly on simple equilibrium binding at single, individual antibody combining regions of substances that are either similar or dissimilar to the eliciting antigenic determinant. It is selectivity in single-site binding that is of interest to us in the context of affinity chromatography and in giving us a true assessment of the specificity of antibodies. It is interesting, therefore, to speculate that antibodies are not monospecific as we have been led to believe from the behavior of complex, secondary phenomena.

Reversible binding is described by a decrease in the Gibbs' free energy attending the mixing of free antibody sites and haptenic determinants in aqueous solution under standard conditions. The standard free energy change, ΔG, is related to the mass action association constant, K, by the familiar relationship.

$$-\Delta G = RT\ln K = RT(\ln 10)\cdot(\log K)$$

where R is the gas constant and T is the absolute temperature ($^{\circ}K$). It is also understood that the observed free energy change involves, importantly, both components residing for some fraction of time in a definite relative configuration or conjunction at a binding site or region. There is no other requirement for simple "binding" to occur aside, perhaps, from defining a threshold. A measurable binding takes place whenever there is sufficient decrease of standard free energy on mixing antibody and ligand. Nothing needs to be stipulated about spatial complementarity of binding surfaces, etc. Instead, a great many factors contribute to the free energy decrease, such as, formation of a set of short- and long-range noncovalent interactions between atoms, changes and restrictions in conformation of ligand and antibody, changes in binding of water and small ions to each component, changes in solvent structure on removal of ligand from free solution, etc. The dynamic character of hapten binding by an antibody is revealed in the recent study of Zavodszky et al. [1].

It can be reasonably assumed that a ligand-binding protein and any potential ligand chosen completely at random, without regard for structural resemblance to a known ligand, will interact in such a way as to minimize the free energy of the system. A certain discrete energy minimum (or perhaps several minima) should be found and will be associated with a definite relative configuration(s) of the two principal components. That is to say, "binding" will have occurred. In most cases, these trials would lead to unobservably weak binding, but nothing fundamentally different from stronger, measurable associations will have occurred. If trials with mutually unrelated ligands are continued, circumstances occasionally should conspire to permit an appreciable minimization of free energy, and a measurable binding will occur. Less frequently, a ligand will be found that gives a fairly substantial association constant. A simplified drawing of two such ligands in their respective bound configurations is shown in Figure 1. This diagram is meant to illustrate just a few things: That there is (1) usually a little more space in the receptor binding cavity (or "region") than is needed to accommodate any one ligand; (2) there is considerable overlapping of the spaces occupied by the several ligands, and thus, their binding would be competitive in a mixed system; and (3) the configuration of microscopic events corresponding to the minimal free energy state is different in each case, that is, the disparate-structure bindings are multimodal with respect to the receptor. The figure can depict sets of short-range interactions and crude morphology but not the varied and dynamic changes involved. Finally, it should be pointed out that some degree of topological complementarity involving favored conformations

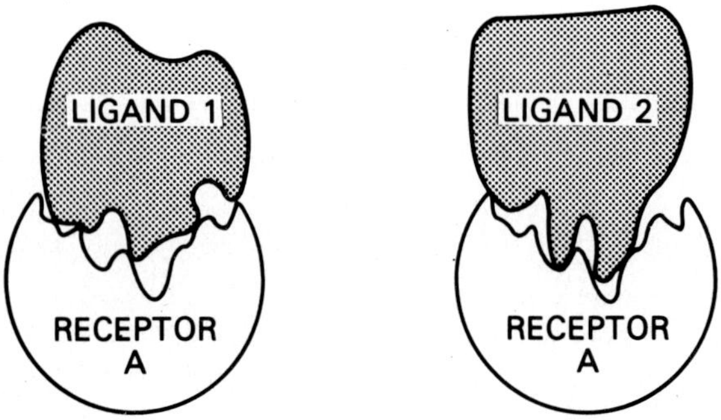

Fig. 1. A simple illustration of the binding configuration of two structurally unrelated ligands in the binding region of the same receptor in separate experiments.

will tend to increase the number of short-range interactions; yet, exact complementarity and thereby unique fit should not be necessary in order for good binding to occur. Permitting some leeway in "fitting" can allow for marked variation in modes of complementation.

Evidence for the multispecificity of antibodies has been reviewed by Inman [2]. Some disparate-structure crossreactions for restricted, normal antibody populations have been reported more recently. Nutt et al. [3] found anti-azophenylphosphorylcholine antibodies that could bind the 3-nitro-5-iodo-4-hydroxyphenylacetyl hapten. Cameron and Erlanger [4] studied anti-adenosine-5'-monophosphate rabbit antibodies that showed varied crossreactive profiles with such diverse substances as 2,4-dinitrophenylglycine, menadione, hydralazine and caffeine. Unfortunately, there have been too few studies along these lines to establish the generality of multispecificity.

AN EXPECTED FREQUENCY DISTRIBUTION OF ASSOCIATION CONSTANTS

In the discussion that follows, attention is focused on monoclonal (homogeneous), anti-hapten antibodies; these substances provide numerous, structurally varied and accessible models of ligand-binding proteins. Consider a hypothetical experiment wherein a number of antibodies are paired individually with each of a considerable number of haptens having varied and diverse structures; then, the equilibrium association constant, K, is measured for each pair tested. We will assume that the experimenter has the means for measuring K regardless of how small it happens to be. Each measurement is called a "test;" thus, 10 randomly selected antibodies and 10 diverse haptens yields 100 tests. How will values of K be distributed?

Several years ago I presented [2] a statistical model that described the expected probability that any such test will result in an association constant equal to or greater than a stipulated minimum value, K_{min}, or, simply, the frequency of tests resulting in K's equal to or greater than K_{min}. The solid curve in Figure 2 is a mathematically derived plot from the model. The model was based upon (1) a purely a priori distribution of weak, short-range (e.g., van der Waals) interactions, (2) reasonable assumptions about the likelihood of singular, more energetic events also occurring, and (3) some information from a few preliminary screening experiments. The circles represent data collected and reported by Eisen et al. [5] on about 350 myeloma immunoglobulins screened against the epsilon-dinitrophenyl-lysine hapten. In other studies, screening results on one myeloma protein with nearly 400 compounds [6], and on 10 myeloma proteins against 19 radioactive ligands [7], yielded several points falling very close to the solid curve. It is assumed that myeloma immunoglobulins behave statistically in the same manner toward diverse ligands as do normally raised, individual antibodies.

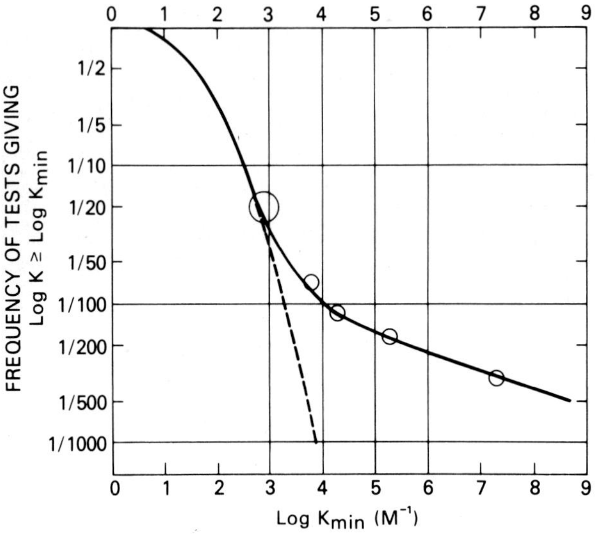

Fig. 2. A theoretically derived tail-accumulative distribution curve showing expected results from many hypothetical binding tests of monoclonal antibodies against disparately structured haptens. The solid curve is a plot of the expected frequencies of tests giving single-site association constants, K, equal to or greater than any stipulated value, K_{min}, as a function of log K_{min}. The statistical model employed [2] is based in part on limited, actual screening experiments (results indicated as circles). The dashed curve represents the calculated contribution of weak, short-range interactions in the model.

The number of weak, short-range interactions was treated in the statistical model as a binomial distribution, setting the most likely number to occur at 5, a number that is a little higher than the minimum of 3 contacting points allowed for irregular surfaces in full contact. The dashed curve of Figure 2 represents this distribution when translated into free energy changes expressed as log K. It illustrates what might be observed if only, say, van der Waals interactions were allowed. The solid, curve, by appropriate summations, incorporates the effects of higher energy events and bends upward and away from the dashed line. The first derivative of this composite function shows a maximum near log K = 1.55 (K = 35), the most likely value resulting from trials between randomly selected components, but a value lying in a range of K that is not generally observable. The chances of higher or lower K values occurring become progressively smaller. The Figure 2 curves do not reveal this maximum since they are accumulative type distributions.

The solid curve of Figure 2 shows that, as the lower-limit or threshold binding constant, K_{min}, is raised, the accumulative frequency of binding events falls off progressively. For example, 1 in 100 tests might give an association constant of 1×10^4 or higher and 1 in 200 a $K \geq 10^{5.5}$ M^{-1}.

SIGNIFICANCE FOR AFFINITY AND DYE-LIGAND CHROMATOGRAPHY

Let us suppose that an investigator has a mixture containing a ligand-binding protein behaving like a typical antibody, and he has about 100 diverse ligands or haptens separately coupled to supports in 100 small affinity columns. Suppose further that the natural or recognized ligand is available in very small amount, adequate only for setting up assays. This investigator can screen the mixture with the 100 columns, and if the model and assumptions discussed above are essentially correct, and if an association constant can be measured for each test, then this hypothetical experiment could give results like those shown in Figure 3. Each bar represents the result of a single test, and the results are arranged in random order as they might actually occur. Most binding events would fall below $K = 10^3$ M^{-1}.

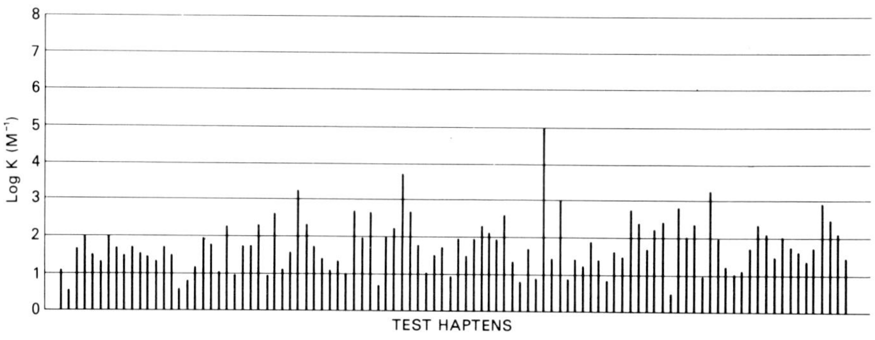

Fig. 3. "Results" from a hypothetical screening experiment with a "typical" antibody versus 100 diverse haptens having little or no structural resemblance to the eliciting hapten or to one another. All single-site association constants, K, are assumed measurable. This random order histogram was constructed from the modeled curve shown in Fig. 2.

A few tests might give values above $K = 10^{3.5}$; the corresponding ligands could serve usefully in affinity separations. In the log K range of 3.5 to 4.2 (roughly), buffer elution would serve to recover the retarded protein. At higher affinities soluble ligand elution would be recommended.

In actual practice, the above hypothetical experiment would be far too tedious for selecting a ligand, but the principle involved is best illustrated this way. The relatively low "success" frequency of 1 or 2 tests per 100 means that high selectivity in the ensuing separation could be achieved. If an extensive panel of affinity adsorbents were at hand, a simpler screening strategy could be devised. However, there is a still easier approach to using the multispecificity principle for affinity separations: First, it should be emphasized that the expected frequencies of binding constants shown in Figures 2 and 3 are averaged values and represent "typical" antibodies and haptens. Many antibodies could be expected to exhibit, by themselves, lower or higher frequencies at a given K_{min} value. A similar variation should occur in binding frequencies for ligands when they are individually tested against many antibodies. Thus, it is quite possible to find and sort out ligands that give higher binding frequencies, that is, ones that generate curves running well above the one plotted in Figure 2. A panel of adsorbents bearing these compounds would simplify the task of screening ligand-binding proteins to find useful alternative ligands. The work in finding this "high frequency" panel can be greatly reduced by a priori considerations: Ligands which possess molecular weights exceeding about 300, appreciable aromatic character, hyperconjugation of double bonds, a rich content of polarizable and electron-withdrawing or donating groups, etc. are likely candidates. Most available compounds filling this prescription are dyes. I would then like to propose that the highly successful, empirical approach to affinity separations called, dye-ligand chromatography, is indeed just such a usefully biased adaptation of the more general, multispecific affinity chromatography. Recognition of this general concept will allow one to adjust the frequency bias in screening ligands in order to balance the labor involved with the degree of selectivity desired, these being somewhat directly related factors. The underlying assumption being made here is that dye-ligand interactions are not fundamentally different, in regard to their selectivity characteristics, from other ligand-receptor interactions. That is, dye-ligands should exhibit fine-specificity binding profiles related to their detailed structural features rather than to gross characteristics, such as, net charge and hydrophobicity. Glazer [8] observed that dye-enzyme interactions were highly specific, and where good binding was observed, "dyes of closely related structure were either not bound or bound very weakly." More recent experiences from other labs appear to confirm this view, yet much well-directed work on this question remains to be done.

CURRENT AND PROSPECTIVE RESEARCH

It has been indicated that existing experimental evidence relevant to testing the validity of general multispecificity is scant and inconclusive. The statistical model and its implications discussed above should be taken only as a guide to further experimentation. The question of whether or not ligand-binding proteins are multispecific in the manner described is one of fundamental importance to biology. It appears to be a reasonable question and is amenable to experimental verification.

Currently, I am setting up a systematic screening program to test various portions of the frequency curve of Figure 2. Zones of radiolabeled, monoclonal antibodies (myeloma and hybridoma proteins) and other ligand-binding proteins will be passed through many small affinity columns with or without soluble ligands and the elution profiles will be studied. The ligands are large haptens, many of which are being synthesized for this study. A few systematic approaches to these syntheses have been reported [9, 10] which yield haptenic reagents that are reacted with amine supports through their reactive acyl azide groups. In my laboratory at present, haptens are being prepared that have chloroacetyl groups that can form stable thioether linkages upon reaction with sulfhdryl functions on supports [11].

It is feasible to relate zonal retention measurements from affinity columns to the equilibrium association constants for the interactions occurring in the column. Quantitative treatments of this problem, based on the assumption that equilibrium binding occurs throughout the run, have been reviewed by Chaiken [12] and developed for the case of mono- and bivalent antibodies by Eilat and Chaiken [13]. Recently, DeLisi and Hethcote [14] have treated the theory of affinity chromatography for zonal development where nonequilibrium is assumed. I have successfully applied their results in determining a single-site association constant for a bivalent myeloma protein with known reactivity with the 2,4-dinitrophenyl hapten. With a few simplifying assumptions, it was possible to develop from their treatment an expression that includes the measurable parameters of experiments. Thus, one can obtain the directly useful relationship

$$r = \frac{\bar{V} - V_d}{V_d} = A \cdot \frac{KN}{1 + KL} + B \cdot \left(\frac{KN}{1 + KL} \right)^2$$

for the case of a bivalent antibody applied zonally to an affinity column and developed with buffer containing soluble ligand of molar concentration, L. The elution volume of the antibody, $\bar{V}$, and the distribution volume, V_d, which is the elution volume of a similar, non-interacting antibody, are both expressed as the first moment of the elution peak, $\Sigma c_i v_i / \Sigma c_i$, where c_i is the concentration (actually, net counts per minute from a tritium label) of the ith fraction whose elution volume at the midpoint of its collection is v_i. A and

B are empirical coefficients that are derived from experimental runs. A allows for an essentially constant relationship between (1) void volume and V_d, (2) the effective support-bound ligand concentration and the actual, chemically determined value, N (molar), and (3) the single-site association constant for soluble (K) and support-bound hapten. In addition to these relationships, B includes a factor relating the K value for binding of the first site and the second step association constant in the case of a bivalent binding. The first and second terms in the above equation express the contributions to the relative retention, r, of mono- and bivalent binding processes, respectively. The coefficient, A, should be fairly constant for different antibodies of the same molecular size and for different haptens; B could vary somewhat between different pairs of components.

After a series of runs had been completed where L and N were varied, I obtained A, B and K simultaneously by an implicit solution involving averaging data via a linear regression analysis (method to be published). The obtained K closely matched the known value.

It should be possible with this method to screen with affinity columns and also to estimate single-site binding constants for values between $10^{2.5}$ and 10^6 M^{-1} when such bindings occur. Only a few micrograms of antibody per run are needed. Above $K = 10^6$ the method becomes insensitive to K and an alternative method, such as, the more tedious equilibrium dialysis techniques, would be required. This need would occur infrequently.

In conclusion, convenient methods appear to be at hand for directly exploring the generality of multispecific binding by soluble ligand-binding proteins even though the labor involved is quite large. Validation of this principle would be important for understanding the evolution and behavior of ligand-binding systems, antibody diversity, drug actions, and many other interesting questions. In particular, results of practical benefit may be obtained that will permit us to enlarge the scope of affinity chromatography.

ACKNOWLEDGMENTS

The author wishes to express his appreciation to Dr. Charles DeLisi and Dr. Herbert W. Hethcote of the National Cancer Institute, National Institutes of Health, for their helpful collaboration in applying their theory of affinity chromatography to a practical experimental approach for obtaining association constants of antibodies.

REFERENCES

1 P. Zavodszky, J-.C. Jaton, S.Yu. Venyaminoy and G.A. Medgyesi. Mol. Immunol., 18(1981)39-46.
2 J.K. Inman, in G.I. Bell, A.S. Perelson, G.H. Pimbley, Jr (Eds.), Theoretical Immunology, Marcel Dekker, New York, 1978, Ch. 9, pp.243-278.
3 N.B. Nutt, A.L. Grossberg and D. Pressman, Immunochem., 13(1976)559-564.
4 D.J. Cameron and B.F. Erlanger, Nature (London), 268(1977)763-765.
5 H.N. Eisen, M.C. Michaelides, B.J. Underdown, E.P. Schulenburg and E.S. Simms, Federation Proceedings, 29(1970)78-84.
6 F.F. Richards, L.M. Amzel, W.H. Konigsberg, B.N. Manjula, R.J. Poljak, R.W. Rosenstein, F. Saul and J.M. Varga, in E.E. Sercarz, A.R. Williamson and C.F. Fox (Eds.), The Immune System: Genes, Receptors, Signals, Academic Press, New York, 1974, pp.53-67.
7 J.M. Varga, S. Lande and F.F. Richards, J. Immunol., 112(1974)1565-1570.
8 A.N. Glazer, Proc. Natl. Acad. Sci. (USA), 65(1970)1057-1063.
9 J.K. Inman, B. Merchant and S.E. Tacey, Immunochem., 10(1973)153-163.
10 J.K. Inman and S.B. Shukla, in E. Gross and J. Meienhofer (Eds.), Peptides-Structure and Biological Function, Pierce Chemical Co., Rockford, IL. (USA), 1979, pp.949-951.
11 J.K. Inman, this volume.
12 I.M. Chaiken, Anal. Biochem., 97(1979)1-10.
13 D. Eilat and I.M. Chaiken, Biochemistry, 18(1979)790-795.
14 C. DeLisi and H.W. Hethcote, this volume.

T.C.J. Gribnau, J. Visser and R.J.F. Nivard (Editors),
Affinity Chromatography and Related Techniques

A THEORY OF COLUMN CHROMATOGRAPHY FOR SEQUENTIAL REACTIONS IN HETEROGENEOUS NONEQUILIBRIUM SYSTEMS: APPLICATION TO ANTIGEN-ANTIBODY REACTIONS

C. DeLISI and H.W. HETHCOTE*
Laboratory of Theoretical Biology, National Cancer Institute, National Institutes of Health, Bethesda, Maryland 20205

ABSTRACT

A theory of column chromatography has been developed to describe a heterogeneous, nonequilibrium system in which diffusion, transport and chemical reaction are all occurring simultaneously. The variables entering the equations are bead bound antigens, mobile antibodies that react specifically with the antigens but which have a distribution of rate and equilibrium constants for them, and mobile antigens which inhibit the antibody-bead bound antigen interaction. Characteristics of the affinity and rate constant distributions are related to characteristics of the elution profile by relatively simple expressions even for systems in which neither chemical equilibrium nor a local steady state has been established. The effect of movement in and out of the bead under nonideal conditions (activity coefficient different from unity) is included. The theory allows characterization of multistep reactions in heterogeneous systems. An important aspect of the development is the theoretical relation between the affinity of antibody for a surface bound antigen as opposed to the affinity for a free antigen. The theory suggests a method to obtain the affinity constants for both reactions.

1. INTRODUCTION

Column chromatography is used extensively for fractionating heterogeneous populations of molecules and for molecular weight determinations (ref. 1,2). It has been much less widely used as a quantitative method for studying chemical reactions, i.e., for determining equilibrium constants, rate constants and--when reactants are heterogeneous--their distributions.

*Permanent address: Department of Mathematics, University of Iowa, Iowa City, Iowa, 52242

Quantitative studies of the reaction between solution phase ligand and molecules bound to surfaces are assuming increasing importance because of developments in cell biology indicating that such reactions play a role in the regulation of cellular activity (ref. 3). Simply knowing that a ligand-receptor reaction must occur is not however, sufficient to understand the physical basis of regulation. Indeed cells ma be triggered to different types of activity merely by changing physical parameters such as ligand valence, ligand affinity and receptor number (ref. 4,5). Moreover, binding often involves multiple steps that may trigger competitive processes, and when this happens, knowledge of the rate constants for each step is important. In addition, receptors are sometimes heterogeneous in their affinity for ligand, as in the case of B-cell bound immunoglobulin, and the distribution must be known if one wishes to understand how antibody affinity is regulated. Finally, theory predicts that the equilibrium and rate constants for cell bound receptors may differ substantially from their values when the receptors are dispersed in solution (ref. 6), even when the reaction mechanism is the same. Systematic experimental studies of the relation between the two will be of considerable importance if we wish to reliably extrapolate--as must often be done--from dispersed to cell bound systems.

These brief and incomplete remarks are intended to provide some biological perspective for the importance of quantitative chemical reaction studies, and consequently for the necessity of having widely available, simple, fast and reliable methods for carrying out such studies. In this paper we show how column chromatography can be used to obtain information that bears upon the above topics, especially in relation to antigen-antibody interactions. More specifically, we distinguish between equilibrium constants of ligand for free hapten and bead bound hapten, and show how both can be found when the ligand is monovalent and the population homogeneous. Generalized equations applicable to bivalent ligands having a distribution of affinities for hapten are then presented. Finally we illustrate by way of a simple example, how the rate constants for reaction are related to the dispersion in the elution profile, thus suggesting the possibility of obtaining rate constant information from second and perhaps higher order moments of the profile. We will begin with a nonequilibrium theory of molecular sieving.

2. MOLECULAR SIEVING

2.1 General Description

The literature abounds with detailed descriptions of the experimental procedure (ref. 1,2). Briefly, we will be considering a cylindrical column packed to a height h with beads that are composed of cross-linked polymers. On a molecular scale the interior of a bead can be thought of as a network of tortuous channels of various sizes.

At time t = 0, a small amount of the ligand to be studied is introduced at the top of the column in a layer whose thickness is negligible compared to h. The length of the column is oriented along the direction of some external field (usually gravitational) so that the ligands move through the column by transport and diffusion. Movement down the column is, however, delayed for various periods of time as a consequence of diffusing out of the mobile phase (i.e., the void volume between the beads) and into the stationary phase (i.e., the penetrable or interior volume of the beads to which the ligand has access). Since the rates for entering and leaving a bead will depend upon, among other things, ligand size and geometry, the time taken to traverse the column will also depend upon ligand size and geometry.

2.2 The Model

Since we are discussing conditions under which the ligand concentration in any local region of the column can be considered relatively dilute, only a very small fraction of the bead interiors will be occupied by ligand. Consequently the general nonlinear problem becomes essentially a linear one, i.e., terms higher than first order in concentration do not appear in the equations. Thus let C be the number of ligands per unit void volume (mobile phase) and let B be the number per unit penetrable volume. With A_o and A_p representing respectively the average void and penetrable cross-sectional areas, CA_o and BA_p are the number of ligands per unit distance in each phase. The diffusion-reaction-transport equations are, in terms of these variables (ref. 7)

$$\frac{\partial}{\partial t}(CA_o) = D\frac{\partial^2}{\partial z^2}(CA_o) + u\frac{\partial}{\partial z}(CA_o) - k_1CA_o + k_{-1}BA_p \qquad (1)$$

$$\frac{\partial}{\partial t}(BA_p) = k_1(CA_o) - k_{-1}(BA_p) \qquad (2)$$

where k_1 and k_{-1} are the rate constants for transition between the two phases, u is the transport velocity, D the diffusion coefficient of the ligand and z the distance from the bottom of the column. The top of the column is at z = h.

Equation 1 describes the rate of change of the mobile concentration in terms of contributions from diffusion (the first term on the right), transport (the second term on the right) and interchanges between the two phases (the last two terms). Equation 2 simply says that when a ligand enters the stationary phase, its average residence time is $1/k_{-1}$; the details of what it may be doing inside the bead being on average unimportant to its large-scale movement. The residence time is, of course, a function of the size and geometry of the ligand. In general k_1 and k_{-1} are concentration dependent, but under the ideal conditions indicated above (dilute ligand concentration), they can be considered essentially constant. Equations 1 and 2 also neglect the possibility of bead size heterogeneity and nonuniform packing. Both these restrictions can be relieved somewhat (ref. 8).

It will be convenient to write eqs. 1 and 2 in terms of density functions; i.e., probabilities per unit distance of finding a ligand at a particular position in the column. This is accomplished by dividing the equations by I, the total number of ligands initially present. Thus with

$$p \equiv CA_o/I \tag{3}$$

and

$$q \equiv BA_p/I \tag{4}$$

eqs. 1 and 2 become

$$\frac{\partial p}{\partial t} = \frac{D\partial^2 p}{\partial z^2} + \frac{u\partial p}{\partial z} - k_1 p + k_{-1} q \tag{5}$$

$$\frac{\partial q}{\partial t} = k_1 p - k_{-1} q \tag{6}$$

These equations are to be solved subject to the conditions that the mobile phase concentration is initially zero everywhere except at $z = h$ where a very thin layer exists. Exact solutions can be obtained for $D = 0$ and excellent approximations can be obtained for the general diffusion-reaction-transport problem (ref. 7). However, solution to the partial differential equations is not necessary for obtaining expressions for the quantities of primary experimental interest: the mean of the elution profile and the dispersion around the mean.

2.3 Mean Time for Passage Through the Column

We are interested in the mean time required for the average ligand to reach a position z in the column and, more specifically, in the mean time for elution from a column of height h.

We assume that above $z = h$, there is solvent but no beads, and that ligands are prohibited from moving upward through h. The latter condition is exactly satisfied when diffusion can be neglected, but an exact treatment with diffusion included requires relaxation of this condition. To keep the development simple, we restrict the presentation to situations in which the contribution to movement due to diffusion can be neglected relative to the contribution due to transport. An exact solution can also be obtained when diffusion is important and we will comment on this below.

The mean passage time $T(z)$ for a particle starting at h at $t = 0$ to arrive at z is, by definition (ref. 9)

$$T(z) = -\int_z^h dz' \int_0^\infty t \frac{\partial}{\partial t} [p(z',t) + q(z', t)]\, dt \tag{7}$$

One can show by using eqs. 5-7 that T satisfies the ordinary differential equation (Appendix A)

$$\frac{dT}{dz} = -\frac{(1 + K)}{u} \tag{8}$$

where $K \equiv k_1/k_{-1}$. See (ref. 7) for a alternate definition of the mean passage time and derivation of (8). The solution to eq. 8 subject to the condition $T(h) = 0$ (from eq. 7) is

$$T(z) = (1 + K)(h-z)/u \tag{9}$$

so that the mean time for arrival at the bottom of the column is

$$T(0) = (1 + K)h/u \tag{10}$$

These equations can be rewritten in terms of the readily measurable experimental quantities, V_p and V_o, the penetrable and void volumes. From eqs. 3, 4 and 6 we find the equilibrium constraint

$$K = V_p/V_o \tag{11}$$

Therefore eq. 10 becomes

$$T(0) = (1 + V_p/V_o)h/u \tag{12}$$

The result of course assumes that the equilibrium distribution between the void and penetrable volumes is determined only entropically; i.e., nonspecific interaction between ligand and bead is assumed negligible. In the presence of nonspecific binding a multiplicative constant would precede V_p.

When diffusion along the length of the column is included, the derivation is more complicated, but the result changes in a numerically trivial manner. In particular, if u_1 and D_1 are the transport velocity and diffusion coefficient in the solvent above the beads, and f is the total length of the column, including th solvent above the beads, then (ref. 7)

$$T(0) = (1 + K)(h/u) \left\{ 1 - (D/uh)(1 - e^{-uh/D})[1 - \frac{D_1 u}{Du_1(1+K)} (1 - e^{-u_1(f-h)/D_1})] \right\} \quad (1$$

Thus if $D/uh << 1$, diffusion is unimportant. As an example with $D = 10^{-6}$ cm^2/sec, $h = 25$ mm, $u = 0.04$ mm/sec, $\frac{D}{uh} = 10^{-4} << 1$. More generally if diffusion is to contribute less than 1 percent to the mean time (clearly less than the experimenta errors), a reasonable criterion for neglect of diffusion is

$$D/u << .01\ h \qquad (14)$$

2.4 The Mean Eluted Volume: Definition and Implications

It is often convenient to record experimental results in terms of the volume eluted at a particular time, rather than the time required to elute a particular volume. Since the volume eluted at time t is $uA_o t = (uV_o/h)t$, then the volume elu at T(0) is

$$V_e = V_o + V_p \qquad (15)$$

For simplicity we will refer to V_e, the volume eluted during the mean time require by a molecule to traverse the column, as the *mean (eluted) volume*. It is to be carefully distinguished from the *peak* (eluted) *volume* which uses the time at which the profile peaks.

It is important to notice *that eq. 15 was derived without equilbrium assumption* The result is *generally valid under nonequilibrium conditions* subject only to the assumptions underlying eqs. 6 and 7 as a model for sieving and the recognition tha the contribution of diffusion to the mean value of the elution profile is small compared to the contribution of transport.

Equation 15 is *formally* identical to a standard equation used in molecular siev However its interpretation is somewhat different. The standard equation says that the *peak* eluted volume varies linearly with the penetrable volume; eq. 15 says tha the *mean* eluted volume varies linearly with the penetrable volume. The mean is of course a weighted volumetric average over the entire profile and in general is equ to the mean only for symmetric profiles. For the model represented by eqs. 6 and the peak and mean are exactly equal only when the initial layer moves down the col as a sharp front. This happens under local equilibrium conditions, but then there no dispersion in the profile due to nonequilibration.

The distinction between the mean volume and the peak volume is important in principle and in practice. According to eq. 15 with the void volume fixed, the penetrable volume is uniquely determined by the mean eluted volume. This is not true for the peak eluted volume which in general depends on V_p and either k_1 or k_{-1} (ref 7). The distinction is important when molecular weight is low ($\lesssim$ 40,000): mass determinations may then be in error by as much as a factor of two. Moreover--and this will be especially important in the case of bivalent antibodies--eq. 15 holds under nonequilibrium conditions, wheras the expression using volume eluted at the peak holds only under local equilibrium. These results have been generalized by making k_1 and k_{-1} linearly dependent on concentration, and solving the resulting equations by a perturbation method (ref. 8).

2.5 Heterogeneity in Bead Residency Times ($1/k_{-1}$)

The results of the last section indicate a simple linear relation between the average eluted volume and the penetrable bead volume. The relation assumes complete uniformity in bead size and packing and in the size, shape and weight of the molecules. Here we consider the effect on the results when the uniformity condition is not fulfilled. The development will also be applicable to certain types of heterogeneous reactive systems, e.g., inpenetrable beads that have haptens covalently bound to their surfaces, reacting with a heterogeneous monovalent antibody population. To focus ideas we will suppose that K is distributed and that heterogeneity arises solely as the result of the distribution in residency times $1/k_{-1}$. We denote by n(K)dK the fraction of constants in the interval (K,K + dK).

The quantity of interest is now the mean time to reach any position in the column averaged over n(K). We denote this average by an overbar so that

$$\overline{T}(z) = \int_0^\infty dKn(K) \int_z^h dz' \int_0^\infty t\frac{\partial}{\partial t}(p + q)dt = \int_0^\infty dKn(K)T(z) \tag{16}$$

From eqs. 9 and 16

$$\overline{T}(z) = (1 + \overline{K})(h - z)/u \tag{17}$$

so that

$$\overline{V}_e = V_o(1 + \overline{K}) \tag{18}$$

For molecular sieving in which molecules move in and out of beads without attac ments, eq. 18 can be written in terms of the average penetrable volume. Thus with

$$\overline{K} = \overline{V}_p/V_o \tag{19}$$

$$\overline{V}_e = V_o + \overline{V}_p \tag{20}$$

The effect of heterogenity, therefore, is simply to replace K by its mean. Although such an inconsequential change is to some extent desirable, it is also unfortunate since information about the width of the distribution (or higher momen does not appear and is therefore not obtainable by mean time measurements. Howeve moments other than the mean eluted volume can be used to obtain additional informa as shown in the next section.

2.6 Dispersion in the Elution Profile

The fact that the thin layer of ligand at the top of the column is not eluted a a thin layer can be attributed in large part to nonequilibrium effects, and perhap also to heterogeneity in bead packing and size, and in the ligands. In the last section we have developed an expression for the average over the elution profile. In this section we comment on the dispersion in the elution profile.

If we define the average variance as

$$\overline{S(z)} = \overline{T_2(z)} - \overline{T_1(z)}^2 \tag{21}$$

where

$$\overline{T_j(z)} = \int_0^\infty n(K)dK \int_z^h dz' \int_0^\infty t^j \frac{\partial}{\partial t}(p + q)dt \tag{22}$$

Then a differential equation can be obtained for $\overline{S}$ in the same way it was found for $T = T_1$. In particular (ref. 8)

$$u\frac{\partial \overline{S}}{\partial z} = \frac{-2\overline{K^2}}{k_1} \tag{23}$$

with solution

$$\overline{S(0)} = \frac{2}{k_1}\frac{h}{u}\overline{K^2} = \frac{2}{k_1 V_o^2}\frac{h}{u}\overline{V_p^2} = \frac{2}{k_1 V_o^2}\frac{h}{u}[\overline{V}_p^2 + (\overline{V_p^2} - \overline{V}_p^2)] \qquad (24)$$

Thus the width of the elution profile varies linearly with the second moment of the distribution of the beads and the ligands. Note that the first term in the brackets is the square of the mean of V_p while the second term is the variance of V_p.

Although dispersion in bead size, ligand size or packing uniformity affects the dispersion in the elution profile, it has only a trivial effect on the mean eluted volume and therefore does not obscure the parameters of interest. Explicit averages taken over such distributions will be neglected in section 3.

3. CHEMICAL REACTIONS

In this section we extend the preceding development to antibody-hapten reactions. We shall, however, in this necessarily brief presentation, focus on information obtainable from the mean eluted volume, leaving aside information obtainable from the dispersion in the eluted volume. The interested reader can consult the references or contact either of the authors for additional information.

We consider an experiment in which hapten is either covalently bound throughout the beads or free, but it is in either case uniformly distributed throughout the column. Antiserum is introduced as a thin layer on top of the column, with an initial concentration such that the total amount of antibody at any time at any position in the column is small compared to either the bound or free hapten concentration. This condition again permits us to ignore any depletion in hapten, either bead-bound or free, and thus to consider a linear system of equations.

3.1 Homogeneous, Monovalent Antibodies

Let p and p_1 be the probability densities for free and bound antibodies outside the bead, and let q, q_1 and q_2 be the probability densities of antibodies inside the bead that are either free, bound to mobile hapten or bound to matrix hapten, respectively. Further define k_2 and k_{-2} as the forward and reverse rate constants for reactions between antibody and soluble hapten and k_3 and k_{-3} as the forward and reverse rate constants for reaction with matrix hapten. Then the equations of the model are

$$\frac{\partial p}{\partial t} = u\frac{\partial p}{\partial z} - k_1 p + k_{-1} q - k_2 Hp + k_{-2} p_1 \tag{25}$$

$$\frac{\partial p_1}{\partial t} = u\frac{\partial p_1}{\partial z} - k_1 p_1 + k_{-1} q_1 + k_2 Hp - k_{-2} p_1 \tag{26}$$

$$\frac{\partial q}{\partial t} = k_1 p - k_{-1} q - k_2 Hq + k_{-2} q_1 - k_3 Nq + k_{-3} q_2 \tag{27}$$

$$\frac{\partial q_1}{\partial t} = k_1 p_1 - k_{-1} q_1 + k_2 Hq - k_{-2} q_1 \tag{28}$$

$$\frac{\partial q_2}{\partial t} = k_3 Nq - k_{-3} q_2 \tag{29}$$

where H is the concentration of soluble hapten, and N the concentration of hapten bound to the bead matrix that is accessible to the antibody.

The distinction between (k_2, k_{-2}) and (k_3, k_{-3}) may be important. Theory predicts that the rate and equilibrium constants governing the reaction between soluble antibodies and hapten bound to a surface can change substantially when the surface bound haptens are dispersed in solution (ref. 6). In particular, the equilibrium constant in the former case may be four to five orders of magnitude higher than in the latter, even when the reaction mechanisms are identical. The extent to which this prediction bears on the present situation in which hapten is bound throughout the bead is not clear *a priori*, but may be answerable experimentally (see below).

The relation between the mean eluted volume and the equilibrium constants governing the reactions can again be found by calculating the mean time for an antibody to pass through the column. The result for the mean eluted volume is (ref. 10)

$$V_e = V_o + V_p[1 + K_3 N/(1 + K_2 H)] \tag{3}$$

where $K_3 \equiv k_3/k_{-3}$ and $K_2 \equiv k_2/k_{-2}$.

Eq. 30 assumes that the reaction between soluble hapten and antibody equilibrates rapidly, but no other equilibrium assumptions are made, i.e., the equation is valid even when reaction with the bead is not in local equilibrium. When antibody-hapten equilibrium is not rapid, the expression for the eluted volume is considerably more complicated, and involves the rate constants k_2, k_{-2}, k_1, k_{-1}. An exact treatment of the full kinetic problem, as well as its applicability, will be presented elsewhere (ref. 10). The equilibrium assumption for the soluble hapten reaction will be valid for dissociation times

$$1/k_{-2} \lesssim h/u \qquad (31)$$

In practice eq. 31 requires dissociation times shorter than minutes. Since typical antibody site hapten dissociation times are seconds or shorter (ref. 11), we expect the approximation to be generally valid for antibodies.

The ratio K_3/K_2 is readily determined. Let ΔV_e denote the difference between the mean eluted volume with $N > 0$ and $N = 0$. Then from eq. 30

$$\frac{\Delta V_e}{V_p} = (K_3N/K_2)/H; \qquad K_2H > 1 \qquad (32)$$

so that a plot of the left side of eq. 32 against $1/H$ will be a straight line passing through the origin with a slope K_3N/K_2. Since N is experimentally controllable, K_3/K_2 can be determined. The hapten concentration required for the validity of eq. 32 is easily determined experimentally since only when $K_2H > 1$ will the left side of the equation be dependent on H.

A determination of the actual magnitudes of K_3 and K_2 may be somewhat less accurate since it requires experiments in the absence of soluble hapten. With $H = 0$ the equation corresponding to 32 is

$$\Delta V_e = K_3N \qquad (33)$$

so that K_3 can be determined from the slope of the left side against matrix bound hapten concentration (N).

3.2 Homogeneous Bivalent Antibodies

In this case, the avoidance of an equilibrium assumption when antibody reacts with matrix bound hapten may be especially important because of the possibility of bivalent attachments. These attachments will likely not be in local equilibrium with free antibody, and consequently a derivation of the elution volume that avoids such an assumption is important. The results (ref. 10) corresponding to eq. 30 and 32 are

$$V_e = V_o + V_p \left[1 + \frac{2K_3N}{1 + K_2H} + \frac{K_3N\ K_4N}{(1 + K_2H)^2} \right] \qquad (34)$$

$$\frac{\Delta V_e}{V_p} = \frac{2K_3}{K_2} \left(\frac{N}{H}\right) + \frac{K_3\ K_4}{K_2^2} \left(\frac{N}{H}\right)^2 ; \ K_2H > 1 \qquad (35)$$

According to eq. 35, if bivalent attachments are not occurring, a plot of the left hand side of eq. 35 against 1/H will be a straight line; if they are occurring, t plot will be a parabola. A least square fit of the data determines K_3/K_2 and K_3K_4 If either K_2 or K_3 can be determined separately as described above, then all three equilibrium constants can be found. A similar result has been obtained by Eilat et al. (ref. 12). The main difference is that here, V_e is the mean, not the peak eluted volume, and we have not assumed an equilibrium controlled reaction between antibody and bead bound hapten.

3.3 Heterogeneous Populations

With the exception of myeloma proteins and a few IgG hybridomas that have rece been developed, the overwhelming majority of antibody populations are heterogeneo in their affinity for ligand. The source of this heterogeneity resides principal in the reverse rate constant distributions. Generalization of the above results take account of heterogeneity follows the development in section 2.5. In the present case one has that, for monovalent antibodies

$$\Delta V_e \cong \overline{(K_3/K_2)}\ (N/H); \qquad K_2H > 1, \text{ all } K_2 \qquad (36)$$

Equation 36 simplifies since, for any particular antibody, K_3 and K_2 are highly correlated, differing only in their diffusive parts (ref. 6). Therefore one can write that

$$K_3 = bK_2 \tag{37}$$

where b is independent of affinity heterogeneity. The right hand side of eq. 36 then reduces to bN/H, and therefore b can be determined. When H = 0, one obtains in place of eq. 36

$$\Delta V_e \cong \overline{K_3}N \tag{38}$$

Therefore the means $\overline{K_3}$ and $\overline{K_2}$ can be obtained. Higher order moments as well as information on the rate constants can be obtained by measuring dispersion along the column (ref. 10).

ACKNOWLEDGEMENT

It is a pleasure to acknowedge a number of stimulating conversations with Drs. Matthew Pincus and Marc Rendell during the early phase of this work.

APPENDIX A.

Differential Equation for the Mean First Passage Time

By definition

$$T(z) = -\int_z^h dz' \int_0^\infty t \frac{\partial}{\partial t}(p + q)\, dt \tag{A.1}$$

Differentiating with respect to z

$$\frac{dT}{dz} = \int_0^\infty t \frac{\partial (p + q)}{\partial t}\, dt \tag{A.2}$$

Integrating the right hand side by parts yields

$$\frac{dT}{dz} = -\int_0^\infty (p + q)\, dt \tag{A.3}$$

Integrate eq. 6 over time to obtain

$$\int_0^\infty q\,dt = K \int_0^\infty p\,dt \qquad (A.4)$$

Substitution of eq. A.4 into eq. A.3 gives

$$\frac{dT}{dz} = -(1 + K)\int_0^\infty p\,dt \qquad (A.5)$$

The integral on the right can be evaluated by integrating eq. 5 with D = 0 over time.

$$-p(z,o) = u\int_0^\infty \frac{dp}{dz}dt - k_1\int_0^\infty p\,dt + k_{-1}\int_0^\infty q\,dt \qquad (A.6)$$

The last two terms on the right vanish because of eq. A.4. We now take

$$p(z,o) = \delta(z-h)$$

and a no flux condition (up(h) = o) at z = h. Then, upon integrating eq. A.6 over z,

$$\frac{1}{u} = \int_0^\infty p\,dt \qquad (A.7)$$

Eq. 8 in the text follow immediately from eq. A.5 and A.7.

Summary of Key Equations

Equation	Comments
$V_e = V_o + V_p$; eq. 15	$D/uh \ll 1$; no chemical reaction; neglects dispersion in bead size and packing, does not assume local equilibrium.
$S(o) = \dfrac{2}{k_1 V_o^2}\left(\dfrac{h}{u}\right)\overline{V_p^2}$; eq. 24	Same as above.
$\dfrac{\Delta V_e}{V_p} = \dfrac{K_3 N}{1 + K_2 H}$; eq. 30	Monovalent ligand; monovalent antibody; homogeneous affinity, $k_{-1}/k_{-2} \ll 1$.
$\dfrac{\Delta V_e}{V_p} = \dfrac{2K_3 N}{1 + K_2 H} + \dfrac{K_3 N K_4 N}{(1 + K_2 H)^2}$; eq. 34	Monovalent ligand; bivalent antibody. If N is known only to within a constant, K_2 and K_3/K_4 can be found, but not the actual magnitudes of K_3 and K_4.

Notation

V_e = Mean value of elution profile.

V_o = Void volume.

V_p = Penetrable volume.

$S(o)$ = Dispersion about the mean (variance).

D = Effective diffusion coefficient of macromolecule.

u = Velocity of macromolecule.

h = Height of column.

k_1 = Rate constant for moving into bead.

ΔV_e = Mean value of elution profile in presence of bead bound ligand minus mean value in its absence.

K_3 = Site-site equilibrium constant for monovalent binding of antibody to bead bound ligand.

K_4 = Equilibrium constant for bivalent binding to bead bound ligand of antibody singly bound to a bead.

K_2 = Site-site equilibrium constant for antibody soluble-ligand interaction.

N = Concentration of bead bound ligand accessible to antibody.

H = Concentration of soluble ligand.

REFERENCES

1 K.H. Altgelt and L. Segal (Eds.), Permeation Chromatography, Marcel Dekker, (1971), New York.
2 G.K. Ackers, The Proteins 1 (1975) 1-94.
3 C. DeLisi and R. Blumenthal (Eds.) Physical Chemistry of Cell Surface Events and Cellular Regulation, Elsevier, North Holland, 1978.
4 C. DeLisi and R. Siraganian, J. Immunol., 122 (1979) 2293-2299.
5 M. Dembo and B. Goldstein, Cell, 22 (1980) 59-67.
6 C. DeLisi, The Effect of Cell Size and Receptor Density on Receptor-ligand Rate Constants. Mol. Immunol., 18, (1981) 507-511.
7 H.W. Hethcote and C. DeLisi, A Nonequilibrium Model of Liquid Column Chromatography I: Exact Expression for Elution Profile Moments, submitted for publication.
8 C. DeLisi and H.W. Hethcote, A Nonequilibrium Model of Liquid Column Chromatography II: Explicit Solutions and Non Ideal Conditions, submitted for publication.
9. G.H. Weiss, in Advan, Chem. Phys., Vol., 13, Wiley, New York, 1967, p. 1.
10 H.W. Hethcote and C. DeLisi, Determination of Equilibrium and Rate Constants by Column Chromatography, in preparation.
11 G.I. Bell and C. DeLisi, Cellular Immunol. 10 (1974) 415-431.
12 D. Eilat, I.M. Chaiken and W.M. McCormack, Biochem. 18 (1979) 790-795.

T.C.J. Gribnau, J. Visser and R.J.F. Nivard (Editors),
Affinity Chromatography and Related Techniques

HORMONE-RECEPTOR INTERACTIONS IN HOMOGENEOUS AND HETEROGENEOUS SYSTEMS: RELATIONSHIP TO AFFINITY CHROMATOGRAPHY

I. PARIKH and P. CUATRECASAS

The Wellcome Research Laboratories
Research Triangle Park, North Carolina 27709, USA

INTRODUCTION

Paul Ehrlich, as early as 1909, introduced the concept of reception and resulting response of hormonal signals through cell surface receptors. It is only during the last ten years that significant progress has been made in the understanding of the nature and function of the cell surface receptors. This progress is primarily due to the successful application of affinity chromatography for isolation of hormone-specific receptor proteins. Since the advent of affinity chromatography and the availability of a wide variety of chemical coupling reactions, together with various hydrophilic neutral solid supports, the isolation and purification of many of the hormone receptors has become experimentally feasible (1-9). However, molecules such as drug and hormone receptors are usually present in such minute quantities in biological tissues that conventional purification techniques such as gel filtration or separation based on electrophoretic mobilities may be inadequate for their purification.

A few hormone and drug receptors have been isolated by affinity chromatography and some of these have been further characterized. Even with a technique as powerful as affinity chromatography, special strategies are necessary for each individual system because of the extremely small quantities of receptor proteins present in biological system.

The biphasic heterogeneous interactions between a soluble hormone receptor (solubilized membranes or cytosolic receptors) and an immobilized ligand in affinity chromatography is well recognized. In physiological systems, the hormone-receptor interactions may be heterogeneous (membrane-bound receptors) or homogeneous (cytoplasmic receptors). In spite of the obvious similarities between the two heterogeneous systems, the overall situation is very different. Among others, the two major qualitative differences between the two heterogeneous systems are: a) the nature of the soluble and/or immobilized components and b) the "local" concentration of the insoluble component.

It is now well recognized that a hormone or a drug upon immobilization may or may not lose its ability to recognize and bind to the specific receptor protein [10]. This may depend on the point of attachment of the ligand to the matrix, chemical and physical nature of the matrix and possible steric hindrance encounter in the binding with the receptor. The concept of interposing a spacer arm between the ligand and the matrix was introduced solely to modify the properties of the immobilized ligand as closely as possible to those of the free ligand [3]. With appropriate selection of the spacer arm, it is possible to almost eliminate the physical and chemical restriction imposed upon by the matrix which are often responsible in lowering the overall affinity of an immobilized ligand (Table I). It is interesting to note that solubilization of a membrane-bound receptor appears to result in an increased affinity to the hormone (Table II). Th

TABLE I

COMPARISON OF RECEPTOR AFFINITIES TO IMMOBILIZED AND FREE HORMONES

Receptor	Hormone	Hormone Free (Kd)	Hormone Immobilized (Kd)	Fold-decrease in Kd	Reference
β-Adrenergic	(±) Alprenolol	10 nM	-	-	11
	Mercapto alprenolol	30 nM	1400[1] nM	37	11
	Mercapto alprenolol	30 nM	3100[2] nM	103	11
Estrogen	Estradiol-hemisuccinate	0.7 μM	9.5[2] μM	13	12
	Estradiol-hemisuccinate	0.7 μM	1.4[3] μM	2	12
	Estradiol-hemisuccinate	0.7 μM	3.0[4] μM	4	12

[1]12 A° spacer; [2]20 A° spacer; [3]Bovine serum albumin as spacer. [4]Poly-L-lysine-L-alanine copolymer as macromolecular spacer

macromolecular spacer first introduced in purification of estrogen receptors has not only provided with an ideal spacer, but also substantially improved the stability of the ligand by multipoint attachment on the matrix [12].

The most serious disadvantage of an improperly immobilized ligand with reduced affinity would be a reduced capacity of the affinity matrix for the hormone. However, the reduced affinity of a ligand upon its immobilization have certain practical advantages. The weakly bound receptor would obviously be easily desorbed and eluted from an affinity column without the need for harsh buffers which could result in an irreversible denaturation of the receptor protein. Thus a compromized balance betweeen reduced capacity and increased ease of elution are often sought in affinity chromatography of hormone receptors.

MEMBRANE ASSOCIATED RECEPTORS

β-Adrenergic Receptor - The catecholamine hormones exert their biological actions in target cells through binding to specific receptors located at the outer side of the plasma membranes. This hormone-receptor interaction results in an increase in cAMP production by adenylate cyclase. This latter enzyme is located at the inner side of the plasma membrane. There may be various components involved in the transmission of the hormonal signal from the receptor to the adenylate cyclase enzyme. The isolation, characterization and reconstitution of the membrane components involving hormonal recognition, signal transmission and activation of the enzyme are essential for understanding the mechanism of action of catecholamines. Although other membrane components of this system are not yet isolated, the catecholamine receptor has been purified from various sources.

The β-adrenergic receptors and adenylate cyclase of turkey erythrocytes were solubilized simultaneously by digitonin in the presence of sodium fluoride [16]. The digitonin extracts were affinity chromatographed on alprenolol-agarose. While the bulk of proteins and the adenylate cyclase were eluted in the breakthrough fractions, the adsorbed receptor was eluted with buffers containing alprenolol and 1 M NaCl in 25-30% yield providing a 2,000-fold purification. The treatment of turkey erythrocytes membranes with digitonin solubilizes only about 30% of the receptor together with about 30% of the adenylate cyclase under the experimental conditions used [11]. Although other detergents, such as Triton X-100, Lubrol PX, Lubrol WX, Nonidet P-40, deoxycholate and lithium diiodosalicylate were ineffective, one cannot rule out that certain detergents do solubilize the receptor sites but then interfere with the binding assay.

Various chemical analogs of the alprenolol-agarose matrix have been used in the affinity chromatography of β-adrenergic receptors. The higher affinity of alprenolol (an antagonist) (I_{50} = 30-45 nM) together with relative ease of derivatization have attracted its use in the affinity chromatography of β-adrenergic receptors in contrast to various agonists such as isoproterenol (I_{50} = 5 μM). The receptor from frog erythrocyte membranes, after solubilization with digitonin, was partially purified on alprenolol-agarose matrix with about 60% recovery [17].

TABLE II

COMPARISON OF HORMONE AFFINITIES TO MEMBRANE-BOUND AND SOLUBILIZED RECEPTORS

Receptor	Ligand	Receptor membrane-bound (Kd in nM)	Receptor Solubilized (Kd in nM)	Reference
β-Adrenergic	Mercapto alprenolol	30	0.15	11
	Dihydroalprenolol	8.2	4	11
	(-) Norepinephrine	780	17	11
	(-) Isoproterenol	120	3.3	11
	(-) Propranolol	5	4	11
	(±) Alprenolol	15	10	11
	(±) Hydroxybenzyl pindolol	0.67	0.87	11
hGH	hGH	0.25	0.45	14
Prolactin	Prolactin	0.3	0.06	13
Thyrotropin	TSH	54	30	15
	LH	540	130	15
	hCG	980	130	15

An earlier attempt to isolate the β-adrenergic receptor from canine ventricular myocardium by affinity chromatography was unsuccessful primarily due to an undesirable design of the affinity matrix [18]. The apparent lack of specificity of this matrix could be attributed to the fact that norepinephrine was coupled to a carboxyl-agarose via an amide bond thus converting an agonist to a biologically inactive compound [19].

Insulin Receptors - This is a classic example where isolation and purification to homogeneity of a membrane-bound receptor was successfully achieved by affinity chromatography. The successful purification of this followed by many other hormone receptors clearly lies in the introduction as well as continued progress in the refinement of the affinity chromatography techniques.

The insulin receptors are located on the outer surface of plasma membranes of a variety of cells [20]. They have been solubilized from fat and liver cell membranes with non-ionic detergents with complete retention of activity [21]. A variety of insulin-substituted agarose derivatives have been prepared and their efficacy for affinity chromatographic purification of insulin receptors have been described [22-24]. The binding and retention of the receptor protein to the affinity matrix is not affected by the presence of 0.1% Triton X-100. The receptor was eluted from the affinity column with buffers containing 4 M urea and 0.1% Triton X-100 in high yield. An overall purification of about 200,000 fold can be achieved by this procedure.

In contrast to the cytosolic proteins, most cell-membrane proteins are glycoproteins. In light of this fact, appropriately selected plant lectins [25], which have very high selectivity for certain sugars, may find increasing applications in affinity chromatography of membrane-bound hormone receptors. This approach was first exploited for the purification of insulin receptor with wheat germ agglutinin and conconavalin A derivatized agarose adsorbents [26]. Furthermore, these plant lectins have certain biological activities which are similar to those of insulin, show specific interactions with insulin receptor and competitively displace insulin binding to the receptor. These observations led to the successful application of lectin-agarose derivatives in the affinity chromatography of the Triton-solubilized insulin receptors from rat fat and liver cell membranes. Most of the lectin binding glycoproteins of fat and liver cell membranes bind very weakly to the lectin-agarose adsorbents in the presence of the non-ionic detergent. The elution of the receptor protein is achieved under very mild conditions with lectin-specific sugars. Thus, in spite of the relatively non-specific character of such lectin-agarose adsorbents, it is possible to achieve substantial (approx. 3000-fold) purification of the insulin receptor. The use of lectin-agarose as an adsorbent offers certain advantages including the ease of elution with a lectin-specific sugar, higher capacity and isolation of receptor protein in a hormone-free state thus facilitating subsequent binding assays. Various other glycoprotein hormone receptors including TSH receptors [15] and EGF receptors [27] have been purified with lectin-agarose adsorbents.

Epidermal Growth Factor Receptors - Unlike the insulin receptor, but like the muscarinic receptor for acetylcholine [28], the EGF receptor is unable to bind to specific ligands following solubilization with detergents. However, the receptor protein can be specifically and covalently prelabeled (affinity labeling) prior to detergent solubilization [27]. The solubilized receptor can then be purified by affinity chromatography either with conconavalin A- or wheat germ agglutinin-coupled agarose adsorbents [27]. Anti-EGF antibodies coupled to agarose have also been used for the purification of the prelabeled EGF receptor [27].

Acetylcholine Receptors - A large number of studies have been recently undertaken on the cholinergic (nicotinic) receptor since the discovery that the receptor can be selectively labeled with snake venom α-bungarotoxin [29]. The acetylcholine receptor is an integral membrane protein and detergents are needed to release it into solution. Several non-ionic detergents, such as Triton X-100, Tween 80, Brij 35 and negatively charged ones (e.g., deoxycholate or cholate) give efficient solubilization without loss of its ability to bind to nicotinic agonists and antagonists. However, the binding property for muscarinic agonists and antagonists is irreversibly lost upon solubilization of the receptor. Routinely the receptor has been isolated from the electric organs of certain fish in which the receptor content is in the range of 0.1-1.0% of the total proteins. A receptor content as high as 25-50% can be achieved in the purified electroplax membrane preparation. Thus purification of this receptor has provided the least complications. In practice the affinity matrices used for the purification of this receptor have been either agarose-immobilized quarternary ammonium functions [30,31] or certain snake venom neurotoxins [32,33]. The use of neurotoxins in the affinity purification of the cholinergic receptor, in general, results in lower overall recoveries than with immobilized quarternary ammonium salts.

Since about 1% of the total electrogenic organ protein of Torpedo fish consists of acetylcholine receptor, only 100- to 300-fold purification is necessary to obtain a homogeneous receptor. In the case of the eel, electrophorus, a somewhat greater purification is necessary to obtain homogeneous receptor [34]. Comparatively rich sources for any mammalian acetylcholine receptor have not yet become available. However, an acetylcholine receptor requiring 48,000-fold purification has recently been isolated from rat skeletal muscle by affinity chromatography [35]

CYTOPLASMIC AND EXTRACELLULAR STEROID RECEPTORS

In addition to the recently discovered membrane-associated binding sites for certain steroids [40,41], steroid receptors are all located in the cytoplasm of the target cells. Whereas the membrane-associated binding sites for the steroids have not yet been purified, the cytoplasmic receptors for most of the steroids have been purified to homogeneity [12,42-44]. The steroid (and also thyroid) hormone-receptor complex that forms in the cytoplasm of target cell is translocated through the nuclear membrane and interacts with elements of the gene expression machinery and modifies transcription.

TABLE III

MEMBRANE-ASSOCIATED RECEPTORS

Receptor	Source	Solubilizing Agent	Ligand	Elution	Reference
Insulin	Rat liver, fat cells	Triton X-100	Insulin	Urea, pH 6.0	22,24
	Rat liver, fat cells	Triton X-100	Con A	α-methyl-mannoside	26
	Rat liver, fat cells	Triton X-100	Wheat germ agglutinin	N-acetylgluco-samine	26
β-adrenergic	erythro-cytes	digitonin	Alprenolol Aprenolol	± 1M NaCl	11,17
Acetylcho-line	Electric fish	non-ionic detergents, deoxycholate	Quarter-nary ammonium	NaCl	30,31
	Electric fish	non-ionic detergents, deoxycholate	Naja Naja toxin	Carbamyl-choline	33,36
	Electric fish	non-ionic detergents, deoxycholate	cobratoxin	Carbamyl-choline	37
Growth Hormone	Rabbit liver	Triton X-100	hGH	5 M $MgCl_2$	14
hCG-LH	Rat testies	Triton X-100	hCG	Acetic acid	38
Prolactin-hGH-Placental lactogen	Mammary glands	Triton X-100	hGH	5M $MgCl_2$	13
TSH	Thyroid membranes	Li-diiodo-salicylate	TSH	Acetate, pH 2.5	15
EGF	Placenta	Nonidet P-40	Plant-lectins	Specific sugars	27
	Placenta	Nonidet P-40	Anti-EGF antibodies	Formic acid	27
IgE	rat baso-philic cells	Nonidet P-40	IgE	Acetate + NP-40	39

Estrogen Receptors - Various forms of estrogen receptors have been identified in uterine and other target tissues. The "native form" of the receptor, localized in the extranuclear space of the cell [45], is characterized by the reversible change of its sedimentation on sucrose gradient from 8 to 4S upon increasing the ionic strength of the buffer [46]. Recently, a "derived form" of the cytosolic receptor has been identified which sediments at 4S in low or high salt. This form is obtained from the "native" receptor upon activation of an endogenous Ca^{++}-dependent enzyme [12]. Furthermore, a "nuclear" receptor represents a form of the receptor which migrates from the cytosol into the nucleus after formation of the hormone-receptor complex [47].

Purification of the estrogen receptors has been very difficult because of their scarsity in biological tissues. The reversible nature of ligand interaction and the stereochemical specificity and high affinity for estrogens suggest that the estrogen receptor may be ideally suited for purification by affinity chromatography The "derived" (low salt stable, 4S) form of the receptor was purified to homogeneity from calf uterus cytosol by affinity chromatography [12,49]. A similar procedure was applied for the purification of the "derived" form of the receptor from human uterus [48]. The relative effectiveness of a large variety of affinity matrices has been described for the purification of the "derived" form of the receptor [12,49,50].

The extremely small quantities and the relatively labile nature of the estrogen binding proteins from different target tissues present special practical difficulties in the handling and isolation of these receptors. It has been estimated that complete purification of the receptor proteins of calf uterine preparations would require 20,000- to 100,000-fold purification [12]. Virtually all polymers used as solid supports in affinity chromatography exhibit marked adsorptive properties for free estradiol. Desorption of free estradiol during chromatography of samples containing estrogen receptors can result in "inactivation" or apparent removal of the estrogen-binding activity from the sample. Although free estradiol may not be present when the derivatized gel is washed with simple buffers, application of protein-containing solutions, e.g., albumin or uterine cytosol, alter the partitioning properties of estradiol between the gel and the aqueous medium and can thus markedly enhance the release of adsorbed hormone from adsorbents washed only with simple buffers. This serious problem can generally be avoided by exhaustive washing of the adsorbent with organic solvents [12,51].

The ester and azo bonds used in coupling the ligand to agarose or to other polymers are relatively unstable and may slowly release the ligand during chromatography. The lability of the bonds may depend on the pH and ionic strength of the buffer used, and the presence of reducing substances. Similarly, nucleophiles in the tissue extracts containing the estrogen receptors may catalyze the hydrolysis of the immobilized ligand.

The release of some free estradiol from the gel during chromatography of receptor-containing samples does not necessarily mean that the adsorbent will be ineffective in selectively extracting the receptor from the sample. If the affinity of the free hormone released from the gel is not very different from that of gel-bound hormone, the small amount of free hormone will not compete effectively with the much greater amount of immobilized hormone, and the receptor will thus preferentially bind to the solid support. If, however, the free hormone has a much greater affinity, e.g., by a factor of 1000, a very small fraction, 0.1%, of the total hormone will interfere with selective adsorption if it is present in free form. For this reason the release of adsorbed hormone (see above) is generally a more serious problem. Instability of the cyanogen bromide bonds formed is less likely to lead to the release of hormone in a form that effectively competes with the matrix-bound material. The problem of ligand leakage [49,51,52] has been largely responsible for the previously reported unsuccessful attempts for purification of estrogen receptor by affinity chromatography [53,54].

Since the release of adsorbed estradiol during chromatography of the sample may lead to the erroneous conclusion that the column is removing the receptor from the sample, it is essential that steps be taken to examine the chromatographed samples for free estradiol as well as for the presence of estradiol-receptor complexes. Estradiol-agarose adsorbents containing macromolecular spacer arms increase the chemical stability of the agarose-bound estrogen due to their multipoint attachments to the matrix and provide other advantages in affinity chromatography [12,51].

The estradiol derivatives which have proved to be most useful in the purification of receptors are those in which the hormone is attached to agarose through position 17 of the estradiol molecule. Although such estradiol derivatives demonstrate lower affinity than native estradiol for the receptor, they retain sufficiently high affinity to be useful in affinity chromatography. The affinity of 17-β-estradiol-17-hemisuccinate, the estradiol derivative which has proved to be the most useful, is only 300 times lower than that of 17-β-estradiol. Immobilization of these derivatives on certain of the agarose gels does not cause a further, serious decrease in the affinity for the receptor. For example (Table I), attachment of 17-β-estradiol-17-hemisuccinate to diaminodipropylamine agarose results in a 13-fold fall in afffinity (Ki of about 10^{-5} M), wherease substitution on albumin-agarose only leads to a 2-fold fall in affinity (Ki of about 10^{-6} M) [12].

The specific estrogen derivative used has important implication for the subsequent steps of receptor elution from the gel. Because the estradiol derivatives which are used have substantially lower affinity than 17-β-estradiol for the receptor, it is not necessary that the amount of free estradiol which is added to the eluting medium be in great excess compared to the gel-bound ligand. Since the "excess" of free estradiol required for effective competitive exchange is related to both the concentration and the affinity of the particular gel derivative used, the specific ratio used in most work is adapted to reflect the particular conditions of the experiment. It is thus possible to use very low concentrations of estradiol of very high specific activity in the exchange reaction used in elution. This permits the use of the same radioactive estradiol for the subsequent assay of the binding activity present in the eluted sample, and it avoids contamination with large amounts of free, native estradiol.

Most gel derivatives which contain estradiol linked to agarose by bonding to the A-ring of estradiol possess very low affinity for the receptor [12]. This is indicated by the total inability of the very stable 3-O-ether derivatives to bind estradiol receptors. This low affinity, coupled with the chemical instability of some of the other A-ring derivatives (which release free estradiol) probably explains the failure of these adsorbents. The position 7 of the B-ring of estradiol is amenable for derivatization without appreciable loss of binding activity [50].

In contrast to the "derived" form, the "native" form of the estrogen receptor has resisted all attempts at purification because of its tendency to form large and irreversible aggregates with small variations in buffer conditions. A property, characteristic of the native form of the receptor, to specifically interact with heparin, was successfully exploited for the affinity chromatography of this receptor [42]. The "native" form of the receptor was thus purified to homogeneity on heparin-agarose followed by gel filtration and ion exchange chromatography [42]. The purified native form of the receptor consists of four identical subunits of a single polypeptide chain each of about 69,000 daltons. The "derived" or the "nuclear" forms of the receptor do not interact with heparin. More recently, an elegant two stage affinity chromatography was used for purification of the native receptor. The technique of two successive affinity chromatographic steps consisted of heparin-agarose followed by estradiol-agarose, which eliminated the need for additional gel filtration or ion exchange chromatography [55].

<u>Progesterone Receptor</u> - These cytoplasmic receptors are specific for progesterone and its biologically active metabolite, 5α-pregnane-3,20-dione. The tissues where progesterone receptor have been detected also contain estrogen receptor. There is no cross reactivity between the two receptors, although the synthesis of progesterone receptor appears to be mediated by the action of estrogen. The progesterone receptor appears to function by complexing with progesterone in the cytoplasm followed by translocation into the nucleus, where the complex associates with acceptor sites located on the nuclear chromatin. In case of hen oviduct receptor, one end product of this series of interactions is the induction of mRNA synthesis for the egg white protein, avidin [56].

The cytoplasmic progesterone receptor of hen oviduct has been purified in 8% overall yield by affinity chromatography followed by ion exchange chromatography

[44]. The receptor is a protein of about 227,000 daltons and consists of two dissimilar subunits. The affinity matrix, deoxycorticosterone coupled to Sepharose via the macromolecular spacer, bovine serum albumin, binds to the receptor with about the same affinity (Kd = 0.8 nM) as the free steroid (Kd = 0.5 nM). The above affinity matrix provided a greater than 20,000-fold purification of the receptor.

As in case of estrogens, the solid supports used in affinity chromatography exhibit very marked adsorption of progesterones and deoxycorticosterone. It is obvious that the release of such non-covalently bound steroids will compete with the receptor and prevents its adsorption to the affinity matrix. This serious problem, which has often been emphasized [12,49,52], can generally be avoided by very exhaustive washing procedures.

Glucorticoid Receptors - The glucocorticoid and adrenocorticoid steroids are secreted in the adrenal cortex by the action of ACTH. These class of steroids have numerous and diversified physiological functions and pharmcological effects. Among other effects, they influence carbohydrate, lipid, protein and purine metabolism and electrolyte/water balance and possess anti-inflammatory properties. It is only recently that a glucocorticoid receptor was purified. An elegent two stage affinity chromatography procedure was used to purify the 90,000 dalton receptor from rat liver cytosol. The affinity matrix for the first column consisted of a covalent agarose conjugate, either of dexamethason-17β-carboxylic acid or dexamethasone-21-methane sulfonate [43] via a disulfide bond [57]. The fascile cleavage of the dilsulfide bond by reducing agents, such as thiols, allowed an easy elution of the receptor with 8,700-fold purification. The partially purified receptor obtained from the steroid-affinity column was further purified to homogeneity by a DNA-cellulose column [43].

The existence of a protein that binds to corticosteroids with higher affinity than albumin has been demonstrated in the serum of many animals. This protein has been called corticosteroid-binding globulin (CBG) or transcortin. It is interesting to note that cortisol (a hydrophilic steroid) and progesterone (a hydrophobic steroid) have very similar affinity to human transcortin [58]. However, a progesterone-binding globulin (PBG) isolated from the serum of pregnant guinea pig possesses high affinity for progesterone but not for cortisol.

Vitamin D Receptors - Vitamin D (Calciferol), followed by its metabolic conversion to 1,25-dihydroxy vitamin D_3, acts by stimulating intestinal calcium absorption and mineralization of bone and cartilage. The 25-hydroxycholecalciferol (25-OH-D_3) binding proteins from chick serum and from chick kidney have been extensively purified by affinity chromatography [59].

TABLE IV

CYTOPLASMIC RECEPTORS

Receptor	Source	Ligand	Elution	Reference
Vitamin-B12	Gastric juice	Vit. B12	guanidine·HCl	60
	Human plasma	Vit. B12	guanidine·HCl	60
	Human granulocytes	Vit. B12	guanidine·HCl	60
cAMP	Skeletal muscle	cAMP	cAMP	61
cAMP	Skeletal muscle	cAMP	cAMP	61
Progesterone	Chick oviduct	deoxy cortisterone	Progesterone	44
Estrogen	Calf uterus	Estrogen	Estradiol	12,49,50
	Calf uterus	Heparin	Heparin	42
Glucoc-corticoid	Rat liver	dexamethasone	β-mercapto-ethanol	43
Vitamin D	Chick kidney	Hydroxy cholecali-ferol	Ammonium acetate	59

In almost all the hormone-receptor systems described above, the general initial approach has been to study in detail the binding interaction between a radio-labeled hormone and a tissue or its cytosol that contains the putative receptor structures. The hormone binding must show absolute specificity, the affinity of the hormone must be consistent with the biological activity of the ligand and the number of binding sites must be consistent with the physiological mechanisms operative in the intact system. As specific receptors have been purified, and as more information is gained concerning the nature of membrane receptors and their interaction with hormones, greater focus will be placed on the nature of the molecular events which occur in the membranes following the formation of the initial complex.

REFERENCES

1 P. Cuatrecasas, M. Wilchek and C.B. Anfinsen, Proc. Nat. Acad. Sci. USA, 61(1968)636-643.
2 R. Axen and J. Porath, Nature, 210(1966)367-369.
3 P. Cuatrecasas, J. Biol. Chem., 245(1970)3059-3065.
4 P. Cuatrecasas and C.B. Anfinsen, Ann. Rev. Biochem., 40(1971)259-278.
5 W. Jacoby and M. Wilchek (Eds.), Methods in Enzymology, Academic Press, New York, 1974.
6 I. Parikh and P. Cuatrecasas, in M.Z. Atassi (Ed.), Immunochemistry of Proteins, Plenum Press, New York, 1977, pp 1-44.
7 S. Jacobs and P. Cuatrecasas, in S. Jacobs and P. Cuatrecasas (Eds.), Receptor and Recognition Series, Chapman and Hall, London, Vol. 11, 1981, pp 61-86.
8 S.K. Sharma and P.P. Mahendroo, J. Chromatogr., 184(1980)471-499.
9 C.R. Lowe and P.D.G. Dean, Affinity Chromatography, Wiley, London, 1974.
10 E. Steers, P. Cuatrecasas and H.B. Pollard, J. Biol. Chem., 246(1971)196-200.
11 G. Vauquelin, P. Geynet, J. Hanoune and A.D. Strosberg, Eur. J. Biochem., 98(1979)543-556.
12 V. Sica, I. Parikh, E. Nola, G.A. Puca and P. Cuatrecasas, J. Biol. Chem., 248(1973)6543-6558.
13 R.P.C. Shiu and H.G. Friesen, J. Biol. Chem., 249(1974)7902-7911.
14 C. McIntosh, J. Warnecke, M. Nieger, A. Barner and J. Köbberling, FEBS Lett., 66(1976)149-154.
15 R.L. Tate, J.M. Holmes, L.D. Kohn, and R.J. Winand, J. Biol. Chem., 250(1975) 6527-6533.
16 G. Vauquelin, P. Geynet, J. Hanoune and A.D. Strosberg, Proc. Nat. Acad. Sci. USA, 74(1977)3710-3714.
17 M.G. Caron, Y. Srinivasan, J. Pitha, K. Kociolek and R.J. Lefkowitz, J. Biol. Chem., 254(1979)2923-2927.
18 R.J. Lefkowitz, E. Haber and D. O'Hara, Proc. Nat. Acad. Sci. USA, 69(1972) 2828-2832.
19 P. Cuatrecasas, G.P.E. Tell, V. Sica, I. Parikh and K.J. Chang, Nature, 247(1974)92-97.
20 P. Cuatrecasas, Ann. Rev. Biochem., 43(1974)169-214.
21 P. Cuatrecasas, Proc. Nat. Acad. Sci. USA, 69(1972)318-322.
22 P. Cuatrecasas, Proc. Nat. Acad. Sci., USA, 69(1972)1277-1281.
23 P. Cuatrecasas and I. Parikh, Methods in Enzymol., 34(1974)670-688.
24 S. Jacobs, Y. Schechter, K. Bissel and P. Cuatrecasas, Biochem. Biophys. Res. Comm. 77(1977)981-988.
25 T. Kristiansen, Methods in Enzymol., 34(1974)331-341.
26 P. Cuatrecasas and G.P.E. Tell, Proc. Nat. Acad. Sci. USA, 70(1973)485-489.
27 R.A. Hock, E. Nexo and M.D. Hollenberg, Nature, 277(1979)403-405.
28 C.M.S. Fewtrell and H.P. Rang in H.P. Rang (Ed.), Drug Receptors, University Press, Baltimore, 1973, pp. 211-224.
29 J.P. Changeux, M. Kasai and C.Y. Lee, Proc. Nat. Acad. Sci. USA, 67(1970) 1241-1247.
30 J. Schmidt and M. Raftery, Biochem. Biophys. Res. Comm., 49(1972)572-578.
31 R.W. Olsen, J.-C. Meunier and J.P. Changeux, FEBS Lett., 28(1972)96-100.
32 G.I. Franklin and L.T. Porter, FEBS Lett., 28(1972)101-106.
33 E. Karlsson, E. Heilbronn and L. Widlund, FEBS Lett., 28(1972)107-111.
34 A. Karlin, Life Sci. 14(1974)1835-1415.
35 S.C. Froehner, C.G. Reiness and Z.W. Hall, J. Biol. Chem., 252(1977)8589-8596.
36 D.E. Ong and R.N. Brady, Biochemistry, 13(1974)2822-2827.
37 J.P. Brockes and Z.W. Hall, Proc. Nat. Acad. Sci. USA, 72(1975)1368-1372.
38 M.L. Dufau, D.W. Ryan, A.J. Baukal and K.J. Katt, J. Biol. Chem., 250(1975) 4822-4824.
39 A. Kulczycki and C.W. Parker, J. Biol. Chem., 254(1979)3187-3193.
40 I. Parikh, W.L. Anderson and P.J. Neame, J. Biol. Chem., 255(1980)10266-10270.
41 T. Suyemitsu and H. Terayama, Endocrinology, 96(1975)1499-1508.

42 A.M. Molinari, N. Nedici, B. Moncharmont and G.A. Puca, Proc. Nat. Acad. Sci. USA, 74(1977)4886-4890.
43 M.V. Govindan and B. Manz, Eur. J. Biochem., 108(1980)47-53.
44 R.W. Kuhn, W.T. Schrader, R.G. Smith and B.W. O'Malley, J. Biol. Chem., 250(1975)4220-4228.
45 I. Parikh, M. Sar, B. Moncharmont, N. Medici, G.A. Puca and P. Cuatrecasas, manuscript in preparation.
46 G.M. Stancel, M.T. Leung and J. Gorski, Biochemistry, 12(1973)1230-1236.
47 G.L. Greene, L. Closs, H. Fleming, E.R. DeSombre and E.V. Jensen, Proc. Nat. Acad. Sci. USA, 74(1977)3681-3685.
48 A.J. Coffer, P.F.D. Milton, J.P. Davis and R.J.B. King, Mol. Cell. Endocrinol., 6(1977)231-246.
49 I. Parikh, V. Sica, E. Nola, G.A. Pucca and P. Cuatrecasas, Methods in Enzymol., 34(1974)670-688.
50 R. Bucourt, M. Vignau, V. Torelli, H. Richard-Foy, C. Geynet, C. Secco-Miller, G. Redeuilh and E.-E. Baulieu, J. Biol. Chem., 253(1978)8221-8228.
51 I. Parikh, S. March and P. Cuatrecasas, Methods in Enzymol., 34(1974)77-102.
52 J.H. Ludens, J.R. DeVries and D.D. Fanestil, J. Biol. Chem., 247(1972)7533-7538.
53 B. Vonderhaar and G.C. Mueller, Biochim. Biophys. Acta, 176(1969)626-631.
54 P.W. Jungblut, I. Hatzel, E.R. DeSombre and E.V. Jensen, Colloq. Ges. Physiol. Chem., 18(1967)57-86.
55 E. Nola, A.M. Molinari, N. Medici, B. Moncharmont, R. Piccoli, G.A. Puca, I. Parikh and P. Cuatrecasas, in Proceedings of VIII Meeting of the International Study Group for Steroid Hormones, Rome, 1977.
56 B.W. O'Malley, G.C. Rosenfeld, J.P. Comstock and A.R. Means, Nature, 240(1972) 45-48.
57 F. Sweet and N.K. Adair, Biochem. Biophys. Res. Comm., 63(1975)99-105.
58 S.D. Straupe, G.B. Harding, M.W. Forsthoefel and U. Westphal, Biochemistry, 17(1978)177-182.
59 R. Sharpe, C.J. Hillyard, M. Szelke and I. Macintyre, FEBS Lett., 75(1977) 265-271.
60 R.H. Allen and P.W. Majerus, J. Biol. Chem., 247(1972)7695-7701.
61 J. Ramseyer, C.B. Kanstein, G.M. Walton and G.N. Gill, Biochim. Biophys. Acta, 446(1976)358-370.

T.C.J. Gribnau, J. Visser and R.J.F. Nivard (Editors),
Affinity Chromatography and Related Techniques

LIGAND/LIGATE INTERACTIONS IN HETEROGENEOUS SYSTEMS: THEIR INFLUENCE ON OPERATIONAL PROPERTIES OF AFFINITY CHROMATOGRAPHIC AND BINDING ASSAY ADSORBENTS.

V. KASCHE and B. GALUNSKY
Biology Department, University of Bremen, NW II,
D-28oo Bremen 33, F.R.G.

1. INTRODUCTION

The specificity of molecular interactions is mainly determined by the life-time of the complex. The more this life-time exceeds the life-time of the collision complex ($\sim 10^{-5}$ sec. for a typical protein-protein interaction) the more specific is this interaction (1). It was early proposed that the long life-times of the specific interactions involving biomolecules - biospecific interactions - can be used to isolate these molecules from a mixture (2). These should be more specific than isolation procedures based on purely physical chemical properties. The development of the experimental methods based on biospecific interactions was, however, after some early studies (3-5), slow until spherical chromatographic media had been developed that were suitable for easy and reproducible procedures to prepare the biospecific adsorbents required for these isolation procedures. Now affinity chromatography is an established analytical and preparative - at least on a laboratory scale-procedure (6,7). In this procedure one of the components is modified by immobilization on/in an insoluble support. This ligate binds the other component - the ligand - in the biospecific interaction. Modified components of biospecific interactions also find widespread use in analytical applications as EIA, FIA and RIA (8-1o) for diagnostic purposes. They are, however, also of great interest for basic studies in cell biology as receptor-ligand interactions, localization and internalization of binding sites for specific ligands (11). These different applications of modified components in biospecific interactions are illustrated in Fig.1. They can be subdivided in equilibrium and non-equilibrium methods or according to the nature of the modified component in homogeneous and heterogeneous methods. In these methods

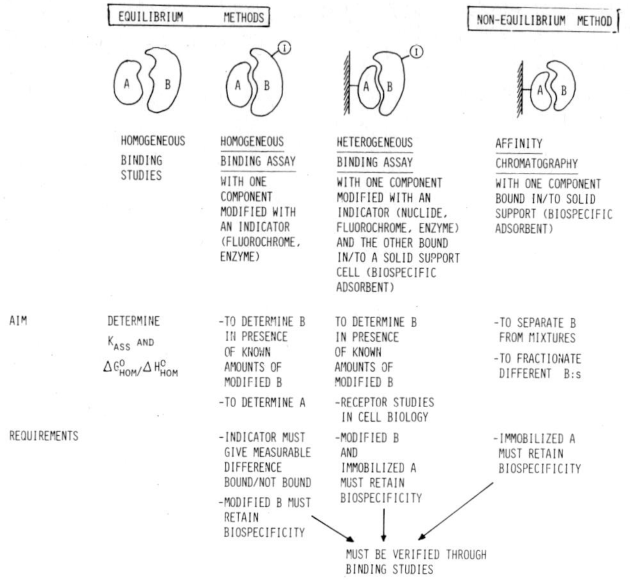

Fig.1. Applications of biospecific interactions where one component is modified with an indicator or by immobilization.

the modified component must retain its biospecific function. The modification is generally not homogeneous i.e. different groups on the modified component may have been labelled or used for immobilization. This heterogeneity may result in different associatio constants that are used as a measure for the activity of the biospecific function. For all applications shown in Fig.1, knowledge of the association constants and their variation for the different modified ligands is essential in quantitative studies.

In this communication procedures to do this and to determine other properties of modified components in biospecific interactions that are important for their various applications will be analyzed. The importance of some of these factors for the resolution in high-pressure liquid affinity chromatography (HPLAC) is studied. Non-stationary (affinity chromatography) and stationary (biospecific sample application in isoelectric focusing) methods to determine association constants for ligand/ligate interactions in biospecific adsorbents will also be compared.

2. THEORY OF BIOSPECIFIC INTERACTIONS

2.1. Basic relations for equilibria with biospecific adsorbents and modified components

For the heterogeneous system with immobilized A, native ligand B differently modified ligand B_i and desorbing ligands X and Y that may interact with B and A, respectively, we have the following equilibria

$$A + B \underset{\bar{k}_{d,1}}{\overset{\bar{k}_{a,1}}{\rightleftharpoons}} AB \quad \text{Association constant} \quad \bar{K}_A \tag{1}$$

$$A + B_i \rightleftharpoons AB_i \quad \bar{K}_{A,i} \tag{2}$$

$$X + B \rightleftharpoons XB \quad K_X \tag{3}$$

$$X + B_i \rightleftharpoons XB_i \quad K_{X,i} \tag{4}$$

$$A + Y \rightleftharpoons A_jY \quad K_Y \tag{5}$$

Average rate constants (small $\bar{k}$) and association constants (large $\bar{K}$) are given for the equilibria involving an immobilized component. Here we assume that the A:S are immobilized at different functional groups yielding a distribution in association constants. At equilibrium the following relation for the ratio of the concentrations of bound (to the ligate) to free ligand B, the distribution coefficient for B can be derived (12, 13)

$$K_D = \frac{\bar{K}_A \cdot n_o (1+K_X \cdot c_B + \sum_i K_{X,i} \cdot c_{B_i})}{(1+\bar{K}_Y \cdot C_Y + \bar{K}_A \cdot c_B + \sum_i K_{A,i} \cdot c_{B_i})(1+K_X \cdot C_X + K_X \cdot c_B + \sum_i K_{X,i} c_{B_i})} \tag{6}$$

where c and C give the free and total concentrations of the species given by the subscript, and n_o is the operational content of the immobilized component that has retained its biological function. The distribution coefficient is a concentration ratio. In chromatography K_D is sometimes defined so that molecules B that are free inside the biospecific adsorbent are counted as bound B. This yields a more complicated relation that is not suitable to analyze equilibrium systems. Therefore the definition of K_D given by Eq.(6) is used here.

For equilibria in heterogeneous systems free and bound B may be distributed in different volumes. In this case another ratio - a mass ratio - the distribution ratio (or capacity factor) k' is defined as (14)

$$k' = \frac{V_s}{V_o} \cdot (K_D + 1) \tag{7}$$

where V_s is the volume of the stationary phase volume where B is bound and V_o the volume of the system (mobile phase volume = void volume) outside the biospecific adsorbent. It gives the mass ratio for the distribution of B in the stationary and mobile phase. The relations (6) and (7) will be used later to describe the equilibrium and non-equilibrium methods shown in Fig.1.

2.2. Pertubation of association constants when one component is modified

The association rate constant for a bimolecular reaction is proportional to the diffusive rate

$$k_+ = 4 \cdot \pi \cdot a \cdot D \cdot f \tag{8}$$

where a is the effective complex radius, D the effective diffusion coefficient and f a geometry factor (<1) (1). When a molecule ist modified or immobilized steric hindrance reduces f when compared with the non-modified molecule. Physically the space angle where collisions may occur is reduced when one of the components is modified or immobilized. This effect is not so marked for the dissociation rate constant. This reduces the association constant as does dipole orientation effects when a charged support is used for immobilization (15). In this case local pH-changes at the support may also change the association constant in either direction depending on how the acid-base equilibria that are changed influence the biospecific interaction. The conformation changes induced by the modification or immobilization may change the association constant in either direction. No theory has yet been developed to analyze the latter in a predictive manner.

Thus the free energy for the biospecific interaction with one component modified or immobilized at constant pH is

$$\Delta G^o_{mod} = \Delta G^o_{hom} + \Delta(\Delta G_{Dip}) + \Delta(\Delta G^o_{pH}) + \Delta(\Delta G^o_{ster}) + \Delta(\Delta G^o_{conf}) + \Delta V \cdot p \tag{9}$$

where

ΔG^o_{hom} = free energy change for the unperturbed interaction (at normal pressure) in a homogeneous system

$\Delta(\Delta G^o_{Dip})$ = free energy change due to dipole orientation (>O)

$\Delta(\Delta G^{o}_{pH})$ = free energy change due to local pH changes ($\gtrless 0$)

$\Delta(\Delta G^{o}_{ster})$ = free energy change due to steric hindrance (for association) (>0)

$\Delta(\Delta G^{o}_{conf})$ = free energy change due to conformation changes ($\gtrless 0$)

ΔV = change in molar volume upon complex formation

p = pressure

For protein-protein interactions ΔV may be >0 or <0 (16). The relative importance of these free energy changes can be evaluated from determinations of apparent equilibrium constants and the relation

$$\Delta G = - RT \ln K_{app} \qquad (10)$$

only when the pK:s for the acid-base equilibria coupled to the biospecific interaction are unperturbed by the modification (17).

2.3. The pH-dependence of biospecific interactions

The pH-dependence of an apparent association constant is given by a relation derived by Wyman (18)

$$\frac{d \log K_{app}}{d\ pH} = - dX \qquad (11)$$

where $-dX$ is the number of protons liberated as a consequence of the biospecific interaction. The quantity $-dX$ is generally $\neq 0$. Thus for X and Y in Eq. (3)-(5) protons can be used as desorbing ligands where a large change in K_{app} is desired provided that the pH-stability of the proteins allows this.

3. APPLICATION OF BIOSPECIFIC INTERACTIONS

3.1. Affinity chromatography

Affinity chromatography has the following main fields of application:

- preparative use, where one biopolymer or a group of biopolymers with same function but of different activity are isolated from biological samples. Here the operational capacity of the biospecific adsorbent (n_o in Eq. (6)) is the essential property;

Table I: Summary of operational properties of biospecific adsorbents used for proteolytic and NAD (NADP)-dependent enzymes and serum albumin

Enzyme(s)	Immobilized Ligand (MW)*	Matrix	Binding yield of ligand (%)	Binding yield of active ligand (%)	Operational Capacity (mM/1 wet support)	resolution	reusability** pure enzyme solutions	tissue homogenates	Ref.
Chymotrypsin Trypsin	Soybean-trypsin inhibitor (H)	Sepharose 4 B	> 90	40-75[1)] -90[2)]	0.3 §	resolves trypsins and chymotrypsins	20	5	a
Chymotrypsin Trypsin	Soybean-trypsin inhibitor (H)	Lichro-sphere	50	50	1[3)]	— " —	> 100*		
Trypsin	ovomucoid (H)	Sepharose 4 B		20	0.3	resolves α/β trypsin	-	-	b
Trypsin	benzamidine(L)	Cellulose	96	-	0.02	-	-	-	c
Trypsin	benzamidine(L)	Sepharose 4 B	2-90	< 1	0.14	-	-	-	c
Trypsin	phenylbutyl-amine (L)	Sepharose	95	-	0.003	-	-	-	c
Trypsin	ovomucoid (H)	Spheron 300	-	-	0.06	-	-	5-10	d
Chymotrypsin	antilysine(H)	Spheron 300	-	3.8	0.06		-	5-10	d
Trypsin	benzamidine(L)	Spheron 300	-	0.4	0.05	-	> 10	-	d
Trypsin	Soybean-trypsin inhibitor (H)	Aminohexyl-cellulose	-	-					e
LDH lactate dehydro-genase	5'-AMP (L)	Sepharose 4 B		0.1	0.1	resolves LDH isozymes			f, g
ADH Alcohol dehydro-genase	NAD^+ (L)	Sepharose		0.1					h
	N^6-5'-AMP(L)	Sepharose		0.3					h
	Procion Red (L)	Sepharose		2	0.1				h
	Cibacron Blue(L)	Sepharose		1	0.1				h
Serum Albumin	Cibacron Blue(L)	Sepharose CL b B		-	0.1				g
	Cibacron	Agarose (5%) (Matrex)		-	0.5				i

H=high molecular weight
L=low molecular weight

1) operational
2) static
3) flow-rate dependent
§ largest reported values

* HPLAC
** cycles for 50% reduction in capacity

References
a) V. Kasche, Biochem. Biophys. Res. Commun 38 (1970) 875; V. Kasche et al. Biochem. Biophys. Acta 490 (1977) 1.
b) N.E. Robinson et al. Biochemistry 10 (1971) 2743.
c) G.W. Jameson et al. Biochem.J., 141 (1974) 555.
d) J. Turkova et al. J. Chrom. 148 (1978) 293.
e) Information from E. Merck, Darmstadt, FRG.
f) K. Mosbach in Affinity Chromatography, Pergamon Press(1978), p.55.
g) Information from Pharmacia Fine Chemicals, Uppsala, Sweden.
h) P. D. G. Dean et al. ref. f, p.25.
i) P. D. G. Dean, personal communication.

- analytical use, where a group of similar biopolymers are separated from each other in order to determine their content in biological samples. In this application the resolution of the adsorbent is important. The analytical use also includes determination of $\overline{K}_A$, K_x and $K_{x,i}$ in Eq.(6).

The operational capacity in affinity chromatography, n_o, is generally smaller than the amount of ligate bound, N, determined

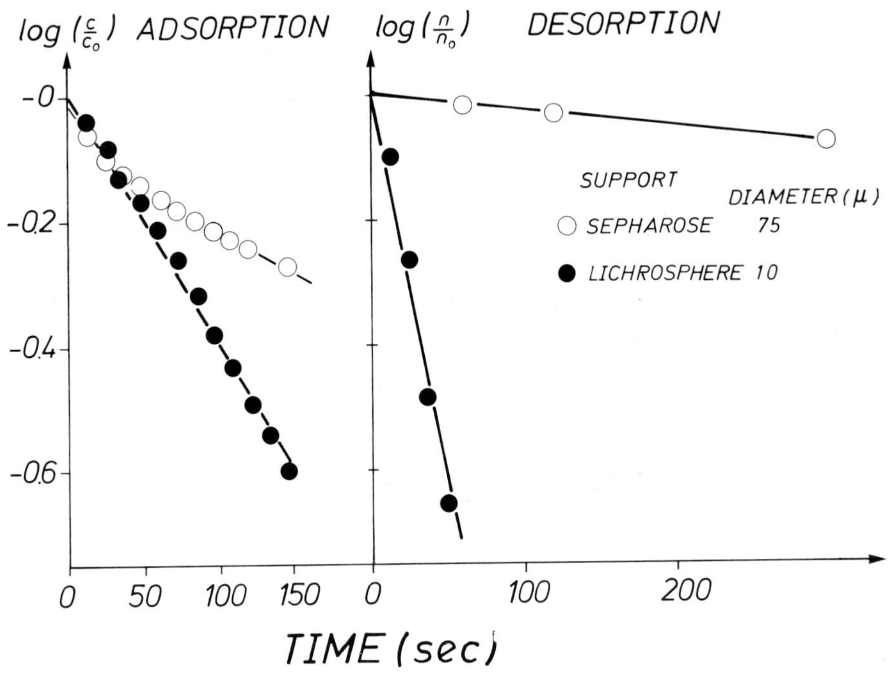

Fig.2 Adsorption and desorption of a ligand to a biospecific adsorbent as a function of time.

A.) Particle size dependence of the adsorption. The same amount of α-chymotrypsin (initial concentration after addition c_o) was given to 1o ml of a stirred suspension in pH 8.o buffer (o.2 M NaCl, Tris-HCl (I=o.o5))of 32 mg soybean trypsin inhibitor(STI)-Lichrosphere (●)(d=1oμ) or 25o μl STI-Sepharose (O)(d=75μ) at time zero. Both adsorbents had the same total stationary capacity. The latter adsorbent had a broad particle diameter distribution. The concentration of free enzyme (c) was measured as a function of time. Total suspension volume 11 ml.

B.) As A, but the desorption of the enzyme was measured after adjusting the pH in the suspension to 4.9 with acetic acid.

from the mass balance, and the static capacity, n_S, determined in equilibrium binding experiments. This is shown in Table I. The maximum operational capacity observed is about 1 mM/(ℓ support). That $n_o > n_S$ is probably caused by the particle size dependence of the rate of ligand mass transport to the stationary phase that is the rate limiting step in the adsorption process. Then some potential binding sites are practically unaccessible in non-equilibrium experiments as affinity chromatography (Fig.2).

Here we will mainly discuss the analytical uses of affinity chromatography. For this aim this method can be considered as a sorption chromatographic procedure for which there exists a well developed theory (14,19). It can easily be extended to cover affinity chromatography. This has, however, not always been done to analyze the operational performance of analytical affinity chromatography (7,15,17,2o,21).

3.1.1. Resolution in analytical affinity chromatography

In analytical affinity chromatography the resolutions is determined by the plate height. The experimental plate height is determined from experimental data using the relation

$$H_{exp} = \frac{L}{16\left(\frac{V_e}{\Delta V_e}\right)^2} \qquad (12)$$

where L is the column length, V_e elution volume and ΔV_e peak base width (14).

The theoretical analysis of plate heights in sorption chromatography yields the following relation for a ligand B, $X = H^+$ and $C_{B_i} = C_y = O$ (Fig.1, Eq.(1)-(5) (14,19)

$$H_{theor} = H_{Disp} + H_{e,diff} + H_{i,diff} + H_{kin} \qquad (13)$$

at low flow rates the axial dispersion due to diffusion in the mobile phase predominates than

$$H_{theor} \approx H_{Disp} = \frac{2\gamma \cdot D_B}{u_e} \qquad (14)$$

and at high flow rates the peak broadening due to the adsorption - desorption process predominates, then (19)

$$H_{theor} \approx H_{kin} = \frac{2\bar{k}'_B \; u_e}{(1+k_o)(1+\bar{k}'_B)^2 \cdot \bar{k}_{d,1}} \qquad (15)$$

where u_e is the interstitial flow rate and γ and k_o column constants. The validity of the relations (14) and (15) has been verified by experimental data (19). They can therefore be used to analyze the factors that determine the resolution in affinity chromatography. The resolution for the separation of similar ligands B' and B is defined as (14)

$$R_S = \frac{1}{4} \cdot \left(\frac{\bar{k}'_{B'}}{\bar{k}'_B} - 1\right) \cdot \sqrt{\frac{L}{H}} \cdot \left(\frac{\bar{k}'_B}{\bar{k}'_B+1}\right) \qquad (16)$$

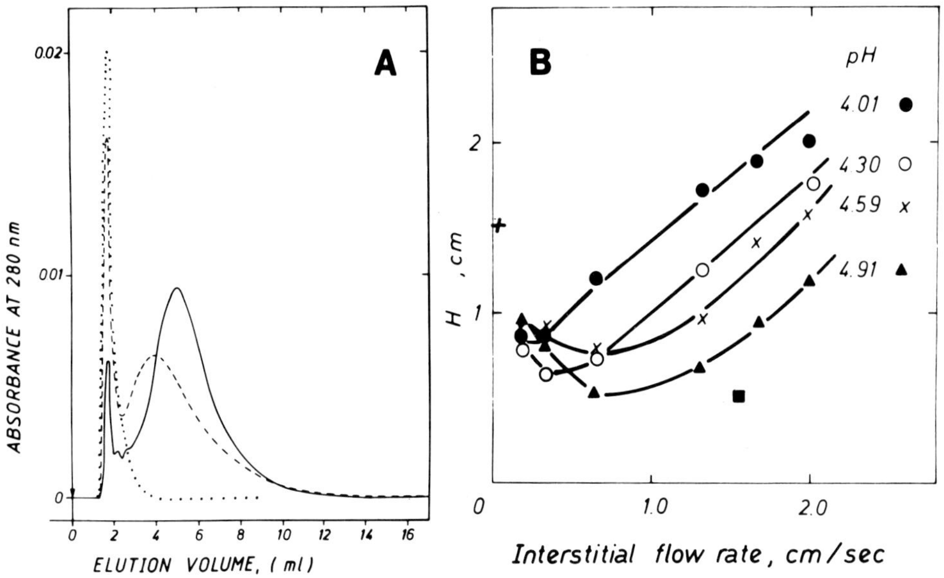

Fig.3. High pressure liquid affinity chromatography (HPLAC) of α - chymotrypsin using a column with soybean trypsin inhibitor bound in Lichrosphere (diameter 1o μ). The elution was performed at constant pH using 0.1 M KH_2PO_4 adjusted to the desired pH.

A.) Elution profiles for α-chymotrypsin and chymotrypsinogen at different flow rates and pH
...... chymotrypsinogen flow rate o.4 cm/sec
------ α-chymotrypsin flow rate 2.o cm/sec
——— α-chymotrypsin flow rate o.4 cm/sec

B.) Experimental Plate height as a function of flow rate and pH. For comparison data from HPLAC of other enzymes (LDH ■ ,ref.23) and low pressure affinity chromatography of α-chymotrypsin at pH 4.8 + , ref.17) are included.

Thus the resolution decreases with increasing H. From Eq.(19) follows that at constant flow rate, H increases with decreasing $\overline{k}$':s (decreasing $\overline{K}_A$) Eq.(6). When the association constant K_A is not homogeneous for all bound ligates the lowest values will consequently reduce the resolution. Thus a narrower distribution in K_A-values should improve the resolution. This can be achieved by varying the binding conditions. When neutral lysine side chains are involved in the immobilization reaction this can be done by reducing the pH during the reaction without loss in binding yield (17). In this way biospecific adsorbents for low pressure affinity

Table II. Association constants (K_A), and dissociation rate constants (k_d) for (bio)-specific interactions in homogeneous systems.

INTERACTING MOLECULES A	B	K_A (M^{-1})	k_d (sec^{-1})	REF
PROTEIN - SMALL LIGANDS				
PLASMACYTOMA PROTEIN X-24	(1-6)-β-D GALACTOTRIOSE	1.10^5	10	(A)
α - CHYMOTRYPSIN	PROFLAVIN	$1.5.10^5$	8.10^3	(B)
LACTATE DEHYDROGENASE	NADH	10^5	10^4	(B)
MALATE DEHYDROGENASE	NADH	10^7	50	(B)
PROTEIN - PROTEIN				
α - CHYMOTRYPSIN	α - CHYMOTRYPSIN	5.10^3	0.7	(B)
β - TRYPSIN	SOY-BEAN TRYPSIN INHIBITOR	10^{11}	10^{-5}	(C)
TRYPSIN	BASIC PANCREATIC TRYPSIN INHIBITOR	2.10^{15}	7.10^{-8}	(B)
CONCANAVALIN A	CELL SURFACE RECEPTOR	2.10^6	$\ll 1$	(D)

A) S. VUK-PAVLOVIC, Y. BLATT, C.D.J. GLAUDEMAUS, D. LANCET AND I. RECHT (1978) BIOPHYS. J. 24, 161-170

B) A. FERSHT (1977) ENZYME STRUCTURE AND MECHANISM, W.H. FREEMAN, READING P. 130-131.

C) I.A. MATTIS, M. LASKOWSKI JR. (1974) IN "PROTEINASE INHIBITORS" (TSCHESCHE, ED.), SPRINGER, BERLIN, P.389.

D) M. MONSIGNY, C. SENE, A. OBERENGWITCH, (1979) EUR.J.BIOCHEM. 96, 295.

chromatography with increased resolution for proteolytic enzymes have been prepared (23).

In high-pressure liquid affinity chromatography (HPLAC) separation times are reduced and higher resolutions (lower H_{exp}) should be possible to achieve. This is demonstrated in Fig.3. The plate height, however, is much higher than for HPLC of low molecular weight compounds $H_{exp} \sim 10^{-2}$ cm (19). In these experiments k'_B was found to have a maximum with varying flow rate. At the highest flow rates some of the active enzyme passed the column without being adsorbed (and k' decreases). Then not all enzymes may diffuse from the mobile to the stationary phase. In the latter they will preferably bind to adsorption sites with high association rate constants $k_{a,1}$ in the stationary phase. When more enzymes enter the stationary phase also adsorbent sites with lower association rate constants - and association constants - will be used provided that the adsorbent sites are non-homogeneous. This was the case for the adsorbent used in these experiments as non-linear Scatchard plots were obtained in equilibrium binding experiments (Fig 5) (28). Therefore k'_B decreases at lower flow rates. This decreased resolution at lower flow rates should be possible to avoid by changing the immobilization condition as used in (23) or by using pellicular supports where steric hindranc is minimized. The observed plate heights in HPLAC are about an order of magnitude larger than in normal HPLC even for similar capacity factors. This is mainly due to the much lower dissociation rate constants; k_d, for biospecific interactions (Table II and Eq.(15)).

The biospecific adsorbent used in the HPLAC experiments was used for more than 2oo runs with protease solutions without marked change in resolving capacity. This higher reusability and speed of separation favours the further development of HPLAC procedures (Table I).

3.1.2. Determination of association constants

The mobility of an adsorbed ligand B in any position of an affinity chromatography column is proportional to the fraction of B in the mobile phase or

$$v_B = v_o \frac{n_{m,B}}{n_{s,B}+n_{m,B}} = \frac{v_o}{1 + k'_B} \qquad (17)$$

where v_B, v_o are the mobilities of the adsorbed and a non-adsorbed solute moving only in the mobile (void) phase, $n_{m,B}$ and $n_{s,B}$ are the total moles of B in the mobile and stationary phase respectively, and k'_B is the capacity factor (mass ratio) given by Eq.(7). The mobility is generally concentration dependent as k'_B varies with the concentration of B.

From Eq.(17) a simple relation for the elution volumes in an affinity chromatographic experiment can be derived for the case where k'_B is constant in the column. Then

$$k'_B = \frac{V_B - V_o}{V_o} \qquad (18)$$

where V_B and V_o are the elution volume of B and the void volume respectively. From Eq.(6) and (7) follows that k'_B is independent of c_B and other free concentrations only when $c_{B_i} = 0$ and

(i) $c_B << C_X, C_Y = 0$, and $K_X \cdot c_B << 1$ (i.e. elution of a zone of B with a buffer with constant content of X) then

$$k'_B = \frac{V_s}{V_o} \left(1 + \frac{\bar{K}_A \cdot n_o}{(1+K_X \cdot C_X)(1+\bar{K}_A \cdot c_B)}\right) \qquad (19)$$

(ii) $C_Y = C_X = 0$, c_B constant in column (i.e. a frontal analysis experiment) then

$$k'_B = \frac{V_s}{V_o} \cdot \left(1 + \frac{\bar{K}_A n_o}{(1+\bar{K}_A \cdot c_B)}\right). \tag{2o}$$

The equations (19) and (2o) can be used to determine association constants from affinity chromatographic experiments provided that the assumptions made in (i) and (ii) are fulfilled. Then the experimental conditions can be practically considered to represent an equilibrium system (in B).

For <u>elution of zones</u> with desorbing ligands X in the buffer the following relation can be derived from Eq.(18) and (19), provided that the assumptions (i) are valid,

$$\frac{1}{V_B-(V_o+V_s)} = \frac{1+\bar{K}_A \cdot c_B}{\bar{K}_A \cdot V_s \cdot n_o} + \frac{K_x \cdot C_x(1+\bar{K}_A \cdot c_B)}{\bar{K}_A \cdot V_s \cdot n_o} \tag{21}$$

that equalsthe expression originally developed by Dunn and Chaiken (24), considering the corrections indicated in (13). The quantity (V_o+V_s) is the elution volume of B without X and the biospecific adsorbent. Thus, V_B, V_s, n_o and C_x can be directly measured. To determine $\bar{K}_A$ and K_x from Eq.(21) $\bar{K}_A \cdot c_B$ must be << 1 as c_B is not constant when C_x-varies. Then Eq.(21) linear in C_x and $\bar{K}_A$ and K_x can be determined when $1/(V_B-(V_o+V_s))$ is plotted as a function of C_x from the slope and slope/intercept ratio.

A similar method to determine $\bar{K}_A$ and K_x has been developed for an analogeous procedure when the biospecific adsorbent is used as an electrophoretic medium (25). This affinity electrophoretic procedure (Fig.4) is based on the assumption that the volumes where ligands are bound to the biospecific adsorbent equals the volume where the ligands can move in the electric field.

For <u>frontal elution</u> (case (ii)) Eq.(18) and (2o) can be written

$$V_B-(V_o+V_s) = \bar{K}_A(V_s \cdot n_o-(V_B-(V_o+V_s))c_B) \tag{22}$$

V_o+V_s is the elution volume for the front when $n_o = 0$.

From Eq.(22) $\bar{K}_A$ can be determined from a plot of $(V_B-(V_o+V_s))$ as a function of $(V_B-(V_o+V_s))\cdot c_B$.

The operational capacity, n_o, used in Eq.(21) and (22) is, as indicated above (Table I, Fig.3), smaller than the stationary active ligate content, n_s. When $\bar{K}_A$ is determined using these methods n_o

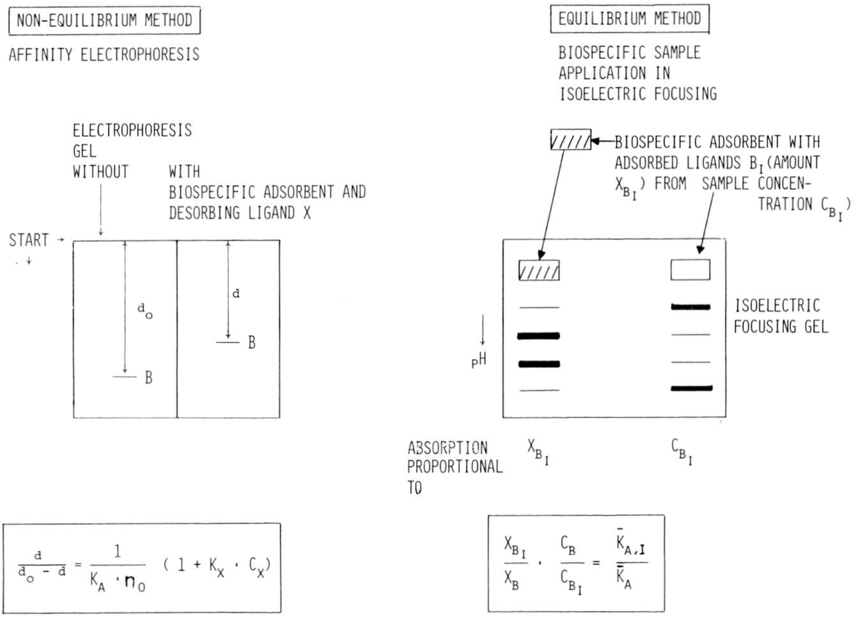

Fig.4. Different electrophoretic procedures with biospecific adsorbents used to determine ligate/ligand association constants.

has to be carefully determined.

In the above procedure the average association constant $\bar{K}_A$ is determined. When the association constants are different for differently bound ligates the plots of the experimental data should deviate from linearity. This can however also be due to the assumptions made for (i) and (ii)not being fulfilled.

Another difficulty in the above methods is that they are based on the determination of a small difference, by separate measurements, of the large quantities V_B and (V_o+V_s) or distances of migration (affinity electrophoresis). The experimental relative error of the single measurements may exceed 5% and the differences observed are less than 5o% of the maximum value (24,25,26). The relative error in association constants determined by such methods has been shown to be > 2o% (17,27). Thus these methods are not ideal for a more detailed analysis of association constants of ligates on biospecific adsorbents including the heterogeneity in these constants. Equilibrium binding studies should therefore be preferred in determination of association constants.

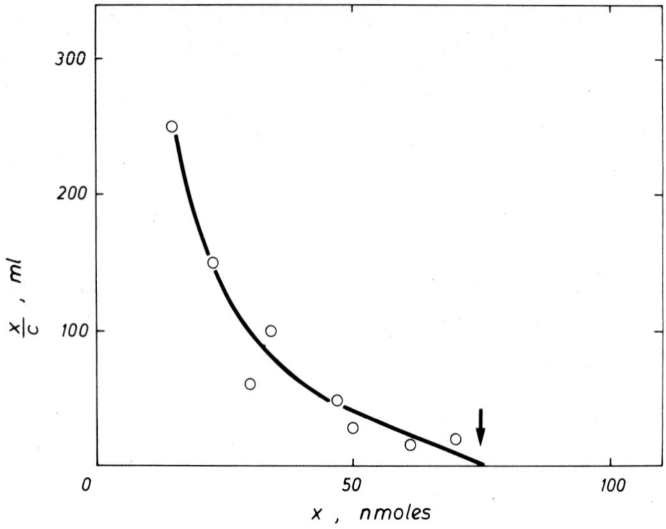

Fig.5. Scatchard plot for the binding of α-chymotrypsin to soybean trypsin inhibitor immobilized to aminosilanized Lichrosphere. Conditions as in Fig 2. The arrow gives the static capacity of the biospecific adsorbent as determined by the mass balance after immobilization.

3.2. Binding studies in equilibrium systems

3.2.1. Determination of heterogeneity in ligate association constant

Through equilibrium binding studies using a homogeneous ligand B (Fig.2) $\overline{K}_A$ can be determined from the linear relation derived from Eq.(6) for $C_X = C_Y = C_{B_i} = 0$

$$K_D = \frac{X_B}{C_B} = \overline{K}_A\ (n_s - X_B) \tag{23}$$

where X_B is bound B-content. In Eq.(23) the stationary ligate capacity n_s content must be used instead of the operational capacity n_o. Such binding curves are given in (15,17) and Fig.5. The deviation from linearity is a measure for the distribution in K_A-values for the ligate. They can be used to study the conditions required to reduce the heterogeneity in ligate association constants yielding biospecific adsorbents with improved resolution as was done in (15,23). They can also be used to estimate the perturbation of the ligate association constant in the heterogeneous system due to steric hindrance, conformation changes and the electric field at a charged support (15,28). $\Delta(\Delta G^o_{Dip})$ and $\Delta(\Delta G^o_{ster} + \Delta G^o_{conf})$ were found to be ∿ 4 and ∿ 2 kJ/mol for the binding of α-chymotrypsin to soybean trypsin inhibitor immobilized in Sepharose of high and low stationary charge density, respectively. In this case linear Scatchard plots were observed as the binding was performed at pH 7.

In these experiments long equilibration times (> 6o min) are required so that all ligate sites are participating in the equilibria. This is apparent from kinetics of adsorption and desorption shown in Fig.2 that shows that the accessibility of some ligates in the heterogeneous system is markedly reduced. The latter will probably not participate as adsorbents in affinity chromatographic experiments. This also demonstrates the difficulty to obtain an unambiguous measurement of the operational capacity, n_o that must be used to analyze affinity chromatographic data (Eq.(2o) and (21)).

3.2.2. Determination of average association constant for several modified ligands B_i simultaneously

Biospecific sample application in isoelectric focusing (29,3o) provides a tool to determine the number of modified ligands B_i (with nuclides, fluorochromes or enzymes as indicators) and their biological properties (association constants) (Fig.4). As the isoelectric points of the modified ligands are normally changed by the covalent binding of the indicator the different conjugates should be separated by isoelectric focusing. If a biospecific adsorbent is equilibrated with a sample of different modified ligands the ratio of K_D for the unmodified ligand B and modified ligand B_i derived from Eq.(6) is ($C_y = C_x = O$).

$$\frac{K_{D,B}}{K_{D,B_i}} = \frac{X_B}{X_{B_i}} \cdot \frac{c_{B_i}}{c_B} = \frac{\bar{K}_{A,B}}{\bar{K}_{A,B_i}} \qquad (24)$$

The ratio X_B/X_{B_i} can be determined after desorption of the biospecific adsorbent placed on the isoelectric focusing gel. The ratio c_{B_i}/c_B is determined from the supernatant of the biospecific adsorbent by isoelectric focusing. Thus in this simple experiment the average association constants of several modified ligands relative to the unmodified ligand can be determined simultaneously in one experiment. This method can be used to evaluate biospecific adsorbents for affinity chromatography or modified ligands for binding assays provided that biospecific adsorbents are available for the latter. It can, however, not be used to evaluate ligate heterogeneity.

4. INFLUENCE OF LIGAND HETEROGENEITY ON LIGAND BINDING ASSAYS

In homgeneous and heterogeneous ligand binding assays an unknown amount of unmodified ligand is determined by two measurements using a biospecific adsorbent for the ligand and modified ligands (8-1o). The amount of bound modified ligand is measured in the presence ($x_i^{*\prime}$) and absence (x_i^*) of the unmodified ligand. From the measured ratio

$$F = \frac{\sum_i \alpha_i x_i^{*\prime}}{\sum_i \alpha_i x_i^{*}} \qquad (25)$$

where α_i is:

(i) heterogeneous binding assay: molar property (fluorescence quantum yield or depolarization, enzyme activity or nuclide activity on separated biospecific adsorbent) of the modified bound ligand B_i;

(ii) homogeneous binding assay: the difference in the molar property between the bound and free ligand B_i when difference measurement with and without A are performed;

that is the basis of the experimental measurement. The unknown ligand content, c_B, is determined from a calibration curve of F vs. c_B at constant modified ligand content. From Eq.(6) the following relation for F can be derived

$C_y = C_x = O$

$$F = \frac{1 + \sum_i \overline{K}_{A,i} \cdot c_{B,i}}{1 + \overline{K}_A \cdot c_B + \sum_i \overline{K}_{A,i} \cdot c'_{B_i}} \cdot \frac{\sum_i \alpha_i \cdot \overline{K}_{A,i} \cdot c'_{B_i}}{\sum_i \alpha_i \cdot \overline{K}_{A,i} \cdot c_{B_i}} \qquad (26)$$

This relation is independent of the heterogeneity in the modified ligand only when the term

$(\sum_i \alpha_i \overline{K}_{A,i} \cdot c'_{B_i}) / (\sum_i \alpha_i \cdot \overline{K}_{A,i} \cdot c_{B_i})$ is

$= c'_{B_i} / c_{B_i}$ i.e. only o n e modified ligand is used in the binding assay; or

= 1 or $\bar{K}_{A,i} \cdot c'_{B_i} = \bar{K}_{A,i} \cdot c_{B_i}$ then all concentrations in Eq.(26) must be total contents i.e. the number of bound ligands << free ligands.

We have used biospecific sample application in isoelectric focusing to analyze the heterogeneity - in α_i and $\bar{K}_{A,i}$ - for fluorochrom-modified proteins (3o). The product $\bar{K}_{A,i} \cdot \alpha_i$ was found to vary almost by an order of magnitude for different mono- and di-labelled conjugates. Unfortunately there exist few data on the heterogeneity of modified ligands used in ligand binding assays. It is, however, expected that the heterogeneity observed in (3o) is a general property for modified ligands. This can be studied by the method shown in Fig.4 that was used to determine the data in (3o).

For quantitative studies using modified ligands especially in cellbiology and for standardization purposes, the use of only <u>o n e</u> modified ligand should thus be preferred. Then the measurements are independent on the heterogeneity in α_i and $\bar{K}_{A,i}$ of the modified ligand.

5. SUMMARY

The operational properties: operational and stationary ligate capacity; resolution; association constants as a measure for the biological activity that is used for fractionation mass transfer limitations in the matrix of biospecific adsorbents are important for the design of affinity chromatographic procedures.

They have been analyzed based on ligand/ligate interactions in heterogeneous systems and two basic ratios used to characterize chromatographic procedures. The latter are the distribution coefficient K_D (<u>a concentration ratio</u>) and the distribution or capacity factor k' (<u>a mass ratio</u>). This has been used to analyze experimental data on the resolution in high performance affinity chromatography. These quantities have also been used to analyze equilibrium systems (ligand binding assays). In both cases the observed properties depend on the heterogeneity in association constants for the different ligate/ligand interactions. It has been shown that they can only be unambiguously determined in equilibrium binding studies. This is mainly due to the flow-rate dependent mass transfer of ligands from the mobile to the stationary phase in non equilibrium (chromatography) procedures. An equilibrium method to study ligands modified with marker groups (using biospecific adsorbents) and isoelectric focusing has been shown to be a valuable tool to study the heterogeneity in the biological activity of the differently labeled ligands.

REFERENCES

1 C. de Lisi, Quart. Rev. Biophys.,13(198o)2o1-23o.
2 R. Röber, Physikalische Chemie der Zellen und Gewebe, Verlag Staempfli & Cie., Bern, 1947, p.156.
3 D.H. Campbell, E. Luescher and L.S. Lerman, Proc. Nath. Acad. Sci. USA, 37, 575-579.
4 C.C. Curtain, Brit. J. exp. Pathol., 35(1954)255-263
5 G. Manecke, Naturwiss.,42(1954)212-213.
6 W.B. Jacoby and M. Wilchek (eds.), Methods in Enzymology, Vol.34(1974).
7 J. Turková, Affinity chromatography, Elsevier Amsterdam, 1978.
8 R.S. Yalow, Science, 2oo(1978)1236-1245.
9 B.K. van Weenen and A.H.W.M. Schuurs, Febs Letters, 24(1972)77-81.
1o Immunofluorescence and related staining techniques (W.Knapp, K.Holubar and G.Wich, eds.) Elsevier/North-Holland Biomedical Press, Amsterdam, 1978.
11 J. Niedel, I. Kahane, P. Cuatrecasas, Science 2o5(1979)1412-1414.
12 V. Kasche, Biochem. Biophys. Res. Commun., 38(197o)875-881.
13 L.W. Nichol, A.G. Ogston, D.J. Winzor and W.H. Sawyer, Biochem. J. 143(1974) 435-443.
14 L.R. Snyder and J.J. Kirkland, Introduction to modern liquid chromatography, John Wiley & Sons, New York, 1979.
15 V. Kasche, studia biophysica, 35(1973)45-56.
16 R. Jaenicke, Ann. Rev. Biochem., 1981, in press.
17 V. Kasche, Acta Univ. Upsaliensis, 2(1971)1-132.
18 J. Wyman Jr., Adv. Prot. Chem. 19(1964)223-287.
19 C. Horvath and H.-J. Lin, J. Chromatogr., 149(1978)43-7o.
2o P.C. Wankat, Anal. Chem. 46(1974)14oo-14o8.
21 D.J. Graves and Y.-T. Wu, Adv. in Biochem. Eng., vol.12(1979) 219-253.
22 S. Ohlson, L. Hanson, P.-O. Larsson and K. Mosbach, Febs Lett. 93(1978)5-8.
23 D. Gabel, H. Amneus and V. Kasche, J. Chromatogr. 12o(1976)391-397
24 B.M. Dunn and I.M. Chaiken, Proc. Nat. Acad. Sci. USA, 71(1974) 2382-2385.
25 V. Horejsi, J. Chromatogr., 178(1979)1-13.
26 S.J. Johnson, E.C. Metcalf and P.D.G. Dean, Anal. Biochem. 1o9(198o)63-66.
27 O.A. Deranlean, J. Am. Chem. Soc., 91(1969)4o44-4o49.
28 V. Kasche, K. Buchholz and B. Galunsky, submitted for publication
29 H. Amneus, L. Näslund, D. Gabel and V. Kasche, Biophys. Biochem. Res. Commun. 9o(1979)1313-132o.
3o V. Kasche and I. Büchtmann, Hoppe Seyler's Z. physiol. Chem., in press.

CHAPTER II

POLYMERIC MATRICES AND LIGAND IMMOBILIZATION

INTERNATIONAL SYMPOSIUM
ON AFFINITY CHROMATOGRAPHY
AND RELATED TECHNIQUES

T.C.J. Gribnau, J. Visser and R.J.F. Nivard (Editors),
Affinity Chromatography and Related Techniques

PROPERTIES AND INTERACTIONS OF POLYSACCHARIDES UNDERLYING THEIR USE AS CHROMATOGRAPHIC SUPPORTS

J.K. MADDEN and D. THOM

Unilever Research, Colworth Laboratory, Colworth House, Sharnbrook, Bedford MK44 1LQ

ABSTRACT

For consideration of the conformations and interactions of polysaccharides underlying their use as chromatographic matrices they can conveniently be divided into two classes on the basis of their covalent structure; namely periodic and interrupted periodic types.

In the periodic type fibrous cellulose was among the first supports used for affinity chromatography. Its inherent advantages and disadvantages can be directly related to its native ordered structures. Chemically cross-linked random coils have also been extensively evaluated. Archtypes are those formed from dextran, a bacterial polysaccharide with minimal opportunities to adopt ordered structures. Optimum cross-linking requires a compromise between conflicting requirements for (a) ideal physical form and (b) optimum pore size. More recently cellulose and starch matrices have re-emerged where the natural ordered structures must be disrupted by chemical treatments prior to cross-linking.

In the interrupted periodic type the seaweed polysaccharide agarose is by far the most commonly used. Detailed investigations using a wide range of physical techniques have shown that the key properties of the matrices arise from (a) intermolecular association of long, regular chain sequences into double helices and (b) further association of these helices into distinct supramolecular "domains". This molecular understanding can be extended to explain further improvements achieved by chemical modification or by combining agarose with synthetic polymers.

A broader range of polysaccharides have also found recent application for the insolubilisation of whole cells. Prominent examples include; (a) carrageenans, which are structurally closely related to agarose, (b) polyuronates such as alginates and pectins where non-covalent association into highly ordered "egg-box" like structures are triggered by divalent cations.

INTRODUCTION

Polysaccharides are hydrophilic polymers which play important roles in the structural assemblies which characterize the extracellular organisation of biological tissues. Their properties (reviewed 1-4) are widely exploited to control the structure and texture of hydrated systems in industrial applications. They also make certain polysaccharides particularly effective supports for the rapidly growing science and technology of affinity chromatography.

Conformational analysis has shown (3-5) that for D sugars their shape may generally be regarded as fixed in a $^{4}C_{1}$ chair conformation with C(6) equatorial. Overall chain geometry is predominantly determined by the relative orientation of adjacent residues around their glycosidic linkages. Polysaccharides contain large numbers of such linkages and in aqueous solution their favoured conformation cannot be assumed to be ordered, since the chain flexibility arising from the continuous fluctuation of the polymers about these linkages provides a strong entropic drive (1,3) to disordered random coil forms. Under particular circumstances, however, favourable, non-bonded energy terms (hydrogen bonding, dipolar and ionic interactions and solvent effects) can combine to overcome this conformational entropy and fix the macromolecules in ordered shapes (1-3). Interactions between pairs of residues is insufficient to do this and it occurs by co-operative interactions which reinforce each other within long extended sequences of the polymer chains.

A key step in the development of insoluble polysaccharide supports for affinity chromatography is often the manipulation and control of the factors controlling the balance between highly ordered structures and disordered shapes in aqueous solution to create the most appropriate physical form for network formation. This is frequently combined with controlled chemical cross-linking to either form the basic 3-dimensional matrix or to reinforce networks formed by non-covalen association of polysaccharide chains. For detailed consideration of the approache which have been explored it is convenient to divide the polysaccharides into two classes on the basis of their covalent structure, namely 1. simple periodic structures and 2. interrupted periodic structures.

1. MATRICES FORMED FROM SIMPLE PERIODIC STRUCTURES

(i) Shapes and interactions

These polysaccharide chains contain only one type of residue linked through identical positions and glycosidic configurations. Conformational analysis confir (6) that such sequences can generate ordered conformations which will fall into one of four distinct types (Fig. 1). Such shapes can be considered as a helix and defined by their symmetry (n) and their projected residue height (h). The important variable in relating primary structure to final shape is the relative orientation of the glycosidic bonds to each sugar ring, and not the interactions between adjacent residues since these can be very similar.

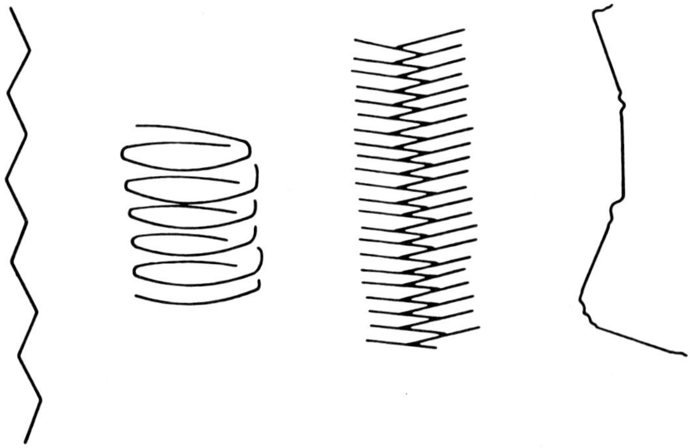

Fig. 1. Schematic representation of the subclasses of ordered conformations predicted by conformational analysis for periodic structures containing only one type of residue.

(a) Extended ribbons. For these structures the bonds connecting each sugar residue to its two glycosidic oxygens form a zig-zag shape, the values of n are in the range $\pm$ (2-4) and h is close to the maximum length for a sugar residue. The archetype is the β-1,4 glucan, cellulose, the major skeletal component of plant cell walls which was among the first supports used for affinity chromatography, Campbell et al. (7) purifying rabbit anti-bovine serum albumin antibodies with a column containing albumin coupled to diazotised p-amino benzyl cellulose. Mushroom tyrosinase was subsequently purified using p-azophenol substituted cellulose (8) and liver flavokinase on flavin-cellulose columns (9,10).

In the solid state cellulose adopts (11,12) an extended ribbon form known as "Herman's bent chain" conformation. This has n = 2 and h = 5.15, close to the maximum possible for sugar residues. It is stabilised by hydrogen bonding of each successive residue to its two nearest neighbours through O(3) and O(5). Such flattened ribbons are characteristic of the structural components which are most difficult to extract from plant cell walls. X-ray diffraction also shows that they have the highest degree of ordered chain packing, a further indication that flat ribbon structures are most easily aligned and bonded into crystalline arrays. In such structures chains can be packed in either a parallel or an anti-parallel arrangement. It has been shown by x-ray diffraction studies that in its natural organisation extended ribbons of cellulose probably occur in parallel rather than anti-parallel arrays (13). Flattened sheets of chains lie side by side and are joined by hydrogen bonds. These sheets are laid on top of one another rather in the way in which bricks are staggered to give strength and stability to a wall.

The high strength, insolubility and inertness to solvent systems that are so characteristic of cellulose and lead to its choice as one of the first supports for affinity chromatography can be readily understood for such compact, tightly bonded aggregates. This organisation is also directly responsible for the fibrous and non-uniform structure and for the poor flow rates and restricted penetration of larger ligands which ultimately lead to its replacement by superior supports. The considerable non-specific absorption frequently found for cellulose can also be rationalised, since the crystalline aggregates have two hydrophilic faces and two which are considerably more hydrophobic.

(b) Coiled Springs. In these structures the bonds from each sugar unit to its glycosidic oxygen define a u-turn, the allowed values of n cover a much wider range, ($\pm$ 2-10) and more significantly h can be close to zero. The structure therefore resemble a coiled spring in various states of extension. The supreme example is starch which is a mixture of two polysaccharides with covalent structure containing 1,4 linked α-D-glucopyranose units. Amylose is essentially linear whereas amylopectin has chains of ~20 units joined by α-1,6 linkages to form a branched structure. Amylose can adopt a wide variety of ordered conformations in the solid state depending on its environment. Its best known ordered states have the spring very compressed with its geometry varying between 6,7 and 8 residues per helical turn (14-17) but more extended forms also occur, the most extreme being a potassium bromide complex (18) which is believed to have n = 4 and h = 4.03 Å. Such chains also adopt highly ordered structures in their native starch granules. A crystalline form known as the "A" form is found in cereals whereas a "B" form is found in tubers. There is now convincing evidence that both structures are double helices (19,20) which are virtually identical, with the observed differences in x-ray diffraction patterns arising from their crystal packing and water content.

(c) Crumpled Ribbons. In this case the relevant bonds have a twisted arrangement and extended sequences would be characterised by steric clashes between non-adjacent residues. Few examples of this class are found in nature.

(d) Flexible Coils. This group consists of 1,6 linkages and are therefore characterised by an extra bond between each sugar residue. The only common example is dextran, an extracellular glucan produced by microorganisms such as *Leuconostoc mesenteroides*, which contains >90%, α 1-6 linkages and can be also lightly branched by 1,2; 1,3 or 1,4 linkages (21). The increased flexibility of the extra glycosidic bond imparts substantially greater conformational entropy, and ordered states are not to be expected in the hydrated state, a property which proves particularly useful in the preparation of uniform matrices for chromatography (described below). In fact no ordered structures of this type have been established although such sequences have been induced to form microcrystalline arrays in solid films (22).

(ii) Chemically cross-linked matrices

Dextran is readily soluble in aqueous solutions due to its flexibility and hydrophilic nature. Uniformly cross-linked beads with a low non-specific absorption can be prepared by emulsifying dextran solutions in organic solvents and chemically cross-linking with epichlorohydrin under alkaline conditions. Such matrices have found wide applications for gel chromatography, since the covalent "tie points" coupled with the very limited tendency of the dextran chains to adopt ordered structures in the solid state and the drive of the flexible chains to dissolve in aqueous solution means that these matrices can be dried and re-swollen almost indefinitely. They are also stable to autoclaving.

Unfortunately the flexibility of dextran chains also has undesirable consequences (Fig. 2). Lightly cross-linked matrices with desirable pore sizes and permeabilities for large ligands are too soft for affinity chromatographic uses, giving compression and very variable flow rates (23). Heavily cross-linked matrices can overcome this, but only at the expense of pore size and ligand permeability. Optimum

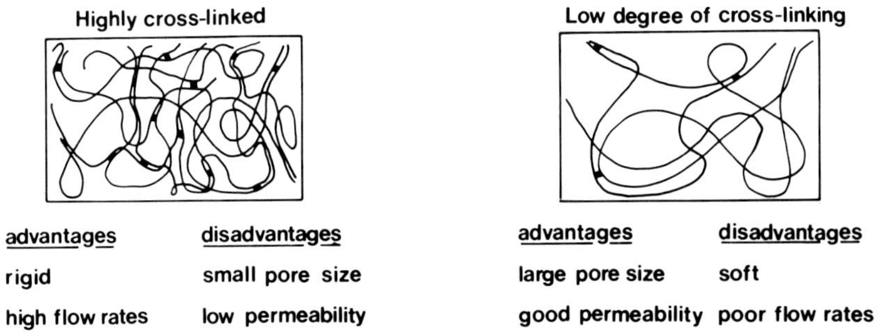

Fig. 2. Comparison of the properties of dextrans with different extents of chemical cross-linking.

cross-linking density requires a compromise between these and in practice means that, in general, these supports cannot compete effectively with agarose. A similar approach has also been taken with a range of polysaccharides and prominent examples include guar (24,25), glycogen (26) and pectic acid (27). However, their use has mainly been restricted to specialist applications involving the isolation of enzymes or plant lectins with specific binding sites for elements of the polysaccharide structure.

More recently the understanding of Sephadex gels has been used to produce improved matrices from cellulose and starch. For cellulose two broad approaches have been applied (Fig. 3) both involving controlled disruption of the native ordered structure as the initial step. In the first (28), carboxymethylcellulose (CMC) was used as the starting material. A high degree and uniformity of substitution ensured that the chains were completely in the flexible coil form. This CMC was converted into the active azide intermediate and reacted with excess, W-diaminoalkanes to simultaneously cross-link the matrix and introduce primary amino groups. Reactive carboxyl groups can subsequently be introduced by reaction with succinic anhydride. This approach provides an alternative chemically cross-linked matrix where the charged substituents prevent the cellulose chains from readopting their native ordered structures during use or drying stages.

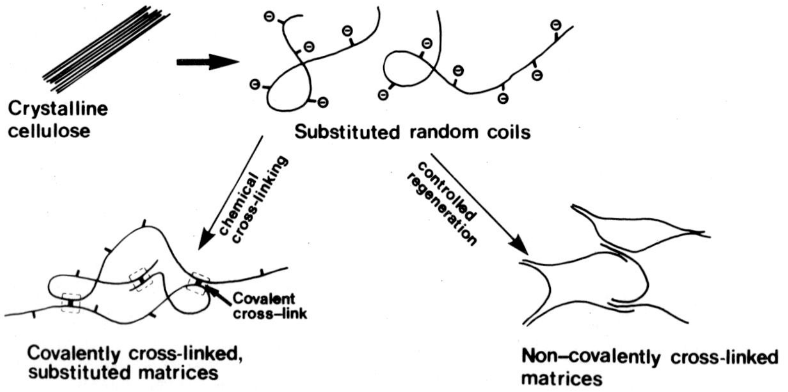

Fig. 3. Schematic representation of the formation of cellulose matrices by controlled disruption of the microcrystalline arrays followed by cross-linking.

Similar approaches can be adopted with starch although there have been few concerted attempts to use it as an affinity matrix. Limited examples include a comparison between cross-linked starch and cellulose (29) and the use of dialdehyde starch for enzyme immobilisation (30) after condensation with p,p' diaminodiphenylmethane. Highly permeable and stable matrices based on amylose have also been reported (31). Very similar investigations have also been carried out by industrial concerns interested in producing "superabsorbent gels" capable of quickly and effectively absorbing aqueous fluids. For natural polymers most approaches have been along very parallel lines to those outlined for cellulose, but with little attention being paid to the production of uniform beads with good flow properties and considerably more to the production of networks with very fast and efficient absorption and retention of fluids. Limited work (32) along these lines has again been carried out with starch; the first step involving controlled disruption of the crystalline starch granule with simultaneous cross-linking. In a second the polysaccharide chains are derivatised to increase their hydrophilic character and prevent ordered structures reforming on drying.

The second approach with cellulose has been covered in detail by Dr. Štamberg (33). Again the first stage is disruption of crystalline cellulose, this time to form cellulose xanthate. In later stages, emulsified droplets of the polymer solution have the xanthogenate groups removed in a controlled process. The cellulose chains reassociate non-covalently (Fig. 3) to provide rigid, beaded (20-1000μm) cellulose with properties rivalling agarose based matrices. In this regenerated cellulose the carbohydrate chains are almost certainly in the cellulose II form (34) as opposed to the native state (cellulose I) since there is at present no known way of regenerating the natural organisation. The chain conformation is essentially the same in both forms, the significant difference being that in cellulose II the sheets of extended ribbons are packed in anti-parallel arrays and laid directly on top of one another rather like planks in a timber yard rather than bricks in a wall.

2. MATRICES FORMED FROM INTERRUPTED PERIODIC STRUCTURES

These chains also contain periodic sequences capable of adopting ordered conformations but these are separated or "interrupted" by deviations from regularity which force chains to leave ordered associations and combine with more than one partner. The resultant balance of ordered and "soluble" regions leads to the highly hydrated gel state which is so important for many industrial applications including affinity chromatography.

(i) Interrupted helices - Agarose and Carrageenans

Agarose is undoubtedly the most widely used support matrix for affinity chromatography. It is closely related to the carrageenan family of polysaccharides and our understanding of this whole class of natural polymers will be used to illustrate the shapes, interactions and properties which have lead to this dominant position. Both agarose and the carrageenans (ι and κ) are isolated from red seaweed (Rhodophyceae) and both classes undergo reversible sol $\leftrightarrow$ gel transitions on cooling and heating in aqueous solution. All are, primarily, alternating $(A-B)_n$ copolymers containing varying levels of substituents and the regular sequences which form the ordered structures consist of 1,3-linked β-D-galactose and either 1,4 linked 3,6 anhydro-α-L- or D-galactose residues for agar and the carrageenans, respectively. In either case this regular repeating sequence of sugar residues is "interrupted" by a proportion of sequences ("soluble kinks") where the regular 3,6 anhydro residues are replaced by galactose. These kinks can be removed by a Smith degradation (35) to create shorter blocks or "segments" which retain the ability to adopt ordered structures, but cannot gel.

Double helices have been characterised for agarose and ι-carrageenan in the solid state by X-ray diffraction (36,37). Agarose and its derivatives form 3-fold left-handed helices with a pitch of 1.90 nm. The influence of the 3,6 anhydro D-galactose configuration in carrageenans is to reverse the sign of the helix; ι-carrageenan adopting right-handed helices with a pitch of 2.66 nm. Although the X-ray diffraction data for κ-carrageenan is less well defined, it has also been interpreted in terms of right-handed double helices with a pitch slightly shorter (2.46 nm) than for ι-carrageenan.

The native members of both the agar and carrageenan families show temperature dependent sigmoidal changes in optical rotation (OR) but the extent of hysteresis varies quite markedly. The important factor appears to be the extent of substitution by negatively charged groups. Unsubstituted agarose shows a much greater degree of hysteresis than a natural pyruvic acid substituted agar from *Gracilaria compressa* and this in turn is greater than a natural sulphated agar from *Gliopeltis cervicornis* (37). Similar differences are observed (38,39) in the OR and spin-spin (T_2) relaxation behaviour of the carrageenan series, with i-carrageenan showing less hysteresis than κ-carrageenan and both being significant less than for agarose. The sign and magnitude of the OR changes (37,40,41) correspond closely to those predicted from the left-handed and right-handed helix geometry found for agarose and carrageenan, respectively, in the solid state.

The mechanism of gelation has been explored in more detail using isolated segments. For ι-carrageenan, segments again show a sigmoidal OR change (Fig. 4) with a sign and magnitude close to that predicted from the double helix geometry in the condensed phase (40). The changes are concentration dependent (at constant temperature and ionic strength) and correspond to a co-operative dimerisation process (41). Low angle X-ray scattering also shows ordered rods corresponding to the expected double helical dimensions. Corroborative evidence (Fig. 4) for conformational ordering is provided by ^{13}C and ^{1}H NMR measurements (41). Sharp, well resolved spectra are obtained at 80°C when the polysaccharide exists as a random coil. All these peaks collapse dramatically on cooling to give the rigid ordered form and comcomitantly a large decrease in the relaxation time for the carbon nuclei (T_2) was observed in the pulse spectrometer. No sign of broadening or shifting of peaks was observed when the transition was partially completed which suggests no time averaging between ordered and disordered states and this is consistent with a co-operative ["two state all-or-none"] mechanism.

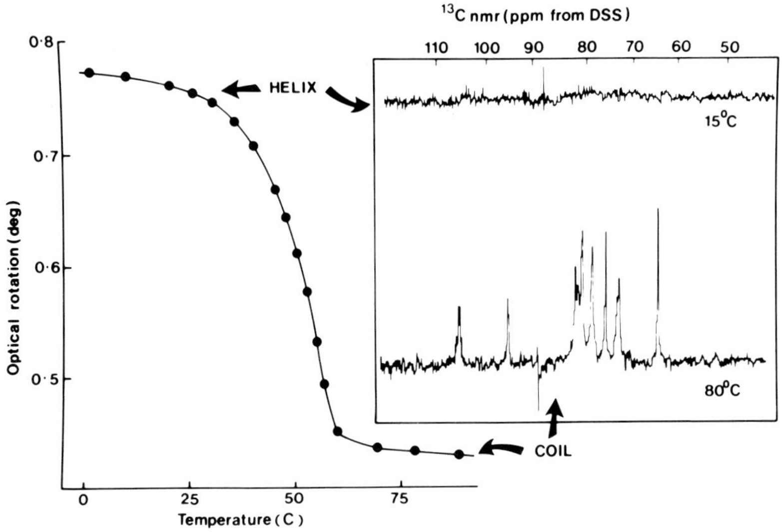

Fig. 4. The temperature dependent sigmoidal increase in OR and comparison of [^{13}C] NMR spectra at 80°C and 15°C for ɩ-carrageenan segments. The concentration was 6 % (wt/vol) in 0.1M sodium chloride solution.

Optical rotation measurements for segmented κ-carrageenan show a co-operative concentration dependent transition (38) on heating and cooling which provides good evidence that a similar ordered conformation occurs in solution. The transition shows hysteresis behaviour which can be correlated with aggregation of the ordered conformation. Recent evidence (42), from the dynamics of salt (K^+) induced disorder-order transition of native κ-carrageenan using a polarimetric stopped flow technique, strongly support the proposal that the aggregates of ordered chains in the junction zones of the gel are double helical entities. The approach relies on the finding that κ-carrageenan does not form its ordered conformation, and does not gel, in the sodium salt form, but that on addition of potassium salt ordering occurs. When the salt jump experiment was made at temperatures far from the hysteresis region the results indicate that the formation of the ordered state is a second order process.

In the case of segmented agarose, aggregation of the ordered conformation is so great on cooling, that the transition cannot be monitored by OR because precipitation occurs at the onset of conformational ordering (38). In contrast to unsubstituted agarose, segments of a highly sulphated agarose show co-operative OR transitions indicative of conformational ordering. Dynamics of salt-induced disorder-order transitions of these sulphated segments (43) indicate that the formation of this ordered state is again a second order process. There is therefore strong evidence that the aggregates of ordered chains in the junction zones of agarose and agar gels are also double helical.

The evidence thus suggests a common mechanism of gel formation for ι-carrageenan κ-carrageenan and substituted and unsubstituted agarose in which chains are associated into double helices separated by soluble sequences involving "kinks" and there is increasing aggregation of the helices as the degree of substitution decreases. Striking evidence for the importance of the high levels of structural organisation within aggregates has recently been demonstrated by detailed studies of the gelation behaviour of carrageenans. It has been known for many years that gelation of carrageenans is extremely sensitive to the nature of the counterion. Thus optimum gelation is observed with K^+ or Rb^+, somewhat weaker gels with Cs^+ or NH_4^+ and no cohesive structure with Li^+ or Na^+, although significant thickening and ultimately precipitation occur at high ionic strength. With tetramethylammoniu[m] as sole counterion there is no evidence of any long range structure and solutions remain freely mobile at all temperatures. All these cationic forms, however, show essentially the same degree of conformational ordering, both by OR and NMR.

The molecular origin of this behaviour is indicated by light scattering studies (44,45-47) of structurally regular iota carrageenan chain segments. In the tetramethylammonium salt form (47) conformational ordering is accompanied by an exact doubling of M_w, entirely consistent with a simple 2 coil $\rightarrow$ double helix transition. In the presence of potassium, however, a molecular weight increase of $\sim$5-6 fold is observed (46), while sodium gives a value somewhat in excess of 2 ($\sim$ 2.1 - 2.2), suggesting different extents of cation-induced helix-helix aggregation. The disorder-order transition of intact (i.e. kinked) iota carrageena[n] in the non-gelling tetramethylammonium salt form is accompanied (46) by an approximately ten-fold increase in M_w. The results for structurally regular chains show that this cannot arise from helix aggregation, but rather indicates the formation of limited chain clusters or 'domains', crosslinked through double helices, with exchange of partners at 'kink' points. This mechanism is evidently not sufficient for gelation, and further association of 'domains' by helix-helix aggregation (46,47) appears to be essential for development of a continuous network. The cation specificity of aggregation suggests incorporation of counterions within the ordered structure (Fig. 5).

The gelation behaviour of the related but less highly sulphated kappa carrageena[n] shows a similar dependence on counterion. In this case, however, conformational ordering, as monitored by optical rotation, is observed only under ionic condition[s] which promote aggregation in iota (46), and is accompanied by very extensive aggregation as shown by light scattering studies on structurally regular segments. It therefore appears that the kappa carrageenan double helix is stable only when aggregated. Further support for this interpretation comes from comparison of the temperature course of conformational change for iota and kappa under different conditions of ionic environment. In the non-aggregating Me_4N^+ salt form, iota

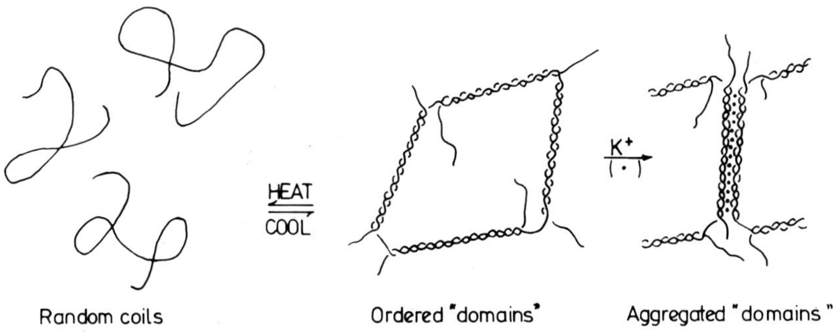

Fig. 5. Schematic representation of the "domain" model of carrageenan gelation.

carrageenan shows a single thermally reversible transition, which has the form expected for a simple two state all-or-none process. Kappa carrageenan in the (aggregating) K^+ salt form also shows a single transition, but shifted to higher temperature by ~15°C, and showing substantial hysteresis, consistent with aggregation stabilising the ordered conformation to temperatures above those at which it will form spontaneously (46,48). Iota carrageenan in the K^+ salt form shows evidence of two conformational transitions, one without hysteresis and centred at the same temperature as the single transition for the non-aggregating Me_4N^+ salt form, and the second centred at a higher temperature, close to that of the kappa (K^+) transition, which it further resembles in displaying thermal hysteresis. These two processes, which are more clearly resolved (48) by DSC, are attributed to coil-helix and coil-aggregated helix respectively. The bulk properties of ι- and κ-carrageenan gels are therefore determined by "quaternary" interactions between ordered tertiary structures (double helices), rather than by direct interchain association through double helical junction zones. As we have seen, however, both show second ordered transition kinetics, indicating that the rate limiting step in the coil-aggregate transition involves the double helical entities.

(ii) Manipulation of helix matrix properties

The two stage "domain" mechanism of gelation has so far only been demonstrated for carrageenan. Since it removes the severe topological constraints associated with the formation of continuous networks solely through double helices, however, it seems likely that similar structures could arise for agar and agarose. The observed differences in gel strength at constant concentration (or minimum concentrations required for gelation) between helix forming polysaccharides are entirely consistent with this suggestion (49). Thus agarose, which is essentially uncharged and has a strong tendency to aggregation, gels at low polymer levels, while iota carrageenan, which is highly charged and aggregates only slightly (presumably due to inter-molecular electrostatic repulsion) requires far higher (~ 10x) concentrations for development of a continuous network, with kappa carrageenan occupying an intermediate position in charge density, extent of aggregation, and concentration requirements for gelation.

We can, therefore, conclude that the bulk properties of agarose gels are determined by extensive aggregation of ordered double helices. The result is a rigid, brittle network with large pores which can accommodate both molecules to be coupled to the matrix and macromolecular ligands which are the target of the affinity methods. This combined with its hydrophilic character, stability and ease of coupling using the cyanogen bromide procedure has made beaded agarose the most widely used matrix for affinity chromatography, fulfulling many of the requirements of an ideal matrix.

However, it is not perfect. Although there is a large hysteresis between setting and melting temperatures, the beads will melt at high temperatures and certainly cannot be autoclaved. As we have seen, the more highly substituted the agarose the lower this melting point will become. The occurrence of these sulphate and carboxylate substituents can also lead to undesirable non-specific binding effects. In addition, beaded agarose has limited stability to acid, alkali and certain solvents and the gel structure can be extensively damaged by freezing or drying procedures.

The rigidity and stability of agarose supports can be considerably improved by chemical cross-linking (50,51) with bifunctional reagents such as epichlorohydrin, 2,3-dibromopropan-1-ol, bis-epoxides and divinyl sulphone (DVS). Cross-linking of Sepharose is carried out on swollen beads and unlike dextran gels, where additional cross-links decrease the effective pore size, no decrease in permeability is observed, thus suggesting that the cross-links serve to reinforce the existing, aggregated network structure and not to introduce additional constraints on the effective pore size. In a detailed study comparing DVS with dibasic acid chlorides (52) it was established that (a) sulphone groups were not responsible for the increased rigidity and (b) optimum cross-linking occurs when there are five atoms in the covalent bridge. The disaccharide repeating unit of agarose contains four unsubstituted hydroxy groups which might, in principle, be involved in chemical cross-linking, at O(2) of 3,6-anhydro-L-galactose, and O(2), O(4) and O(6) of D-galactose. In the ordered conformation, however, O(2) of galactose is buried within the double helix, and would therefore be inaccessible for substitution.

Inspection of published co-ordinates from x-ray analysis shows that the closest approach of accessible hydroxy groups on separate strands of the double helix structure (between O(6) of galactose on one chain and O(2) of the nearest anhydrogalactose residue on the other chain) is 7.4 Å (53), which corresponds to a bridge of 7.5 Å which would be formed by DVS. It is very likely that the observed increase in the rigidity of the agarose beads also depends on reinforcing cross-links between aggregated helices and, possibly most importantly, between the more flexible "soluble kinked sequences" linking the rigid aggregated double helices.

After chemical cross-linking agarose can also be treated to remove ionic groups responsible for non-specific absorption (50). If the necessary stereochemical conditions are fulfilled (54), sulphate esters are readily removed by alkali in the presence of sodium borohydride to prevent chain "peeling" β-elimination reactions. Pyruvate groups can subsequently be reduced with lithium aluminium hydride.

Freeze-thaw, solvent and drying damage (55) to agarose beads can also be understood in terms of the molecular mechanisms underlying gel properties. Freezing produces large ice crystals which force aggregated helices to associate further to form much larger crystalline zones. Solvent and drying treatments can do the same, often more extensively. On thawing or rehydration intramolecular forces within these crystalline zones overcome the drive of the "soluble sequences" to dissolve in the aqueous fluid and, unlike Sephadex, agarose beads usually do not regain their original form and properties.

One approach to overcome this is to incorporate a second polymer system into the agarose network. This will inhibit further aggregation of the original junction zones on freezing or drying and will also provide an additional drive to reswelling on thawing or rehydration. Gels composed of mixed agarose/polyacrylamide networks have been described (56) and are available commercially from LKB (Ultrogel). Rigidity, uniformity and the presence of different functional groups have been reported but we might also expect them to be more stable to freeze-thaw, solvent and drying damage. Similar effects might also be expected for chemically cross-linked agarose gels, particularly where significant amounts of the cross-linking agent have only reacted with one hydroxyl on the agarose chain to provide substituent groups which can inhibit further packing of the aggregated helices on freezing or drying (Fig. 6).

(iii) Interrupted Ribbons - alginates and pectins

More recently the increasing interest in the applications of immbolised whole cells has lead (57,58) to the investigation of the carrageenans themselves as appropriate support materials. Another major class of polysaccharide which is also under investigation in these systems (59-61) are the polyuronates, alginates and pectins, where gels can be formed by controlled release of divalent cations.

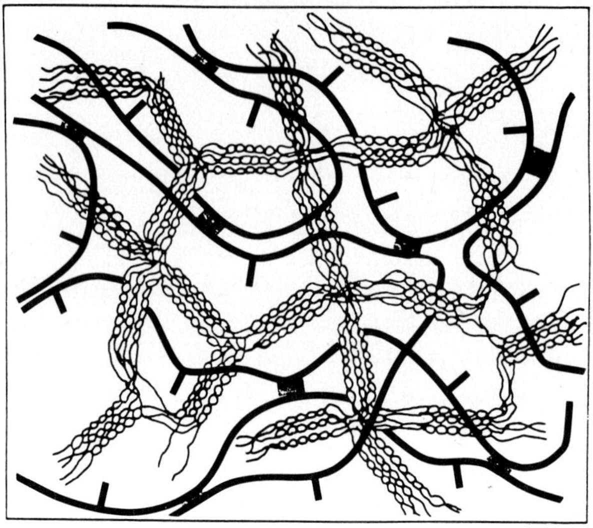

Fig. 6. Schematic representation of interpenetrating networks of aggregated agarose helices and chemically cross-linked chains.

Both families are examples of interrupted ribbon structures. For alginates homopolymeric, buckled ribbons of poly α-L-guluronate sequences play the most important role in junction formation with poly β-D-mannuronate sequences and sequences which approach, albeit irregularly (62), alternating sequences functioning mainly as interrupting soluble sequences. In pectin poly α-D-galacturonate sequences are almost the exact mirror image of polyguluronate except at C(3) and show similar cation binding behaviour. In this case methyl esterified sequences, 1,2 linked α-L-rhamnose kinks and neutral sugar side chains all play interrupting roles.

The gelation of both alginates and pectins has been interpreted (63-65) in terms of an "egg-box" model involving co-operative binding of divalent cations by polyguluronate and polygalacturonate sequences, respectively. The model is based on the physiochemical properties of alginate and pectin gels (64) and on the co-operativity and stoichiometry of their interactions with Ca^{++} ions (65). The primary association involves dimerisation of either polyguluronate or polygalacturonate sequences, with Ca^{++} ions sandwiched in the interstices formed by the packing of these buckled ribbons and efficiently co-ordinated by both carboxylate anions and other oxygen atoms on the polysaccharide chains. This specific site binding leads to dramatic changes in the $n \rightarrow \pi^*$ transition which are equivalent in extent but, as expected, opposite in sign for the polyguluronate and polygalacturonate sequences of alginate and pectin, respectively.

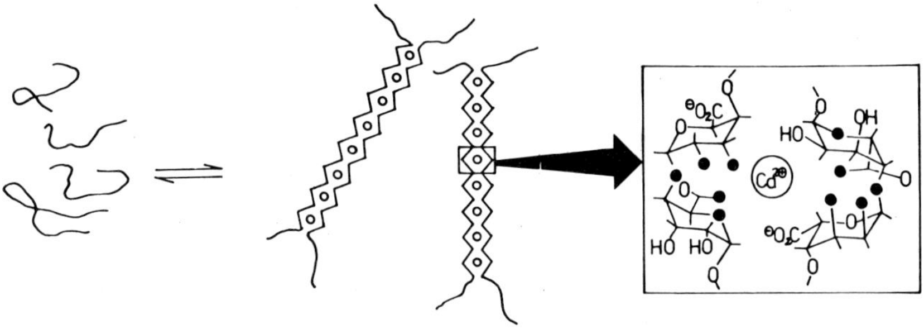

Fig. 7. Schematic representation of the "egg-box" model for the sol → gel transition in alginates. The disposition of groups coordinating the divalent cations is also shown and for clarity chains have been moved apart.

The coprecipitation of cells with alginate has been used as one method for whole cell immobilisation (59,60). However, recent results suggest that many of the processes for entrapping cells have diffusion problems of both oxygen and nutrient supply and also preclude the use of macromolecular substrates or products. A more recent approach (61) uses pectin which has been chemically cross-linked to form an insoluble support. Bacterial cells are then immobilised on the surface of the cross-linked pectin by the formation of specific complexes involving iron. Polydentate ligands are used to increase metal binding capacity of the pectin (61). Similar approaches have also been investigated for the coupling of enzymes to other extended ribbon polysaccharides such as cellulose, alginate and chitin (66,67).

ACKNOWLEDGEMENT

We thank our colleague Dr. E.R. Morris for helpful advice and discussions.

REFERENCES

1. D.A. Rees, in G.O. Aspinall (Ed.), Carbohydrates, MTP International Review of Science, Organic Chemistry Series One, Vol. 7, Butterworths, London (1973) p251.
2. D.A. Rees, in W.J. Whelan (Ed.), Biochemistry of Carbohydrates, MTP International Review of Science, Biochemistry Series One, Vol. 5, Butterworths, London, (1975) p1.
3. D.A. Rees, Polysaccharide Shapes, Outline Series in Biology, Chapman & Hall, London 1977.

4. D.A. Rees, E.R. Morris, D. Thom and J.K. Madden, in G.O. Aspinall (Ed.), The Polysaccharides, Academic Press, New York, in press.
5. J.F. Stoddart, Stereochemistry of Carbohydrates, Wiley, New York, 1971.
6. D.A. Rees and W.E. Scott, J. Chem. Soc., B (1971) 469.
7. D.H. Campbell, E. Luescher and L.S. Lerman, Proc. Nat. Acad. Sci. USA, 37 (1951) 575.
8. L.S. Lerman, Proc. Nat. Acad. Sci. USA, 39 (1953) 232.
9. C. Arsenis and D.B. McCormick, J. Biol. Chem., 239 (1964) 3093.
10. C. Arsenis and D.B. McCormick, J. Biol. Chem., 241 (1966) 330.
11. R.H. Marchessault and A. Sarko, Advan. Carbohyd. Chem., 22 (1967) 421.
12. P.R. Sundararajan and R.H. Marchessault, Canad. J. Chem., 50 (1972) 792.
13. K.H. Garner and J. Blackwell, Biopolymers, 13 (1974) 1975.
14. R.E. Rundle, J. Amer. Chem. Soc., 69 (1967) 1769.
15. D. French, A.O. Palley and W.J. Whelan, Starke, 15 (1963) 549.
16. R.R. Bumb and B. Zaslow, Carbohyd. Res., 4 (1967) 98.
17. A. Hyble, R.E. Rundle and D.E. Williams, J. Amer. Chem. Soc., 92 (1970) 5834.
18. J.J. Jacobs, R.R. Bumb and B. Zaslow, Biopolymers, 6 (1968) 1659.
19. H. H-C Wu and A. Sarko, Carbohyd. Res., 61 (1978) 7.
20. H. H-C Wu and A. Sarko, Carbohyd. Res., 61 (1978) 27.
21. C.R. Ricketts, Prog. Org. Chem., 5 (1961) 73.
22. E.R. Ruckel and C. Schuerch, Biopolymers, 5 (1967) 515.
23. J. Porath and T. Kristiansen, in H. Neurath and R.L. Hill (Eds.), The Proteins, 3rd edn., Vol. 1, Academic Press, New York, 1975 p.95.
24. N.M. Young and M.A. Leon, Carbohyd. Res., 66 (1978) 299.
25. J. Lönngren, I.J. Goldstein and R. Bywater, FEBS. Lett., 68 (1976) 31.
26. L. Rexová - Benková and V. Tibenský, Biochim. Biophys. Acta, 268 (1972) 187.
27. J.J. Marshall and W. Woloszczuk, Carbohyd. Res., 61 (1978) 407.
28. W. Brümmer, Kontakte (Merck) 1/74, p23; 2/74, p3.
29. P. Luby, L. Kuniak and D. Berek, J. Chromatogr., 59 (1971) 79.
30. L. Goldstein, M. Pecht, S. Blumberg, D. Atlas and Y. Levin, Biochemistry, 9 (1970) 2322.
31. M. Serban, H.D. Schell and M.A. Mateescu, Rev. Roum. Biochim., 12 (1975) 187.
32. British Patents. GB 1508570, GB 1508123, GB 1519949.
33. J. Štamberg, Analytical Chemistry Symposia Series, this volume.
34. F.J. Kolpack and J. Blackwell, Macromolecules, 9 (1976) 273.
35. D.A. Rees, J. Chem. Soc., (1961) 5168; (1963) 1812.
36. S. Arnott, W.E. Scott, D.A. Rees, C.G.A. McNab, J. Mol. Biol., 90 (1974) 253.
37. S. Arnott, A. Fulmer, W.E. Scott, I.C.M. Dea, R. Moorehouse and D.A. Rees, J. Mol. Biol., 90 (1974) 269.
38. I.C.M. Dea, A.A. McKinnon and D.A. Rees, J. Mol. Biol., 68 (1972) 153.
39. S. Ablett, P.J. Lillford, S.M.A. Baghdadi and W. Derbyshire, Nato Adv. Study Inst. Ser., Ser. C, 61 (1980) 687.
40. D.A. Rees, W.E. Scott and F.B. Williamson, Nature, 227 (1970) 390.
41. T.A. Bryce, A.A. McKinnon, E.R. Morris, D.A. Rees and D. Thom, Faraday Discuss. Chem. Soc., 57 (1974) 221.
42. I.T. Norton, D.M. Goodall, E.R. Morris, D.A. Rees, J. Chem. Soc. Chem. Commun., (1979) 988.
43. I.T. Norton, PhD Thesis, Univ. of York 1980.
44. R.A. Jones, E.J. Staples and A. Penman, J. Chem. Soc., Perkin II, (1973) 1608.
45. G. Robinson, E.R. Morris and D.A. Rees, J. Chem. Soc. Chem. Commun., (1980) 152
46. E.R. Morris, D.A. Rees and G. Robinson, J. Mol. Biol., 138 (1980) 349.
47. I.T. Norton, D.M. Goodall, E.R. Morris and D.A. Rees, J. Chem. Soc. Faraday I, submitted for publication.
48. E.R. Morris, D.A. Rees, I.T. Norton and D.M. Goodall, Carbohyd. Res., 80 (1980) 317.
49. D.A. Rees, Chem. Ind., (1972) 630.
50. J. Porath, J.C. Jenson and T. Låås, J. Chromatog., 60 (1971) 167.
51. T. Låås in H. Peeters (Ed.) Protides of the Biological Fluids, Brugge Pergammon Press Ltd., Oxford (1975) p495.
52. J. Porath, T. Låås and J.C. Janson, J. Chromatog., 103(1975) 49.
53. E.R. Morris, Personal Communication.

54. E.G.V. Percival, Quart Rev, 3 (1949) 369.
55. T.C.J. Gribnau, C.A.G. Van Eekelen, C. Stumm and G.I. Tesser, J. Chromatog., 132 (1977) 519.
56. J. Uriel, J. Berges, E. Boschetti and R. Tixier, C.R. Acad. Sci. Paris 273 Serie D (1971) 2358.
57. T. Tosa, T. Sato, R. Mori, K. Yamamoto, I. Takata, Y. Nishida and I. Chibata, Biotechnol. Bioeng., 21 (1979) 1697.
58. T. Sato, Y. Nishida, T. Tosa and I. Chibata, Biochim. Biophys. Acta, 570 (1979) 179.
59. P.O. Larsson and K. Mosbach, Eur. J. Appl. Microbiol. Biotechnol., 7 (1979) 103.
60. P. Brodelius, B. Deus, K. Mosbach and M.H. Zenk, FEBS. Lett., 103 (1979) 93.
61. M.A. Vijayalakshmi and A. Marcipar, J. Polymer. Sci., Polymer Symp., 68 (1980) 57.
62. A. Haug, B. Larsen, O. Smidsrød and T. Painter, Acta Chem. Scand., 23 (1969) 2955.
63. G.T. Grant, E.R. Morris, D.A. Rees, P.J.C. Smith and D. Thom, FEBS Lett., 32 (1973) 195.
64. E.R. Morris, D.A. Rees, D. Thom and J. Boyd, Carbohyd. Res., 66 (1978) 145.
65. M.J. Gidley, E.R. Morris, E.J. Murray, D.A. Powell and D.A. Rees, J. Chem. Soc. Chem. Commun., (1979) 990.
66. W.H. Hanisch, P.A.D. Rickard and S. Nyo, Biotechnol. Bioeng., 20 (1978) 96.
67. J.F. Kennedy and C.E. Doyle, Carbohyd. Res., 28 (1973) 89.

T.C.J. Gribnau, J. Visser and R.J.F. Nivard (Editors),
Affinity Chromatography and Related Techniques

BEAD CELLULOSE

J. ŠTAMBERG[+], J. PEŠKA[+], H. DAUTZENBERG[++] and B. PHILIPP[++]
[+]Institute of Macromolecular Chemistry, Czechoslovak Academy of Sciences, 162 06 Prague 6 (Czechoslovakia)
[++]Institut für Polymerenchemie, Akademie der Wissenschaften der DDR, Teltow (G.D.R.)

ABSTRACT

Experience with classical fibrillar and powdered celluloses as adsorbents led to enhanced efforts to develop new cellulosic materials composed of porous spherical particles. The various methods use different principles, vary in the equipment and labour needed and yield products of different quality. Bead cellulose, based on an original Czech procedure, is a versatile product of this type. Its properties are reviewed, with emphasis on possible modifications of the primary product, including changes in porosity, cross-linking and derivatization. The successful use of bead cellulose materials as matrices for enzyme immobilization and adsorbents for affinity chromatography are discussed.

INTRODUCTION

Cellulose is one of the most widespread polymeric substances. Billions of tons are created each year through photosynthesis, and mankind utilizes more than 500 millions of tons in the form of wood, paper, textiles, films, plastics, coatings, fuel, etc. (refs. 1,2). Its significance has recently increased as reserves of oil and coal are being reduced.

Particularly interesting applications of cellulose occur in the field of functional polymers, based on interactions between active sites in the polymeric structure and the components of the liquid or gaseous phases. Functional polymers include adsorbents, ion-exchange resins, chemically reactive polymers, catalysts, redoxites and carriers of various biological functions. A wide range of functional celluloses have been described earlier and their applications were generally successful. However, limitations were recognized, derived from the unsuitable physical structure (lack of porosity) and from the unsatisfactory geometrical shape of the individual particles. Recently, most problems related to the use of the old fibrillar and powdered celluloses have been solved by introducing a novel form, which is both porous and spherical (ref. 3).

PREPARATION OF SPHERICAL CELLULOSE

Since the first attempt by O´Neill and Reichardts to prepare spherical cellulose (ref. 4), a number of other procedures have appeared. All are based on a common principle, which includes following steps:

(a) formation of a liquid phase containing cellulose or a cellulose derivative;
(b) shaping the liquid to form spherical objects (droplets);
(c) solidification of the liquid droplets;
(d) regeneration of cellulose into the solid spherical particles;
(e) final treatment, including washing.

Table 1 summarizes known procedures for the preparation of spherical renegerated cellulose and spherical cellulose derivatives, particularly hydroxyethylcellulose, cellulose acetate and ion-exchange celluloses.

In order to distribute cellulose into the individual liquid particles, either extrusion through apertures of suitable sizes (refs. 4-8) or dispersion in a medium which does not mix with the cellulose-containing liquid phase (refs. 5, 9-21) is used. The particle size is controlled by the conditions of passage through the nozzle, by the efficiency of mixing during dispersion, or by the addition of surface-active compounds.

An important step in the preparation is the solidification of liquid particles, i.e., completion of the sol-gel transition. This is carried out so as to avoid deformation of the spherical shape and adhesion of the individual particles to yield agglomerates. The procedures used so far to achieve the sol-gel transition involved salt and acid regeneration baths as used in the production of cellulose fibres; also, the composition of the dispersed phase was changed so as to reduce the solubility of the cellulose component, the temperature was decreased in order to achieve solidification of the melt (in the case of cellulose acetate) or to reduce the solubility, and chemical cross-linking was used, especially by using epichlorohydrin in an alkaline medium. The preparation of spherical celluloses was completed by various additional procedures resulting in a porous structure and in the completion of the regeneration of cellulose and involving also removal of decomposition products by washing.

In our procedure for the preparation of spherical cellulose (ref. 19), where spherical cellulose is called "bead cellulose", the original solution consists of cellulose xanthate, i.e., viscose; the particles are formed by employing the dispersion procedure, the drops are solidified by the thermal sol-gel transition, cellulose is regenerated in spherical gel particles by known procedures and chemical derivatives are prepared by additional polymer-analogous transformations (ref. 22).

TABLE 1

Preparation of spherical celluloses

(a) Ejection processes

Reference	Principle	Solution	Solidification
4 (O´Neill, 1951)	forming in inert liquid	viscose	salt or acid precipitation
5,6 (Determann, 1968)	spraying into air	viscose	acid precipitation
7 (Yasui, 1973)	forming in air	viscose	acid precipitation
8 (Gensrich, 1980)	forming in inert liquid	viscose	heating

(b) Suspension process

Reference	Dispersed phase	Solidification	Product
5,6,9 (Determann, 1968)	Schweizer´s soln. viscose Fe tartrate soln. cadoxen soln.	acid precipitation	cellulose
10 (Chitumbo, 1971)	viscose	cross-linking	cellulose
11-13 (Andreassen, 1972)	cellulose poly-electrolyte in water	cross-linking	ion exchanger
14 (Brown, 1972)	HEC in water[a]	cross-linking	HEC[a]
15 (Satake, 1974)	cellulose poly-electrolyte in water	cross-linking	ion exchanger
16 (Satake, 1975)	HEC in water[a]	cross-linking	HEC[a]
17 (Chandler, 1975)	cellulose acetate in DMSO/water	cooling	cellulose acetate
18 (Li Fu Chen, 1976)	cellulose acetate in organic solvent	dispersed in water	cellulose acetate[b]
19 (Peška, 1976)	viscose	heating	cellulose
20 (Motozato, 1978)	melt of cellulose acetate	cooling	cellulose acetate[b]
21 (Kuga, 1980)	solution in $Ca(SCN)_2$	diln. with MeOH	cellulose

[a]HEC = hydroxyethylcellulose.
[b]hydrolysis as after-treatment to yield cellulose also included.

The difference between our procedure and those described earlier consists mainly in the type of sol-gel transition, which is subject to the requirements of a suspension procedure derived by analogy with suspension polymerization. We chose a sol-gel transition that starts as a result of merely increasing the temperature (thermal sol-gel transition, TSGT process). Owing to the presence of xanthate groups, cellulose xanthate is a polyelectrolyte that is soluble in an aqueous sodium hydroxide solution. It should be pointed out, however, that xanthate groups decompose spontaneously even at room temperature, and the decomposition is reflected in an increase in viscosity followed by gel formation (Fig. 1). We investigated (ref. 23) the dependence of viscosity changes on time and found that if the temperature rises from 25 to 95°C, the time of gelation is reduced from several days to a few minutes. At 95°C the sol-gel transition proceeds very quickly, as required by a successful suspension process. Under these conditions, the spherical shape may also be preserved after solidification; no noticeable adhesion of the individual particles to yield agglomerates takes place. An advantage of the thermal sol-gel transition is its simplicity and the complete fulfilment of the requirements of the suspension process. It occurs under stable hydrodynamic conditions, and immediately in the total volume of the particle rather them starting from the drop surface as when the precipitant is added to the dispersed phase (the latter procedure leads to an undesirable cuticle formation "skin effect").

The new procedure for the preparation of spherical cellulose has been optimized with respect to the composition of the liquid phases during the dispersion process, to the requirement of suitable additives and to the simplificatio and improvement of the regeneration procedures, bearing in mind the economies, quality of the product and pollution of the environment. The production procedure is easy to carry out and is now in the stage of experimental pilot-plant operation.

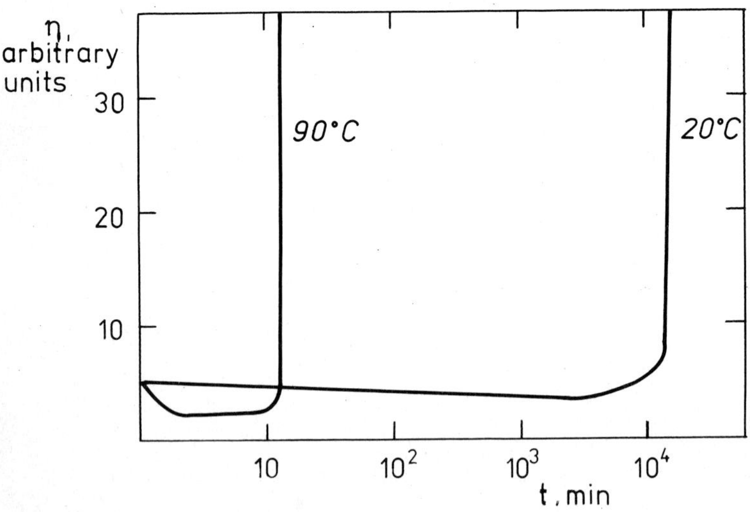

Fig. 1. Dependence of viscosity on time for viscose at different temperatures.

PROPERTIES OF BEAD CELLULOSE

The suspension process utilizing the thermal sol-gel transition (TSGT process) yields regular spherical beads, the geometry and particle size distribution of which have been described elsewhere (ref. 3). Bead cellulose is a highly porous spherical product which, in the centrifuged state, is composed of pure regenerated cellulose (15%) and water. The sulphur content in the dry matter is <0.01% and the content of carboxylic groups is <20 µmol/g (compared with 5-20 µmol/g for native cellulose and up to ca 50 µmol/g for regenerated celluloses). The incombustible residue is <0.15% and on one-stage acid extraction decreases to <0.01%.

The porosity value calculated from water regain in the centrifuged state is P=90% (where P=volume fraction of pores in the swollen matter). The pore size distribution was determined from permeability measurements on dextrans of various molecular mass (ref. 24). Accessibility for dextrans (with 10^5-10^6 m.w.) was observed which allows us to estimate the diameter of the main fraction of pores from 10 to 60 nm and rest of them - about 10-20 vol.% - above the said upper limit.

With its porous structure, the swollen bead cellulose resembles vinyl copolymers of the macroporous (macroreticular) type. In the coagulation and regeneration of cellulose from xanthate solution, a microheterogeneous structure is formed consisting of regions of high order interconnected by amorphous material. Owing to this microheterogeneous structure, spherical macroparticles of bead cellulose are much more rigid than particles of homogeneous polysaccharide gels which arise, e.g. by the chemical cross-linking of the individual polymer chains with epichlorhydrin. The entry of cross-linking groups or other substituents into macromolecules may reduce the formation of crystalline domains. For this reason, bead cellulose is mechanically more stable than, e.g. dextran gels of comparable porosity.

The porous structure of bead cellulose may be modified by simple aftertreatments (ref. 25), producing tailor-made spherical cellulose carriers and adsorbents. Changes in the porous structure occur during drying by direct removal of water. Simple drying in air reduces the water content only slowly, so that at room temperature it decreases to about 70-80% of its initial level after 70 h (at R.H. of about 40-60%). Repeated swelling gives an equilibrium water regain of 3-5 g H_2O/g dry matter. Additional drying in a stream of air (fluidization drier, 95°C, 30 min) makes the water content decrease to as little as to 5-30% of the initial level; after reswelling, the water regain becomes ca 1 g H_2O/g. Further drying in the vacuum of a water pump (80°C, 48 h) allows the water content to be decreased to as little as to 1% of the initial value or lower. Glassy beads are obtained, which swell in water to only 0.9 g H_2O/g. Hence, drying gradually reduces the water content which cellulose can again retain in its pores after subsequent reswelling. This is due to the contraction of the skeleton and to the formation of new hydrogen or other structure-forming bonds (ref. 26). Drying leads to the formation of larger or new microcrystalline domains.

The porous structure may be varied by the solvent replacement method followed by drying or reswelling. Pores filled with water in the initial state can be filled with any solvent by exchange without shrinkage and a corresponding decrease in porosity. According to the type of solvent from which cellulose is eventually dried, products of various true porosity(dry-state porosity) are formed with a characteristic macroreticular structure (Fig. 2). Fig. 3 gives

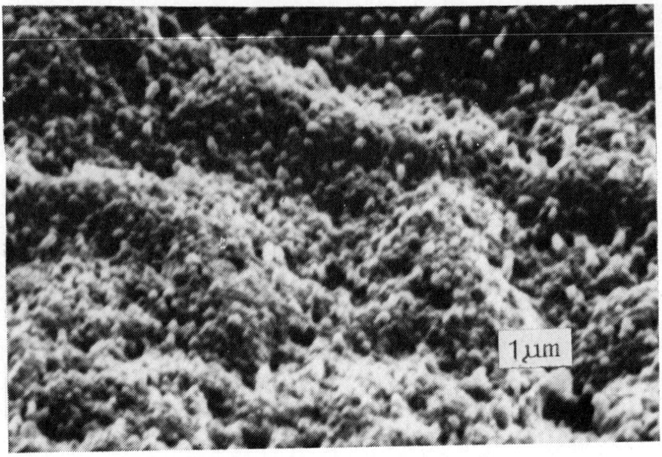

Fig. 2. SEM photograph of solvent-dried bead cellulose with true porosity (ref. 27).

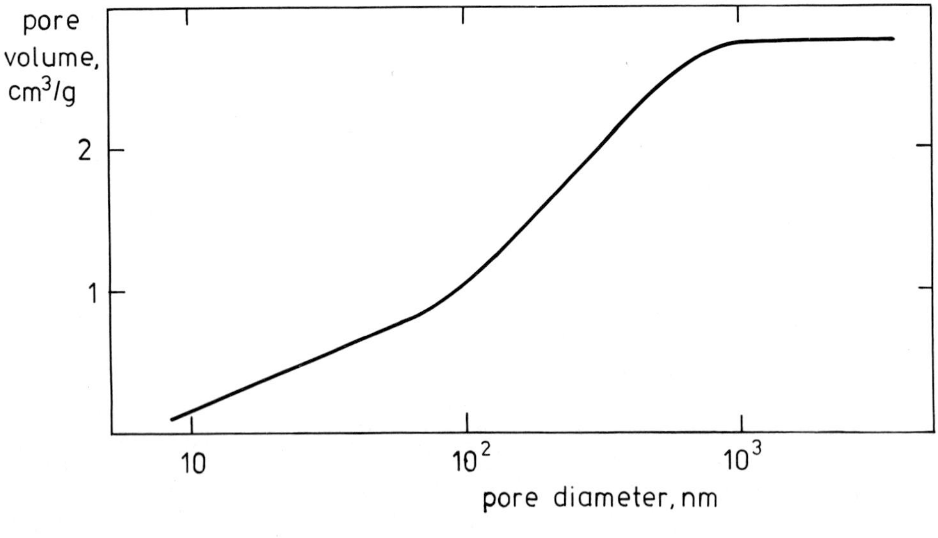

Fig. 3. Pore distribution in solvent-dried bead cellulose with true porosity.

the results of a Hg-porosimetric evaluation of bead cellulose dried from benzene (ref. 28). A porosity value 83% (vol.% of pores) was calculated from cyclohexane regain for this product; surface area 340 m^2/g was determined by BET method. Drying from solvents of higher polarity yielded products with lower porosity; no porosity was demonstrated in the dry state for the product dried from water.

After repeated water swelling of cellulose samples dried from solvents of various polarity, different water regain values were obtained. It can be concluded that depending on the conditions of drying, it is possible to prepare products with adjustable porosity values in the dry state from 0 to 83% and in the water-swollen state from 50 to 90%.

It is known that cellulose exceeds some other synthetic polymers in its chemical reactivity. By employing polymer-analogous transformations, many derivatives were prepared that could be used as functional polymers (refs. 29-31). We found that bead cellulose could be derivatized without damaging the original spherical particle shape (ref. 22). Many reactions proceed with particular ease owing to the porosity of the skeleton and the accessibility of reaction sites. In some instances it is advisable to modify the reaction conditions with respect to high porosity, which means a high content of the swelling liquid and undesirable dilution of the reaction system. For these reasons, in some instances it is appropriate to fill the inter-particle volume with an inert medium, e.g., a hydrocarbon, thus directing the transformation process into the intra-particle domain (ref. 32).

APPLICATIONS OF BEAD CELLULOSE

Bead cellulose can be widely used as a functional polymer and as the starting material for the preparation of other functional polymers. Some applications are listed here:

- filtration of water from mechanical impurities and removal of colloidal compounds
- fixation of organic compounds on anion-exchange derivatives (humic acids)
- collection of trace elements (metals) on chelate derivatives (analysis of water)
- obtaining of proteins from natural materials and waste solutions
- use in pharmacology and medicine

The immobilization of enzymes and affinity chromatography are particularly important; Table 2 summarizes papers published so far on immobilization on bead cellulose. Using some particular examples, it was shown that advantages over other carriers can mainly be achieved by employing a suitable geometrical shape, high porosity with good mechanical stability and a pronounced hydrophilic character of the skeleton with only low non-specific sorption of proteins.

Bead cellulose was activated by means of cyanogen halides, especially cyanogen bromide, periodate oxidation, reaction with 4-nitrophenyl and N-hydroxysuccinimidyl chloroformates, introduction of a primary amine group by reaction with epichlorohydrin and ethylenediamine or hexamethylenediamine, or etherification with chloroethylamine followed by reaction with glutaraldehyde, activation with benzoquinone and introduction of isothiocyanate, thiol, disulphide and hydrazide groups.

TABLE 2

Immobilization of enzymes on bead celluloses

Matrix	Binding/activation	Enzyme	Reference
cellulose	cyanogen bromide	steroid dehydrogenase	33
	periodate	trypsin	34,35
		chymotrypsin	35
		glucose oxidase	36
	diisocyanate	amyloglucosidase	37
	activated ester	trypsin	38
		amyloglucosidase	39,51
	isothiocyanate	pepsin	40
		rennin	40
		penicillase	40
	benzoquinone	galactosidase	41
DEA-ethylcellulose	ionic	glucose isomerase	37
DEA-hydroxypropylcellulose	ionic	invertase	42,43
aminoethylcellulose	glutaraldehyde	glucose oxidase	36
carboxymethylcellulose	carbodiimide	glucose oxidase	36

Ion-exchange bead celluloses were used for immobilization directly (by ionic bonds) or after modification (by covalent bonds). In the former case a derivative with diethylaminohydroxypropyl (DEAHP) groups was used, and in the latter case AE-cellulose,(cellulose with aminoethyl groups), after modification with glutaraldehyde or CM-cellulose (cellulose with carboxymethyl groups) using carbodiimide, was employed.

Chen, Tsao and Dickensheet have described the immobilization of enzymes (glucoamylase, glucose isomerase, invertase) on spherical cellulose activated with isocyanate or diethylaminoethyl groups (refs. 18,37).

The examples described so far do not cover all the fields of applicability of bead cellulose. It could be used not only in studies on enzymes and in laboratory preparations, but also, and with more success than with other carriers, on an industrial scale for technological and economic reasons.

The advantageous flow properties of layers of bead cellulose should lead to successful applications in affinity chromatography. The examples described so far are summarized in Tables 3 and 4. The first attempts concerned the introduction of the imidazole group for the sorption of haemoglobin. A further paper was devoted to the immobilization of DNA. Bead cellulose has been used in the binding of anthraquinone dyes of the triazine and Remazole type; the sorbents prepared were used in the binding of lactate dehydrogenases of various origins.

A novel, simple procedure for the utilization of bead cellulose in the isolation of lectins has also been described (ref. 50). Bead cellulose has been modified with aminosaccharides. Further data are summarized in Table 4.

TABLE 3

Affinity chromatography on bead cellulose for isolation and purification of proteins

Affinity Ligand	Protein	Reference
histamine	methaemoglobin	44,45
DNA	DNase	39
reactive dyes	lactate dehydrogenase	46,47,48,49
saccharides	lectins	50, cf. Table 4

TABLE 4

Recovery of lectins on bead celluloses

Ligand	Lectin from	Capacity (mg/ml affinant)	Increase in Activity
D-glucose	Canavalia ensiformis	36	64x
	Pisum sativum	25	128x
	Perca fluviatilis (ova)	21	64x
D-mannose	Canavalia ensiformis	44	64x
L-fucose	Ulex europaeus	25	512x
D-galactose	Ricinus communis	32	64x
	Glycine soja	18	256x
N-Acetyl-D-glucosamine	Triticum aestivum	35	256x
	Helix pomatia (alb.glands)	28	128x
N-Acetyl-D-galactosamine	Dolichos biflorus	33	128x

CONCLUSIONS

Bead cellulose has not yet achieved its full potential in the field of functional polymers, including the immobilization of enzymes and affinity chromatography. Its production by the TSGT process can be accomplished by employing simple equipment and cheap raw materials, under economical conditions, and may be integrated into the traditional production of viscose rayon and cellophane.

When applying bead cellulose in the field of immobilized enzymes and affinity chromatography, it is possible to take advantage of the regular spherical shape, high porosity, hydrophilicity, chemical reactivity and mechanical strength of the individual particles. The procedures developed under laboratory conditions can be transferred without further difficulty to the process scale. It would be a pity if the lack of interest on the part of manufacturers of functional polymers in spherical cellulose, which unfortunately has lasted for almost

30 years (since the publication of the patent by O´Neill and Reichardts), continued to restrict developments. We believe that the results achieved in the fields of enzyme immobilization and affinity chromatography using bead cellulose will promote greater enthusiasm for the large-scale manufacture of spherical celluloses in the next few years.

REFERENCES

1 E. Ott and H.G. Tennent, in E. Ott, H.M. Spurlin and M.W. Grafflin (Eds.), Cellulose and Cellulose Derivatives, 2nd edn., Part I, Interscience Publishers, New York, 1954, p.1.
2 B. Philipp, Textiltechnik, 30(1980)750.
3 J. Štamberg, J. Peška, D. Paul and B. Philipp, Acta Polymerica, 30(1979) 734-739.
4 J.J. O´Neill Jr., R. Reichardt and E.P. Reichardt, U.S.Pat.No 2,543,928 (1951).
5 H. Determann, H. Rehner and T. Wieland, Makromol. Chem., 114(1968)263.
6 H. Determann and T. Wieland, U.S.Pat.No. 3,597,350 (1971).
7 K. Yasui, Y. Isome, K. Sugimori and S. Katsuyama, Japanese Kokai Patent No. 73 43,082 and 73 60,753 (1973).
8 H.J. Gensrich, V. Gröbe, D. Bartsch, J. Štamberg, J. Peška, D. Paul, M. Holtz, D. Berek and I. Novák, Czech.Patent Appl.No. PV 3634-80 (1980).
9 H. Determann and T. Wieland, Swedish Patent No. 382,066 (1976).
10 K. Chitumbo and W. Brown, J.Polymer Sci., C36(1971)279-292.
11 O.H. Edlund and B.A. Andreassen, German Offen.No. 2,138,905 (1972).
12 B.A. Andreassen, Swedish Patent No. 343,306 (1972).
13 B.A. Andreassen, Swedish Patent No. 382,329 (1976).
14 W. Brown and K. Chitumbo, Chemica Scripta 2(1972)88-92.
15 T. Satake, I. Kano, Y. Tsutsui and K. Yokota, Japanese Kokai Patent No. 74 91,977 (1974).
16 T. Satake and K. Hata, Japanese Kokai Patent No. 75 151,289 (1975).
17 B.V. Chandler and R.L. Johnson, German Offen.No. 2,507,551 (1975).
18 Li Fu Chen and G.T. Tsao, Biotechnol.Bioeng., 18(1976)1507-1516.
19 J. Peška, J. Štamberg and Z. Blače, U.S.Pat.No. 4,055,510 (1977); Czech.Patent No. 172 640 (1976).
20 Y. Motozato, Japanese Kokai Patent No. 78 07,759 and 78 86,749 (1978).
21 S. Kuga, J.Chromatogr., 195(1980)221-230.
22 J. Peška, J. Štamberg and J. Hradil, Angew.Makromol.Chem., 53(1976)73-80.
23 O. Quadrat, P. Pavlík, J. Peška and J. Štamberg, Acta Polymerica, to be published.
24 F. Rypáček, private communication.
25 J. Peška, J. Štamberg and Z. Pelzbauer, Cell.Chem.Technol., 21(1978)419-428 (corrigenda in 1/1979).
26 H. Ruck, Papier, 33(1979)14-21.
27 Z. Pelzbauer, to be published.
28 M. Bleha, personal communication.
29 N.M. Bikales and L. Segal (Eds.), Cellulose and Cellulose Derivatives, Part V, Wiley-Interscience, New York, 1971.
30 L.S. Gal´braikh and Z.A. Rogovin, Fortschr.Hochpolym.-Forsch., 14(1974) 87-130.
31 K. Takemoto, SEN-I GAKKAISHI, 31(1975)P105-P109.
32 M. Beneš, personal communication.
33 R. Bovara, G. Carrea, P. Cremonesi and G. Mazzola, Italian Congress on Clinical Analytical Chemistry, Milano, 1979.
34 J. Turková, J. Vajčner, D. Vančurová and J. Štamberg, Collection Czechoslov.Chem.Commun., 44(1979)3411-3417.
35 J. Turková, L.V. Kozlov, L.Ya. Bessmertnaya, L.V. Kudryavtseva, V.M. Krasilnikova and V.K. Antonov, Bioorganicheskaya Khimiya, 6(1980) 108-115.

36 F. Švec, O. Valentová, M. Marek, J. Štamberg and Z. Vodrážka, Biotechnol. Bioeng., in press.
37 G.T.Tsao, Li Fu Chen and P.A. Dickensheets, Biotechnol.Bioeng., 19(1977) 365-375.
38 J. Drobník, J. Labský, H. Kudlvasrová, V. Saudek and F. Švec, Biotechnol. Bioeng., in press.
39 V. Saudek, J. Drobník and J. Kálal, Polymer Bulletin 2(1980)7-14.
40 J. Augustin and B. Skárka, Poster Abstracts, Symposium - Modern Approaches in Applied Enzymology, Reinhardsbrunn, GDR, May 13-15, 1980, p.15.
41 Moonjin Chun, Theses, Justus-Liebig-Universität, Giessen, 1979.
42 J. Hradil and F. Švec, Enzyme Microbial Technol., to be submitted.
43 J. Hradil and F. Švec, Enzyme Microbial Technol., in press.
44 K. Pommerening, W. Jung and B. Neumann, Poster Abstracts, Symposium - Modern Approaches in Applied Enzymology, Reinhardsbrunn, GDR, May 13-15, 1980, p.13.
45 K.Pommerening, M. Kühn, W. Jung, K. Buttgereit, P. Mohr, J. Štamberg and M. Beneš, Int.J.Biolog.Macromolecules, 1(1979)79-88.
46 D. Mislovičová, P. Gemeiner, L. Kuniak and J. Zemek, J.Chromatogr., 194(1980)95-99.
47 P.Gemeiner, D.Mislovičová, J.Zemek and L. Kuniak, Collection Czechoslovak Chem.Commun. 46(1981)419-427.
48 D. Mislovičová, P. Gemeiner, L. Kuniak and J. Zemek, Kvasný průmysl, 26(1980)163-164.
49 D. Mislovičová, P. Gemeiner, L. Kuniak and J. Zemek, in Collection of Papers (Nové trendy výskumu a využitia biopolymerov),Institute of Chemistry, Slovac Academy of Sciences, Bratislava, 1980, p.41.
50 V. Hořejší, J. Drobník, J. Štamberg and F. Švec, Czechoslovak Patent No. 203,767 (1980).
51 F. Švec, H. Kudlvasrová, I. Konečná, I.I. Menyaylova and L.A. Nachapetian, Czech.Patent Appl.No. PV 2921-80 (1980).

T.C.J. Gribnau, J. Visser and R.J.F. Nivard (Editors),
Affinity Chromatography and Related Techniques

SYNTHETIC POLYMERS APPLIED TO MACROPOROUS SILICA BEADS TO FORM NEW CARRIERS FOR INDUSTRIAL AFFINITY CHROMATOGRAPHY

J. SCHUTYSER, T. BUSER, D. VAN OLDEN and H. TOMAS
Akzo Research - Corporate Research Department, Velperweg 76, 6800 AB Arnhem, the Netherlands
and
F. VAN HOUDENHOVEN and G. VAN DEDEM
Diosynth b.v., P.O. Box 20, 5340 BH, Oss - the Netherlands

SUMMARY

New carriers have been developed for industrial affinity chromatography by combining the known outstanding mechanical and hydrodynamic characteristics of macroporous inorganic material, such as silica, with the high versatility of hydrophilic (co)-polymers having different kinds of functional groups.

Heparin and antithrombin III were immobilized in one step only on the new carriers and the resulting conjugates were successfully used to purify antithrombin III and heparin, respectively. The latter purification was performed on a semi-industrial scale.

Considering these results and the excellent flow properties, the ease in packing columns, the almost complete inertness in common organic solvents and aqueous solutions ranging from pH 2 to 9.0, as well as the possibility of regeneration by pyrolysis to blank silica, the hydrophilic tailor-made carriers should have a promising future as carrier for industrial affinity chromatography.

INTRODUCTION

Porous inorganic particles with different particle and pore sizes such as Controlled Pore Glass available from Corning Glass Works and Sphérosil, porous silica, supplied by Rhône Poulenc are frequently reported to have, unlike agarose, excellent mechanical and hydrodynamic properties for affinity chromatographic applications on a large or industrial scale [1].

A serious deficiency, however, which limited their use as a support for affinity chromatography is the non-specific adsorption or denaturation of proteins (predominantly basic and neutral ones) on the silica surface even after derivatisation. Silanol groups present on the surface act as weak ion-exchanging groups and are responsible for this disturbing adsorption phenomenon. Not surprisingly, in the past several efforts have been directed to eliminating this adsorption tendency. In 1976 Regnier reported the derivatization of controlled pore glass with 3-glyceryl-propyl-siloxy groups, resulting in a significant but not always complete deactivation of the silica surface [2]. Also glass or silica treated with 3-aminopropyltriethoxy silane was shown to be a suitable starting support for the immobilization of enzymes[3] and affinity ligands [4].

Since commercially available derivatives of controlled pore glass are rather expensive for large scale application a simple chemical modification of relatively inexpensive and readily available porous silica gel seemed us more obvious.

To that end, we have investigated the possibilities to synthesize tailor-made carriers by coating the surface of macroporous silica with a thin skin of cross-linked hydrophilic co-polymers carrying several amounts and kinds of functional end groups linked either directly or indirectly via a spacer group to the backbone of the polymer.

SYNTHESIS OF THE NEW CARRIERS.

A large quantity of hydrophilic vinyl monomers provided with different kinds of functional groups are commercially available and/or described in the literature. Some of them have been utilized for the preparation of synthetic polymer supports for the immobilization of enzymes or affinity ligands [4, 5].

As suitable functional monomers we selected for our study: N-hydroxy-succinimidyl, 6-acrylamido-hexanoic acid ester 1; N-hydroxy-succinimidyl acrylic acid ester 2; 6-acrylamido-hexanoic acid 3; allylamine 4 and N-allyl, 1-bromo-acetamide 5. While the monomers 1 - 3 were prepared according to known routes [5], monomer 5 was synthesized by reaction of the commercially available allylamine with N-hydroxy-succinimidyl, 1-bromo-acetic acid ester; see Fig. 1.

Fig. 1 Selected functional monomers

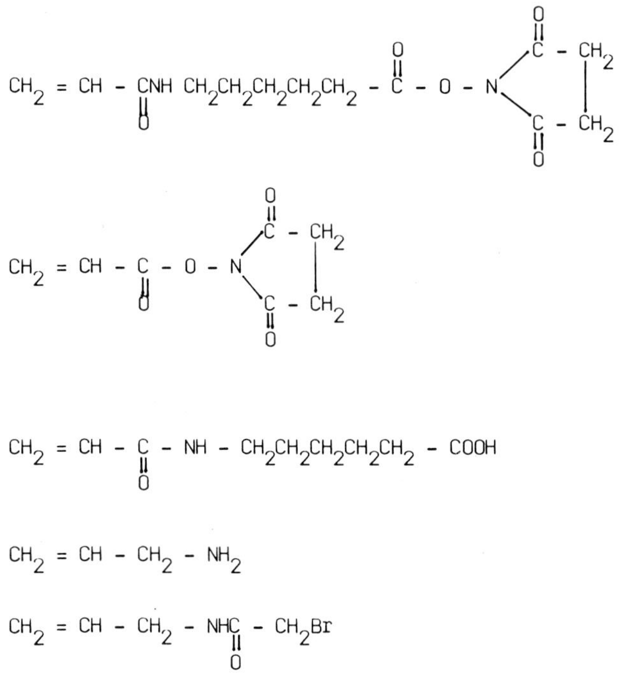

```
                                              O
                                              ‖
                                      O      C - CH₂
                                      ‖     ╱     |
CH₂ = CH - CNH CH₂CH₂CH₂CH₂CH₂ - C - O - N         |           1
           ‖                              ╲     |
           O                               C - CH₂
                                           ‖
                                           O


                      O
                      ‖
                      C - CH₂
              ╱     |
CH₂ = CH - C - O - N              |                                       2
           ‖        ╲     |
           O         C - CH₂
                     ‖
                     O


CH₂ = CH - C - NH - CH₂CH₂CH₂CH₂CH₂ - COOH                                3
           ‖
           O


CH₂ = CH - CH₂ - NH₂                                                      4


CH₂ = CH - CH₂ - NHC - CH₂Br                                              5
                    ‖
                    O
```

As may be seen from Fig. 1 the monomers 1 and 3 comprise ε-amino-caproic acid, a molecule which is frequently utilized in affinity chromatography as a spacer group.

N-methylolacrylamide 6 (abbreviated to MAAM) was chosen as the hydrophilic comonomer for the copolymerization with the above mentioned functional monomers. This polar monomer (6) contains besides a vinyl function a reactive methylol function, which is capable of reacting under heating -possibly in the presence of a trace of acid - with hydroxyl containing compounds such as alcohol, thus resulting in ether formation. The methylol functions can also react with each other under the formation of a stable methylene bridge. When MAAM monomers are incorporated into polymers this reaction generates crosslinking of the polymer chains (Fig. 2).

Fig. 2 Chemistry of N-methylolacrylamide

$$CH_2 = CH - \underset{\underset{O}{\|}}{C}NHCH_2OH \xrightarrow{\text{radical initiator}} \{CH_2 - \underset{\underset{NHCH_2OH}{|}}{\overset{\overset{\textstyle |}{}}{CH}}\}_n$$

$$\underline{6}$$

$$\xrightarrow[H^+/\Delta]{ROH} CH_2 = CH - \underset{\underset{O}{\|}}{C} - NH-CH_2 - OR + H_2O$$

$$CH_2 = CH - \underset{\underset{O}{\|}}{C} - NHCH_2OH \xrightarrow{H^+/\Delta} CH_2 = CH - \underset{\underset{O}{\|}}{C} - NH - CH_2 - NH - \underset{\underset{O}{\|}}{C} - CH = CH_2 + H_2O + CH_2O$$

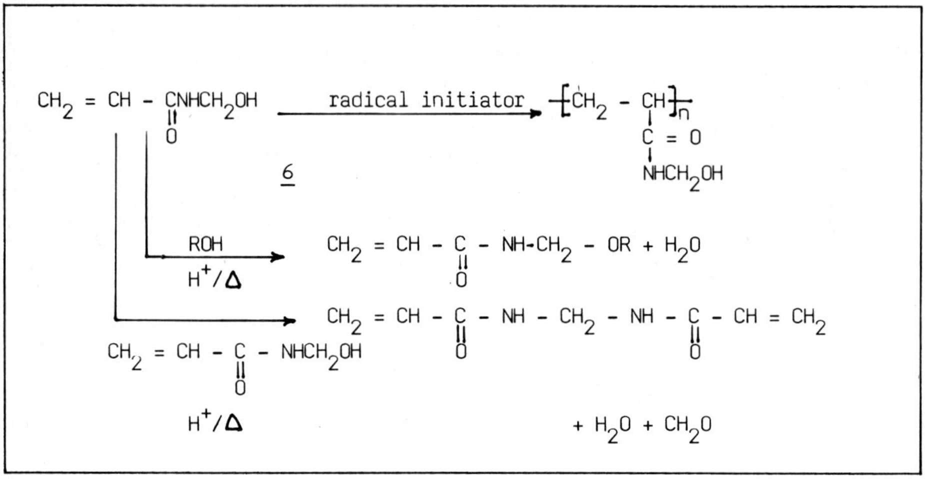

The origin and characteristics of the macroporous inorganic silica materials utilized in our study are given in Table I.

Table I Nominal characteristics of Sphérosil.

Type	Specific surface area m^2/g	Average pore diameter nm	Pore volume cm^3/g
XOB030	50	60	1.0
XOB015	25	125	1.0
XOC005	10	300	1.0

Particle sizes: 40 - 100 μm; 100 - 200 μm; 200 - 800 μm;
Apparent bulk density: about 0.45 g/cm^3
Manufactured and supplied by Rhône Poulenc Industries

Covering the silica surface with a thin skin of crosslinked polymer was performed by the following procedure of synthesis:

1. The first step consisted of adding the inorganic beads to a clear solution of the monomers in a volatile polar solvent. According to a method utilized in the gas chromatography[7] the volatile solvent is carefully evaporated under reduced pressure and with gentle stirring on a rotavapor in order to retain a thin film of monomers on the silica surface. Due to their low molecular weight the monomers penetrate rapidly into the pores.

2. The second step comprises a heterogeneous polymerization. Therefore, a solution of a radical initiator in an apolar solvent, in which the polar monomers are not or very slightly soluble, is added to the monomer coated beads. Under a nitrogen atmosphere the stirred suspension is heated resulting in polymerization and crosslinking.

3.In a third step the polymer coated beads are thoroughly washed with polar solvents to remove unreacted monomers and soluble material. After the washing operation the particles are dried by heating under vacuum.

Research into the hydrolytic stability of the prepared samples in aqueous buffer solutions showed the necessity to link (or graft) the cross-linked polymers covalently to the silica surface. Therefore, as described in the literature [1], the silica surface is treated with 3-glycidoxy-propyl-trimethoxysilane. By hydrolyzing the epoxy groups 3-glyceryl-propyl-siloxy silica is obtained.
Etherification of the hydroxyl function of the glyceryl moiety with the methylol groups of the MAAM-copolymer results in a chemical anchoring of polymer to silica.

In Table II some examples of the new carriers are summarized. It is evident that the content of reactive groups can easily be varied by offering several amounts of reactive monomers. Incorporation of the offered amount of monomers occurs at a fairly constant level.

In order to show the suitability of the new carriers we immobilized heparin and antithrombin III and tested the two affinity columns.

Table II Synthetic polymers applied to silanized Sphérosil

Sample	Elemental C%	analysis N%	Functional monomer used in combination with MAAM and μmol offered per gram	Amount of functional monomer incorporated per gram carrier (μmol) (a)	Amount of incorporate MAAM per gr carrier (μm
1	3.15	0.57	no	-	400
2(b)	5.08	1.09	1 (400)	217	409
3	7.53	1.18	1 (400)	202	376
4(c)	3.12	0.54	1 (200)	88	207
5	6.36	0.92	2 (400)	190	460
6	4.81	0.71	2 (200)	80	430
7	4.34	0.65	2 (50)	20	444
8	5.95	1.03	3 (400)	230	500

(a) Measured by microtitration.
(b) Copolymer applied directly to the blank silica surface.
(c) Sample 4 based on Sphérosil XOB015; other samples derived from Sphérosil XOB03C

ISOLATION OF ANTITHROMBIN III (AT III) FROM BOVINE PLASMA WITH HEPARIN-SILICA.

Heparin, a mucopolysaccharide, has been used in affinity chromatography as a group specific ligand for the purification of a wide range of proteins including coagulation proteins, lipases, lipoproteins and DNA or RNA polymerases[8]. It is an anticoagulant which acts selectively by binding to the plasma protease, antithrombin III, thereby increasing the normal rate at which antithrombin III inhibits thrombin.

We immobilized heparin on our new inorganic carriers containing the ε-amino-caproic acid spacer groups with activated carboxyl end groups by simply adding the heparin solution at pH 7.7. Heparin originating from Diosynth with an activity of 180 U.S.P. units per mg was used. The amount of aminogroups per gram heparin being about 25 μmol per gram was determined by a fluorimetric method according to Jozefsson[9]. Assuming an average molecular weight of 13.000 for heparin and one amino group (belonging to serine coupled to heparin) per molecule heparin it can be calculated that about one out of the 3 heparin molecules possesses an amino group. After the coupling remaining active carboxyl groups were blocked with ethanolamine.

Firstly the influence of the amount of heparin coupled per gram adsorbent on the adsorption capacity expressed as the amount of AT-III activity adsorbed per ml bed volume, was investigated batchwise under the conditions outlined in Table III. The samples studied contained quantities of heparin in the range of 5 to 20 mg heparin per ml bed volume (i.e. 10 to 40 mg per gram). Within this range no clear influence of the incorporated amount of heparin was detected on the adsorption capacity, which was of the order of 10 up to 15 AT III units per ml bed volume. Neither the bead size (100 - 200 μm; 200 - 400 μm) nor the mean pore size (110 nm; 320 nm) appeared to have any influence on the capacity.

In a second series of experiments the adsorption capacity and the rate of adsorption of AT III was studied batchwise for two heparin-silica samples and heparin-Sepharose® CL-6B purchased from Pharmacia. Table III reveals an adsorption capacity for heparin-silica which is four times higher than for heparin-Sepharose. A comparison between the two inorganic samples containing the same loading of heparin showed the dependence of the adsorption rate on the bead sizes in combination with the pore size.

The results of a chromatographic experiment are listed in table IV. Nearly quantitative elution is obtained with a purification factor for the AT III of about 120. The use of gradient elution as suggested in the literature[8] is expected to improve the purification factor to about 500. The operational adsorption capacity was about 8 U per ml bed volume.

SEPARATION OF HIGH AND LOW ACTIVITY HEPARIN BY MEANS OF ANTITHROMBIN III--SILICA ON A SEMI-INDUSTRIAL SCALE.

Heparin was separated by affinity chromatography on antithrombin--substituted silica into two distinct fractions, one with high affinity, the other with low affinity for the protein.

The affinity column was made by coupling purified bovine antithrombin via its amino groups to the new carrier containing ε-amino-caproic acid with activated carboxyl end groups, in the presence of heparin. The heparin (previously treated with acetic anhydride in order to acetylate any free amino group present) was added in an attempt to mask the heparin-binding site of the

Table III Batchwise adsorption of AT III from bovine plasma[a]

Adsorbent	Heparin-silica[b] Mean pore size 1100 Å Bead sizes: 200-400 μm	Heparin-silica[b] Mean pore size 3200 Å Bead sizes: 100-200μm	Heparin-Sepharose CL-6B
Starting AT III U/ml plasma[c]	0.87	0.81	0.77
Percentage of adsorption after			
1 hour	0	3	1
3 hours	0	13	2
19 hours	4	-	3
22 hours	16	18	5
24 hours	16	22	5
Capacity i.e. adsorbed U/ml bed volume after 24 hours	12	14	3

(a) Before use the adsorbent was equilibrated with 10 mmol Trisbuffer at pH 7.4. 5 ml adsorbent was added to 400 ml bovine plasma, defibrinated and adjusted at pH 7.5 with acetic acid. After degassing the suspensions were gently stirred on a rotavapor at 5°C. Occasionally a small quantity of the supernatent was analysed in order to determine the adsorption percentage.
(b) Amount of heparin was about 40 mg per gram (i.e. about 20 mg/ml bed volume)
(c) Antithrombin was assayed for its suitability to inhibit thrombin in the presence of heparin by measuring the hydrolysis of Tos-Gly-Pro-Arg-p-nitroanilide, i.e. Chromozym TH from Boehringer[10]. AT III activity was calculated relative to an AT III standard obtained from Kabi.

The quantity of heparin in the elution fractions was measured by analyzing the uronic acid content by the carbazole reaction[11]. Heparin was assayed for its suitability to inhibit in the presence of antithrombin factor Xa by measuring the hydrolysis of N-benzoyl-L-isoleucyl-L-glutamyl-glycyl-L-arginine-p-nitroanilide hydrochloride and its methylester, substrate S-2222 from Kabi Diagnostica, Sweden[12].

Table IV Chromatography of bovine plasma on heparin-silica[x]

Step	Volume (ml)	U AT III/ml	Total AT III activity	Spec.act. U/A^{280}	Yield of activity (percent)
Plasma	416	0.75	312	0.009	(100)
Breakthrough activity and washings	461	0.50	231	-	74
Desorption	16	4.86	78	1.2	25

x Defibrinated plasma was applied at 25°C on a column of 1.0 x 13.5 cm at a flow rate of 5 ml per hour. The column was packed with heparin-silica, containg 17 mg heparin per ml and based on Sphérosil XOB015 with bead sizes of 200 to 400 μm, and equilibrated before loading with plasma with 10 mmol Tris containing 0.15 M NaCl at pH 7.4. After washing with the equilibration buffer, the adsorbed acitivity was eluted with 10 mmol Tris containing 4 M NaCl at pH 8.4.

AT III from binding to the carrier. Remaining active carboxyl groups after the coupling were blocked with ethanolamine. Determination of the offered amount of protein and amount of protein recovered after washing by the biuret method revealed that 5.4 mg AT III per gram adsorbent (or 2.5 mg per ml bed volume) was bound.

700 ml of the antithrombin-silica, based on Sphérosil XOB015 with bead sizes of 200 up to 600 μm, was packed in a column with an internal diameter of 5.0 cm. The bed height was 35 cm. Applied flow rate was 1105 ml per hour i.e. about 60 ml hr^{-1} cm^{-2}. The fractionation of heparin (170 U.S.P. units per mg) per cycle is illustrated in Fig. 3.

To obtain an efficient separation the column equilibrated with 0.02 M phosphate buffer at pH 7.4 containing 0.15 M NaCl, was overloaded with 510 mg heparine in 51 ml equilibration buffer. Elution was further performed with the same buffer containing amounts of sodiumchloride as indicated below. The breakthrough fractions and washings (0.15 M NaCl) yielded 324 mg heparin comprising 26 % of the offered total heparin activity. 43.5 mg low active heparin (100 U/mg) containing 5 % of injected heparin activity emerged subsequently during elution with 0.5 M NaCl. Per run 151 mg high active

Fig. 3: Separation of high and low activity heparin by means of antithrombin III - Silica.

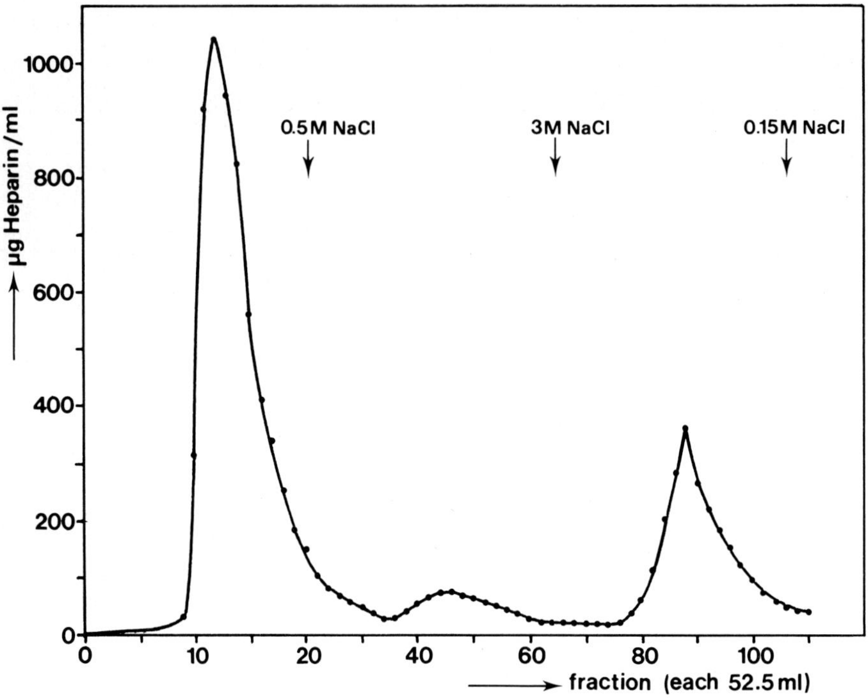

heparin (400 U/mg) representing 70 % of applied activity was eluted with 3.0 M NaCl . Before starting the following cycle the column was re-equilibrated. One whole cycle of fractionation took four hours, which means a daily production capacity of 810 mg high affinity heparin powder (400 U/mg). The powder was obtained by a tenfold concentration of the high activity elution fraction by means of an Amicon PM-10 and by precipitation with methanol.

The capacity of the described column was 210 µgram high activity heparin per ml bedvolume which is on the same level as described by Günzel(13) for antithrombin-Sepharose.

CONCLUSION

We may say that our new carriers are very well suited for affinity chromatography on an industrial scale. They possess excellent flow properties, an almost complete inertness in common organic solvents and aqueous solutions from pH 2 to 9.0 and are easy to handle. Regeneration by pyrolysis at 600°C to blank silica is possible without damaging the pore morphology. Finally, the procedure of synthesis permits coating of the silica surface with a wide versatility of hydrophilic (co)-polymers carrying functional groups. Thus tailor-made carriers are obtained on which affinity ligands can be immobilized in one step only.

ACKNOWLEDGEMENTS

We want to express our gratitude to Mr. D. van Houwelingen and Mrs. M. Cramer for the microtitration and to Mr. W.J. Buis (Organisch Chemisch Instituut TNO, Utrecht) for the elemental analysis of our new carriers.

REFERENCES

1 a C.R. Lowe, Int. J. Biochem., (1977), 8, 177
  b W.H. Scouten, Int. Lab., (1974), nov. dec., 13
  c J.L. Tayot et al, Chrom. Synth. Biol. Polym., (Lect. Chem. Soc. Int. Symp.), (1978), 2, 95
  d L. Jervis, Chrom. Synth. Biol. Polym., (1978), 2, 231
2 a F.E. Regnier and R. Noel, J. Chromatogr. Science (1976), 14, 316
  b G.P. Royer et al, Biochem. and Biophys. research comm., (1975), 64, 478
3 R.A. Messing (Ed), (1975), Immobilized Enzymes for industrial Reactors, Academic Press.
4 G.R. Gray, Anal. Chem., 1980, 52, 9R.
5 a R.L. Schnaar, Y.C. Lee, Biochemistry, (1975), 14, 1535
  b E. Brown et al, Int. Symp. on affinity chromatogr. and molecular interactions, (1979), 37, Ed. Inserm.
6 E. Nyquist (Ed), Functional monomers, (1973), 1, 394, Marcel Dekker, Inc., New York.
7 H. Purnell, gas chromatography, (1962), Wiley & Sons, New York.
8 T.H. Finlay et al, Anal. Biochem., (1980), 108, 354.
9 B. Jozefsson et al, Anal. Chim. Acta, (1977), 89, 21.
10 U. Abildgaard et al, Thromb. Research, (1977), 11, 549.
11 Z. Dische, Journ. of Biol. Chem., (1947), 167, 189.
12 A.N. Teien et al, Thromb. Res., (1976), 8, 413.
13 A.S. Bhargava, G. Freihube, and P. Günzel, Arzneim. - Forsch./Drug Res., (1980), 30 (II), 1621.

T.C.J. Gribnau, J. Visser and R.J.F. Nivard (Editors),
Affinity Chromatography and Related Techniques

REACTIVE CARRIERS FOR THE IMMOBILIZATION OF BIOPOLYMERS

G. MANECKE, H.-G. VOGT and D. POLAKOWSKI
Institut für Organische Chemie der Freien Universität Berlin,
Takustr. 3, D-1000 Berlin 33

ABSTRACT

Various reactive carriers for the immobilization of enzymes are described. As insoluble starting materials polymeric products which contain poly(vinyl alcohol) were used: Poly(vinyl alcohol)-gels crosslinked with terephthalaldehyde, hydrolyzed beads of crosslinked poly(vinyl acetate), poly(vinyl acetate-co-ethylene)-tubes coated with poly(vinyl alcohol) and mainly poly(vinyl alcohol)-containing synthetic wood pulp. Various reactive groups were introduced into these carriers. The influence of some selected parameters on the immobilization and the properties of these immobilized products is shown. The formation of the trypsin-soy bean trypsin inhibitor-complex on the immobilized products is described.

INTRODUCTION

Already in the beginning of the fifties Isliker (ref. 1), Grubhofer and Schleith (ref. 2) and Manecke and Gillert (ref. 3) have prepared serologically specific adsorbents binding antigens or antibodies to reactive carriers. Manecke, Singer and Gillert have also prepared hapten specific adsorbents (refs. 4, 5). They immobilized antibodies specific to artificially conjugated antigens.

It was shown here that it is possible to prepare specific adsorbents for physiologically active proteins and also for small molecular mass compounds. Immobilizing enzymes we continued these investigations of covalent binding of physiologically active proteins.

For the immobilization of enzymes many suitable polymeric supports in the form of particles, tubes, hollow fibres, fibres or membranes (refs. 6-9) can be used.

For the covalent immobilization of enzymes, among others, many polymers based on polysaccharides were extensively investigated. Because of the high content of hydroxyl groups the polysaccharides are hydrophilic. The hydroxyl groups also allow to introduce many different kinds of reactive groups. So cellulose, starch, dextran, Sephadex, agar and agarose have been activated and used as reactive carriers for the immobilization of enzymes very successfully.

Also many totally synthetic reactive carriers have been synthesized and used.

These carriers contain reactive groups and for better wettability, they are also fitted out with hydrophilic groups, which are very often ionogenic groups (refs. 6-

We have recently investigated carriers without ionogenic groups starting from different polymeric products based on poly(vinyl alcohol) (PVA). This polymer contains also many hydroxyl groups, which give the same above mentioned advantages. We think that these polymeric carriers should also give interesting possibilities for the affinity chromatography, especially for the preparation of immuno adsorbent With this report a survey of our own research work concerning the use of PVA for th preparation of reactive carriers for the immobilization of enzymes will be given. Different polymeric products based on PVA were used as starting materials.

RESULTS

Insoluble carriers of gel like structure were obtained by reaction of PVA with terephthalaldehyde in an acidic medium. By the hydrolysis of crosslinked poly(vinyl acetate) beads, commercially available under the trade mark Fractogel (Merck), another PVA-carrier was obtained. Also tubes consisting of poly(vinyl acetate-co-ethylene) coated with PVA on their inner surface have been used for the immobilization of enzymes. Into these polymeric materials by polymeranalogous reactions reactive groups were introduced. These carriers were mechanically stable and besides of thei reactive groups for the immobilization reaction and the hydrophilic hydroxyl groups they were chemically inert. Not only the chemical structure of the carriers but als their morphology is of great importance for the immobilization reaction and the properties of the immobilized products. Very often the diffusion is the rate determini step for the activity of the immobilized enzyme. Therefore it is of great advantage to use carrier materials, which allow the immobilization of enzymes only on their surface. Carriers with a big surface and an inert core, which supplies the carrier with a sufficient mechanical stability are very useful. They should also be useful for the preparation of adsorbents for the affinity chromatography.

Ideally such carrier properties shows synthetic wood pulp (SWP). This synthetic fibre material was originally developed by Crown-Zellerbach for the paper making industry (ref. 10).

The SWP is a new form of polyethylene which is obtained by shear-induced cristallization (ref. 11). Under such circumstances polyethylene crystallizes in the shape of so-called "shish-kebabs". SWP can be prepared in the presence of PVA which acts as an hydrophylizing agent. Though the PVA is physically bound to the surface of the polyethylene fibrides, it cannot be washed off. Such polyethylene-fibrides are commercially available e.g. as Hostapulp® (Hoechst). In our investigations we used the SWP of Hoechst company of the type SWP-E400 and SWP-R 830 with a PVA-content of 1.55 resp. 4.2 %. Enzymes, immobilized on the reactive derivatives of the PVA-content of the synthetic wood pulp, show a high stability of the enzyme-carrier-lir kage. Such immobilized enzyme products differ from other enzyme-carrier-conjugates principally by the presence of an inert core of the carrier, bearing a thin surface layer containing the immobilized enzyme. The absence of pore structure should be favourable for the diffusion of reactands. The derivatization of the PVA-content of the SWP and the immobilization of the enzymes was performed by methods which are partially tested with PVA-carriers of other morphology (ref. 12).

All activation methods, which allow the introduction of reactive groups into one kind of PVA-product are also applicable for the other PVA-materials.

The activation reactions of PVA-products, starting with the reaction of 2-(3-aminophenyl)-1,3-dioxolane with PVA forming an amino derivative, were extensively investigated. The amino-groups were converted into reactive groups by diazotization, by reaction with thiophosgen, or by activation with glutardialdehyde.

The reactive carriers were characterized according to their chemical composition. In the case of the diazotized carriers and the carriers with isothiocyanato-groups the carriers could be prepared with different contents of reactive groups, as the derivatization of PVA-products by the reaction with 2-(3-aminophenyl)-1,3-dioxolane is a good controllable reaction. The determination of the content of the reactive diazonium groups was performed by coupling with tyrosine as a model compound. The results of these our investigations with the SWP are demonstrated in Table 1.

TABLE 1

Preparation of amino groups-containing SWP by means of 2-(3-aminophenyl)-1,3-dioxolane and coupling of the diazotized derivatives with tyrosine and trypsin

Starting material	Preparation of amino groups-containing SWP. Applied amount of 2-(3-aminophenyl)-1,3-dioxolane in mmole/g SWP	Coupling by the diazotized derivatives. Bound amount of tyrosine in µmole/g	trypsin[a] in µmole/g	trypsin[a] in mg/g
SWP-E 400	1.46	1.5	0.6	15
SWP-R 830	1.46	7.5-8	2.0	47
SWP-R 830	0.73	3	1.5	35
SWP-R 830	0.15	2-3	0.5	12

[a] Coupling conditions: 25 mg trypsin; 10 cm^3 phosphate buffer pH 8.0 (μ = 0.15); 100 mg carrier; 2 h; 0°C.

At the chosen conditions the highest content of reactive diazonium groups was 8 µmmole/g SWP i.e. 2 % of the theoretical maximal content for the complete acetalization of the PVA-part of the SWP. This low substitution extent shows that most of the PVA in the SWP is not available for surface reactions.

In Table 2 is shown how the composition of the reactive carriers with diazonium groups influences the immobilization of papain. A content of 1.4 mmole re-

TABLE 2

Immobilization of papain on diazonium groups-containing carriers based on cross-linked PVA-gels

Content of amino groups in mmole/g starting polymer	Immobilized papain	
	bound amount in mg/g carrier	retained activity at pH 6 in %
1.8	615	7.7
1.6	635	9.5
1.4	685	12.8
1.1	335	14.8
0.6	305	20.8
0.3	300	21.5

active groups per gramm of starting polymer seems to be optimal for the binding ability of the carrier. The carrier with the lowest content of reactive groups led to an immobilized product with the highest retained activity. These results can be explained by the following effects: With the number of reactive groups the binding ability of the reactive carriers increases, but with the increasing number of the reactive groups the hydrophobic character of the carrier increases too. This is infavourable for the immobilization: Lower amounts of immobilized enzyme and lower retained activities were found. When the density of reactive groups on the carrier is low, so that the multiple bonding between the carrier and the enzyme is minimized, the retained activity is usually higher.

Similar results have been found with the immobilization of papain on isothiocyanato derivatives of crosslinked PVA. Smaller amounts of papain were bound as the carrier in this case was much more hydrophobic than the respective diazonium form of the polymer. Azo-coupling of trypsin on derivatives of SWP also led to higher amounts of bound enzyme than the coupling by isothiocyanto-groups, when the reactiv carriers were prepared from polymers with the same content of amino-groups. The azo coupling also led to better retained activities.

Binding the enzyme molecule to the carrier by a single bond and if all reactive groups are accessible and do react, one mmole tyrosin i.e. 23,400 mg should be immobilized per mmole of reactive groups. As it can be seen in Tab. 3 the carriers whic were prepared with terephthalaldehyde crosslinked PVA or from hydrolyzed Fractogel immobilized only up to a few 100 mg of trypsin per mmole of the reactive groups of the carrier. This shows that most of the reactive groups are not accessible to the enzyme. In the case when the content of the immobilized trypsin of the carriers crosslinked with therephthalaldehyde was far below the maximum binding capacity because of the formation of more multiple bonds between the carrier and the enzyme molecule the retained activity of the immobilized enzyme decreases. The derivatives of SWP contained lower amounts of reactive groups (0.003 mmole/g carrier). As these groups are located on the surface of the carrier they are easily accessible. There-

TABLE 3

Immobilization of trypsin on different PVA-products by azo-coupling

Reactive diazonium derivative of	Content of reactive groups in mmole/g carrier	Conditions for trypsin immobilization		Immobilized trypsin		Retained activity in %
		offered trypsin[a] in mg/g carrier	time, temperature	in mg/g carrier	in mg/mmole reactive groups (in % of max. binding ability)	
with terephthalaldehyde crosslinked PVA	1.40[b]	100	3h, 4°C	87	60 (0.3)	1.6[d]
	1.40[b]	1000	3h, 4°C	321	230 (1.0)	5.0[d]
hydrolysed Fractogel	1.25[b]	1000	2h, 0°C	88	70 (0.3)	2[e]
synthetic wood pulp (SWP-R 830)	0.003[c]	250	2h, 0°C	35	11650 (50)	11[e]
	0.003[c]	250	2h, 25°C	55	18300 (78)	5.5[e]

[a] In phosphate buffer, pH 8.0.

[b] Determined by nitrogen analysis of the respective amino derivatives under the assumption of a complete conversion upon diazotization.

[c] Determined by reaction with tyrosine as model compound.

[d] Assayed with $2.5 \cdot 10^{-3}$ M D,L-BAPA; phosphate buffer pH 7.8 (μ = 0.15); 25°C.

[e] Assayed with 0.01 M BAEE; 0.2 M NaCl; 0.1 M tris buffer pH 8.5; 25°C.

fore big amounts of trypsin per mmole of reactive groups are immobilized. At the immobilization temperature of 0°C 50 per cent of the theoretical maximum amount of immobilized trypsin could be bound and at 25°C 78 per cent. At the higher temperature the autodigestion of the dissolved enzyme is also faster, therefore the retained activity of the immobilized enzyme was lower here.

Methods for direct activation of the PVA-products glutardialdehyde could be used (ref. 13) as it reacts with the hydroxyl-groups of polymers. PVA-products were reacted with BrCN (ref. 14) giving carriers for the immobilization of enzymes. PVA-carriers were also modified by reaction with 2,4,6-trichloro-s-triazine according to a method described by Kay and Crook (ref. 15).

Reactive PVA-carriers were also obtained by the derivatization with p-benzoquinone (ref. 16).

PVA-carriers with a specificity towards SH-compounds were prepared by reacting PVA-products with 3-maleimido-benzaldehyde (ref. 17). They could also be obtained b a sequence of reactions giving a mixed polymeric disulfide, which contained 2-thiopyridyl groups (ref. 14). The latter type of reactive disulfide-groups containing carrier has been investigated by Axen et al. on agarose base (ref. 18).

In the following some further parameters which influence the immobilization reaction or the properties of the immobilized enzyme will be discussed.

The ability of the reactive SWP-carrier to bind enzymes is not depending only on its content of reactive groups but also on the coupling conditions. For the immobilization of enzymes by azo-coupling often a weak alkaline pH and a low coupling temperature is chosen.

TABLE 4
Temperature dependence on the immobilization[a] of trypsin and α-chymotrypsin by azo-coupling on SWP derivatives

Coupling Temperature °C	Immobilized trypsin: bound amount in mg/g SWP	Immobilized trypsin: ret. activity[b] in %	Immobilized α-chymotrypsin: bound amount in mg/g SWP	Immobilized α-chymotrypsin: ret.activity[c] in %
0	35	11	20	4.8
25	55	5.5	31	2.8

[a] Coupling conditions: 25 mg enzyme in 10 cm^3 phosphate buffer pH 8.0 (μ = 0.15); 100 mg diazotized SWP-R 830 derivative, 2 h.
[b] 0.01 M BAEE; 0.2 M NaCl; 0.1 M tris buffer; at the pH-optimum.
[c] 0.02 M ATEE; 0.2 M KCl; 5 vol-% ethanol; at the pH-optimum.

In Table 4 the influence of the coupling temperature on the immobilization of trypsin resp.α-chymotrypsin by diazotized SWP is demonstrated. Trypsin as well as α-chymotrypsin were coupled in higher amounts at higher temperature. The decomposition of the diazonium groups during the chosen 2 h at 25°C is neglible at the coupling temperature of 0°C although the retained activities were higher in both cases. This can be explained by the higher stability of the enzyme at the lower temperature.

TABLE 5

Enzyme immobilization on SWP derivatives

Reactive group	Immobilized protein	Offered protein in mg/g SWP-derivative	Bind.ability in mg prot./ g SWP-derivat. (Immob.yield in %)	Ret.activity in %
$-N_2^+$	Trypsin	250	47(19)	19.6
	Trypsin	100	34(34)	12
	α-Chymotrypsin	250	35(14)	4.8
	Papain	240	90(36)	18.3
	Urease	90	50(56)	5
	Soy bean trypsin inhibitor	50	38(76)	14
	Catalase	25	22(88)	0.1
	Glucose oxidase	12.5	11(88)	0.4-0.8
-NCS	Trypsin	250	12(4.8)	5.8
>C=NH	Trypsin	100	7(7)	1
$-N=CH(CH_2)_3CHO$	Trypsin	100	25(25)	14.7
Dichloro-s-triazinyl	Trypsin	100	62(62)	23
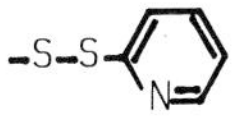	Catalase	25	20(80)	0.01

As it can be seen in Tab. 5 the amount of immobilized enzyme is strongly influenced by the offered amount of the enzyme at the immobilization reaction. Usually at lower amounts of offered enzyme effective carriers give a 100 % immobilization yield. At higher amounts offered, naturally the amount of the bound enzyme increases but the yield of the immobilization decreases. Tab. 5 also indicates that carriers with different reactive groups and also different enzymes with the same reactive carrier naturally show different immobilization characteristics.

As an example for possible properties of an immobilized enzyme some typical characteristics of immobilized trypsin-products bound to SWP-derivatives by azo-coupling are described: The thermostability of the immobilized trypsin can be investigated by different ways. In the investigated case an aqueous solution of the trypsin (2 mg trypsin/cm^3) or a suspension of immobilized trypsin were incubated at definite temperatures, then the activities of the enzymes were determined (ref. 19).

It was found that the immobilized trypsin is thermally more stable than the native trypsin in solution. This is mainly caused by the hindering of the autolysis of the immobilized enzyme. While the time dependence of the decrease of the activity of trypsin in solution follows a law of second order, indicating herewith an autolysis it was not possible to assign a whole numbered reaction order for the immobilized trypsin. The thermal inactivation should be a reaction of first order. The basic requirement for this would be however, that all enzyme molecules are bound to the matrix in the same way and have the same microenvironment. The inactivation of the immobilized trypsin indicated that this is not the case here.

Investigating the temperature dependence of the activity of trypsin and of trypsin immobilized by azo-coupling on SWP, the activation energies of their reaction with BAEE were determined. They were almost the same (10 kcal/mole resp. 11 kcal/mole). This shows that for the activation of the immobilized trypsin the diffusion is not rate determing. The diffusion is only little temperature depending. Therefore one should find for diffusion controlled reactions much smaller apparent activation energies compared with those of the free enzyme (ref. 19).

The investigation of immobilized enzymes often includes the determination of the pH-activity profiles. When in enzymatic reactions protons are involved the immobilized enzymes often show strongly shifted pH-optima compared to the native enzyme. In the case of trypsin which was immobilized on SWP by azo-coupling the pH-activity profiles were investigated with BAEE as substrate in the presence of different concentrations of buffers.

It was found that in the absence of buffer a strong shift of the pH-optimum to higher pH-values was observed. The presence of 0.4 M NaCl in the substrate solution did not improve the activity. Adding a buffer the pH-activity profile showed a similar shape and position to that of native trypsin. This can be explained by the static and dynamic properties of the buffer (ref. 20).

The Michaelis-Menten constants (K_M) of trypsin and the $K_{M(app)}$-values of trypsin immobilized on diazonium derivatives of SWP were determined with L-BAPA as substrate. The kinetic constants were obtained by Lineweaver-Burk-plots. $K_{M(app)}$ of the immobilized trypsin was found to be distinctly lower than the values of the native enzyme ($K_{M(app)}$= 0.7 mM; K_M = 4.0 mM). This might indicate a higher concentration of BAPA in the microenvironment of the immobilized enzyme compared to the bulk solution.

For this audience the adsorption behaviour of immobilized trypsin towards soybean-trypsin-inhibitor (STI) and that of immobilized STI towards trypsin are of special interest. STI is a protein, which forms a 1:1 complex with trypsin. 1 mg of STI can inhibit 1.15 mg of free trypsin (Merck) totally. Since diazonium derivatives of SWP allowed to immobilize trypsin as well as STI the formation of the trypsin-STI-complex of both immobilization products was investigated (ref. 19).

For the adsorption the immobilized STI was suspended in aqueous phosphate buffer (pH 8.0; μ = 0.15) and a freshly prepared solution of trypsin (Merck) in phosphate buffer of the same pH was added. The desorption was performed with 1 mM HCl aqueous solution. The time dependence of these both processes is shown in Fig. 1.

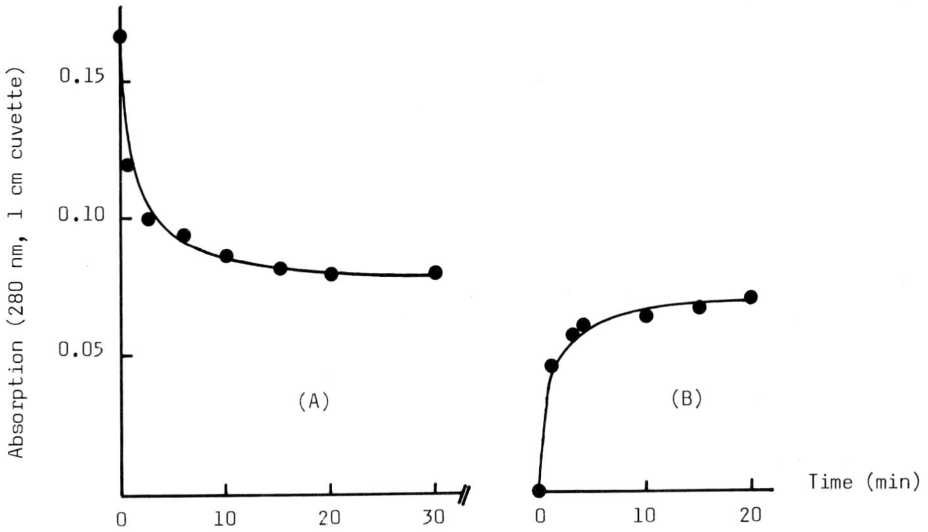

Fig. 1 (A) Time dependence of the adsorption of trypsin on immobilized STI.
(B) Time dependence of the desorption.

In Tab. 6 the complexation properties of the two immobilized adsorbents are summarized.

TABLE 6

Properties of trypsin and soy bean trypsin inhibitor (STI), immobilized on SWP-derivatives by azo-coupling

	Immobilized protein	
	Trypsin	STI
Mole mass	23,400	22,000
Covalent-bound amount		
in mg/g conjugate	32.6	36.3
in μmole/g conjugate	1.39	1.65
Activity of covalent-bound trypsin		
towards BAEE[a] in U/g conjugate	149	
in % (rel.)	12	
towards STI in % (rel.)	1.5	
Activity of covalent-bound STI		
in BAEE-units/g conjugate		240
towards trypsin in % (rel.)		14
STI adsorbed in mg/g trypsin-conjugate	0.5	
Trypsin adsorbed in mg/g STI-conjugate		6

[a] 0.01 M BAEE; 0.2 M NaCl; 0.1 M tris buffer pH 8.0; 25°C.

The adsorption activities of the immobilized adsorbents are not reduced by repeated cycles of adsorption and desorption.

It was the aim of this short and very selected outline of our research work performed in the field of enzyme immobilization to demonstrate the influence of some parameters which might also affect the preparation of immobilized immuno adsorbents and to draw the attention to some reactive carriers which were developed in our laboratories and which might be also useful for the preparation of adsorbents for the affinity chromatography.

ACKNOWLEDGEMENT

We thank Prof.Dr. H. Cherdron (Hoechst AG) for supplying us with synthetic wood pulp and Prof.Dr. V. Kasche (University Bremen) for his cooperation in the investigations with soy bean trypsin inhibitor.

REFERENCES

1 H.C. Isliker, Ann.New York Acad.Sci., 57 (1953) 225.
2 N. Grubhofer and L. Schleith, Naturwiss., 19 (1953) 508.
3 G. Manecke and K.-E. Gillert, Naturwiss., 42/8 (1955) 212.
4 G. Manecke, S. Singer and K.-E. Gillert, Naturwiss., 45/18 (1958) 440.
5 G. Manecke, S. Singer and K.-E. Gillert, Naturwiss., 47/3 (1960) 63.
6 O. Zaborsky, Immobilized Enzymes, CRC Press, Cleveland, 1973.
7 H.H. Weetall, Enzymology, Vol. I, Immobilized Enzymes, Antigens, Antibodies, and Peptides, Dekker, New York, 1975.
8 K. Mosbach, Methods in Enzymology, Vol. XLIV, Immobilized Enzymes, Academic Press, New York, 1976.
9 L. Goldstein and G. Manecke, The Chemistry of Enzyme Immobilization, in L.B. Wingard, Jr., E. Katchalski-Katzir and L. Goldstein (Eds.), Applied Biochemistry and Bioengineering, Vol. I, Academic Press, New York, 1976, p. 23.
10 Ger.Offen. 2249 604 (1973), Crown Zellerbach International, C.A. 79 (1973) 7064p
11 W. Gordon, H.J. Leugering and H. Cherdron, Angew.Chem., 90 (1978) 833.
12 G. Manecke and H.-G. Vogt, Angew.Makromol.Chem., 78 (1979) 21.
13 P. Monsan, G. Puzo and H. Mazarguil, Biochimie, 57 (1975) 1281.
14 H.-G. Vogt and G. Manecke, Angew.Makromol.Chem., 88 (1980) 37.
15 G. Kay and E.M. Crook, Nature, 216 (1962) 514.
16 G. Manecke and W. Beier, Angew.Makromol.Chem., in press.
17 G. Manecke and H.-J. Middeke, Angew.Makromol.Chem., 91 (1980) 179.
18 R. Axén, H. Drevin and J. Carlsson, Acta Chem.Scand., B29 (1975) 471.
19 G. Manecke and H.-G. Vogt, Angew.Makromol.Chem., 85 (1980) 41.
20 J.M. Engasser and C. Horvath, in D. Thomas and J.P. Kernevez (Eds.), Analysis and Control of Immobilized Enzyme Systems, North Holland Publishing Company, Amsterdam, 1976, p. 187.

T.C.J. Gribnau, J. Visser and R.J.F. Nivard (Editors),
Affinity Chromatography and Related Techniques

MACROPOROUS SPHERICAL HYDROXYETHYL METHACRYLATE COPOLYMERS, THEIR PROPERTIES, ACTIVATION AND USE IN HIGH PERFORMANCE AFFINITY CHROMATOGRAPHY

J. ČOUPEK
Laboratory Instruments Works, 162 03 Prague 6, Czechoslovakia

ABSTRACT

Hydrophilic polymers of 2-hydroxyethyl methacrylate with ethylene dimethacrylate (SPHERON/SEPARON HEMA) were synthesized in the form of rigid macroporous spherical particles. Conditions of the polymerization reaction control the particle and pore diameters, inner surface area, pore volume and other physical and chemical properties of copolymers formed, which can be used as efficient carriers of reactive functional groups, molecules or cells.

Copolymers are characterized by extremely high hydrolytic and mechanical stability and can be used under pressure conditions of HPLC. The average binding capacity for small molecules amounts to 2.2 mmol per gram of dry carrier. Due to the high rigidity of particles, only negligible changes in swelling with variations of pH and ionic strength could be observed.

Many activation reactions were successfully applied to the carriers. The majority of known types of ion exchangers have been prepared proving the superiority of the synthetic materials to the natural carriers in respect to their mechanical stability. Highly hydrophilic derivatives formed by covalent binding of mono- and oligosaccharides were used for the fast size exclusion chromatography of proteins and in the affinity chromatography of lectins. Oxirane activated SPHERON carriers are promissing for technical and analytical high performance liquid chromatography applications. Results obtained with some high performance techniques in separation of biopolymers are discussed.

INTRODUCTION

Development of high performance liquid chromatography related to the progress achieved in the design of instrumentation especially of precise pumps for the mobile phase and of highly sensitive and selective detectors, made possible extension of this separation technique to virtually all branches of the chemical and biochemical analysis.

Results obtained in the research of the synthesis and technology of sorbents and columns for liquid chromatography,an essential increase in their efficiency and resolution,facilitated studies involving separation and analytical determination in complicated systems of natural and synthetic components appearing mainly in modern organic synthesis,pharmacology,biochemistry,biology and in clinical practice.Macroporous sorption materials and carriers of active functional groups defined,along with their physicochemical and chromatographic parameters, also by a considerably high exclusion limit of molecular mass,permitte to extend the validity of the generally acknowledged liquid chromatography separation principles to biopolymers and their fragments.

With increasing demands on the efficiency and reduction of the separation time,working on chromatographic columns necessitates conditions of an increased pressure drop along the column which automatical rules out application of some soft gels in a strongly swollen state. Suitable packings must possess a rigid or semirigid character,a suffic ent porosity enabling macromolecules to penetrate into internal domain of the sorbent structure,and an adequate specific surface area for rea ing a high loading capacity in the adsorption,partition,ion exchange and affinity chromatography.Spherical particles have many advantages in attaining the highest regularity in column packing,and the minimum pressure drop at a given flow rate and mobile phase viscosity.The column efficiency is inversely proportional to the particle size of the packing (refs.1,2).After the proper technique of packing the columns has been solved,sharply fractionated microparticular packings with the particle size 5 or 10 µm ± 1.5 µm reach theoretical values of the separation efficiency.

Stability against effects of reagents,the required interaction properties and reactivity of carrier in modification reactions rank among other required properties of a good packing material.The phase boundar at which interactions occur in macroporous materials must be sufficien ly hydrophilic to prevent denaturation of biopolymers or their irrever sible adsorption.Particle size must not vary with varying pH and ionic strength of aqueous solutions.Carriers should be stable in organic sol vents,also in the modification reactions proper,and must not be subjec ted to the action of enzymes and microorganisms.Good thermal stability is advantageous especially in the modification reactions at elevated temperatures,similarly to the sterilization of carriers by heating. On the other hand,however,the basic carriers should be sufficiently reactive to make possible their chemical transformations for covalent binding of further required functional groups,small molecules and of synthetic or natural polymers.

The requirements just outlined represent properties of an ideal carrier useful not only for chromatographic purposes.Some of the like properties are much needed also in their utilization in enzyme catalysis and in applications on a technological scale.

Mechanical and hydrodynamical requirements on a good sorbent are fulfilled by macroporous silica (refs.3,4) or by controlled pore glass (refs. 5,6).On the other hand,however,their sorption interactions with proteins,which are difficult to check,are undesired.To some extent, they can be eliminated by coating the inner surface of the carrier with a chemically bonded hydrophilic stationary phase (refs.7-9) which can be further chemically modified,if needed (ref.10).Similarly,carriers of the TSK-gel SW type are surface-hydrophilized silicagels too (ref.11). On the contrary,hydrophobic adsorption effects have recently been used in the chromatography on alkyl modified macroporous silicagels (refs. 12-14).All these inorganic carriers cannot,however,be applied at elevated pH values and their lifetime is comparatively limited also in neutral and acid salt solutions,where the covalently bonded phase may be gradually split off.Adsorption of cations and exclusion of anions due to silanol groups released after hydrolysis cause further complications met in the practice.

Our investigation of various chromatographic techniques used in the separation of biopolymers and search for suitable sorbents were based on a somewhat different conception of the carrier matrix.We concentrated on macroporous rigid organic copolymers,which by their mechanical properties could successfully compete with silica,would not possess some of its disadvantages outlined above,and which would exceed silica not only in their chemical stability,but also in a larger variety of modification reactions and the range of applications in chromatography and possibly even beyond its scope.Macroporous copolymers can be prepared beside the form of perfect spherical particles of an arbitrary size from 3 to c.2000 μm also in the form of membranes,plates,rods, blocks etc.This contribution,however,deals only with particular materials for chromatographic purposes and for the immobilization of biologically active compounds.

Requirements regarding a good carrier suitable for laboratory and industrial uses have to large extent been satisfied by copolymers of 2-hydroxyethyl methycrylate with ethylene dimethacrylate,manufactured under the name of SPHERON by Lachema Brno.For high performance liquid chromatography,the same copolymers are produced by the firm Laboratory Instruments Works Prague,under the trade mark SEPARON HEMA.

SYNTHESIS

The suspension radical copolymerization of 2-hydroxyethyl methacrylate with ethylene dimethacrylate is carried out in an aqueous dispersion in the presence of inert organic solvents,the type and concentration of which affect the eventual macrostructure of copolymers (ref.15 A combination of poly(vinylpyrrolidone) and poly(vinyl alcohol) acts as stabilizer of the suspension.The total concentration of the mixed stabilizer in the disperse phase,its composition and hydrodynamic condition of stirring determine the mean size and shape of the size distribution curve of spherical particles of the copolymer.Similarly to other physicochemical properties,the content of accessible (and thus modifiable) hydroxyl groups depends on the initial monomer-to-crosslinking agent ratio and on the microstructure of the carrier;it varies around 2.2 mmol OH-groups per gram.On completion of the polymerization reaction,the eventual product is centrifuged,repeatedly washed with water and extracted with organic solvents in order to remove last traces of unreacted monomers and admixtures.After drying,the copolymer are fractionated to the particle size required for subsequent application (Fig.1.)The values required for high performance liquid chromatography techniques are $d_p = 5 \pm 1.5$ μm and 10 ± 2.0 μm,that for the usua column chromatography is 100 - 200 μm and those needed for technological purposes are 200-300,300-500 and 500 - 1000 μm.

PROPERTIES

Copolymers SPHERON were thoroughly investigated with respect to the properties in chromatographic applications by Mikeš et al.(ref.16). The selected fundamental physicochemical parameters determined by the usual methods are summarized in Table 1.

Copolymerization yields strongly crosslinked microparticles of the xerogel which with the proceeding reaction undergo aggregation and for the macroporous structure of beads.Due to this structure the hydroxyethyl methycrylate carriers possess mechanical prperties analogous to those of silica.The dynamic properties of a column packed with SPHERON are characterized by a flat dependence of the height equivalent to the theoretical plate on the flow rate (ref.16).The macroporous structure of the carrier can be seen in Fig.2.

At the usual buffer concentrations the SPHERON particle practicall does not change its volume if pH or the ionic strength of solution is changed (refs.17,18).This fact makes possible their utilization especi ly in the high performance ion exchange chromatography of proteins.

Fig.1.SEPARON HEMA 1000 for HPLC.Mean particle size 10 /um

Fig.2.Macroporous structure of a broken SPHERON particle.

TABLE 1

Parameters of some macroporous SPHERON supports

Quantity	Unit	SPHERON		
		100	300	1000
Inner surface area (B.E.T.)	m^2/g	56	48	94
Exclusion limit[+]	dalton x 10^3	100	501	1000
Specific pore volume	cm^3/g	0.534	0.601	1.69
Specific unpenetrable volume[x]	cm^3/g	0.724	0.752	0.761
Mean pore diameter (mercury)	nm	22	25	--
Mean pore diameter (N_2desorpt.)	nm	--	25	37
Cation exchange capacity	mequiv/g	0.04	0.04	0.04

[+]Determined by GPC with polydextrane standards Pharmacia Uppsala
[x]For water, n_{H_2O}

In the case of columns pre-packed with microparticular ion exchangers based on SPHERON,gradients can be run to a wide extent without any loss in the column efficiency.Only negligible changes in the particle volume were observed in water-organic mixed phases used in adsorption (hydrophobic) chromatography of proteins.The synthetic character of copolymers and their structure containing ester bonds of acid with the tertiary alpha-carbon atom guarantee their stability against the action of the majority of organic solvents and against enzymatic hydrolysis. In addition,the carriers are very resistant against the effects of temperature.In the dry state their decomposition by depolymerization sets in (ref.15) at temperatures above 200°C.

DERIVATIVES AND ACTIVATION

Surface treated carriers

With respect to the internal structure of copolymers composed of very densely crosslinked submicroscopic particles agglomerated into beads of spherical shape, the modification reactions proceed predominan ly on the internal surface of macropores. This finding has been indirect ly confirmed by the diffusion and chromatographic properties of sorben prepared by modification reactions. For copolymers transformed to a hig degree of conversion, however, one cannot rule out the formation of an a tive layer on the pore surface. With increasing thickness of this layer the capacity of the sorbent increases. The mass transfer is being decre sed, slowing down the establishment of sorption and reaction equilibria The negative consequences which are likely to occur in HPLC (peak broa ning caused by diffusion effects in the stationary phase) are evident. Hydroxyl groups of the copolymer SPHERON (SEPARON HEMA), the structure of which is shown in Fig.3., can be chemically transformed by reactions known from the organic chemistry of primary alcohols. A number of polym have been prepared which contain very reactive groups, such as alkoxide (ref.19), carboxyls (ref.18), anhydrides (ref.20), isocyanate (ref.21), phosphoryl chloride (ref.22) etc., or on the contrary, functional groups able to react under „physiological" conditions required in the immobi- lization of enzymes (ref.23).

A special attention was devoted to chemical modifications of the SP RON carriers with ionogenic groups. This made possible the preparation of high performance ion exchangers possessing an almost theoretical ca pacity towards small ions characterizing the high conversion of hydro- xyl groups (refs.17,18,24) to the corresponding ionogenic groups.

SPHERON derivatives containing bound salicyl (ref.25), 8-hydroxyqui lin (ref.26) or thiol (ref.27) are also of interest. These groups were found to be useful in adsorption studies and in chromatography of ino nic ions.

A special group of modifications of the SPHERON carriers consists sorbents prepared by the hydrophobization of the matrix with alkyl, ar nitro- or dinitrophenyl groups by means of an ester (ref.28) bond (re action of SPHERON with chloride of the respective acid) or of an ethe (ref.29) bond (reaction of alkoxide-SPHERON) with a halogen derivativ Sorbents containing aromatic nitro-groups covalently bonded on the or ganic matrix exhibit strong charge-transfer interactions.

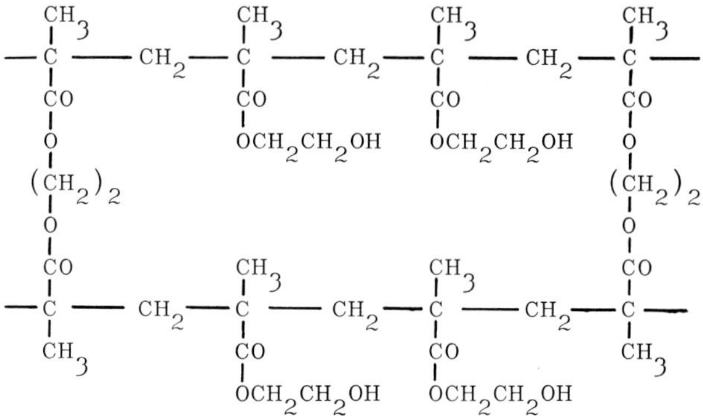

Fig.3. Schematic formula of 2-hydroxyethyl methacrylate-co-ethylene dimethacrylate copolymers SPHERON (SEPARON HEMA).

The SPHERON (SEPARON HEMA) copolymers can be further hydrophilized by a covalent attachment of mono- and oligosaccharides with dry HCl,BF_3 or their mixtures used as catalysts in an aprotic solvent (ref.30). If a thiol derivative of SPHERON is used as the initial carrier in the immobilization of saccharides according to this method,S-glycosidic derivatives are readily obtained (ref.31),possessing a somewhat higher stability towards hydrolysis.O-Glycosidic derivatives containing up to 16% (w/w) of saccharide combine good mechanical and hydrodynamic properties of SPHERON with (for proteins) very advantageous interaction parameters of polysaccharides (e.g.agarose).They may serve as initial carriers in further modifications and subsequent applications.

For the activation reactions of SPHERON/SEPARON carriers occuring in the immobilization of biologically active compounds it holds,similarly to the general organic-chemical reactions of the primary hydroxyl group,that all the activation reactions known for modifications of polysaccharides (SEPHADEX,SEPHAROSE) can be utilized virtually without any rest.The slightly hydrophobic character of the SPHERON matrix, which however may be eliminated to a great extent by glycosylation mentioned above,must of course be borne in mind in the applications.

Binding onto carriers activated with cyanogen bromide (ref.32) proceeds at alkaline pH values.If the binding reaction must take place in an acid solution (in respect to the activity of enzyme immobilized), NH_2-SPHERON (hexamethylenediamine modified) or COOH-SPHERON (ε -aminocaproic acid) can be employed,and the activation may be carried out with water soluble carbodiimide (ref.33).Hydrazide derivatives,diazonium salts and anhydride derivatives of SPHERON also yield active immobilized enzymes (refs.20,34).Similarly chymotrypsin and serumalbumin were successfully immobilized in the activation with benzoquinone (ref.35).By oxidizing covalently bonded saccharide molecules on the hy droxyethyl methacrylate matrix with $NaIO_4$,one can obtain reactive alde hyde groups which were utilized in the immobilization of lectins (ref. 31)and in the subsequent affinity chromatography of glycoproteins and synthetic glycosubstances,as well as in the immobilization of trypsin (ref.36).

A hydroxyethyl methacrylate copolymer activated on the inner pore surface with epoxypropyl functional groups (SEPARON E) seems to be very promissing carrier for affinity chromatography.Using oligopeptidic synthetic ligands,sorbents for the high performance separation of several proteinases from accompanying inactive proteins were obtained (ref.37).Epoxide activated carriers were used in theoretical studies dealing with the effect of the matrix on the denaturation of trypsin and chymotrypsin based on the measurement of thermodynamic data (ref. 38).They were employed also in the immobilization of selected heterocyclic compounds (ref.39),like imidazole,pyridine purine,histidine and their derivatives,for the selective adsorption of metalloporphyrines, metallophtalocyanines and chlorophyll and in the immobilization of heparin (ref.40).

Small spherical particles of SEPARON HEMA activated with epichlorohydrin and hexamethylenediamine were used in the immobilization of cel of Kluyveromyces lactis by means of glutaraldehyde with the beta-alanine spacer (ref.41).

Copolymers with monomers containing reactive functional groups or thei precursors

Of the precursors containing primary aromatic amino groups,the preparation (ref.42) and copolymerization (ref.43) of p-acetaminophenylethoxy methacrylate (APEMA)

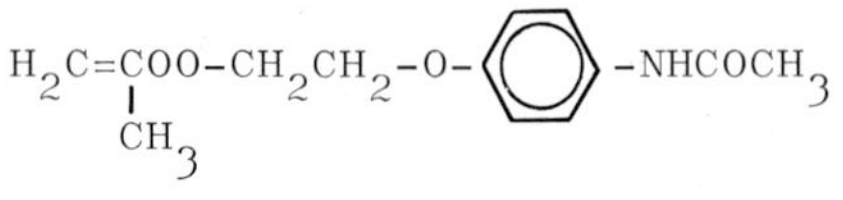

were studied in greater detail.The aromatic p-amino groups formed by the hydrolytic splitting of the acetate groups in the polymer can be utilized in a great number of activations such as diazotization,reaction with glutaraldehyde,phosgene,thiophosgene or mercuri acetate (ref 43).

Ternary copolymerization with reactive monomers containing functional groups were investigated in the preparation of glycidyl methacrylate copolymers (ref.44) and particularly of carriers with reactive p-nitrophenyl esters of methacrylic acid (NPAC-SPHERON) (ref.45).Using the immobilization of glycine and phenylalanine,the dependence of the binding capacity of NPAC-SPHERON on the length of the spacer and the dependence of hydrolytic and aminolytic reaction of p-nitrophenyl groups on pH were studied.A number of enzymes,such as chymotrypsin,trypsin and papain were bonded onto this carrier.

A disadvantage of macroporous copolymers with reactive monomers consists in the deep incorporation of reactive functional groups in densely crosslinked regions of the matrix,which greatly impedes and often makes quite impossible complete elimination of reactive functional groups after immobilization of the ligand.If the groups involved give rise to hydrophilic and readily solvated functional groups after hydrolysis,it can be expected that such carriers will be slowly mechanically degraded in contact with water or aqueous solutions.The microregions of the xerogel released by the hydrolysis and solvated with water may swell,and the arising local pressures tear the structure,thus laying bare further,still unreacted reactive groups.Such polymers can never be completely saturated.These phenomena were indeed observed with both copolymers under study,and their further investigation was therefore abandoned in our laboratory as nonpromising.

HIGH PERFORMANCE LIQUID CHROMATOGRAPHY ON SPHERON/SEPARON PACKINGS

Steric exclusion chromatography

The exclusion limit of macroporous hydroxyethyl methacrylate copolymers may be controlled by physical and chemical conditions of their synthesis between 10^4 and $5x10^6$ daltons at an advantageous inner pore volume to void volume ratio V_i/V_o = 1.1 to 1.3.The importance of this ratio (column permeability) for steric exclusion chromatography was explained by Regnier and Gooding (ref.46).A typical calibration dependence between the elution volume and logarithm of molecular mass is obtained using standard polydextrane series (Pharmacia Uppsala) shown in Fig.4,with water as the mobile phase.Dynamic properties of a typical HPLC column for size exclusion and hydrophobic chromatography packed with SEPARON HEMA 1000 can be seen in Fig.5.

Immobilization of glucose on the inner surface of the carrier gives a linear dependence of log M_w vers. V_e for a series of protein standards manufactured by Boehringer Mannheim (Fig.6). This demonstrates a pure mechanism of separation by steric exclusion.An example of a fast separation of a natural mixture of proteins obtained from human plasma using glucose modified carrier SEPARON 300 Glc is shown in Fig.7.

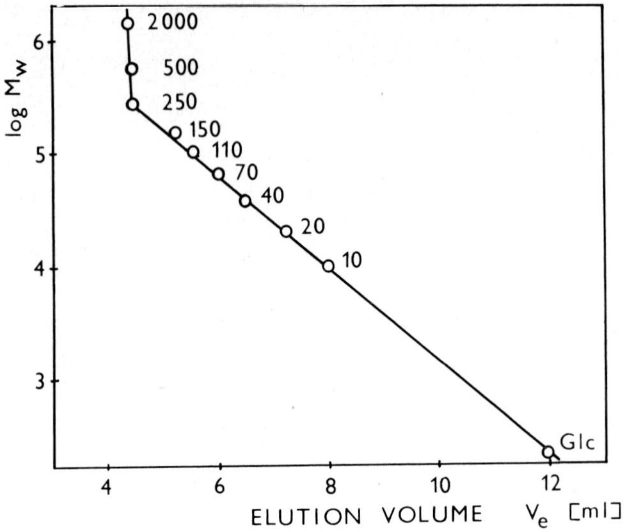

Fig.4. Calibration plot of SEPARON 300 Glc with polydextrane series. Mobile phase:water,column: 500 x 6 mm,particle diemeter: 10$\pm$2 μm, detector: RI

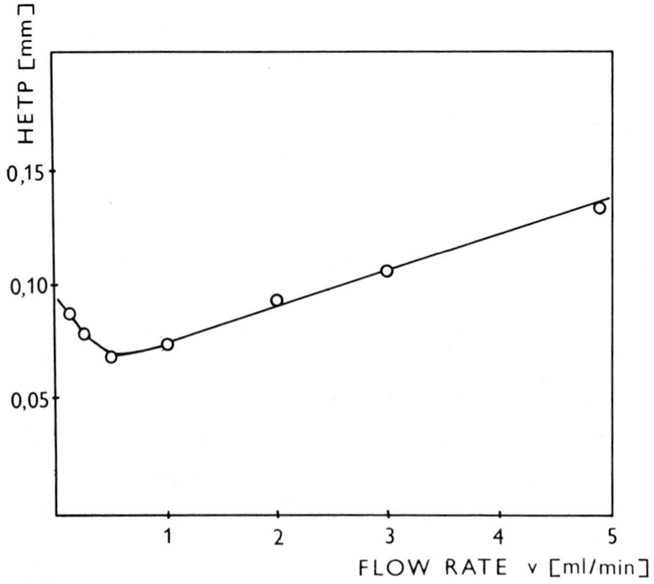

Fig.5. Separation efficiency of SEPARON 1000 column in dependence on the flow rate of mobile phase.
Column: 250 x 6 mm,particle diameter 10$\pm$2 μm,mobile phase:water, detector RI,solute:glucose.

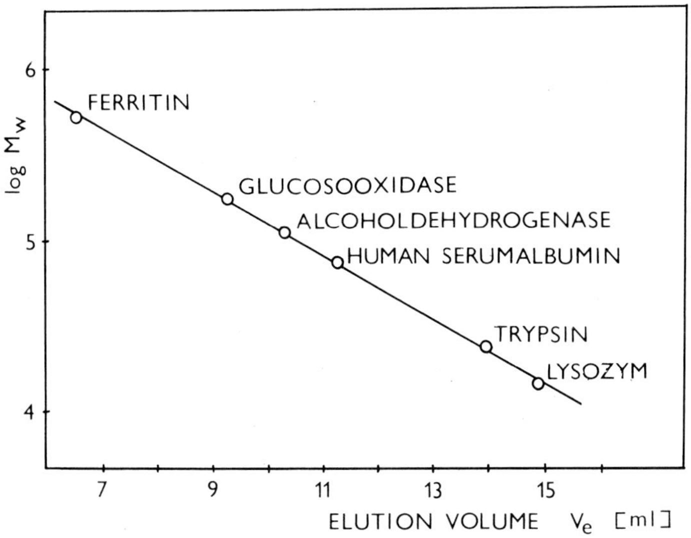

Fig.6. Fast size exclusion chromatography on SEPARON 300 Glc.Dependence of elution volume on molecular mass in a series of standard proteins. Mobile phase:water,column 500 x 8 mm,particle diameter 10±2 μm,detector:UV 235 (Pye Unicam LC-3).

Hydrophobic (adsorption) chromatography

Adsorption interactions and „normal phase" chromatography on unsubstituted SPHERON were studied first using a number of model aliphatic and aromatic hydrocarbons (ref.47) with n-heptane as the mobile phase.The retention behaviour of hydrocarbons as a function of their structure has confirmed the view that sorption interactions correspond rather to the mechanism of liquid-liquid chromatography.Hydrophobic „reversed phase" chromatography in aqueous solutions (ref.48) on unmodified SPHERON indicates the fact that numerous proteins are salted in onto the copolymer matrix at higher ionic strength values and desorbed at a lower ionic strength.Similarly to low-molecular weight chromatography with reversed phase,desorption is facilitated by an addition of alcohol.The problems investigated involved separation of peptides from the tryptic digest of lysozyme,separation of human serumalbumin,chymotrypsinogen and lysozyme, human serumproteins and of α , β and γ -trypsin (ref.49).

Sorbents with the covalently bound octadecyl phase (SEPARON 40-S) are now being intensively studied.Their main advantage in comparison to silica ODS sorbents consists in the possibility of using them also in media with pH 9-11.

Ion exchange chromatography

Separation of proteins by the method of high performance ion exchange chromatography on SPHERON/SEPARON carriers has been investigated in detail by O.Mikeš and his collaborators (refs.16-18,24,50-53).The results obtained in his laboratory provide conditions for the application of SPHERON ion exchangers not only on an analytical scale,but also in high performance preparative chromatography.

Compared with macroporous ion exchangers based on silica,a great advantage of organic carriers can be seen in their outstanding hydrolytic stability in a wide range of pH and ionic strength,and also in the considerably higher values of loading capacity.The exchange capacity values for small ions of various SPHERON ion exchangers are listed in Table 2.

TABLE 2

SPHERON ion exchange derivatives

Matrix: SPHERON 300	(specific pore volume 0.6 cm^3/g,mean pore diameter 25 nm)	
CM-SPHERON 300	carboxymethyl derivative	2.0 meq/g
P -SPHERON 300	phosphate	4.0 meq/g
S -SPHERON 300	sulfate	1.5 meq/g
DEAE-SPHERON 300	diethylaminoethyl	2.0 meq/g
QAE -SPHERON 300	quaternary amonium	2.0 meq/g
Matrix: SPHERON 1000	(specific pore volume 1.69 cm^3/g,mean pore diameter 37 nm)	
SPHERON C 1000	carboxyl	2.0 meq/g
SPHERON Phosphate	phosphate	3.5 meq/g
SPHERON S 1000	sulfopropyl	1.5 meq/g
SPHERON DEAE 1000	diethylaminoethyl	1.5 meq/g
SPHERON TEAE 1000	triethylamonium derivative	1.5 meq/g

All ion exchangers listed above can be regenerated with 2N HCl and 2N NaOH solutions.

Affinity chromatography

The predominating majority of reactions discussed in Chapter DERIVATIVES AND ACTIVATION have been utilized in the immobilization of ligands intended for affinity chromatography.Fundamental research concerning SPHERON use in this field has been carried out by J.Turková and coll.,whose attention was mainly concentrated on the investigation of the affinity chromatography behaviour of proteinases and their inhibitors(refs.23,32,34,37,53-58).Many papers have been dealing with the affinity chromatography separations in systems antibody-antigen (refs.59-64).Fast affinity chromatography of lectins (ref.30) was successfully performed on sorbents with covalently bonded saccharides,such as glucose,galactose,N-acetylglucosamine,N-acetylgalactosamine,mannose and fucose.Concanavalin A,immobilized in oxidized SEPARON 1000-Gal, acted as a selective sorbent for glycosylated polyacrylamide used as a model polymer and for glycoproteins (ovalbumin)(ref.31).

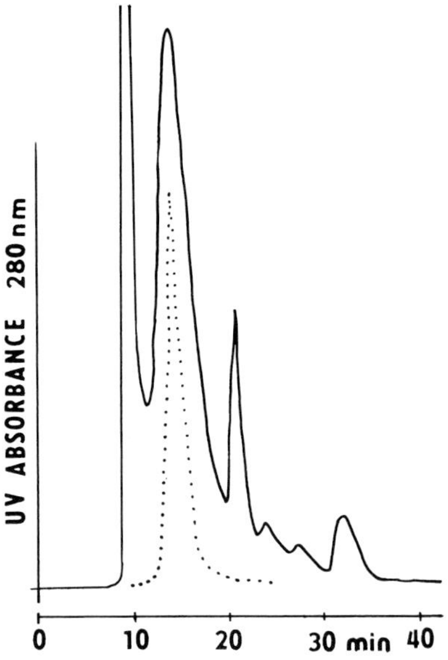

Fig.7. Preparative size exclusion chromatography of human serum proteins with indicated esterolytic activity (·····).
Column:1000x8 mm,SEPARON 300 Glc, d_p = 30 µm,mobile phase:isotonic solution,detector:UV 280 nm (Pye-Unicam).

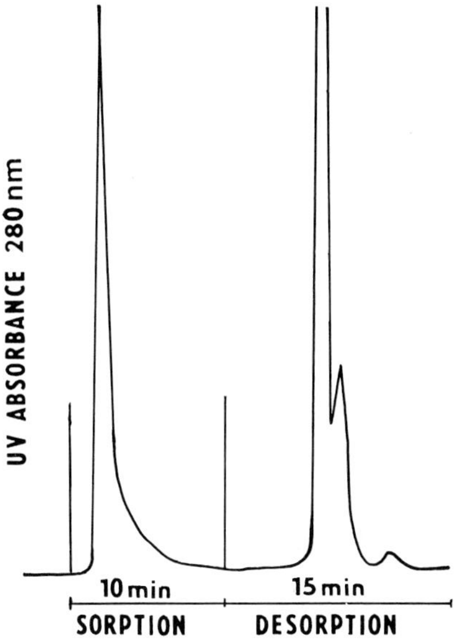

Fig.8. High performance affinity chromatography of trypsin preparation.
Column:100x4 mm,SEPARON 1000 modified with -aminocaproyl-L-Phe-D-Phe-OMe (ref.37),d_p = 10 µm
sorption:0.1M NaAc (pH 4.5)
desorption:0.1M NaAc + 1M NaCl
detector:UV 280 nm (Optronica)

In several Czechoslovak laboratories,high performance affinity chromatography separations are successfully performed using short columns packed with SEPARON HEMA 1000 carriers with covalently bonded ligands, according to the purpose of their application.Column length 100 mm,internal diameter 4 mm,flow rate of the mobile phase 0.5 - 1.0 ml/min, particle size of sorbents d_p=10$\pm$ 2 µm.An example of fast affinity chromatography of a partly autolyzed trypsin sample using a 280 nm UV detection is illustrated in Fig.8.

For purposes of high performance liquid chromatography in all working modes,but particularly for work with aggressive mobile phase and for the separations in the systems readily denaturated in contact with metal surfaces,high pressure,high performance glass cartridges have be developed in cooperation of Laboratory Instruments Works with Institu of Macromolecular Chemistry,Czechoslovak Academy of Sciences Prague. Chemically hardened glass tubes used for manufacturing of the cartrid makes possible working up to pressure of 40 MPa (400 atm) with visuall observed elution in the column.We believe that such glass cartridges protected mechanically with metallic jacket and packed with microparticular HPLC sorbents SEPARON find a great number of applications especially in the field of biochemical and clinical analysis.(ref.65).

REFERENCES

1 J.J.Van Deemter,F.J.Zuiderweg and A.Klinkenberg,Chem.Eng.Sci.,5(19 271.
2 J.C.Giddings,J.Chem.Phys.,31(1959)1462.
3 K.K.Unger,J.Schick-Kalb and K.F.Krebs,J.Chromatogr.,83(1973)5.
4 J.J.Kirkland,J.Chromatogr.,125(1976)231.
5 W.Haller,Nature (London),206(1965)693.
6 W.Haller,J.Chem.Phys.,42(1965)686.
7 C.Persiani,P.Cukor and K.French,J.Chromatog.Sci.,14(1976)417.
8 P.Roumeliotis and K.K.Unger,J.Chromatogr.,185(1979)445.
9 H.Engelhardt and D.Mathes,J.Chromatogr.,142(1977)311.
10 H.Chang,K.M.Gooding and F.E.Regnier,J.Chromatogr.125(1976)103.
11 T.Hashimoto,H,Sasaki,M.Aiura and Y.Kato,J.Chromatogr.160(1978)301.
12 P.Roumeliotis and K.K.Unger,J.Chromatogr.149(1978)211.
13 W.Mönch and W.Dehner,J.Chromatogr.,147(1978)415.
14 M.Rubinstein,Anal.Biochem.,98(1979)1.
15 J.Čoupek,M.Křiváková and S.Pokorný,J.Polymer Sci.,Symp.No.42(1973)1
16 O.Mikeš,P.Štrop and J.Čoupek,J.Chromatogr.,153(1978)23.
17 O.Mikeš,P.Štrop,J.Zbrožek and J.Čoupek,J.Chromatogr.,180(1979)17.
18 O.Mikeš,P.Štrop,M.Smrž and J.Čoupek,J.Chromatogr.,192(1980)159.
19 L.Lochmann,J.Čoupek and J.Trekoval,Czech.Pat.No.170019.
20 J.Hradil,J.Čoupek,M.Křiváková,J.Štamberg,A.Stoy and J.Turková,Czech Pat.No.167593.
21 O.Wichterle and J.Čoupek,Czech.Pat.No.179484.
22 J.Kálal,J.Čoupek,S.Pokorný,F.Švec and K.Bouchal,Czech.Pat.No.16826
23 J.Turková,Affinity chromatography,Elsevier,Amsterdam 1978.
24 O.Mikeš,P.Štrop,J.Zbrožek and J.Čoupek,J.Chromatogr.,119(1976)339.
25 Z.Slovák,S.Slováková and M.Smrž,Anal.Chim.Acta,87(1976)149.
26 Z.Slovák and J.Toman,Z.Anal.Chem.,278(1976)115.
27 Z.Slovák,M.Smrž,B.Dočekal and S.Slováková,Anal.Chim.Acta,111(1979)2
28 J.Hradil,J.Čoupek,M.Křiváková and J.Štamberg,Czech.Pat.No.159990.
29 K.Pecka (in preparation).
30 K.Filka,J.Čoupek and J.Kocourek,Biochim.Biophys.Acta,539(1978)518.
31 K.Filka,J.Čoupek and J.Kocourek,Protides Biol.Fluids,27(1979)375.
32 J.Turková,O.Hubálková,M.Křiváková and J.Čoupek,Biochim.Biophys.Acta 332(1973)1.
33 O.Valentová,J.Turková,R.Lapka,J.Zima and J.Čoupek.Biochim.Biophys. Acta,403(1975)192.
34 J.Turková,in Methods in Enzymology (Ed.K.Mosbach),Vol.44,Academic Press,New York,1976,p.66.

35 N.Stambolieva and J.Turková,Coll.Czech,Chem.Commun.45(1980)1137.
36 D.Vančurová,J.Turková,A.Frydrychová and J.Čoupek,Coll.Czech.Chem. Commun.,44(1979)3405.
37 J.Turková,K.Bláha,J.Horáček,J.Vajčner,A.Frydrychová and J.Čoupek, J.Chromatogr.,(in press).
38 J.Turková,L.V.Kozlov,L.Ya.Bessmertnaya,L.V.Kudryavtseva,V.M.Krasilnikova and N.K.Antonov,Bioorg.Khim.,6(1980)108.
39 M.Kühn,P.Mohr and J.Čoupek,DDR Pat.No.136269.
40 M.Rybák and J.Čoupek,Czech.Pat.Appl.No.PV-4665-80.
41 V.E.Gulaya,J.Turková,V.Jirků,A.Frydrychová,J.Čoupek and S.N.Ananchenko,Eur.J.Appl.Microbiol.Biotechnol.,8(1979)43.
42 O.Wichterle,Czech.Pat.Appl.No.PV-3574-72.
43 M.Jelínková,S.Pokorný and J.Čoupek,Angew,Makromol.Chem.,52(1976)21.
44 F.Švec,J.Hradil,J.Čoupek and J.Kálal,Angew.Makromol.Chem.,48(1975)135.
45 J.Čoupek,J.Turková,O.Valentová,J.Labský and J.Kálal,Biochim.Biophys. Acta,481(1977)289.
46 F.E.Regnier and K.M.Gooding,Anal.Biochem.,103(1980)1.
47 M.Minárik,R.Komers and J.Čoupek,J.Chromatogr.,148(1978)175.
48 P.Štrop,F.Mikeš and Z.Chytilová,J.Chromatogr.,156(1978)239.
49 P.Štrop and D,Čechová,J.Chromatogr.,207(1981)55.
50 O.Mikeš,Int.J.Peptide Protein Res.,14(1979)393.
51 O.Mikeš,P.Štrop and J.Sedláčková,J.Chromatogr.148(1978)237.
52 O.Mikeš,J.Sedláčková,L.Rexová-Benková and J.Omelková,J.Vhromatogr., 207(1981)99.
53 J.Turková,Enzyme Engineering,4(1978)451.
54 J.Turková,K.Bláha,O.Valentová,J.Čoupek and A.Seifertová,Biochim.Biophys.Acta,427(1976)586.
55 J.Turková and A.Seifertová,J.Chromatogr.,148(1978)293.
56 J.Turková,S.Vavreinová,M.Křiváková and J.Čoupek,Biochim.Biophys.Acta, 386(1975)386.
57 J.Turková,O.Valentová and J.Čoupek,Biochim.Biophys.Acta,420(1976)309.
58 H.Malaníková and J.Turková,J.Solid Phase Biochem.,2(1978)237.
59 T.Linnas,R.Mikelsaar,H.Nutt and O.Kirret,Eesti NSV Tead.Akad.Toim. Keem.,27(1978)46.
60 H.Tlaskalová,L.Tučková,M.Křiváková,J.Rejnek and J.Čoupek,Immunochemistry,12(1975)801.
61 H.Tlaskalová-Hogenová,J.Čoupek,M.Pospíšil,L.Tučková,J.Kamínková and P.Mančal,J.Polymer Sci.,Symp.No.68(1980)89.
62 J.Rejnek,L.Tučková,J.Trávníček,H.Tlaskalová and F.Kovářů,Molecular Immunology,17(1980)65.
63 L.M.Sirakov,J.Barthová,T.Barth,S.P.Ditzov,K.Jošt and I.Rychlík,Coll. Czech.Chem.Commun.,40(1975)775.
64 J.Vaněčková,J.Barthová,T.Barth,I.Krejčí and I.Rychlík,Coll.Czech.Chem. Commun.,40(1975)1461.
65 P.Špaček,S.Vozka,J.Čoupek,M.Kubín,J.Voslář and B.Porsch,Czech.Pat. Appl.No.PV-4635-81.

T.C.J. Gribnau, J. Visser and R.J.F. Nivard (Editors),
Affinity Chromatography and Related Techniques

CHEMICAL REACTIVITY OF NATURAL AND SYNTHETIC POLYMERS IN RELATION TO THE SYNTHESIS OF AFFINITY SUPPORTS

D C SHERRINGTON

Instituto di Chimica Organica Industriale, Universita di Pisa, Italy[x]

ABSTRACT

The conformational and structural properties of polymer supports are described with reference in particular to crosslinked species. Approachs to chemical modification are outlined and the situation involving linear macromolecules described as a reference point. Compatibility problems are highlighted and then the various models describing the reactivity of resin-bound functional groups are elucidated. In particular pseudo-homogeneous systems are discussed and those involving some diffusion control by the matrix. Cross reference is made to the different structural types of resin available. Chemical effects on the reactivity of bound groups are described. General situations where changes occur in the activation parameters defining the reactivity are presented, along with more specific neighbouring group effects and site isolation effects.

STRUCTURE OF MACROMOLECULAR SUPPORTS

In general organic macromolecules form the basis of most affinity supports. A linear macromolecule is a long chain species, and an individual chain adopts, almost exclusively, a random coil conformation in solution (Fig.1a). As a result of specific electronic and/or stereochemical factors natural polymers such as polypeptides and polysaccharides may retain (from the solid state) a fixed structural conformation, such as a helix, within the overall random coil geometry (Fig.1b).

The author acknowledges the receipt of a Senior Research Fellowship from the Ciba-Geigy Trust.

x : Permanent Address:

Department of Pure and Applied Chemistry, University of Strathclyde, Glasgow.

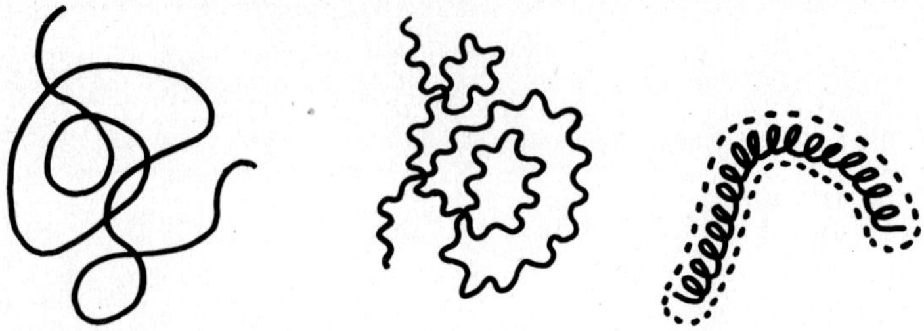

a) b) c)

Fig.1. Conformations of isolated linear polymer chains a) random coil b) random coil with some internal structure c) worm-like structure.

Fig.2. Interpenetrating polymer coils in the solid state and in solutions of concentration>∽2% by weight.

Indeed the flexibility of such macromolecules may be sufficiently limited to produc a wormlike chain structure rather than a full random coil (Fig.1c).(ref.1). A collection of linear macromolecules most often consists of a set of interpenetrat polymer coils in the solid state, and even in quite dilute solution in a good solve polymer coils still overlap and interact (Fig.2). Again strong intramolecular forces can give rise to more precise conformational structures. Polymer coils readily expand in a good solvent and contract in a poor one, finally precipitating in a non-solvent. Similar changes can occur in the case of hydrophilic polymers dissolved in water when the pH or ionic strength is changed.

If the macromolecules are formed in the presence of a crosslinking agent, or are subsequently crosslinked in a post-polymerization process, all the random coils are effectively interconnected to form an infinite network. Such a system can no long form a true molecular solution and may be regarded as insoluble in a strict thermodynamic sense. If during the crosslinking process individual polymer segments remain in a fully solvated state, ie. a solvent is present which is compatible with both monomer and polymer, then a fully expanded network is formed, and thermodynamically this constitutes a true gel. In some instances monomer itself will function effectively as the good solvent, until the level of conversion depletes its availability. Such gel networks or 'gel-type' crosslinked supports allow a good solvent to be reversibly removed and re-introduced, and the gel state re-established quite readily (Fig.3). Polystyrene resins prepared in the presence of toluene, and poly(acrylamide) or poly(hydroxyethylmethacrylate) prepared in the presence of water fall into this category.

If during the crosslinking process, or indeed during polymer chain formation, a phase separation or desolvation of polymer chains occurs, then irreversible collapse and chain clustering occurs. The network formed under these circumstances is no longer a gel in a true thermodynamic sense. Removal of solvent, strictly the diluent, from such a system more often than not yields a residual network of aggregated, or indeed crystallized, polymer chains, with a permanent system of macropores (Fig.4). Such supports are often termed 'macroporous' or 'macroreticular'.

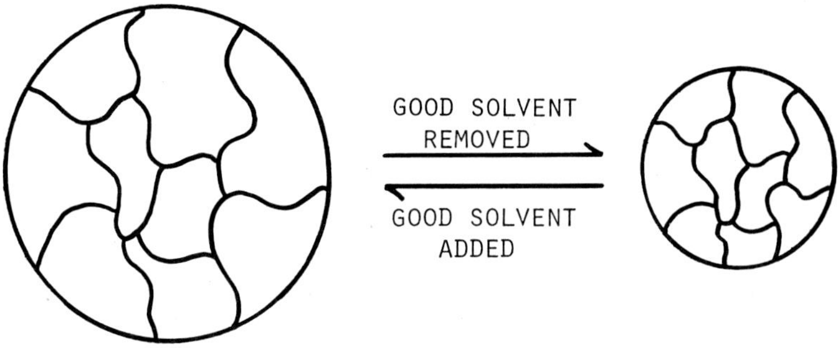

Fig.3. Swelling behaviour of gel-type crosslinked polymer supports.

Fig.4. Macrorecticular polymer support showing rigid permanent macroporous structure. /// = highly entangled (or crystallised) regions.

The total pore volume and pore size distribution will depend upon the type and relative volume of diluent originally present, on the degree of crosslinking of individual polymer chains, and on the reaction conditions such as the temperature. The macropores are often accessible to both solvents and non-solvents alike and in general the whole structure is not susceptible to the dramatic changes observed wit gel-type resins when the nature of the surrounding medium is changed. Polystyrene resins prepared in the presence of aliphatic hydrocarbons or alcohol diluents, and polyacrylamide resins prepared in the presence of methanol/water mixtures represent examples of this type. Agarose 'gels' also can be regarded as fulfilling this description, whereas dextran gels relate more closely to the former true gel system

These two pictures do of course represent ideal and perhaps extreme models, and many preparations give rise to structures with intermediate characteristics. Howev in this paper it is perhaps convenient to think in these terms in order to clarify experimental observations. Further details are available in references 2 and 3.

PHYSICAL CHARACTERISTICS OF SUPPORTS

Dry solid supports can be conveniently characterised in terms of their total surface area (internal and external), total pore volume and average pore diameter (or pore size distribution). Commercially available materials have surface areas typically in the range $\sim$ 5-600$m^2 g^{-1}$, pore volumes of $\sim$0.3-4 ml g^{-1} and mean pore diameters of $\sim$20-2000A. It is also important to appreciate that these physical parameters are not independent of each other but rather are closely interrelated. Assuming some simple model for the pore geometry eg. cylindrical symmetry, then the surface area, S, pore volume, P, and average pore radius, r, are related by the sim geometrical equations : $P = n\pi r^2 l$ and $S = n2\pi rl$, where n and l are the number and average length of pores respectively; and nl is the effective total pore length. These are not in themselves important equations because such a model is over simple, however, they do illustrate some important generalities. Thus, for a given mass of a support (ie. a limited maximum for P) as the mean pore radius is increased the number of pores is likely to fall, as therefore is the total interior surface area, S. Conversely as the total interior surface area rises, the number of pores is also likely to rise, though each pore is likely to have a reduced radius. The optimum choice of physical characteristics for a particular application often therefore involves a compromise. In affinity chromatography emphasis is generally placed on the availability of large pores, and this is one reason why the agarose 'gels' with molecular weight exclusion limits of ∼10,000,000's are particularly popular. It is worthwhile bearing in mind, however, that selection of materials with such large pore diameters is likely to limit the other physical parameters, and for some applications this might not be totally desirable.

Gel-type supports usually have relatively small pore diameters and large effective surface areas. They are advantageous in some circumstances because of their high loading capabilities, up to ∼10mmol g^{-1}. Macroporous supports have large pore diameters but relatively small surface areas. Chemical modification of these species occurs largely on the pore surfaces, and generally the highly crosslinked and entangled (or crystalline) macromolecular chains are not readily available for functionalization. Their loading capabilities are thus lower, but they do vary a lot and can be as high as ∼3mmol g^{-1}, but are often much lower than this. For affinity chromatography applications this is of course more than adequate.

CHEMICAL MODIFICATION OF SUPPORTS

Chemical modification of a macromolecular support may be achieved by i) appropriate copolymerization methods in formation of the support (ref. 4,5), ii) post-polymerization reactions on pre-formed supports (ref. 6,7), and iii) a combination of these. Route i) is exemplified by the copolymerization of acrylamide with labile acrylates such as the N-phthalimidyl derivative. The latter group is easily displaced in any subsequent reactions with a nucleophile (reaction 1). Route ii) is typified by the activation of pre-formed polysaccharide supports with cyanogen bromide (reaction 2). These of course are classic methods of achieving immobilization and indeed there are a wide range of pre-functionalized supports available commercially for exploitation.

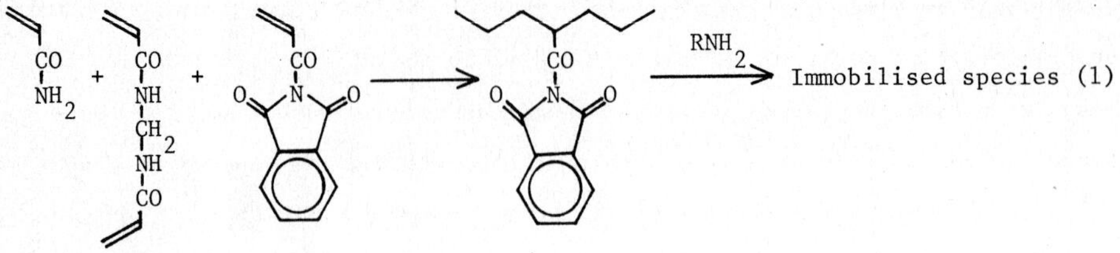

However, it is worth considering how these two approachs may lead to quite differen situations with regard to the reactive group introduced. Route i) has the advanta that the structure and number of functional groups is generally well defined, and therefore the chemistry of immobilization reaction also is clearly defined. In som cases even the distribution of functional groups might be estimated from the react-ivity ratios of the comonomers employed. On the other hand some of the groups so introduced may be relatively inaccessible for any subsequent chemical reaction, and in the case of macroreticular supports this possibility will be increased. With route ii) the groups introduced are generally accessible for further exploitation but their structure, concentration and distribution are likely to be uncertain. It is notoriously difficult to make an adequate structural analysis of any functionali bound to a three dimensional matrix, and this shortcoming remains one common to all applications of supports. Advances in such structural analysis continue to be mad but even today very often this approach relies to some extent on faith that a particular modification has been accomplished. In spite of this, the important advantage that a wide variety of pre-formed supports of high quality are readily available makes this the approach most often used, but the alternative route is worth bearing in mind if detailed knowledge of the chemistry of the immobilization process becomes important.

REACTIVITY OF LINEAR POLYMERS

In principle any chemical reactions between two small molecules in solution should be reproducible when one of the molecules is covalently bound to a macro-molecular backbone, providing the other remains free to execute its normal solution translational motion. Thus individual segments of a linear macromolecule should behave chemically in a similar manner to some appropriate small molecular analogue (ref. 8). Indeed providing the polymer dissolves in the reaction medium in a relatively expanded conformation then this generality holds true. Under these circumstances it is possible to make a quantitative kinetic analysis of such systems bearing in mind the constraint that the bound functional groups are restricted to the volume occupied by the expanded polymer coils. If the overall concentration of polymer molecules is such that the coils interact, then the bound groups effectively occupy the total solution volume, and their concentration can be calculated accordingly. On the other hand if the overall concentration of polymer molecules is sufficiently low that each expanded coil moves independently of the others, then the bound groups occupy a volume equal to the total volume of polymer coils, and significantly less than the total solution volume, the local concentration of bound groups must therefore be based upon this restricted volume. The rate constants for local reactions may therefore be deduced, and from temperature dependence studies, activation parameters may be calculated and a comparison made with a low molecular weight model reaction. Conclusions concerning similarities or differences in mechanisms can then be drawn, and detailed analyses along these lines have been made for many reactions catalysed by linear polybases (ref. 9), for reactions involving polyelectrolytes (ref. 10) and for a number of polymer-supported copper complexes used as oxidation catalysts (ref. 11).

COMPATIBILITY FACTORS

In the case where the polymer coils adopts a tightly coiled conformation and is relatively incompatible with a reacting small molecule then the latter may find considerable difficulty in achieving sufficiently adequate penetration for effective reaction with an immobilised group, even though the polymer is not crosslinked. For example with acetylated polystyrene, oxidation with sodium hypochlorite proceeds slowly where the degree of acetylation is less than ~ 30%. Beyond this value the linear polymer is sufficiently expanded in the aqueous oxidizing solution for rapid oxidation to occur.

Similar effects can arise with hydrophilic supports reacting with hydrophobic molecules. Thus potassium cellulosate suspended in acetonitrile solvents reacts efficiently with polar molecules such as acryloyl chloride and chloromethyl salicilaldehyde but much less efficiently with non-polar ones such as chloromethyl styrene (ref. 12) (reaction 3). A high loading of the glucose residues can also be achieved on reaction with chloroacetylchloride whereas with 11-bromoundecanoyl chloride only poor loadings result (ref. 13).

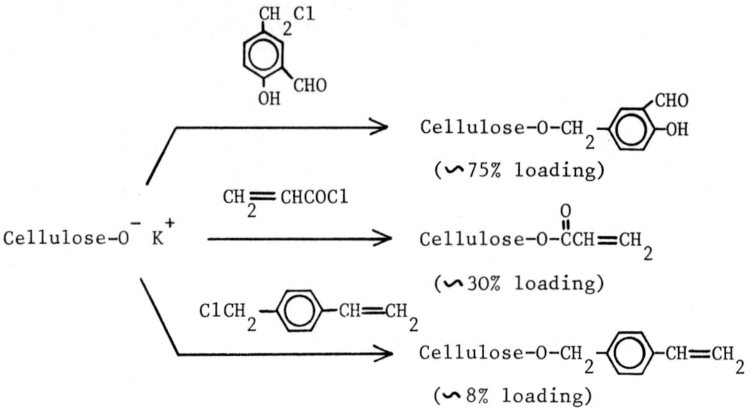

(3)

The latter are probably adequate for affinity chromatography purposes but are total unsatisfactory for many other applications. Almost certainly these low yields are associated with poor compatibility between the support and the more hydrophobic reagents since the inherent reactivities of the functional groups involved are more than adequate. With the long chain bromoacylchloride, however, specific steric factors may also play a significant role.

REACTIVITY OF CROSSLINKED POLYMERS

With crosslinked macromolecular supports the situation can be even more complica and a number of different possibilities can arise.

Pseudo-Homogeneous Systems

These represent perhaps the simplest situation and one which is most often desi able. Here the polymeric network plays no significant role other than to limit t volume in which the bound group is found and molecularly the picture is very simil to the case of linear polymers. Solvent within the polymer matrix may be conside as 'homogeneous' phase within which reaction occurs and wherein reactants and produ are in rapid thermodynamic distribution equilibrium with the external solution, the latter acting merely as an inert reservoir and functioning only to supply reactant and store any free products and/or by-products formed (Fig.5a). The rate controll process is the same as in the corresponding truly homogeneous model reaction, and bimolecular rate constant, k_H, also the same. For N_A molecules of a bound group A

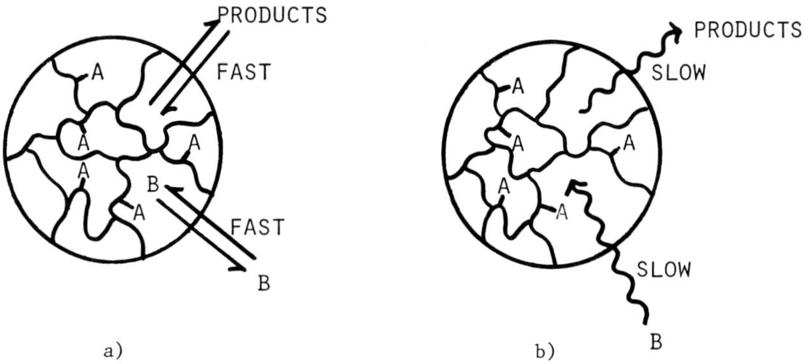

Fig.5. Reaction models for a bound group A reacting with a mobile molecule B. a) pseudo-homogeneous system b) diffusional control by the matrix.

on a polymer matrix of volume, v, reacting with N_B molecules of a mobile species B, in a solution volume, V, (where $V \gg v$) and also distributed uniformly through, v, the local rate of reaction within the support, R_L, will be $R_L = k_H\ (N_A/v)\ (N_B/V)$. In practice volume v is fed by B molecules from the bulk volume, V, into which product and/or by-product molecules emerge. Furthermore supported reactions are generally monitored by following the appearance or disappearance of a species from the bulk volume, V. Hence the experimentally observed rate in the bulk volume, R_B, will be given by $R_L \times \frac{v}{V}$ ie. $R_B = k_H\ (N_A/v)\ (N_B/V) \times \frac{v}{V}$ or $R_B = k_H\ (N_A/V)\ (N_B/V)$. This of course is exactly the same expression for two unbound molecules A and B reacting together in a solution volume, V. Hence in these special circumstances the immobilisation of A on the polymer matrix has no effect on the reactivity and the system behaves kinetically as though it were homogeneous.

Lightly crosslinked gel-type polymer matrices highly swollen with a good solvent are most likely to behave in this way (ref. 14,15) but there are also some experimental data from reactions using macroporous species which fits this model (ref.16). A particularly good example of a reaction which conforms to this picture is the use of resin-supported linear oliogethers as solid-liquid phase transfer catalysts (reaction 4) (ref. 17). If the supported system is truly pseudo-homogeneous then the rate of reactions should not be dependent on the particle size of the support

$$\text{n-BuBr} + K^{+}\ OPh^{-} \xrightarrow[\text{toluene}]{\text{(P)}-CH_2(OCH_2CH_2)_nOR} \text{n-BuOPh} + K^{+}Br^{-} \qquad (4)$$

nor its structure, and for equivalent amounts of reagents, reaction rates should be the same as a model homogeneous reaction. This is clearly the case in this particular instance as shown by the data in the Table. In reactions where there can be no cooperative effects involving the bound group, rates should also be independent of the loading of the group on the support provided the total number of groups is constant.

TABLE

Effect of matrix type and particle size on rate of reaction[a)]

Catalyst (loading)	Particle size(μm)	Initial rate x 10^5 ($M^{-1}s^{-1}$)	Loss of nBuBr(%)[b)]
Ⓟ X2E3Me(61%)	35-75	10.0	93
Ⓟ X2E3Me(61%)[c)]	≪35	10.5	98(2.5h)
Ⓟ X2E3Me(70%)	250-500	10.0	99(2.5h)
Ⓟ X10E3Me(71%)	100-200	11.0	92
Ⓟ X42E3Me(29%)	15-20	10.5	95
HOE3Me(-)	-	9.1	98

a) $[nBuBr]_{initial}$ = 0.24M; K^+OPh^- = 2.0mmol; catalyst = 1.16mmol of O.
b) After 3h unless specified. c) Crushed sample of first catalyst.
(Reproduced from W.M. Mackenzie and D.C. Sherrington, Polymer, 21 (1980) 791, by permission of the publishers, IPC Business Press Ltd. ©)

In reality the assumptions which are necessary to arrive at this result are seld all valid simultaneously. For example if the free molecule, B, is not distributed uniformly between the support phase, v, and the bulk solution, V, ie. if the distribution coefficient, λ , is not unity, then the local concentration of B within the support will be modified, as will the local rate of reaction and also the bulk measured rate. Favourable interactions between the support phase and B may enhance the local concentration of the latter giving rise to a rate enhancement (ref. 18). Conversely where the support and B are relatively incompatible the distribution coefficient of B will fall below unity, and the reaction rate be reduced accordingl Since the polarity of the support phase itself may change significantly during reaction of a bound group so may these accompanying distribution effects, leading to highly complex situation. Likewise changing the reaction solvent (ref. 19), and the pH or ionic strength in the case of aqueous solutions, may also shift any distribution factors.

Diffusional Effects

Very commonly the reactivity of a polymer-immobilized group is substantially altered relative to the situation in homogeneous solution and it is not unusual to find that experimental data varies with the particle size of the support , its porosity and crosslink ratio and with the size of an incoming reacting molecule. Under these circumstances it is usual to suggest that the polymer matrix inhibits the reaction in some rather ill-defined manner.

An important criterion for 'diffusion control' of reactions on polymer supports is an inverse rate dependence on particle size. When diffusion takes place from the geometric exterior of a support particle to its interior (Fig. 5b), then the rate of reaction, R, will be proportional to the total surface area of particles ie. $R \propto 4\pi r^2 N$, where N is the number of particles of average radius, r. The mass of a given sample of a support, m, is of course $4/3\pi r^3 \rho N$, where ρ is the density of the material, ie. for a fixed m; $N \propto 1/r^3$. Hence $R \propto 4\pi r^2 \times 1/r^3$, ie. $R \propto 1/r$ (ref. 20). This process may be pictured as diffusion through a uniform gel network of molecular dimensions, and involving largely local or microenvironmental motions of the macromolecular backbone. Furthermore for a fixed mass of the support rates of reaction can become virtually independent of the loading of the matrix bound group (ref. 21).

One mathematical model which has been developed for reactions involving simultaneous diffusion and chemical reaction and which is particularly convenient uses the concept of an 'effectiveness factor', η, (ref. 22,23). The model assumes that diffusion of a reacting molecule can be characterised by an effective diffusivity D, which is constant throughout a support particle, and that chemical reaction is represented by a true rate constant, k, again constant through the support. For a spherical material radius, r, the normal bimolecular rate, R, is modified by the 'effectiveness factor', η, this being defined as the observed rate divided by the rate which would result if diffusion were so rapid as to have no influence at all. In this instance it can be shown that $\eta = (3\emptyset)\ (1/\tanh \emptyset - 1/\emptyset)$, where $\emptyset = r\ (kD)^{\frac{1}{2}}$. This model has been very successfully applied in some ion exchange resin catalyses (ref. 22).

In the case of gel-type supports, particularly those of higher crosslink ratio or density, this picture seems to be an appropriate one and might be regarded as being associated with inefficient use of immobilised groups deep in the interior of support particles. With macroporous materials, however, where there exists a significant permanent pore volume and a discrete interior surface which is often readily accessible to reacting molecules, poor correlation with this model might arise because of the difficulty in assigning a realistic value for, r, the radius of the support particle, or more accurately the total distance over which diffusion can occur. Indeed in this case to picture diffusion as occurring through a uniform gel matrix would be misleading. Under these circumstances the rate dependence on the inverse of the particle size may also be expected to be invalid (ref. 21). Some of the bound groups, A, may be located on the interior surface of the support and be readi accessible to a small reagent, B. Others may lie below the interior surface and reaction of these is likely to be limited by a diffusional process similar to that already described for gel-type species (Fig. 6).

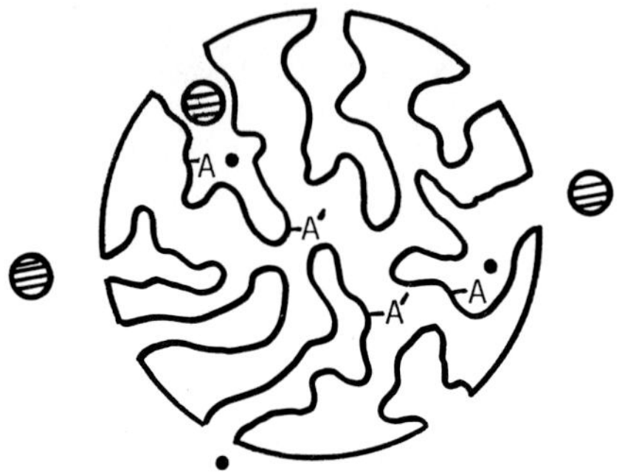

Fig.6. Reactivity of functional groups A bound to a macroporous support. A = surface sites, A' = buried sites, • = small reagent B, ⊜ = large reagent B.

Quite clearly in this case to describe the situation with a simple equation applicable to all sites would be inappropriate and misleading. Where the reagent molecu B, is very large even reaction with bound groups on the interior surface may be inhibited, in which case the simple diffusional model may again become sufficiently adequate to fit the experimental data. In these circumstances, however, the proce would be diffusion through macroscopic pores rather than through a molecular matrix and might involve long range cooperative motion of polymer support chains and not just local motions, and be characterised by a quite different diffusion constant,

To add further to the complexities, as A and B react, the nature of the support itself might change and so therefore may the macromolecular conformations. Pore volume and pore radius may also change and hence the assumption that any diffusion coefficient, D or D^1, is constant may become invalid. In some specific cases attempts have been made to treat such systems mathematically (ref. 23) but the meth are probably not generally applicable, and indeed each system is more likely to require an independent analysis.

The reactions of pyridine residues immobilised on gel-type resins with alkyl halides of varying size (reaction 5) display many of the features of matrix diffusion control.

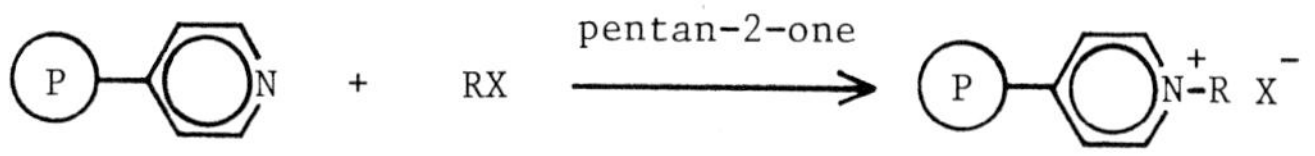

(5)

The relative rate depends on the crosslink ratio of the support, on the size of the reacting halide and is independent of the loading of pyridine groups for a given mass of the support (ref. 21). In spite of this, however, examination under the optical microscope of sectioned samples of materials partially reacted with an alkyl iodide shows the characteristic yellow colouration of the pyridinium iodide charge transfer absorption band across the entirety of the support, with perhaps some suggestion of a higher intensity near the support surface. In addition the simple inverse dependence of rate on particle size is not obeyed. Thus even with apparently uniform gel matrices the picture of diffusion occurring simply from geometric exteriors to the interior is too simple a one in some cases, and shows how these models must be applied with caution.

Hetergeneous Models

Macroporous supports in which the immobilised group is located specifically on the surface of pores might be appropriately represented in terms of a true hetergeneous model (ref. 24,25). Reaction rates in this case would be controlled by the extent of surface coverage and some appropriate adsorption isotherm would be required to relate the surface coverage to the concentration or pressure of reacting molecule (ref.24). Such descriptions have not been widely used for liquid phase reactions but have been applied with some success in gas phase reactions involving immobilised species (ref. 26).

Conceptually, however, there is a close similarity between this heterogeneous model for gas phase reactants and the Michaelis-Menton kinetic scheme describing the reactivity of enzymes. The latter has been widely applied to quantify the reactivity of macromolecular bound species in liquid phase reactions (ref. 9,11), but its application is limited mainly to those systems where the immobilised species is a catalyst present in small concentrations in relation to a reactive substrate. Perhaps even more importantly it can only be applied where the system is effectively homogeneous, and for this reason its use in synthetic systems has been limited largely to linear macromolecular species.

CHEMICAL EFFECTS

So far we have dealt essentially with physical factors which can influence the overall reactivity of a bound group, and have regarded the inherent or intrinsic chemical reactivity to be unaffected by the immobilisation process. There are, however, a number of specific chemical effects which can arise but before these can be characterised and quantified all physical effects must be adequately accounted fo first.

Changes in Activation Parameters

The inherent reactivity of a chemical function with a second molecule is defined by its activation parameters deduced from the temperature dependence of intrinsic rate constants. When the chemical function is immobilised on a matrix, on reactio with a free molecule, it might be expected that a greater loss of internal degrees o freedom might occur in forming the transition state for the reaction, than might occur when both species are completely mobile, ie. the transition state might be mo 'frozen' when the chemical function is an immobilised one. Under such circumstanc their would be a greater loss of entropy. Thus the entropy of activation would be more negative, and the intrinsic rate constant characterising the reaction would be reduced. Furthermore, reactant molecules with many internal degrees of freedom would be expected to be most affected, and this seems to be the case in practice (ref. 27). Conversely favourable specific adsorption of a reacting molecule in the vicinity of an immobilised group may serve to reduce the activation energy for reaction sufficiently to overcome any unfavourable entropy effects, in which case the inherent reactivity might rise.

Neighbouring Group Effects

Perhaps somewhat more obvious and equally important in reactions involving macro molecular-bound groups are a number of potential neighbouring group effects. The simplest example of these is the ionisation, ie. the acidity, of bound carboxylic acid groups. Qualitatively we can see that the ease of removal of a proton from an acid group will decrease as the existing degree of ionisation of other groups increases, simply because the polymeric matrix become progressively more negativel charged. In fact if K and K_o are the ionisation constants for the polymeric acid and an analogous monocarboxylic acid respectively, K/K_o can be as low as $\sim 10^{-4}$ (ref. 28). Thus the acidity of a carboxylic acid group flanked by two unionised groups (I) should be higher than one flanked by one or two ionised groups, (II) and (III), and this seems to be the case in practice (ref. 29).

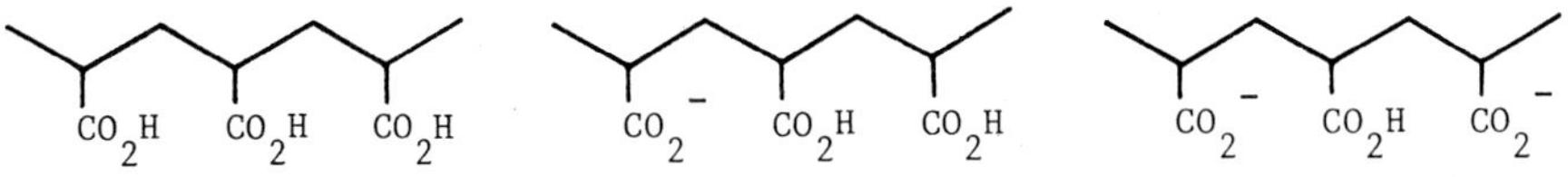

(I) (II) (III)

The neighbouring groups need not be of the same type. Thus the acidity of a carboxylic acid group and a phenol residue can be mutually influenced by hydrogen bond formation between the two (reaction 6). This appears to be an important relationship in some enzymes the presence of the phenol group enhancing the acidity of the carboxylic acid; while its own acidity is simultaneously reduced (ref. 30).

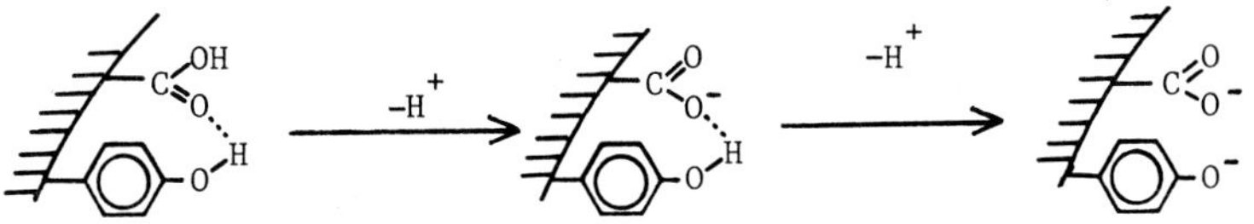

(6)

Closely related phenomena arise in the quaternization of polyvinylpyridines. Three rate constants k_1, k_2 and k_3 are required to quantify these systems (ref. 31). k_1 refers to a pyridine residue flanked by two neutral pyridine groups, while k_2 and k_3 refer to a residue flanked by one and two quaternised groups respectively. The relative values of these can be rationalized on simple electrostatic grounds as in the case of the ionisation of carboxylic acid groups, the outcome of which depends upon whether the quaternizing alkyl halide is itself a neutral or a charged species (ref. 32).

In many cases the term 'neighbouring' must be interpreted very loosely. It may indeed refer to a group on an adjacent polymer segment; whereas it may also refer to a group which finds itself in close proximity simply as a result of the conformation adopted by the macromolecular chain. Such effects therefore need not be limited to highly loaded macromolecular matrices and almost certainly arise very much more often than is generally recognised. Cooperative effects of randomly bound groups have been characterised in detail in the case of the reaction of many immobilised bases.

Isolation Effects

These are somewhat as a corrollary to the previous situation and arise in the main when the loading of an immobilised group is deliberately restricted. Such groups might be considered as analogous to the situation of 'infinite dilution' in a homo geneous solution, and as such, interaction between these groups either directly or via a second mobile is likely to be inhibited. This indeed is more often than no the prevailing situation in affinity chromatography. 'Site isolation' of this ty is also very important in the generation of immobilised coordinately unsaturated transition metal complexes such as titanocene (ref. 33) which have very limited lifetimes when formed as free species in solution. It also allows the selective single binding of symmetrical polyfunctional molecules to an immobilised group. A example of this is the selective monoprotection of symmetrical diols by an immobil trityl chloride residue (ref. 34) (reaction 7). The free unreacted hydroxyl grou (IV) can subsequently be further elaborated and the protected hydroxyl group fina regenerated, to produce overall a specific single ended modification of the symmet diol.

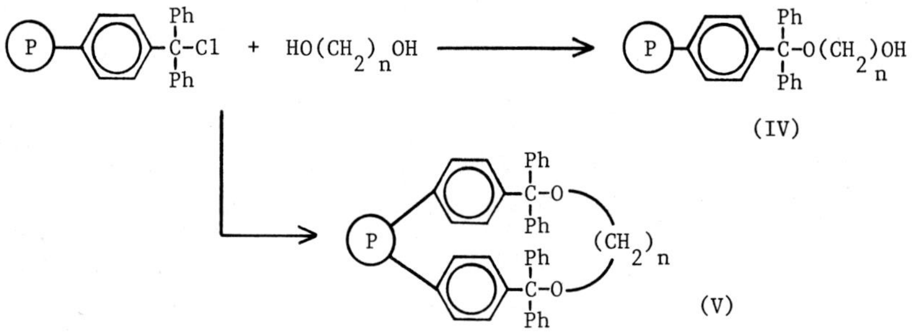

(7)

Providing the loading of immobilised group is kept below ∽ 0.5mmol per gram then satisfactory 'site isolation' can be achieved, but inevitably some 'double binding (V), of bifunctional substrates does occur.

This type of reaction is of course used routinely in the generation of 'spacer arms' in affinity applications (reaction 8) (ref. 35) and while the loading levels are such that site isolation appears to be guaranteed, the distribution of reactiv sites is crucial in controlling the liklihood of 'double binding' and this is more often than not an unknown factor. The occurrence of 'double binding' will also r as the length of the potential spacer group is increased. To what extent side re actions such as these do occur is not clear, but the possibility of such contribut and their effects should be borne in mind.

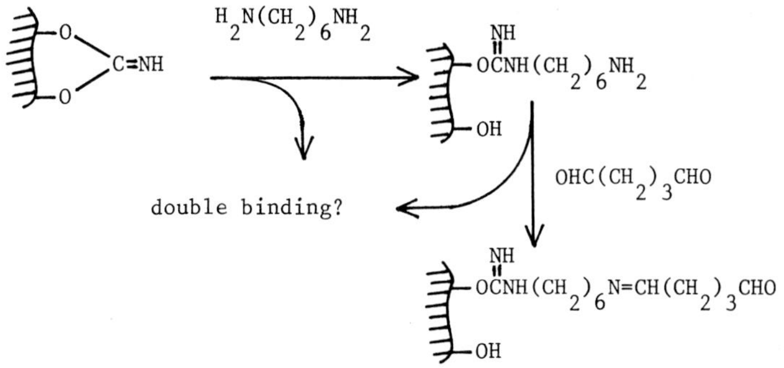

(8)

REFERENCES

1 H.Morawetz, Macromolecules in Solution, 2nd edn., John Wiley and Sons, New York, 1975.
2 D.C.Sherrington, in P.Hodge and D.C.Sherrington (Eds.), Polymer-supported Reactions in Organic Synthesis, John Wiley and Sons, Ltd, London; 1980, Chapter 1.
3 R.Epton and C.Holloway, An Introduction to Permeation Chromatography, Koch-Light Laboratories Ltd., Colnbrook, 1973.
4 e.g. D.C.Sherrington, D.J.Craig, J.Dalgleish, G.Domin, J.Taylor and G.V.Meehan, Europ. Polymer. J., 13(1977)73.
5 e.g. R.L.Schnaar and Y.C.Lee, Biochemistry, 14(1975)1535.
6 J.M.J.Frechet and M.J.Farrall, in S.S.Labana (Ed.) The Chemistry and Properties of Crosslinked Polymers, Academic Press, New York, 1977, p. 59.
7 C.J.Suckling, Chem.Soc.Reviews, 6(1977)215.
8 P.J.Flory, Principles of Polymer Chemistry, Cornell University Press, Ithaca, New York, 1953, Chapter 3.
9 K.Kunitake, See ref.4 Chapter 4.
10 N.Ise. Adv.Polymer Sci., 7(1971)536.
11 G.Challa, A.J.Schouten, G.Ten Brinke and H.C.Meinders, ACS Symposium Series, 121(1980)7.
12 A.Akelah and D.C.Sherrington, J.App.Polymer Sci., "in press".
13 A.Akelah and D.C.Sherrington, "unpublished results".
14 F.Helferrich, J.Amer.Chem.Soc., 76(1954)5567.
15 R.Tartarelli, G.Nencetti,M.Baccaredda and S.Cagnianelli, Ann.Chim. (Rome), (1966) 1108.
16 F.Ancilloti, M.M.Marcello and E.Pescarollo, J.Catalysis, 46(1977)49.
17 W.M.Mackenzie and D.C.Sherrington, Polymer, 21(1980)791.
18 S.L.Regen and A.Nigam, J.Amer.Chem.Soc., 100(1978)7773.
19 R.H.Grubbs, L.C.Kroll and E.M.S.Sweet, J.Macromol.SciChem. A7, (1973)1047.
20 F.Helferrich, Ion Exchange, McGraw-Hill, New York, 1962, Chapter 11. p527.
21 J.A.Greig and D.C.Sherrington, Polymer, 19 (1978) 163.
22 e.g. E.R.Gilliland, H.J.Bixler and J.O'Connell, Ind.Eng.Chem.Fund., 10(1971)185.
23 H.W.Heath and B.B.Gates, Amer.Inst.Chem.Eng.J., 18(1972)321.
24 I.Langmuir, Trans.Faraday Soc., 17(1922)621.
25 C.N.Hinshelwood, The Kinetics of Chemical Change, Oxford University Press, New York, 1940.
26 e.g. A.Martinec, K.Setinek and L.Beranek, J.Catalysis, 51(1978)86.
27 e.g. S.Affrossman and J.P.Murray, J.Chem.Soc., B, (1968)579.
28 e.g. A.M.Kotliar and H.Morawetz, J.Amer.Chem.Soc., 77(1955)3692.

29 V.Gold, C.J.Liddiard and J.L.Martin, J.Chem.Soc., Faraday I, 73(1977)1119.
30 H.A.Scheraga, Biochim. Biophys.Acta, 23(1957)196.
31 J.M. Sauvage and C.Loucheux, Makromol.Chem., 176(1975)315.
32 H.Ladenheim, E.M.Loebl and H.Morawetz, J.Amer.Chem.Soc., 81(1959)20.
33 R.H.Grubbs, C.Gibbons, L.C.Kroll, W.D.Bonds Jr. and C.H.Bruakker, J.Amer.Chem.S
95(1973)2374.
34 J.M.J.Frechet and L.J.Nuyens, Canad.J.Chem., 54(1976)926.
35 P.Mohr, K.Pommerening and M.Kuhn, J.Pol.Sci., Pol.Symp. 68(1980)109.

T.C.J. Gribnau, J. Visser and R.J.F. Nivard (Editors),
Affinity Chromatography and Related Techniques

NEW AFFINITY TECHNIQUES

Sulfonyl halides for the immobilization of affinity ligands and enzymes (A).

K. Mosbach and K. Nilsson

Pure and Applied Biochemistry, Chemical Center, University of Lund, P.O. Box 340, S-220 07 Lund, SWEDEN

(A) A number of immobilization methods are known and applied. Although these procedures have proved successful they have some drawbacks such as leakage under certain conditions. We have found that treatment of supports with sulfonyl chloride derivatives results in conversion of hydroxyl groups of the support into excellent leaving groups, sulfonates (refs. 1-3). Addition of affinity ligands or enzymes to thus activated supports leads to nucleophilic substitution at the matrix carbon giving rise to highly stable, direct linkages between ligand and support.

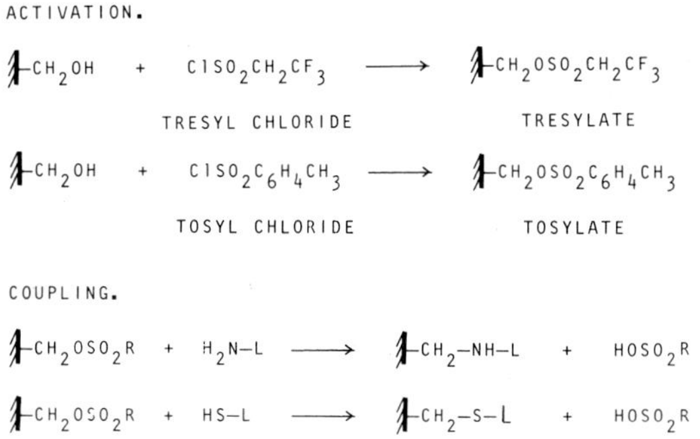

Scheme. L = affinity ligand or enzyme; R = CH_2CF_3 or $C_6H_4CH_3$.

The activation of supports (refs. 1-3) is carried out in dry acetone to avoid hydrolysis of the sulfonyl chloride. Pyridine is used as a catalyst for the reaction. This is followed by gradual transfer of the gel to water for storage. Detailed procedures for coupling and determination of bound ligands are given in refs. 1-3.

It was found that activation with tosyl chloride is rapid, can easily be followed by ultraviolet measurements, gives predictable levels of activation and the activated gel is stable on storage in distilled water for long periods (refs. 1, 2). Coupling results in stable, direct linkages between ligand and support without the introduction of any additional groups, thereby making it possible to devise highly specific affinity chromatographic systems (ref. 2).

By employing a more reactive sulfonyl chloride derivative, i.e. tresyl chloride, a number of enzymes and affinity ligands were immobilized to different hydroxyl gr containing supports, i.e. agarose, cellulose, diol-silica, glycophase-glass and hydroxyethyl methacrylate (refs. 3, 4). After these supports were activated with tresyl chloride high coupling yields were obtained with the general affinity ligan N^6-(6-aminohexyl)-adenosine 5'-monophosphate when coupled under the mild condition of pH 8.2 and at 4 ^oC. Yields and retained specific activities of enzymes were hig than with CNBr activated supports at neutral pH (60-80 and 25-50 %, respectively) (ref. 3).

It was also found that commercially available diol-silica (Lichrosorb Diol, Merc could conveniently and directly be activated with tresyl chloride derivatives. Thu activated supports coupled the above AMP-analogue efficiently, allowing resolution a mixture of various NAD^+-dependent dehydrogenases in HPLAC systems in less than 1 min (ref. 3).

REFERENCES

1. K. Nilsson and K. Mosbach, Eur. J. Biochem., 112(1980)397-402.
2. K. Nilsson, O. Norrlöw and K. Mosbach, Acta Chem. Scand, B35(1981)19-27.
3. K. Nilsson and K. Mosbach, Biochem. Biophys. Res. Commun., Submitted for publication.
4. K. Mosbach and K. Nilsson, Patent pending.

Affinity precipitation (B) and High performance liquid affinity chromatography (C)

K. Mosbach, M. Glad, P.O. Larsson and S. Ohlson
(address: see above)

(B) Over the last few years we have been studying a new affinity technique which we like to call "affinity precipitation" since it strongly resembles the technique of immunoprecipitation (ref. 1). The basic underlying principle is the use of a bi-functional affinity ligand which allows the subunits of oligomeric enzymes to interact, eventually leading to the formation of aggregates that, because of their size and insolubility, precipitate out of a solution. In our initial studies we have used as affinity ligand the bis-NAD compound N_2, N_2'-adipodihydrazido-bis(N^6-carbonylmethyl-NAD) (Fig. 1).

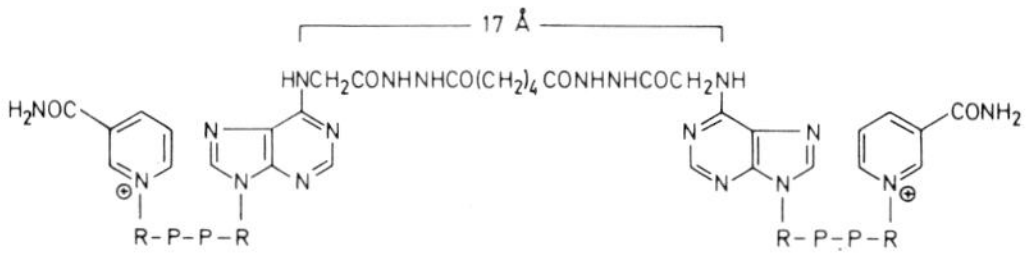

It was obtained by reacting adipic acid dihydrazide dihydrochloride with N^6-carboxy-methyl-NAD, the latter commercially available from Sigma, alternatively, it can be synthesized according to the literature (ref. 2). When bis-NAD is allowed to interact with, for example, the tetrameric enzyme lactate dehydrogenase in the presence of 0.1 M oxalate, the enzyme precipitates out of solution within a few minutes (Fig. 2).

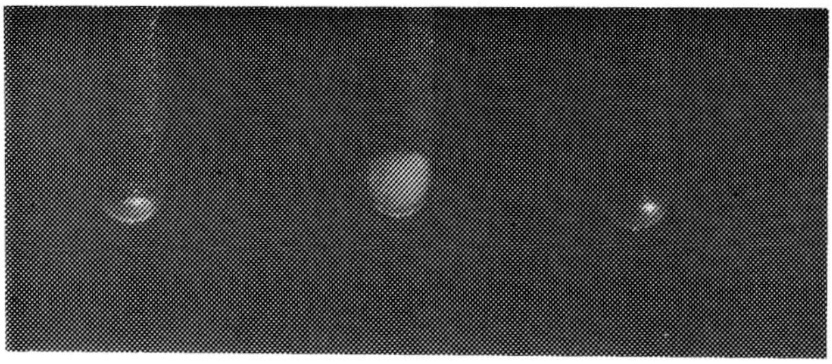

LACTATE DEHYDROGENASE

1	2	3
With no Bis-NAD$^+$ present	With Bis-NAD$^+$ and oxalate present	With no oxalate present

Addition of oxalate, or another third component like pyruvate, is required to obtain such precipitation since only in their presence will stable ternary complexes be formed. When the same bis-NAD compound is used but glutarate is added instead, the enzyme glutamate dehydrogenase will efficiently be precipitated out, whereas for yeast alcohol dehydrogenase pyrazole is required. Thus, the need for a third compo is in fact advantageous as it allows bis-NAD to act as a versatile "general ligand" whereby specificity is given to the system by the third component. The ternary complexes which are formed can be dead-end complexes or complexes with coenzyme a inhibitor or NAD adducts (ref. 2).

Another study confirmed the expected requirement, that for precipitation to occur one-to-one ratio of NAD moiety to enzyme subunit was needed as studied with lactat dehydrogenase. Additions of excess NADH led to the quick dissolution of the precip tate proving the biospecific nature of the complex formed.

The potential usefulness of affinity precipitation as described here for prepar tive enzyme purification was also investigated. After centrifugation of the homoge obtained from ox heart, precipitation of the supernatant led to pure lactate dehyd genase that could be crystallized. The overall purification achieved was 50-fold w 80 % recovery (ref. 3).

A further area in which bifunctional affinity ligands such as bis-NAD may find u lies in basic enzymology. Thus, it may be possible to indirectly measure the depth the active site of an enzyme by applying a series of bis-ligands with increasing l of the interconnecting spacer. It is expected for steric reasons that only from a certain length onwards will precipitation occur. This will provide useful and valu insight of the enzyme's topography.

Recently, a report appeared on the use of a similarly constructed bis-ligand, i. bis-boronate, which was successfully applied in the precipitation of erythrocytes (ref. 4). As in principle most biomolecules that occur in oligomeric form (dimeric and higher) can be made to precipitate provided the affinity binding is strong eno it is expected that affinity precipitation will find wider use in the future.

(C)High Performance Liquid Chromatography (HPLC) and Affinity Chromatography each p an important role in modern separation technology, for analytical as well as for purification purposes. The attractive features of these two separation methods wer combined resulting in a new chromatographic technique which we have given the name "High Performance Liquid Affinity Chromatography (HPLAC)" (ref. 5).

In HPLAC the inherent benefits of HPLC in terms of high resolution and speed of operation paired with the pronounced specificity of affinity chromatography are f utilized. HPLAC has witnessed a rapid growth during the last three years as is evident by the number of published reports in this field (refs. 5-14). In the fol ing various applications of this technique will be discussed.

In our studies on HPLAC, silica has been selected as a suitable carrier for the following reasons: It has chemical and mechanical stability (withstands high pressure falls), has high porosity (allowing free penetration of high molecular weight compounds) and can easily be derivatized. However, a major drawback with the use of silica for the separation of biomolecules in aqueous solvents is the nonspecific adsorption. A hydrophilic layer of glycerolpropyl groups covalently attached to silica or glass, however, minimizes these adverse effects. We used a microparticulate porous silica (5-10 μm, Lichrosphere Si 60-4000, Merck, FRG) which was silanized with γ-glycidoxypropyltrimethoxysilane (Dow Corning, U.S.A.) (ref. 15); the recovery of biomolecules on our HPLAC-supports was in the range of 80-100 %.

In addition, this material exhibits good stability at pH 2-9 and carries functional groups suitable for binding of ligands. Details on the coupling procedure are given in the scheme and in refs. 5, 6). The diol-silica is now commercially available (LiChrosorb diol, Merck, FRG).

SCHEME OF COUPLING PROCEDURES

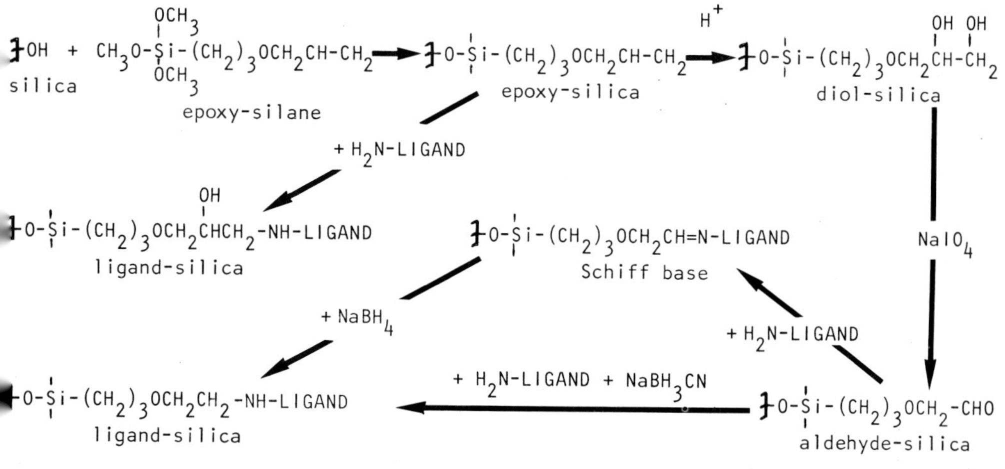

The table gives a summary of the HPLAC applications performed in our laboratory presenting the ligand and coupling method used as well as those biomolecules exhibiting affinity for the different ligands.

TABLE 1

Applications of HPLAC

Ligand	Coupling method	Application	Reference
AMP	Aldehyde	Dehydrogenases including isozymes	5, 6
Anti human serum albumin	Aldehyde	Albumin	5
Anti human creatine kinase	Aldehyde	Isozymes of creatine kinase	16
Benzene boronic acid	Epoxy	Nucleosides, nucleotides, carbohydrates	7
Cibachron Blue F3G-A	Epoxy	Dehydrogenases, kinases, ribo-nuclease	9
Con A	Aldehyde	Glycoproteins	17
NAD^+	Epoxy	Lactate dehydro-genase	18

Separations with antibody-silica

Several applications on the use of antibody-silica in HPLAC have been reported (refs. 5, 8, 16). In order to illustrate the principle of using antibodies immobilized to silica in HPLAC, a model separation of human serum albumin (HSA) from bovine serum albumin (BSA) was carried out with anti-HSA-silica. As shown in Fig. 3 this separation can be completed within 5 min.

A clinically interesting application is the analysis of creatine kinase isoenzymes for the diagnosis of myocardial infarction. An efficient and selective way of separating these isoenzymes is the use of silica-bound antibodies directed towards one of the subunits of the enzyme (ref. 16).

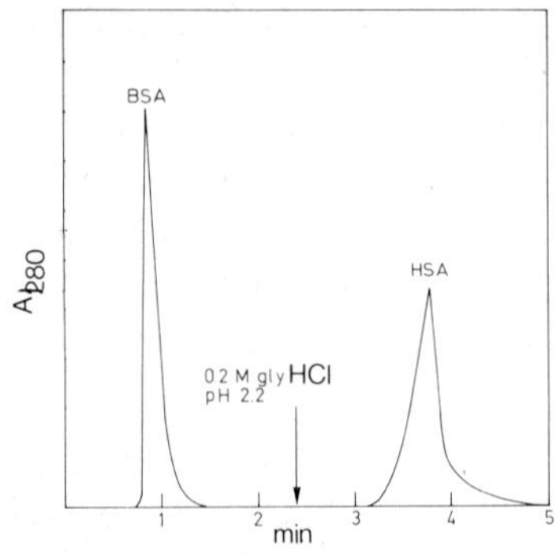

Fig. 3. Separation of BSA and HSA on anti-HSA-silica.

Separations with boronic acid silica

Boronic acids form cyclic boronate esters with vicinal cis-diols at high pH, a fact that has been used in the separation of diol-containing molecules such as nucleosides, nucleotides, carbohydrates and catecholamines (ref. 19). An example of the performance of boronic acid silica (ref. 7) in HPLAC is the separation of a mixture of nucleosides given in Fig. 4. Several of the components were well resolved and the separation was completed within 35 min. as compared with several hours with conventional boronic acid resins.

Boronic acid silica is a highly versatile adsorbent and should be applicable for the separation of a great number of biomolecules including glycoproteins.

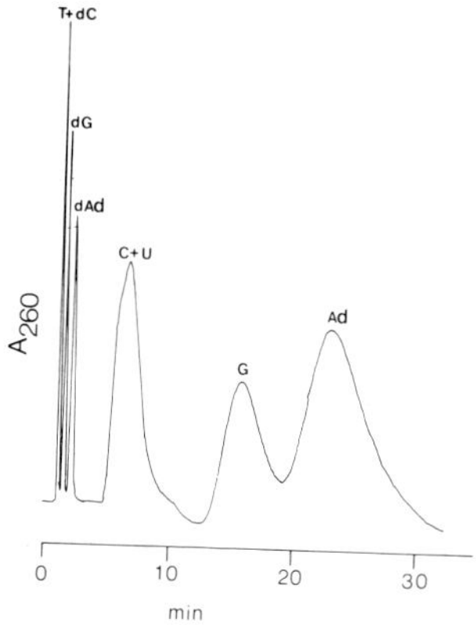

Fig. 4. Separation of nucleosides with boronic acid silica. Adenosine (Ad), deoxyadenosine (dAd), cytidine (C), deoxycytidine (dC), guanosine (G), deoxyguanosine (dG), thymidine (T) and uridine (U).

Separations with dye-silica

In recent years the triazine dye Cibachron Blue F3G-A has been used as a general ligand in affinity chromatography. A variety of nucleotide-dependent enzymes, including kinases and dehydrogenases, can be separated with this ligand. More recently, the use of Cibachron Blue F3G-A (ref. 9) as well as other triazine dyes (refs. 10, 11) bound to silica for the separation of enzymes with HPLAC have been amply documented. In Fig. 5 the purification of hexokinase (HK) and/or 3-phosphoglycerate kinase (PGK) from crude yeast extract was performed on silica-bound Cibachron Blue F3G-A. The elutions were carried out by specific ternary complex formation. The enzyme activity profiles were directly followed on-line by mixing the eluent with an assay medium containing a mixture of substrates for both the enzymes. Thus, the time for analysis was reduced to the actual time of the chromatographic separation.

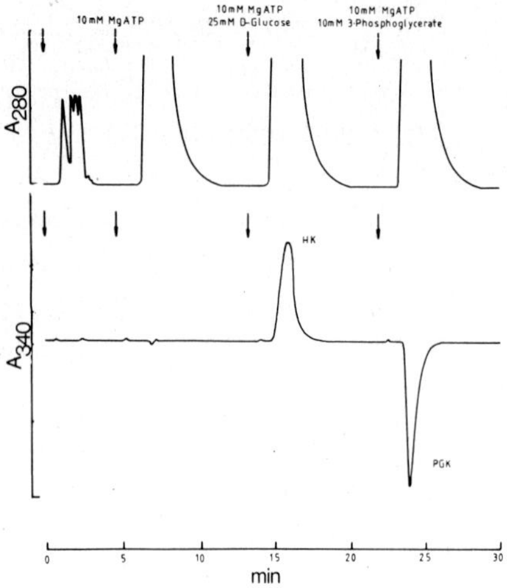

Fig. 5. Separation of kinases from a crude yeast extract on silica-bound Cibachron Blue F3G-A with on-line monitoring of the HK and PGK activities.

REFERENCES

1. P.O. Larsson and K. Mosbach, FEBS Lett., 98(1979)333-338.
2. K. Mosbach, in A. Meister (Ed.), Advances in Enzymology, Vol. 46, John Wiley & Sons, New York, 1978, pp. 205-278.
3. S. Flygare, T. Griffin, P.O. Larsson and K. Mosbach, To be published.
4. T.J. Burnett, H.C. Peebles and J.H. Hageman, Biochem. Biophys. Res. Commun., 96(1980)157-162.
5. S. Ohlson, L. Hansson, P.O. Larsson and K. Mosbach, FEBS Lett., 93(1978)5-9.
6. S. Ohlson, Thesis, Dept. of Pure and Applied Biochemistry, University of Lund, Sweden (1980).
7. M. Glad, S. Ohlson, L. Hansson, M.O. Månsson and K. Mosbach, J. Chromatogr., 200(1980)254-260.
8. J.R. Sportsman and G.S. Wilson, Anal. Chem., 52(1980)2013-2018.
9. C.R. Lowe, M. Glad, P.O. Larsson, S. Ohlson, D.A.P. Small, T. Atkinson and K. Mosbach, J. Chromatogr., in press.
10. D.A.P. Small, T. Atkinson and C.R. Lowe, J. Chromatogr., in press.
11. C.R. Lowe, this volume.
12. V. Kasche and B. Galunsky, this volume.
13. J. Turková, this volume.
14. P.O. Larsson, M. Glad, L. Hansson, M.O. Månsson, S. Ohlson and K. Mosbach, Review, to be published in Advances in Chromatography.
15. F.E. Regnier and R. Noel, J. Chromatogr. Sci., 14(1976)316-320.
16. S. Ohlson, P. Wikström, P.O. Larsson, L. Hansson and K. Mosbach, To be published.
17. A. Borchert et al., To be published.
18. M. Glad et al., To be published.
19. H.L. Weith, J.C. Wiebers and P.T. Gilham, Biochemistry, 9(1970)4396-4401.

T.C.J. Gribnau, J. Visser and R.J.F. Nivard (Editors),
Affinity Chromatography and Related Techniques

SPECIFIC BINDING OF SUBSTANCES TO POLYMERS BY FAST AND REVERSIBLE COVALENT INTERACTIONS+

G. WULFF, W. DEDERICHS, R. GROTSTOLLEN AND C. JUPE
Institute of Organic Chemistry II, University of Düsseldorf (G.F.R.)

ABSTRACT

By a special kind of template polymerization each two binding groups were introduced in synthetic polymers in a defined spatial proximity in cavities of specific shape. The specific interaction of substances with these polymers was investigated. Reversible covalent interactions were used for the binding. The thermodynamic and kinetic properties of possible covalent binding reactions were first studied with low molecular weight model compounds. The interaction of benzene boronic acid with diols in organic solvents was studied in detail. Very fast and selective covalent interactions for diols were observed with the o-dimethylaminomethyl-benzeneboronic acid group and for monoalcohols with the boronophthalide group.

INTRODUCTION

In affinity chromatography usually biologically active substances are coupled as ligands covalently to a solid support. The separation then is based on the ability of these ligands to bind specifically and reversibly other substances. The type of interaction between ligand and ligate mostly is of rather complex nature (see *e.g.refs.*2,3). We have investigated the possibility of using covalent interactions between ligand and ligate, but in contrast to the so-called covalent affinity chromatography the interaction was of a reversible type. It was tried to prepare synthetic polymers in which the binding groups are located in a defined spatial proximity in cavities of specific shape. Such cavities can be considered as models of natural receptors.

+"On the Chemistry of Binding Sites", Part II. For Part I see *ref.*1.

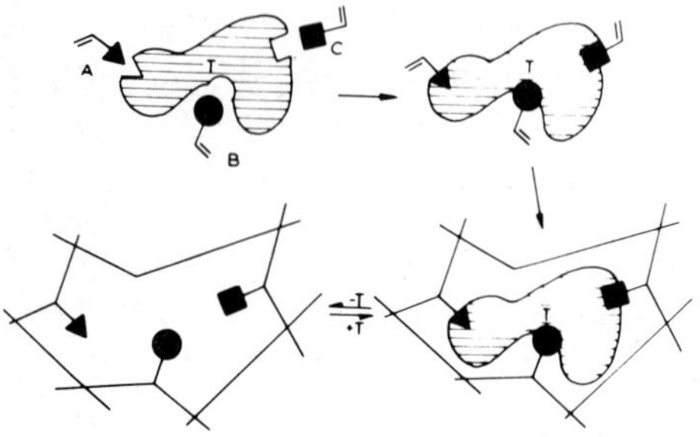

Scheme I

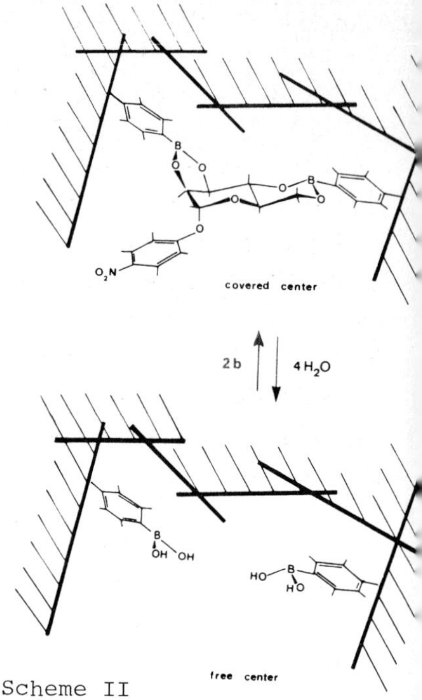

Scheme II

THE PREPARATION OF CHIRAL CAVITIES IN SYNTHETIC POLYMERS

To prepare binding sites in synthetic polymers we introduced some years ago the following approach (*refs.* 4-6). The binding groups to b introduced were bound in the form of polymerizable vinyl derivatives a suitable template molecule (see Scheme I). This monomer then was copolymerized under conditions such that highly crosslinked polymers with chains in a fixed arrangement were formed. After removal of the template, free cavities were formed with a shape and an arrangement of functional groups that corresponds to those of the template. The functional groups in this polymer are located at quite different poin of the polymer chain, they are hold in spatial relationship by the crosslinking of the polymer. This approach to prepare cavities is in c trast to those for example with crown ethers (see *e.g.refs.*7,8) were a low molecular weight part carries the stereochemical information.

As an example for this method the polymerization of 4-nitrophenyl-α-D-mannopyranoside-2,3;4,6-bis-*o*-(4-vinylbenzeneboronate) is shown (*ref.* 5) (see Scheme II). The template is 4-nitrophenyl-α-D-mannopyranoside to which two 4-vinylbenzeneboronic acids are bound by diest linkages to each two hydroxyl groups. The boronic acid was chosen as a binding group, since it is known to undergo an easily reversible interaction with diol groupings. Since we had chosen a chiral template, the accuracy of the steric arrangement in the cavity could be tested by the ability of the polymer to resolve the racemate of the template after splitting off the original template.

This monomer then was copolymerized with radical initiation in presence of an inert solvent with a high amount of a bifunctional cross-linking agent. Under these conditions macroporous polymers were obtained which possess a permanent pore structure and a high inner surface area. By this a good accessibility and a low swellability and therefore a limited mobility of the polymer chains can be expected.

The templates could be split off from this type of polymer with water or alcohol to a degree of 40-80%. If this polymer is treated with the racemate of the template in a batch procedure under equilibrium conditions, preferably that enantiomer is taken up that has been used for the preparation of the polymer. If the specificity is expressed by the separation factor α, that is the ratio of the distribution coefficients between solution and polymer of L and D-form, values in this and similar cases ranging from 1.20 to 3.64 were obtained depending on the equilibration condition and the polymer structure (*refs.* 5,9,10). The highest α-value obtained until now was 3.64. In this case in the simple batch procedure an enrichment of the L-form in the filtrate of 12.8% and of the D-form at the polymer of 40.4% was observed (*ref.* 10). Polymers of somewhat less specifity have been used as chromatographic sorbents for the chromatographic resolution of racemates (see Fig. 1) (*refs.*5,6). 80% of the first eluting peak could be obtained in optically pure form in a separation of the racemate of the template. With a separation factor α of 1.8 a much better separation was to be expected. Apparently the mass transfer was too slow. Responsible for this might be the slow interaction

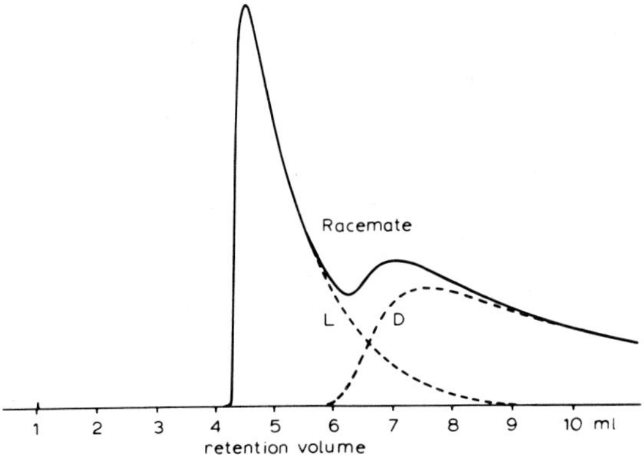

Fig. 1. Chromatography of the racemate of 4-nitrophenyl mannopyranoside. Solvent: methanol/piperidine (98:2) Temp.: 65 °C, α-value: 1.85, optical yield: 87%

of the boronic acids with the diols and the slow embedding of the substrates into the somewhat rigid cavities. In the following the influence of the covalent interaction will be discussed more in detail.

THE COVALENT INTERACTION OF LIGAND AND LIGATE

There are many possibilities for the type of interaction between the ligand and the ligate in affinity chromatography (see *e.g. refs.* 2,3). We wanted to have a fast reversible interaction in order to be able to use this method in h.p.l.c. and to elute substances one after another without changing the eluent. With our method of preparing synthetic receptor sites the binding groups have a twofold function. Firstly, during polymerization a strong interaction between template and binding groups should be present, so that the template molecule can fix the binding groups at the growing polymer chains in a defined stereochemistry. After splitting off the template the binding groups should secondly be able to undergo an easily reversible binding interaction with the template or similar substrates. While strong covalent bonds are preferable for the first purpose the activation energy of the reaction should be low for a reversible binding reaction in a receptor model. The activation energies for most reactions leading to covalent bonds are too high, only a few could be used with some success. It was of interest whether it is possible to influence the kinetics of the covalent binding reactions by changing the reaction conditions or by introducing suitable neighbouring groups into the binding site. These investigations have first been undertaken with low molecular weight model compounds.

Reactions of Benzeneboronic Acid with Diols

The reaction of boronic acid with diols yields different products depending on the medium employed. In organic solvents, neutral esters (*e.g.* **3**) with trigonal boron are formed (*eq.* (1)) whereas in aqueous, alkaline solutions basic esters (*e.g.* **4**) with tetrahedrally coordinated boron atoms result (*eq.* (2)). While studies of the equilibrium in aqueous solution at various pH-values have been reported (*ref.* 11), little is known about the equilibrium of the esterification in organic solvents. Even less details are available concerning the kinetics of the reaction of aliphatic diols with benzeneboronic acid in aqueous or organic phase. Only the very fast reactions of benzeneboronic acid with catechol (*ref.* 12), lactic acid (*ref.* 13) and oxalic acid

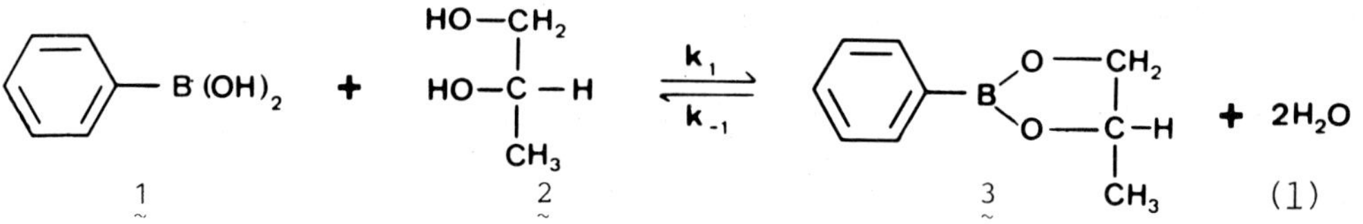

1 2 3 (1)

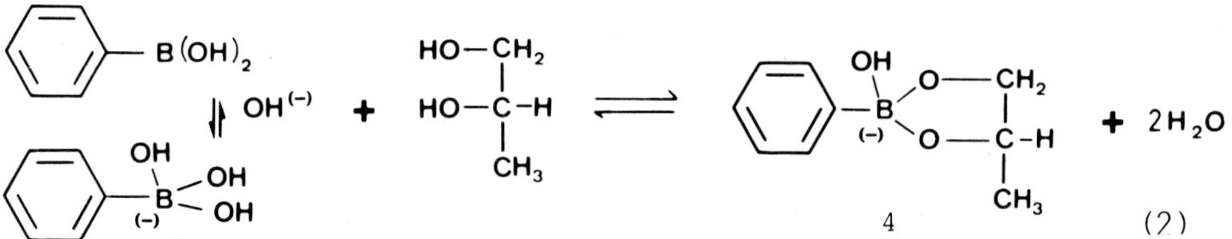

1a 4 (2)

(*ref.*14) as well as the hydrolysis of 2-phenyl-1,3,2-benzodiazaborol (*ref.*15), the kinetics of which obey quite complicated laws, have been more thoroughly investigated.

The reaction of benzeneboronic acid with optically active L-1,2-propanediol was chosen as a model, since the reaction can be readily followed by measurement of the optical rotation. The molar optical rotations of diol 2 ($[M]^{20}_{365}$ + 33.4°) and ester 3 ($[M]^{20}_{365}$ + 109.7°) in dioxane differ by a factor greater than three.

Starting from various concentrations of L-propanediol and benzeneboronic acid or L-1,2-propanediol, benzeneboronic acid and water or L-1,2-propanediol-benzeneboronate and water, a value of K = 19.2 mol · l^{-1} (standard deviation σ = 1.16) was obtained for the equilibrium

$$K = \frac{[3]\ [H_2O]^2}{[1]\ [2]} = 19.2 \text{ mol} \cdot l^{-1} \text{ in dioxane at } 20.0^{\circ}C$$

constant in dioxane at 20.0°C. The fact, that there is no difference between the equilibrium constant determined for the forward and the reverse reaction, reveals that true equilibria were achieved.

The polarimetric method lent itself to the kinetic investigations of reaction (1). It was found that esterification as well as hydrolysis took place very quickly.

The investigations of our substances were limited by the fact that the starting discharge of the reaction, a very important factor in the study of reversible reactions, was not accurately known due to late measurements. In addition, the reactions(the esterification, in particular) clearly revealed an induction period. Thus, so as to estimate the reaction rates, the times, at which half of the possible conversion to equilibrium had occurred, were determined (*cf.* Tab. 1).

TABLE 1

Kinetics of the Establishment of Equilibrium According to *equ.*(1)

Solvent	Starting concentration in $mol \cdot l^{-1}$				Half life time sec	k_1	k_{-1}
	1	2	3	H_2O			
Dioxane	0.0131	0.0153			650	$1.5 \cdot 10^{-1}$	
Dioxane	0.0131	0.0253			210	$2.5 \cdot 10^{-1}$	
Acetonitrile	0.0249	0.0274			170	$5 \cdot 10^{-1}$	
Acetonitrile	0.0249	0.0137			300	$4 \cdot 10^{-1}$	
Dioxane			0.0489	0.743	100		$7 \cdot 10^{-3}$
Dioxane			0.0342	0.650	115		$9 \cdot 10^{-3}$

LEGEND. The rates of reaction were followed polarimetrically. The halflive times are the times for half of the conversion to the equilibrium state. The rate constants k_1 and k_{-1} were calculated for various reaction times, assuming that the forward and reverse reaction are both second order overall, and extrapolated to time t=0.

In order to obtain values, which are independent of the concentrations, it was assumed that the forward and backward reactions were both first order with respect to each reactant; the rate constants for various conversions were thus calculated. Table 1 shows values of k_1 and k_{-1} (see *equ.* (1)) extrapolated to time t=0 ($k_1 \sim 2 \cdot 10^{-1}$ and $k_{-1} \sim 8 \cdot 10^{-3}$ $l \cdot mol^{-1} \cdot sec^{-1}$).

The calculation is surely much simplified, nevertheless, the rate constants obtained for the start reaction should be appropriate for comparison purposes. It can be seen that the esterification at the start occurs approx. 25 times faster than the hydrolysis reaction. On the whole, both the esterification and the hydrolysis are fast compared with the formation and hydrolysis of carboxylic esters, but unfortunately yet too slow for the binding reaction in a fast chromatography. This may well explain the broad peaks obtained in chromatography using organic solvents (*ref.*6).

Control of the reaction kinetics. A rise in temperature does not greatly affect the reaction rate, however, a change of solvent to acetonitrile approximately doubles the rate constant of formation.

Addition of piperidine so strongly accelerates the rate of establishment of equilibrium that it can no longer be followed by our measuring technique. Preliminary investigations of the transesterification reaction of 3 with 2 with the temperature-dependent ^{1}H.n.m.r. spectroscopy show that the rate constants in case of piperidine are several orders of magnitude greater than those in absence of piperidine. The influence of piperidine could probably be due to two different effects. Firstly, in an equilibrium reaction between piperidine and the alcohol, the more nucleophilic alkoxide may be formed and then attack the boron or secondly, piperidine may bind itself to the boron atom to saturate its outer electronic shell. ^{11}B-n.m.r. spectroscopy investigations show that the ^{11}B-signal of 3 at 31 ppm (in acetonitrile) is shifted to 26.2 and 14.5 ppm respectively by addition of piperidine in the molar ratios of 3:piperidine = 3:1 and 3:piperidine = 4:10. The upfield shift of the ^{11}B-signal which appears typically at 31 ppm for trigonal boronic esters, points to a tetrahedral coordination of the boron atom.

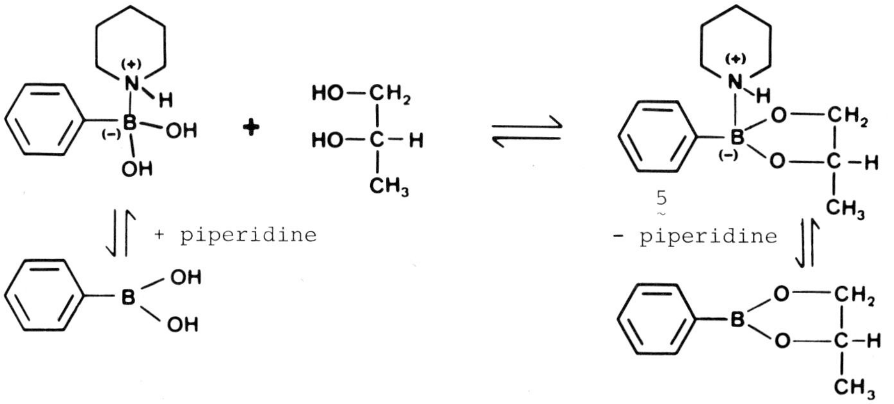

(3)

Thus in equilibrium an adduct with structure 5 is formed, which has already been isolated in crystalline form by other authors (*ref.*16) in the reaction of the pure components in absence of solvent. Also, addition of OH^- to 3 dissolved in water leads to a strong shift of the ^{11}B-resonance signal which may be attributed to the formation of the basic ester 4 (*ref.*17).

The strong acceleration of the rate of attainment of equilibrium for reaction (1) in presence of piperidine might be attributed to its marked ability to form complexes with boronic acid derivatives. Apparently, the equilibrium is reached more rapidly through tetrahedrally coordinated B-atoms than through trigonally coordinated ones, as is also shown by the fast reaction in aqueous alkaline solution (*equ.*(2)) proceeding from compound 1a with tetrahedral coordinated boron atom. In analogy to *equ.*(2) the establishment of equilibrium in the presence of piperidine may be described by *equ.*(3). In accordance with these results triethyl-amine having a similar p_{Ka} value as piperidine but not forming a B-N-interaction as seen from ^{11}B-n.m.r. does not accelerate the rate of attainment of the esterification equilibrium, therefore the acceleration is not caused simply by the basicity of the piperidine.

o-Dimethylaminomethyl-Benzeneboronic Acid as a Binding Group

Chromatography on polymers containing boronic acids in presence of piperidine improves markedly the mass tranfer. Since attempts to polymerize our template monomers in presence of piperidine led apparently to chain transfer reactions, we tried to introduce an intramolecular B-N interaction in the binding group.

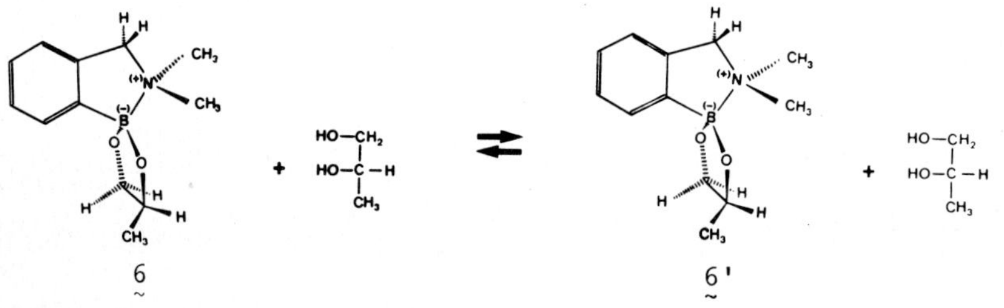

(4)

By the introduction of an ortho dimethylaminomethyl group in the benzeneboronic acid (preparation of the substance see *ref.*18) both the rate of hydrolysis and of formation of the ester was enhanced by several orders of magnitude. By following the exchange equilibrium of *equ.*(4) instead of the esterification reaction the kinetics could be followed by dynamic ^{1}H-n.m.r. spectroscopy. The methyl groups of the propanediol moiety in 6 and 2' ($\Delta\upsilon$=29.4 Hz) showed coalescence at -1 $^{\circ}$C (equimolar amounts 0.42 mol/l each). From these data an enhancement of the rate of exchange by the introduction of the dimethylaminomethyl group of 3-4 orders of magnitude can be estimated. In compounds 6 and 7 a B-N coordination can be detected. The ^{11}B-signal is shifted from δ=36 ppm in 3 to δ=17 ppm in 6. The intramolecular B-N bond in this case leads to a hindrance of rotation around the B-C-axis. A detailed n.m.r. investigation of several esters showed that the B-N bond is frequently broken ($\Delta G^{\neq}$ =40-54kJ/mol) and re-formed (*ref.*1).

8 9 (5)

9a' 9a (6)

TABLE 2

Capacity Factors k' in the Chromatography of Various Substances on Macroporous Copolymers prepared from 7

Substance chromatographed	Eluent 100% MeOH	Eluent MeOH with 0.1% triethylamine
Phenylalanine	0.49	0.26
Benzoic acid	3.40	0.24
Benzylamine	0.76	0.77
Benzylalcohol	0.64	0.62
Resorcinol	0.39	0.36
Hydroquinone	0.36	0.36
Brenzcatechol	0.87	0.87
Phenylethanediol	2.95	2.96
Phenyl-α-D-manno-pyranoside	2.93	2.65

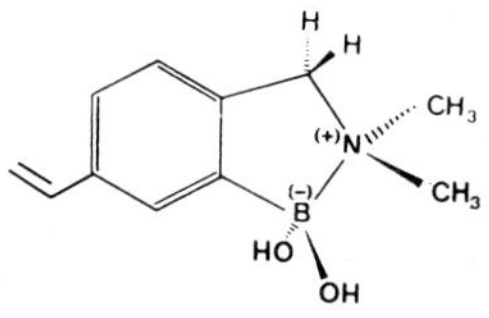

7

In order to study the binding behaviour of this group at a polymer the monomer 7 (with a vinyl group) was prepared (*ref.* 19). This compound in form of its glycolester was copolymerized with divinylbenzene to a macroporous polymer. After splitting off the glycol moiety the binding ability in chromatography was studied for various substrates (see Table 2). It can be seen that this binding group is highly specific for aliphatic diol groupings. Under neutral conditions acids are retarded as well, with addition of 0.1% triethylamine this binding becomes neglible. Amines and monoalcohols as well as aromatic diols show some retardation. The mass transfer in these separations is greatly improved compared to separations at polymers from 4-vinylbenzeneboronic acid.

Other Binding Groups

A suitable binding group for monoalcohols was found to be the boronophthalide 8, a compound that has been prepared before for other purposes (*ref.* 20). In this case without a neighbouring group participation a quick esterification of 8 as well as a hydrolysis of 9 takes place. The kinetics of these reactions could not be followed by conventional methods. So as in case of reaction (4) the exchange reaction (6) was studied by dynamic ^{1}H-n.m.r.-technique. The rate of these reactions is comparable to those in (4). Therefore 8 provides a new and efficient binding group for monoalcohols.

Another covalent interaction used for binding at polymers was the azomethine formation from aldehydes and amines (*refs.* 21,22). In addition to covalent binding electrostatic (*refs.*4,23) and charge-transfer interactions (*ref.*24)have been used in chiral cavities.

LITERATURE

1 T. Burgemeister, R. Grobe-Einsler, R. Grotstollen, A. Mannschreck and G. Wulff, Chem. Ber. in press.
2 W.B. Jakoby and M. Wilchek (Ed.), Methods Enzymol. 34(1974) 1-810.
3 J. Turková, "Affinity Chromatography", Elsevier, Amsterdam 1978.
4 G. Wulff, A. Sarhan and K. Zabrocki, Tetrahedron Lett. (1973)4329.
5 G. Wulff, W. Vesper, R. Grobe-Einsler and A. Sarhan, Makromol. Chemie 178(1977)2799.
6 G. Wulff and W. Vesper, J. Chromatogr. 167(1978)171.
7 J.M. Lehn, Pure Appl. Chem. 52(1980)2303.
8 S.C. Peacock, L.A. Domeier, F.C.A. Gaeta, R.C. Helgeson, J.M. Timko and D.J. Cram, J. Amer. Chem. Soc. 100(1978)8190.
9 G. Wulff, R. Grobe-Einsler, W. Vesper and A. Sarhan, Makromol. Chem. 178(1977)2817.
10 G. Wulff and J. Vietmeier, unpublished results.
11 R.J. Ferrier, Adv. Carb. Chem. Biochem. 35(1978)31.
12 R. Pizer and L. Babcock, Inorg. Chem. 16(1977)1677.
13 S. Friedmann, B. Pace and R. Pizer, J. Amer. Chem. Soc. 96(1974) 5381.
14 S. Friedmann and R. Pizer, J. Amer. Chem. Soc. 97(1975)6059.
15 T. Okuyama, K. Takimoto and T. Fueno, J. Org. Chem. 42(1977)3545.
16 A. Finch and J.C. Lockhart, J. Chem. Soc. (1962)3723.
17 H. Nöth and B. Wrackmeyer, "Nuclear Magnetic Resonance Spectroscopy of Boron Compounds", Springer Verlag, Berlin, 1978, p.74.
18 R.T. Hawkins and H.R. Snyder, J. Amer. Soc. 82(1960)3863.
19 G. Wulff and R. Grotstollen, unpublished results.
20 H.R. Snyder, A.J. Reedy and W.J. Lennarz, J. Amer. Chem. Soc. 80(1958)835.
21 G. Wulff in "25 Jahre Fonds der Chemischen Industrie", Verband der Chemischen Industrie, 1975, p.135.
22 G. Wulff, Nachr. Chem. Tech. Lab. 25(1977)235.
23 G. Wulff and A. Sarhan, Makromol. Chem. in press.
24 G. Wulff and E. Lohmar, Isr. J. Chem. 18(1979)279.

ACKNOWLEDGEMENTS

Thanks are due to "Fonds der Chemischen Industrie" and the "Minist für Wissenschaft und Forschung des Landes Nordrhein-Westfalen" for financial support.

T.C.J. Gribnau, J. Visser and R.J.F. Nivard (Editors),
Affinity Chromatography and Related Techniques

COVALENT ATTACHMENT OF LIGANDS TO SUPPORTS BY USING HETEROBIFUNCTIONAL REAGENTS

J.K. INMAN
National Institute of Allergy and Infectious Diseases, National Institutes of Health, Bethesda, Maryland 20205, U.S.A.

ABSTRACT

In recent years some interesting heterobifunctional reagents have been described and used for heteroligating peptide or protein A to protein B through thioether or disulfide bonds. Improved methods for introducing sulfhydryl functions via amino groups on substrate structures have been reported. These techniques suggest a general approach to covalent attachment of ligands or ligand-binding proteins to affinity supports that offers considerable flexibility and other useful advantages. One such approach, used in the author's laboratory, allowed convenient, modular addition of spacing structures and highly efficient and controllable binding of hapten molecules at very low concentrations to a crosslinked agarose support.

INTRODUCTION

Progress in the methodology of affinity chromatography will depend in part on the development of more efficient and flexible ways for attaching various special molecules to support matrices. Ideally, an affinity adsorbent should be prepared from a matrix that interacts negligibly with mobile species yet offers adequate functionality for covalent attachment of spacers and specific immobilized components. Aside from the latter moieties, there should remain no potentially interacting features such as ionic charges or hydrophobic, aromatic or branched structures that arise as products or by-products of the derivatization process. The adsorbent should bear a known and planned concentration (density) of specific ligands or receptors.

Covalent attachment of specific molecules normally begins with activation of one of their functional groups or, alternatively, of groups attached to the support; a bond-forming reaction then takes place with nucleophilic groups on the other component. Activation is brought about by converting a stable chemical group to a less stable, and more reactive, electrophilic function. Activation of groups bound to the support matrix may lead to undesired crosslinks and/or attached by-products; activated groups may hydrolyze during handling or coupling resulting in bound negative charges. Variable losses of reactive groups also will make it difficult to attain a planned or reproducible ligand density. Activation of the ligand or specific molecule may result in crosslinks (self-polymers) and in other undesirable alterations of structure. Likewise, hydrolysis of activated groups on specific molecules will give diminished control over their density on the adsorbent and may represent an expensive loss of material.

Alternatively, both support and specific component may possess passive nucleophilic groups and can be joined by means of a separately activated, and subsequently purified, bifunctional, crosslinking reagent. If the latter possesses two identical, reactive groups (a homobifunctional reagent), then crosslinks are formed both among support groups and between molecules to be bound. If the two functional groups of the reagents are not the same and have distinctly different reactivities (a heterobifunctional reagent), then it is possible to join ligand or receptor to the support cleanly, efficiently and without self-crosslinking side reactions.

FORMATION OF DISULFIDE AND THIOETHER LINKAGES USING HETEROBIFUNCTIONAL REAGENTS

A general strategy

A general approach to the covalent attachment of ligands or ligand-binding proteins to supports is shown in Figure 1. Amine-bearing supports are generally available; protein ligands usually have amino groups that are not involved in the specific interactions, and synthetic or small molecular ligands often can be prepared with a spaced amino group. From this starting point it should be possible to heteroligate cleanly the ligand to the support using several types of reagent: one bearing electrophilic and nucleophilic functions (E_1 and Nu), and another type possessing two distinctly different electrophilic groups (E_2 and E_3). The division and sequence of reactions allows greater selectivity and permits intermediate purifications or removal of by-products. E_2 encounters only the amino group of the ligand and thus may be the same functional group as E_1. E_3, however, must not react significantly with NH_2 groups during coupling to ligand <u>via</u> E_2. Subsequently, E_3 should combine efficiently with Nu, also under conditions where it would not combine with any free amino groups remaining on the support. From a number of studies on heteroligating methods, it appears that the above requirements can be readily

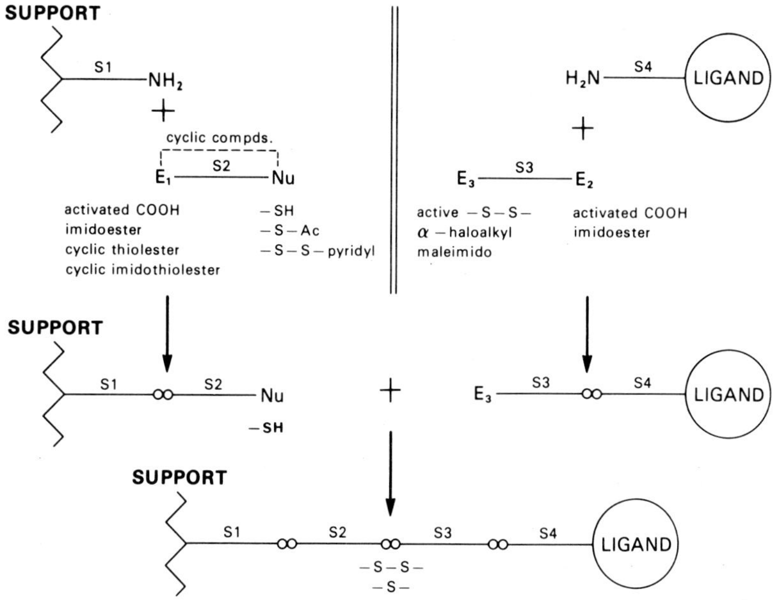

Fig. 1. Use of heterobifunctional reagents for joining ligand (or ligand-binding protein) to an initially functionalized support through disulfide or thioether bonds. E_1 and E_2 are electrophilic groups reactive with amines. E_3 is an electrophilic group selectively reactive with the nucleophilic function, Nu. Interconnecting structures, s1, s2, s3 and s4, which lack special structural features, contribute to the final spacing arm.

met if Nu is sulfhydryl and E_3 is one of several sulfhydryl-specific, reactive functions; their mutual reaction will yield either disulfide or thioether linkages. The latter are quite stable under a wide variety of conditions; the former, although not as stable, may be dissociated with thiols later on if desired.

Activation (formation of E_1, E_2 and E_3) is performed on separate compounds that can be purified and stored for later, multiple uses: It is during activation that serious side reactions are apt to take place. Reactions of E_1 and E_2 with amines may be inefficient, especially in aqueous mixtures, but they can be completed with excess of the heterobifunctional reagents; excesses of the latter are removed prior to the final step. One may simply start out with an available sulfhydryl derivative of the support matrix and incorporate E_3 in the ligand in the course of its synthesis. However, control over sulfhydryl density and spacing can be an important advantage to the investigator. The efficiency of the final reaction (E_3 + Nu) may be increased by employing an appropriate excess of SH groups on the support. Finally, unused SH groups should be removed by reacting them with a simple monofunctional reagent such as iodoacetamide, following fresh reduction, so that disulfide links within the matrix will not remain or form later on.

The support and ligand can be interchanged in the scheme of Figure 1, but certain advantages would be lost. If E_3 were on the support, which should be held in water or solvents, its storage would be somewhat more limited than would be the case for bound Nu and its protected forms. On the other hand, E_3-ligand intermediates can be kept more readily, and longer, in a dry state.

Heterobifunctionals with two electrophilic groups

A rather modest number of heterobifunctional reagents of the type, E_3—— E_2, have been reported. Examples of several kinds of these reagents where E_3 is selectively reactive with sulfhydryl groups are briefly discussed below. In only a few instances have they been used in preparing affinity adsorbents; a few of them are now commercially available.*

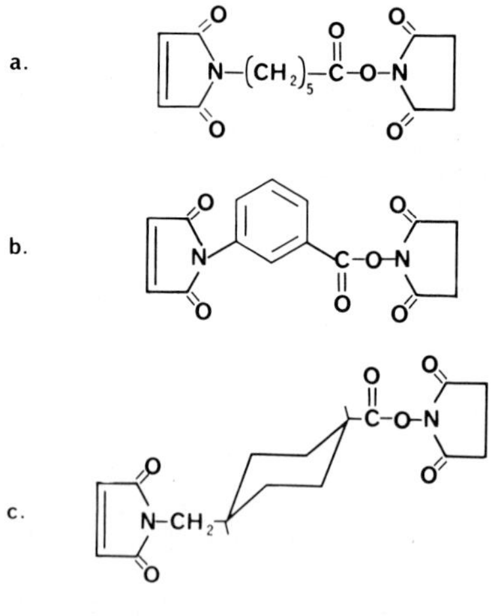

Fig. 2. Heterobifunctional reagents bearing SH-reactive maleimido and NH_2-reactive N-hydroxysuccinimide ester groups. a, succinimidyl 6-maleimidocaproate [1]; b, succinimidyl m-maleimidobenzoate [5]; c, succinimidyl trans-4-(maleimidomethyl)cyclohexanecarboxylate [9].

* Compounds shown in Figures 4a, 2b and 2c and an analogue, succinimidyl 4-(p-maleimidophenyl)butyrate, are available from Pierce Chemical Co., Rockford, IL., USA 61105. Compounds of Figures 2b and 4a are offered by Sigma Chemical Co., St. Louis, MO, USA 63178; the former is also sold by Pharmacia Fine Chemicals. p-Nitrophenyl iodoacetate has been available from Fluka AG.

Keller and Rudinger [1] described the synthesis and use of succinimidyl 6-maleimidocaproate (Fig. 2a). With this reagent, maleimido groups were coupled through amide links to amine-derivatized Sepharose (agarose) [2] or to soluble macromolecular carriers [3] by means of the N-hydroxysuccinimide, active ester function. The carrier-bound maleimide functions were then linked by thioether bonds to SH groups on active peptides, a process representing heteroligation with Figure 1 components interchanged. More recently, Kriwaczek et al. [4] synthesized the p-nitrophenyl ester of 6-maleimidocaproic acid and used it for preparing a maleimidohexanoyl-peptide that was subsequently conjugated to a thiolated TMV virus carrier (in corresponding order of the Fig. 1 scheme).

Another combination of a maleimide plus active ester, namely, succinimidyl m-maleimidobenzoate (Fig. 2b), was synthesized and reported by Kitagawa and Aikawa [5] who described its use in preparing an insulin-β-D-galactosidase conjugate for enzyme-linked immunoassays. This reagent has been employed by other workers for producing enzyme-antibody and other soluble conjugates [6] and for coupling insulin to sheep erythrocytes [7]. In spite of these useful applications, a problem is encountered with the use of N-arylmaleimides in sequential reactions in aqueous media: Hydrolytic opening of the maleimide ring will occur at appreciable rates at pH values above 7. The resulting maleamic acid is markedly less reactive toward SH groups. Machida et al. [8] have shown that this side reaction proceeds somewhat less rapidly with N-alkylmaleimides. Thus, the reagent of Kitagawa and Aikawa will be more difficult to use for heteroligation sequences than the alkylmaleimides of Keller and Rudinger (Fig. 2a, ref. 1) or of Yoshitake et al. [9] who more recently described the synthesis and use of the reagent, succinimidyl *trans*-4-(maleimidomethyl)cyclohexanecarboxylate (Fig. 2c). These latter investigators also found that the maleimido group of their compound was substantially more stable than m-maleimidobenzoic acid in aqueous, pH 7.0, phosphate buffer.

Active esters of the α-haloacetic acids (Fig. 3a) have been employed in a variety of studies, but their use in heteroligation has not been extensive. Rector et al. [10] prepared succinimidyl iodoacetate and used the product without isolation for protein-protein conjugations. Kriwaczek et al. [11] joined thiolated virus particles to peptides that had been bromoacetylated with p-nitrophenyl bromoacetate. Succinimidyl bromoacetate was prepared and used without isolation by Cuatrecasas et al. [12] as an affinity labeling reagent. The synthesis and properties of p-nitrophenyl iodoacetate have been described [13]. Succinimidyl chloroacetate (2-chloroacetoxysuccinimide) has been reported only in the Japanese patent literature. It can be prepared in a straightforward manner from equimolar amounts of chloroacetic acid, N-hydroxysuccinimide and N,N-dicyclohexylcarbodiimide in CH_2Cl_2-dimethylformamide and crystallized from hot 2-propanol (m.p. 121.5°-122.5°) [14].

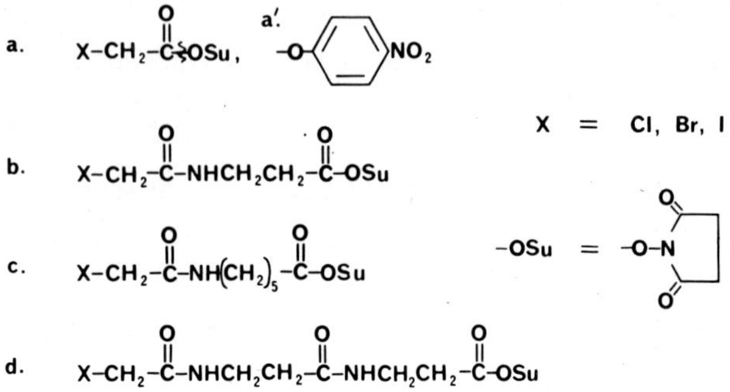

Fig. 3. Heterobifunctional reagents bearing SH-reactive haloacetyl and NH_2-reactive N-hydroxysuccinimide ester groups. a, succinimidyl and p-nitrophenyl haloacetates; b, succinimidyl N-(2-haloacetyl)-β-alaninate; c, succinimidyl 6-(2-haloacetamido)caproate; d, succinimidyl N-(2-haloacetyl)-β-alanyl-β-alaninate.

The order of reactivity of α-haloacetyl compounds toward thiols is I > Br > Cl. The approximate pH ranges for convenient reaction rates with RSH are 7-8, 8-9, and 9-11, respectively. The ratio of S-alkylation to hydrolysis in these pH ranges is quite favorable. Also, selectivity for sulfhydryls is very good within these limits [15, 16]. At higher pH values amines and phenols may be alkylated at significant rates.

Some haloacetyl N-hydroxysuccinimide esters of the type shown in Figure 3b, c and d have been synthesized in my laboratory [14] and are nicely crystallizable compounds. They provide spacing structures, such as -βAla- and -βAla-βAla-, which are appreciably hydrophilic. These compounds are currently being investigated for their utility in preparing reactive ligands, affinity adsorbents and soluble conjugates.

Two heterobifunctional reagents have been described which bear amine-specific groups and "activated" disulfide bonds. The latter groups, after being attached to an amine-bearing component, readily undergo thiol-disulfide interchange with free sulfhydryls on a second component. The two components are thus heteroligated through a disulfide link (Fig. 1). Carlsson et al. [17] described the synthesis of succinimidyl 3-(2-pyridyldithio)propionate (SPDP, Fig. 4a) and its use in the preparation of 2-pyridyldithiopropionyl derivatives of proteins that can be conjugated to other proteins or structures bearing free SH groups. Since 2-pyridyldithio groups are quite stable in aqueous media, they may be placed on adsorbent matrices and used for immobilizing sulfhydryl-containing proteins and ligands [18]. Recently, Jou and Bankert [19] reported the use of SPDP to prepare a 2-pyridyldithiopropionylated protein antigen that subsequently was coupled to thiolated sheep erythrocytes for immunological assays.

a.

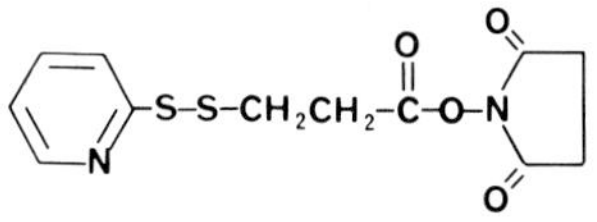

b.

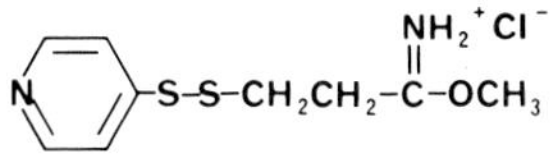

Fig. 4. Heterobifunctional reagents bearing SH-reactive, "activated" disulfide bonds and NH_2-reactive functions. a, succinimidyl 3-(2-pyridyldithio)propionate (SPDP) [17]; b, methyl 3-(4-pyridyldithio)propionimidate hydrochloride [20].

Another similar approach to heteroligation through disulfide linkages was reported by King et al. [20] who synthesized the reagent shown in Figure 4b, namely, methyl 3-(4-pyridyldithio)propionimidate hydrochloride. The N-hydroxysuccinimide ester function was replaced by the highly amine-specific methyl imidate, and the 4-pyridyl isomer was selected. In contrast with SPDP, use of the above reagent yields positively charged amidine linkages which may be considered a less desirable feature on an affinity adsorbent than uncharged amide bonds. Amidine linkages, however, can be dissociated at a later time, if desired, with appropriate nucleophilic agents [21].

Reagents for introducing sulfhydryl groups

The scheme of Figure 1 calls for the use of a heterobifunctional of the type E_1—— Nu or E_1—— Nu-Prot. (where Prot. = a removable protecting group). Although such an electrophilic-nucleophilic compound might not be considered a heterobifunctional "reagent", it will be, in the context of this paper, a thiolating reagent that normally requires a primary amine substrate.

For the past twenty years, two reagents have been used almost exclusively for thiolating polyamines and proteins. Benesch and Benesch [22] introduced the compound, N-acetylhomocysteine thiolactone, which can be aminolyzed as shown in Figure 5a. Direct aminolysis proceeds very slowly except in rather alkaline media (pH 10-11). Although the reaction can be carried out near pH 7 by adding Ag^+ as a catalyst, subsequent removal of the Ag^+ is often troublesome. The acetamido appendage (Figure 5a) on the resulting linkage to an affinity support may interfere with the intended biospecific interaction. This last problem could be more serious in the case of supports or ligands thiolated with the second reagent, S-acetylmercaptosuccinic anhydride (SAMSA) of Klotz and Heiney [23], where the resulting side chain bears a negative charge (see Fig. 5b). However, SAMSA possesses the advantages of being adequately reactive near pH 7 and yielding mainly acetyl-protected thiol groups. At any time after completion of the reaction, S-acetyl groups can be removed easily by treatment with 0.01 to 0.05 M hydroxylamine at pH 7-8 [23, 10] (Fig. 5b).

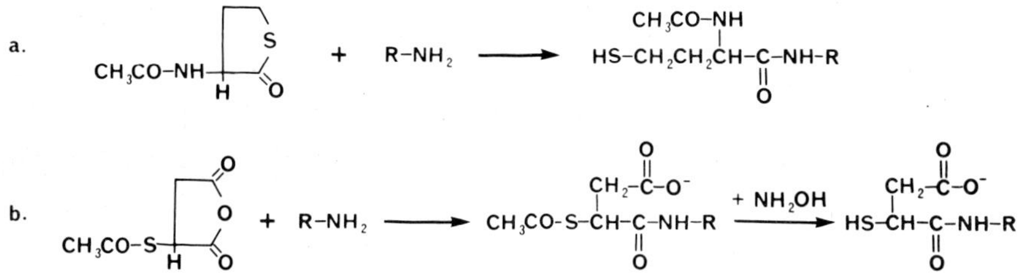

Fig. 5. Thiolation of amines (proteins, supports or ligands) by means of N-acetylhomocysteine thiolactone (a) or S-acetylmercaptosuccinic anhydride (b).

Several imidoesters have been proposed as general reagents for thiolating proteins. Perham and Thomas [24] prepared the free thiol-containing compound, methyl 3-mercaptopropionimidate hydrochloride (Fig. 6a); its synthesis was improved by Traut et al. [25] who also prepared the homologous, 4-mercaptobutyrimidate. A synthesis of the latter compound was likewise described by King et al. [20] who noted that during storage the product gradually cyclized with elimination of methanol to form 2-iminothiolane (Fig. 6b). This cyclic imidothiolester can be synthesized directly and used for thiolation of proteins [20, 26]. Iminothiolane is probably less reactive than the corresponding, open-chain methyl 4-mercaptobutyrimidate, but it can be

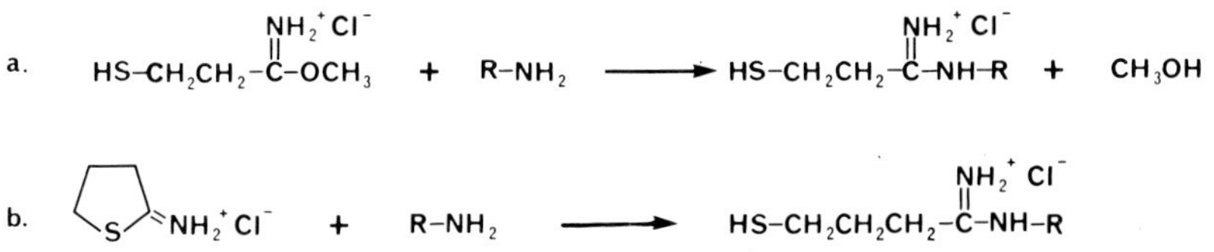

Fig. 6. Thiolation of amines (proteins, supports or ligands) with the imidate ester, methyl 3-mercaptopropionimidate hydrochloride (a), or 2-iminothiolane hydrochloride (b).

stored in the cold for many months and is commercially available (e.g., from Pierce or Sigma Chemical Companies). Methyl 3-mercaptopropionimidate cannot cyclize but slowly decomposes and can be stored for only limited periods. The S-pyridylthio-protected form of this reagent (Fig. 4b) was found to be more stable and can be kept at least two months in a freezer [20]. Thus, it might serve as a useful thiolating agent if used in a manner analogous to SPDP [17] as shown below in Figure 7a. The above reagents are excellent for thiolating proteins since they are especially amine-specific and do not cause a change in net charge when reacting with a protein. When they are used for thiolating supports, the resulting positively charged amidine links could give rise to unwanted electrostatic interactions during later separations.

Proteins and supports can be thiolated with uncharged, 3-mercaptopropionamido groups in several convenient ways as shown in Figure 7; additional spacing is conferred, and no branched structure is introduced. These N-hydroxysuccinimide esters are stable, crystallizable compounds that react cleanly with primary amines. The reagent of of Carlsson et al. [17] (SPDP) serves quite well in the thiolation of amino supports (Fig. 7a), but it is rather expensive for this purpose if purchased. Occasionally, I have used the easily prepared dithiobis(succinimidyl propionate) (DSP)* in the manner shown in Figure 7b. Recently, a similar strategy was employed by Jou and Bankert [19] for thiolating erythrocytes using a mixture of dithiodiglycolic acid ($HOOC-CH_2-S-S-CH_2COOH$) and a water-soluble carbodiimide in place of DSP.

* Available from Pierce Chemical Co.; a synthesis of the radiolabeled reagent has been reported [27].

Their second step also was a reduction with excess dithiothreitol (DTT). In both approaches shown in Figure 7, the same result is obtained. It has been found satisfactory to store supports in the intermediate forms before DTT reduction and to defer the latter step until the SH groups are needed.

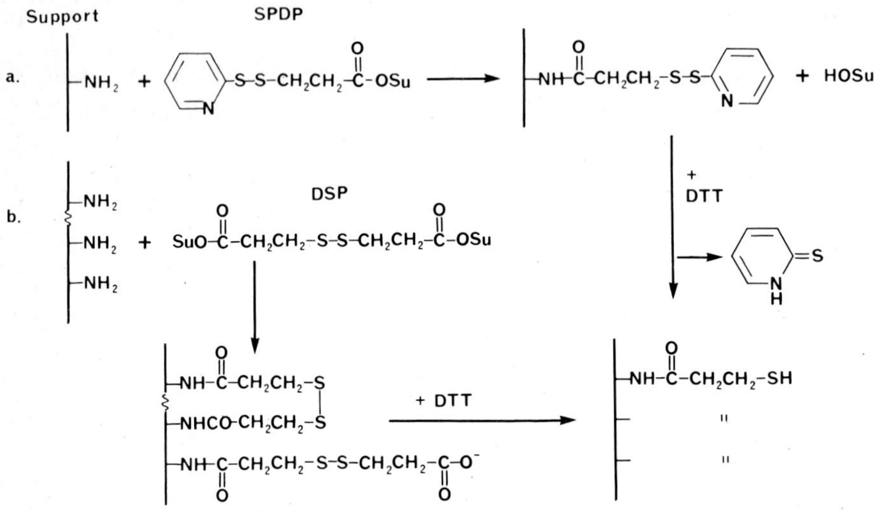

Fig. 7. Two methods for thiolating amine-bearing supports by adding 3-mercaptopropionyl groups. (a.) Succinimidyl 3-(2-pyridyldithio)propionate (SPDP), is added to a suspension of the amine-support in a mixture of 0.15 M HEPES-NaOH buffer, pH 7.6, 1 mM EDTA (3 vols.) and dimethylformamide (DMF, 2 vols.). One milligram (or more as required) of SPDP is added per ml of suspension from a 1% solution in DMF. After 2.5 h (23°) the gel is washed with saline and suspended in the above buffer without DMF and treated for 30 min (23°) with DTT (2 mg/ml suspension) under N_2. (b). Dithio*bis*(succinimidyl propionate) (DSP) was reacted with the amine-support under similar conditions, and reduction with DTT was carried out using 15 mg/ml.

PREPARATION OF A HAPTEN-COUPLED, CROSSLINKED AGAROSE ADSORBENT

AECM-Agarose

Agarose beads have been used widely and successfully in the preparation of affinity supports. The availability of de-sulfated, crosslinked agarose beads in standard porosity ranges has further increased their suitability for this application by providing lower ionic background, lower solubility and greater stability to heat, solvents and alkali [28]. Mechanical stability is excellent since suspensions of such beads (Sepharose CL-4B, Pharmacia) and their chemical derivatives may be frozen and thawed repeatedly without apparent physical or functional damage [14].

I have initially functionalized crosslinked agarose in the manner that I reported earlier for Ficoll (Pharmacia), agarose and soluble dextran [29]. In the first step (Fig. 8), stable carboxymethyl ether linkages were formed with a few of the hydroxyl groups of the poly-D-galactosyl-anhydro-L-galactoside backbone. A large excess of chloroacetate was employed in order to increase the reaction rate at room temperature and minimize exposure of the polysaccharide to strong alkali. The time allowed for this reaction determined the CM and AECM group density. In the second step (Fig. 8), amide bonds were formed between bound carboxyl groups and one end of ethylenediamine molecules to yield the N-(2-aminoethyl)carbamylmethylated (AECM) agarose derivative. The carboxyls were activated by the water-soluble coupling agent, 1-ethyl-3-(3-dimethylaminopropyl)carbodiimide hydrochloride (EDC·HCl). Two-ended reactions (crosslinking) were prevented largely by employing a great excess of the diamine (see discussion in ref. 29).

$$Cl-CH_2-\overset{O}{\underset{\|}{C}}-O^- + HO-A \xrightarrow[2.\ +\ H^+]{1.\ NaOH,\ 2n} HO-\overset{O}{\underset{\|}{C}}-CH_2-O-A$$

$$\downarrow + H_3\overset{+}{N}CH_2CH_2\overset{+}{N}H_3 + 2Cl^-$$
$$ + EDC\cdot HCl$$

$$H_3\overset{+}{N}CH_2CH_2NH-\overset{O}{\underset{\|}{C}}-CH_2-O-A$$

AECM

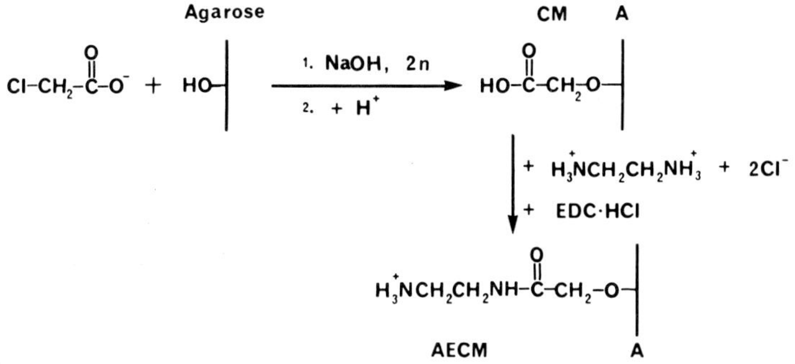

Fig. 8. Formation of carboxymethylated (CM) agarose (A) and N-(2-aminoethyl)-carbamylmethylated (AECM) agarose. The amide linkage in the second step was formed by addition of excess 1-ethyl-3-(3-dimethylaminopropyl)carbodiimide hydrochloride (EDC·HCl).

AECM-Agarose was prepared as follows: Sepharose CL-4B (from Pharmacia; about 250 ml of settled bed) was suspended in 1200 ml of 0.1 M NaCl (saline) and allowed to settle for 60-90 min. Fines were siphoned off. This process was repeated 3 more times, and the gel was washed further on a Buchner funnel

(Coors with paper filter). The gel was suspended in saline to 300 ml volume, and 300 ml of 6.0 N NaOH was added. The suspension was cooled to 15° and 0.60 mole of chloroacetic acid was added (zero time). The suspension was paddled-stirred at 25.0° ± 0.2° until 68 min had elapsed. The mixture was then washed rapidly with saline (on Buchner) to stop the reaction (elapsed time about 75 min). The gel was then washed slowly with 5 l of saline and very slowly with four, 75-ml portions of 1.5 M aqueous ethylenediamine dihydrochloride (EDA•2HCl; 20% w/v, treated with decolorizing charcoal). The suction-drained gel was placed in a beaker and resuspended to 400 ml volume using 1.5 M EDA•2HCl. The pH was adjusted to 4.7 (2 N NaOH), and 6.0 g of EDC•HCl (Fig. 8 legend) was added in portions over 20-25 min at room temperature; the pH was kept near 4.7 (with 2N HCl) during this addition and while the mixture was being stirred for the next 3.5 h. The gel was washed with 5 l of saline, left suspended in saline overnight and washed again slowly with 3 l of saline. The resulting AECM-agarose was washed with and stored in 70% v/v 2-propanol (-20°).

3-Mercaptopropionyl-β-alanyl-AECM-agarose

In order that more non-hydrophobic spacing structure could be placed on the support, a reagent was designed that allows convenient addition of β-alanine residues to AECM-agarose. Accordingly, the crystallizable compound, N-phthalimidyl N'-ethylsulfonylethoxycarbonyl-β-alaninate (Esc-βAla-ON<Pht; Fig. 9, top line), was synthesized briefly as follows: An excess of β-alanine in aqueous dimethylformamide (DMF) was reacted overnight with 2-(ethylsulfonyl)-ethyl p-nitrophenyl carbonate (Esc-ONp) without added base. Esc-ONp was prepared in a manner similar to that described for the methylsulfonyl analogue by Tesser and Balvert-Geer [30]. The ethyl analogue was selected because of the commercial availability of ethylsulfonylethanol. Solvent was removed and the residue was taken up in 1.3% w/v sodium citrate dihydrate. The pH was adjusted to 5.1 with 2 N HCl, and p-nitrophenol was removed by 4 extractions into ethyl ether while keeping the aqueous layer at pH 5.1. The aqueous phase was acidified to pH 1.8, saturated with NaCl, and extracted 3 times with ethyl acetate. Pooled solvent layers were evaporated, and the residue was crystallized from ethyl acetate-ethyl ether 1:1 to yield Esc-β-alanine (m.p. 82.5°-84°).

The corresponding N-hydroxyphthalimide ester (-ON<Pht) was chosen because of its good reactivity and crystallizability [31]. It was prepared from Esc-β-alanine isobutylcarbonic mixed anhydride and N-hydroxyphthalimide and crystallized from absolute ethanol (m.p. 130°-131°). Details of the synthesis of Esc-βAla-ON<Pht and a similarly useful dipeptide derivative, Esc-βAla-βAla-ON<Pht, will be described elsewhere.

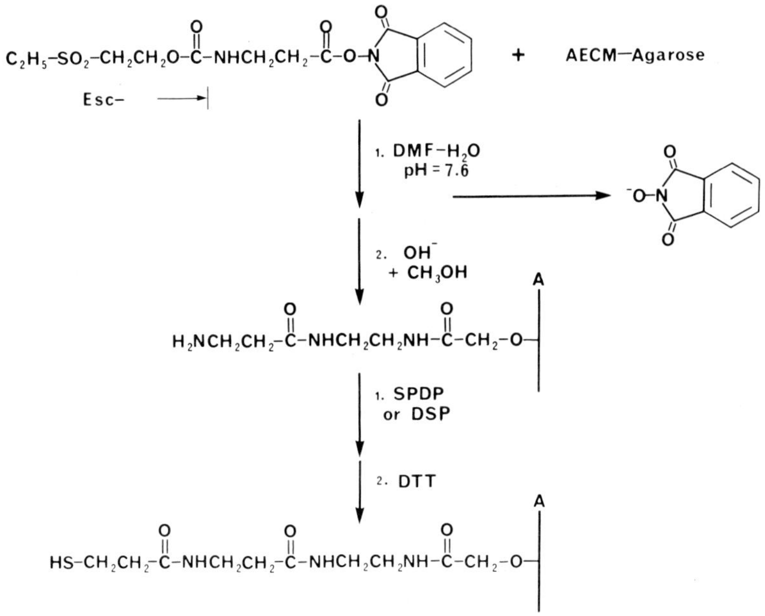

Fig. 9. Extension of the spacing arm of AECM-agarose and thiolation of the terminal amino group. A β-alanine residue is added by means of the active ester, Esc-βAla-ON<Pht (top line, and see text). The alkali-labile Esc group is removed to give β-alanyl-AECM-agarose (middle line). A 3-mercaptopropionyl group is added via SPDP, as described in Fig. 7, to yield 3-mercaptopropionyl-β-alanyl-AECM-agarose.

Esc-βAla-AECM-agarose was prepared as follows: Sixty milliliters of AECM-agarose was suspended to 90 ml in a mixture of 0.15 M HEPES-NaOH buffer, pH 7.6, (3 vols.) and DMF (2 vols.). A solution of 144 mg of Esc-βAla-ON<Pht in 6.0 ml of DMF was added, the suspension was stirred at room temperature for 75 min, and the gel was washed with the above buffer-solvent mixture and with saline. The resulting intermediate follows the first reaction arrow in Figure 9 and is not depicted.

Esc groups were removed by the following washing sequence on a Buchner funnel: (1) saline-methanol 3:1 and 1:1; (2) the composition, 0.05 N NaOH, 0.10 M NaCl, 50% v/v methanol (slowly, for 30 min); (3) saline-methanol 3:1; (4) 0.002 N HCl in saline; and (5) saline alone. The resulting derivative,

H_2^+-βAla-AECM-agarose, is shown in the middle of Figure 9 in the free amine form. The thiolation step was then carried out with SPDP as described in Figure 7 legend to yield 3-mercaptopropionyl-β-alanyl-AECM-agarose as shown on the bottom line of Figure 9.

Thioether attachment of a chloroacetylated hapten

An example of the type of approach I am currently employing for preparing haptenated adsorbents is illustrated in Figure 10. Chloroacetoxysuccinimide (succinimidyl chloroacetate) was made as outlined above and used to chloroacetylate the haptenic amine shown. This reaction is representative of the upper right portion of the general scheme of Figure 1. The support was thiolated as stipulated by the upper left portion of Figure 1 and immediately brought into reaction with the chloroacetylated hapten under nitrogen at a "pH" of 10.9 (readings elevated due to 40% DMF) to form a stable thioether linkage.

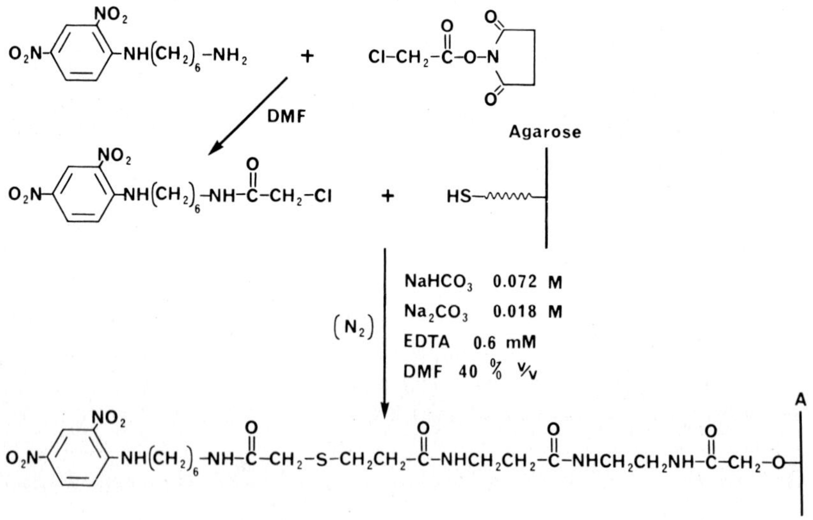

Fig. 10. Reaction of an amine-bearing hapten with the heterobifunctional reagent, succinimidyl 2-chloroacetate. Attachment of the product to the thiolated agarose results in stable thioether linkages and dinitrophenyl-aminohexyl haptenic groups spaced 17 linear atoms from the original polysaccharide matrix.

Freshly prepared 3-mercaptopropionyl derivative of the agarose beads, following the 30-min reduction with DTT (see Fig. 7, pathway a. and Fig. 9, bottom line), was washed thoroughly with a solution having 0.2 M NaCl, 1 mM EDTA, pH 3.7 (with HCl), and then with the composition: 0.072 M $NaHCO_3$, 0.018 M Na_2CO_3, 0.6 mM EDTA, DMF 40% v/v; all solutions had been N_2-equilibrated. The gel was suspended to 1.4 times its settled volume using the latter buffer-solvent mixture, and a solution of chloroacetylated hapten in DMF was added; the amount of hapten did not exceed the SH content of the gel on a molar basis. The mixture was immediately bubbled with N_2 (30 min); the container was capped and rocked at room temperature (23°) overnight (16-20 h). A sample of the supernatant was taken for spectrophotometric analysis, and the gel was washed thoroughly with saline-DMF 3:2 and saline.

A final reduction and alkylation step was performed to tie up unused sulfhydryl groups, some of which had probably been oxidized to disulfide links in the gel: The gel was washed with 0.15 M HEPES-NaOH, 1 mM EDTA, pH 7.6, and suspended in this buffer to 1.5 bed volumes. DTT (2.0 mg/ml of suspension) was added, and the mixture was stirred under N_2 for 45 min. Next, recrystallized iodoacetamide (6.0 mg/ml) was added, and stirring under N_2 was continued for another 45 min. The gel was finally washed with saline and stored frozen in suspension (-20°).

The results of three such preparations are given in Table 1 and illustrate the high efficiency achieved in coupling the dinitrophenyl hapten to an agarose support from solutions of low concentration to give adsorbents with very low

TABLE 1

Coupling efficiency of a chloroacetylated hapten[a] to a thiolated agarose support[b]

Added hapten, μmol/ml gel[c]	Initial conc. of hapten, mM	Hapten bound, μmol/ml gel	Coupling Efficiency, %[d]
0.301	0.215	0.282	93.7
0.616	0.440	0.580	94.1
1.163	0.830	1.073	92.3

[a] 1-(2,4-Dinitrophenylamino)-6-(2-chloroacetamido)hexane.
[b] 3-Mercaptopropionyl-β-alanyl-N-(2-aminoethyl)carbamylmethylated Sepharose CL-4B having approximately 1.2 to 1.3 μmol/ml gel bed of SH groups.
[c] Per ml of settled gel.
[d] As determined by adsorbance of supernatant fluids at 360 nm following 16-20 h of reaction in carbonate-DMF solution of "pH" 10.9 at 23°.

group densities. Coupling efficiencies of over 90% of the added hapten allow a variation range of only a few percent; thus, excellent control over a targeted hapten density can be realized, even at a very low level.

In conclusion, it appears that heteroligating principles and many of the required heterobifunctional reagents being developed for various biochemical and immunological applications can serve usefully in the preparation of affinity adsorbents. These adsorbents should be chemically stable and well-defined and carry specific molecules with readily controllable densities on the support matrix. The general approach outlined here should permit increased flexibility in devising appropriate methods for covalent attachment of ligands and ligand-binding proteins.

ACKNOWLEDGMENTS

I am indebted to the excellent technical assistance of Ms. Barbara Duntley and Ms. Candace Williamson given during the course of this work.

REFERENCES

1 O. Keller and J. Rudinger, Helv. Chim. Acta, 58(1975)531-541.
2 J.L. Fauchere and G.M. Pelican, Helv. Chim. Acta, 58(1975)1984-1994; H.J. Moschler and R. Schwyzer, Helv. Chim. Acta, 57(1974)1576-1584.
3 A.C.J. Lee, J.E. Powell, G.W. Tregear, H.D. Niall and V.C. Stevens, Molecular Immunology, 17(1980)749-756.
4 V.M. Kriwaczek, J.C. Bonnafous, M. Muller and R. Schwyzer, Helv. Chim. Acta, 61(1978)1241-1245.
5 T. Kitagawa and T. Aikawa, J. Biochem., 79(1976)233-236.
6 M.J. O'Sullivan, E. Gnemmi, D. Morris, G. Chieregatti, M. Simmons, A.D. Simmonds, J.W. Bridges and V. Marks, FEBS Letters, 95(1978)311-313; F.T. Liu, M. Zinnecker, T. Hamaoka and D.H. Katz, Biochemistry 18(1979) 690-697.
7 J.A. Schroer, J.K. Inman, J.W. Thomas and A.S. Rosenthal, J. Immunol., 123(1979)670-675.
8 M. Machida, M.I. Machida and Y. Kanaoka, Chem. Pharm. Bull., 25(1977) 2739-2743.
9 S. Yoshitake, Y. Yamada, E. Ishikawa and R. Masseyeff, Eur. J. Biochem. 101(1979)395-399.
10 E.S. Rector, R.J. Schwenk, K.S. Tse and A.H. Sehon, J. Immunol. Methods, 24(1978)321-336.
11 V.M. Kirwaczek, A.N. Eberle, M. Muller and R. Schwyzer, Helv. Chim. Acta, 61(1978)1232-1240.
12 P. Cuatrecasas, M. Wilchek and C.B. Anfinsen, J. Biol. Chem., 244(1969) 4316-4329.
13 E.N. Hudson and G. Weber, Biochemistry, 12(1973)4154-4161; L. Lorand, W.T. Brannen, Jr. and N.G. Rule, Arch. Biochem. Biophys., 96(1962)147-151.
14 J.K. Inman, unpublished observations.
15 C.H. Moussebois, J.F. Heremans, P. Osinski and W. Rennerts, J. Org. Chem., 41(1976)1340-1343.
16 R.C. Parker, S. Stanley and D.S. Kristol, Int. J. Biochem., 6(1975)863-866.
17 J. Carlsson, H. Drevin and R. Axen, Biochem. J., 173(1978)723-737.

18 J. Carlsson, R. Axen and T. Unge, Eur. J. Biochem., 59(1975)567-572; J. Carlsson, I. Olsson, R. Axen and H. Drevin, Acta Chem. Scand., B30(1976) 180-182.
19 Y.H. Jou and R. B. Bankert, Proc. Natl. Acad. Sci. (USA), 78(1981)2493-2496.
20 T.P. King, Y. Li and L. Kochoumian, Biochemistry, 17(1978)1499-1506.
21 J.K. Inman, G.C. DuBois, E. Appella and R.N. Perham, in T.Y. Liu and A.N. Schechter (Eds.), Chemical Synthesis and Sequencing of Peptides and Proteins, Elsevier/North Holland, Amsterdam, 1981, pp.231-237.
22 R. Benesch and R.E. Benesch, Proc. Natl. Acad. Sci. (USA), 44(1958)848-853.
23 I.M. Klotz and R.E. Heiney, Arch. Biochem. Biophys., 96(1962)605-612.
24 R.N. Perham and J.O. Thomas, J. Mol. Biol., 62(1971)415-418.
25 R.R. Traut, A. Bollen, T.T. Sun, J.W.B. Hershey, J. Sundberg and L.R. Pierce, Biochemistry, 12(1973)3266-3273.
26 H.J. Schramm and T. Dulffer, Hoppe-Seyler's Z. Physiol. Chem., 358(1977) 137-139.
27 A.J. Lomant and G. Fairbanks, J. Mol. Biol., 104(1976)243-261.
28 J. Porath, J-.C. Janson and T. Laas, J. Chromatogr., 60(1971)167-177.
29 J.K. Inman, J. Immunol., 114(1975)704-709.
30 G.I. Tesser and I.C. Balvert-Geers, Int. J. Peptide Protein Res., 7(1975)295-305.
31 G.H.L. Nefkens, G.I. Tesser and R.J.F. Nivard, Rec. Trav. Chim. 81(1962) 683-690.

T.C.J. Gribnau, J. Visser and R.J.F. Nivard (Editors),
Affinity Chromatography and Related Techniques

THE DETERMINATION OF ACTIVE SPECIES ON CNBr AND TRICHLORO-S-TRIAZINE ACTIVATED POLYSACCHARIDES

JOACHIN KOHN and MEIR WILCHEK*

Department of Biophysics, Weizmann Institute of Science, Rehovot, Israel and
Fogarty International Center, National Institutes of Health, Bethesda, Maryland.

ABSTRACT

Analytical methods were developed for the quantitative determination of reactive groups on polysaccharides which are formed upon activation with cyanogen bromide (CNBr) and Trichloro-s-triazine (cyanuryl chloride). Cyanate esters and imidocarbonates, the active species on CNBr activated polysaccharide, were determined as follows: Imidocarbonates by selective acid hydrolysis followed by the determination of the liberated ammonia, Cyanate esters and Triazines by the color reaction with pyridine and dimethyl barbituric acid. Using these procedures the upper limit of the coupling capacity of an activated resin can be determined. A titrimetric procedure is also suggested, using the coupling of methionine and derivatives, which facilitates the determination of coupled ligand directly on the resin without prior total hydrolysis.

INTRODUCTION

The immobilization of biologically active compounds on polysaccharides is usually a two step reaction: a) activation of the chemically inert polysaccharide resin and b) coupling of the ligand to the activated resin (Fig. 1).

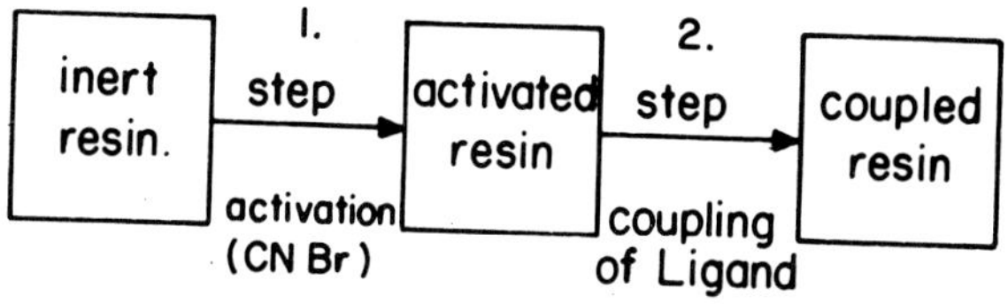

Fig. 1. Steps in immobilization of biologically active compounds

*Fogarty Scholar-in-Residence, NIH, Bethesda, Md.

The most widely used activation procedure employs cyanogen bromide (ref. 1, 2) and in a few studies trichloro-s-triazine was advocated (ref. 3). Most of our pres knowledge on the structure of activated polysaccharides with CNBr (Fig. 2) is deriv

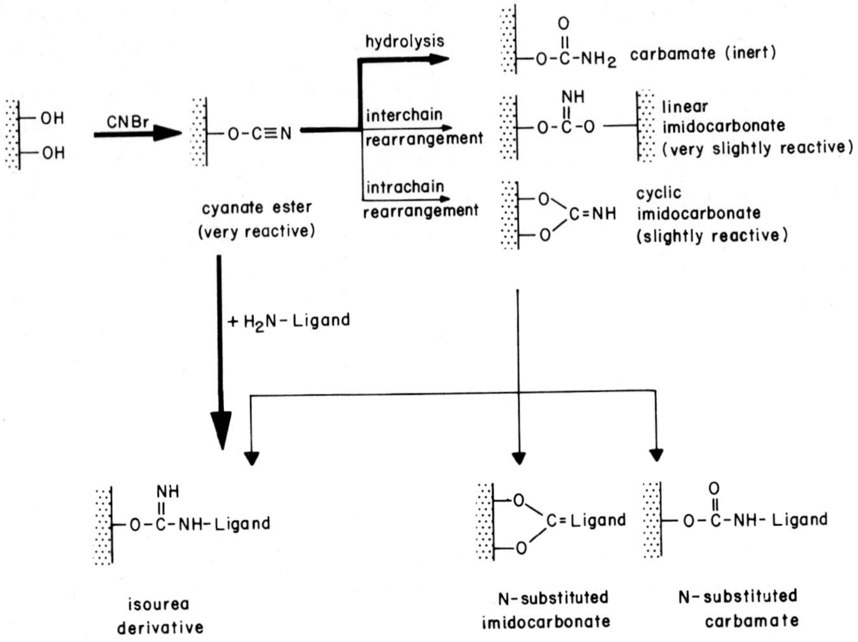

Fig. 2. Mechanism of activation and coupling to Sepharose. Heavy lines indicate m reaction pathways.

from studies of soluble, monomeric model compounds and some studies with polysaccha rides (ref. 4, 5, 6). The nature of the active species, responsible for coupling o the ligand, has only lately been fully identified. Whereas, imidocarbonates were widely described as the active species during coupling (ref. 4), our studies (ref. 7, 8, 9) provided evidence for the cyanate esters as predominantly responsible for coupling of ligands.

Due to the absence of kinetic data regarding the rate of formation and decay o cyanate esters and imidocarbonates, optimisation of the activation and coupling reactions was done on an empirical basis of "trial and error." Although for most practical applications, the amount of covalently coupled ligand is the sole parame of interest, the lack of suitable analytical methods for the determination of the "degree of activity" and the expected "coupling capacity" of activated resins has regarded as a disadvantage of the CNBr activation procedure. Considering the high cost of some ligands (e.g., purified antibodies, hormones, etc.) it may sometime b useful to characterize the activated resin before coupling of ligand is attempted.

In the following we shall describe procedures for the determination of imidocarbonates, cyanate esters and cyanuryl derivatives directly on the activated resins (Sepharose and Sephadex).

Due to the high sensitivity of imidocarbonates to acids, we found it possible to cause the selective hydrolysis of imidocarbonates only. This is the basis of the procedure for the determination of imidocarbonates (ref. 9). Cyanate esters can be determined quantitatively with high sensitivity, using König's Reaction (ref. 7).

The described analytical procedures were developed in such a way as to facilitate their intergration into a single analytical scheme, which makes it possible to characterize activated resins in terms of their content of active groups and in terms of their expected coupling capacity within about 2 hours. The obtained results are accurate and reproducible only, if related to the weight of dry resin. Therefore, whenever possible, the resins were dried prior to analysis. Since a certain fraction of cyanate esters and imidocarbonates may decay during drying, the use of wet samples is sometimes unavoidable. In these instances the resin was collected after analysis and its exact weight was determined after drying it in vacuo over P_2O_5.

Determination of imidocarbonate

As reported previously, activated Sephadex and Sepharose lose a certain fraction of their total nitrogen content during about 2 hours of hydrolysis at pH 3 (ref. 10). Table 1 shows that, upon hydrolysis at pH 1, both Sephadex and Sepharose

TABLE 1

Rate of liberation of NH_4^+ during acid hydrolysis

Time of hydrolysis (min)	Amount of NH_4^+ liberated[a] (in % of total nitrogen content of resin) Sepharose 4B[b]			Sephadex G50 (fine)[c]
	pH 1 - 40°C	pH 1 - 20°C	pH 4 - 20°C	pH 1 - 20°C
2	25.8	21.9	-	-
5	27.2	24.7	16.4	33.1
10	27.4	26.9	20.6	42.9
20	28.5	27.1	23.0	44.4
30	28.8	27.3	24.4	46.0
45	28.7	28.1	-	46.8
60 (1 h)	28.7	28.8	25.4	46.3
120 (2 h)	28.8	28.6	26.0	-
480 (4 h)	-	29.3	28.7	-
1440 (24 h)	28.9	29.8	29.0	46.5

[a] weighted samples of activated resins were hydrolysed in 0.1N HCL or 0.2M sodium acetate (pH 4) respectively. At the indicated times, the suspensions were shortly centrifuged and aliquots of clear supernatant were withdrawn and analysed for NH_4^+ by the ninhydrin reaction.

[b] total nitrogen content according to Kjeldahl: 1350 µmol/g dry resin

[c] total nitrogen content according to Kjeldahl: 2130 µmol/g dry resin

lose a part of their total nitrogen content within about 1 h and prolonged additional hydrolysis does not further decrease the nitrogen content of the resins.

Among the three major products of activation carbamates, imidocarbonates, and cyanate esters (Fig. 2), only the imidocarbonates are known to hydrolyse rapidly at pH 1 according to the following equation (ref. 11).

$$\begin{matrix} R-O \\ R-O \end{matrix}\!\!\Big\rangle C = NH + H^{+} + H_2O \longrightarrow \begin{matrix} R-O \\ R-O \end{matrix}\!\!\Big\rangle C = O + NH_4^{+}$$

In principle, also carbamates are susceptible to acid hydrolysis. However, this reaction seems to be extremely slow at pH 1 and does not interfere with the determination of imidocarbonates. Cyanate esters are hydrolysed to carbamates by acid (ref. 12), and do not cause any liberation of NH_4^+. Imidocarbonates, carbamates and cyanate esters together account for nearly the entire nitrogen content of the resin. Consequently the amount of NH_4^+, liberated during short hydrolysis in 0.1N HCl can be assumed to be equivalent to the amount of imidocarbonates present on the resin.

Procedure for *imidocarbonates*. A sample of activated resin, containing about 1-20 µmol of imidocarbonates is placed in a 25.0 ml measuring bottle. 5ml of 0.1N HCl are added. After 30 min of hydrolysis at 40°C (or alternatively after 60 min hydrolysis at room temperature) the volume is made up to 25.0 ml exactly with 0.2M acetate buffer, pH 5.5 After thorough mixing the suspension is allowed to settle is shortly centrifuged to remove the insoluble resin and aliquots of 1 ml exactly of the clear supernatant are used for determination of NH_4^+ by the ninhydrin react described below or by any alternative method. The amount of NH_4^+ found is equival to the amount of imidocarbonates on the resin.

Determination of ammonia by ninhydrin reaction

To 1 ml of sample solution, containing between 0.05 and 0.8 µmol of NH_4^+ is add 1 ml of 2M acetate buffer (pH 5.5) and 1.5 ml of hydrindantin reagent. (880 mg hydrindantin and 4.0 g ninhydrin are dissolved in 100 ml of methyl - cellusolve) After mixing, the solutions should show a light red color. The stoppered test tub are heated to 100°C for 20 min. The colored contents of each test tube are dilute to 10 ml exactly with cold formaldehyde diluent (0.5% solution of formaldehyde in isopropanol: water). The diluent reacts with excess hydrindantin and within a few seconds the regular blue ninhydrin color is obtained. The "blank" is slightly yel and its absorbance at 570 nm should be below 0.05. In each analysis a "blank" solution (1 ml of 0.2M acetate buffer, pH 5.5) and three standard solutions, containing 0.1 µmol/ml, 0.3 µmol/ml and 0.6 µmol/ml of highly purified, dry NH_4Cl wer included. Absorbance was measured at 570 nm and the amount of NH_4^+ in the sample was determined from a calibration curve in the usual way.

Determination of cyanate esters

The mechanism of color formation by the König reaction and N,N'-dimethylbarbituric acid or barbituric acid as color forming reagent was described earlier (ref. 7). Cyanate esters and probably trace amounts of cyanuric acid derivatives (triazines) are the only species present on activated resin which are capable of cleaving the pyridine ring, thereby initiating color formation. The spectrophotometric determination of cyanate esters by the König reaction requires either the availability of a calibration curve or the value of the molar absorption coefficient ε must be known. In the case of cyanate esters, incorporated into an insoluble resin, the value of ε can only be determined by indirect means, since there is presently no alternative method for the determination of resin-bound cyanate esters. Therefore the apparent ε value for cyanate esters was calculated by determining the decrease in the nitrogen content of the resin due to color formation (ref. 9). ε was found to be 137000 liter $mol^{-1} cm^{-1}$ in aqueous solution containing 1-5% pryidine.[1] This value represents an average of 5 repetitive determinations. Considering the obtained standard diviation (SD = 8700 liter $mol^{-1}cm^{-1}$), the given ε for cyanate esters can be expected to be correct within about $\pm$10%. Therefore, inspite of the fact that the results of the color reaction are highly reproducible (see Table 2), the overall accuracy of the method is limited to $\pm$10%.

Procedure for *cyanate esters*. To 5-20 mg of dry, activated resin or an equivalent amount of wet resin, 5 ml of color reagent is added. (500 mg of N,N'-dimethylbarbituric acid are suspended in 5.0 ml of water, and 45 ml of cold distilled pyridine are added). The mixture is vigorously stirred and warmed to 40°C for 25 min in a closed test tube. After that the mixture is diluted to any convenient volume with distilled water. For 10 mg of freshly activated resin, usually, dilution to 250 - 500 ml is necessary in order to reduce the optical density of the purple solution to 0.5 - 1.0. Aliquots of this diluted solution are filtered and used to measure the absorbance at about 588 nm. Using Beer Lambert's Law, the amount of cyanate esters can be calculated in the usual way, using a value of 137 ml $\mu mol^{-1} cm^{-1}$ as molar absorption coefficient.

[1] The value of ε was found to be very sensitive to solvent effects and increases significantly with increasing concentration of organic solvents.

TABLE 2

Reproducibility of the procedure[a] for the determination of cyanate esters

Amount of sample (mg dry resin)	Repetition	Final volume after dilution (ml)	Measured absorbance at 588 nm	Calculated am of cyanate es (μmol/100 mg
	1	500	0.085	31.0
	2	250	0.167	30.4
1.00	3	250	0.165	30.1
	4	100	0.459	33.5
	1	500	0.249	30.3
	2	250	0.496	30.2
3.00	3	250	0.503	30.6
	4	250	0.497	30.2
	1	1000	0.234	28.5
	2	500	0.497	30.2
6.00	3	500	0.501	30.5
	4	250	1.004	30.5

[a] Samples of commercially activated Sepharose 4B, obtained from Pharmacia (Lot FI 16294) were repeatedly analysed by the procedure described. Both the amount of s and the degree of final dilution were varied, and the analyses were performed on different days. For the calculation ε = 137 ml $\mu mol^{-1} cm^{-1}$.

Determination of Trichloro-s-triazine activated Sepharose

Cyanuryl chloride activated agarose contain two reactive acyl-like chlorines (ref.3). The carbon atoms are strongly electrophilic and react with pyridine sim to the cyanate ester. The pyridine aduct formed on the column is hydrolysed and released as a glutaconic aldehyde which reacts with the dimethylbarbituric reager give the same color as the cyanate ester, with theoretically twice the absorptior coefficient. Preliminary results indicated that the reaction proceeds nicely but do not have yet conditions for obtaining the high ε as expected, even though the reaction on the polysaccharide is quantitative as determined by nitrogen analysis The pyridine reaction can be used to destroy excess chlorides remained after the coupling reaction with ligands. (T. Miron, J. Kohn and M. Wilchek)

Determination of coupling capacity of activated resins

Methionine coupled to polysaccharide can be quantitatively and rapidly oxidi by H_2O_2 to the corresponding sulfoxide (ref. 13). By determining the amount of H consumed by the oxidation of Met, the amount of Met in the sample may be calculat as described below. Met or any Met containing small peptide may be used as model ligand. Cys, His, Tyr, Trp residues interfere with the determination of Met and be absent.

Procedure: To a weighted sample of about 15-50 mg of coupled, dry resin is added 1.0 ml of 1.5M perchloric acid, followed by exactly 200 µl of oxidation mixture. (Mix 96 ml of 60% perchloric acid and 4.0 ml of 30% H_2O_2 and dilute to 300 ml with distilled water). The samples are allowed to react with stirring at 20°C for 25 min exactly. After that time the samples are diluted to about 6 ml with distilled water and nitrogen is bubbled through the reaction mixture for 2 min in order to expel oxygen from the system. Thereafter the stream of nitrogen is interrupted and the iodometric titration of H_2O_2 is commenced by adding 1 ml of KI solution and 0.5 ml of molybdate catalysator solution (ref. 14). 0.05N standard sodium thiosulfate solution is used as titrant in a microcuvet having an accuracy of $\pm$ 1 µl. The end point is reached when the titrated solution remains colorless for at least 1 min. The amount of Methionine on the resin is calculated according to ref. 13. (The blank value is usually zero within the limits of experimental mistake.)

Stability of cyanate esters and inidocarbonates or activited resin

Using the analytical procedures described above, the stability of cyanate esters and imidocarbonates as function of pH was investigated. As expected, the stability of imidocarbonates was found to increase with increasing pH. Interestingly, no stability maximum at pH 7 was apparent. Instead, imidocarbonates seem to be most stable in rather strongly basic medium (pH 12-14), and under typical coupling conditions (pH 9, 4°C) over 50% of all imidocarbonates survive for more than 24 h. Cyanate esters behave in a directly opposite fashion: they were found to be extremely sensitive toward base and most stable in 0.1N HCl. Whereas all cyanate esters can be hydrolysed to carbonates within a few seconds at pH 14 (20°C), over 50% of all cyanate esters remained intact, when a sample of activated Sepharose 4B was hydrolysed for 24 h in 0.1N HCl.

It is well known that activated Sepharose must be washed with acid (10^{-3}N HCl) in order to perserve its coupling capacity, whereas activated Sephadex is best washed with base (0.1M $NaHCO_3$). Since it was assumed that imidocarbonates are the active species in both Sepharose and Sephadex, no satisfactory explanation could be found for the seemingly contradictory behavior of the proposed imidocarbonates. However, if cyanate esters are assumed to be the active species of Sepharose, and imidocarbonates the active species on Sephadex, the stability of those species corresponds perfectly to the stability of activated Sepharose and Sephadex toward acid or base, as shown in a qualitative way by Fig 3.

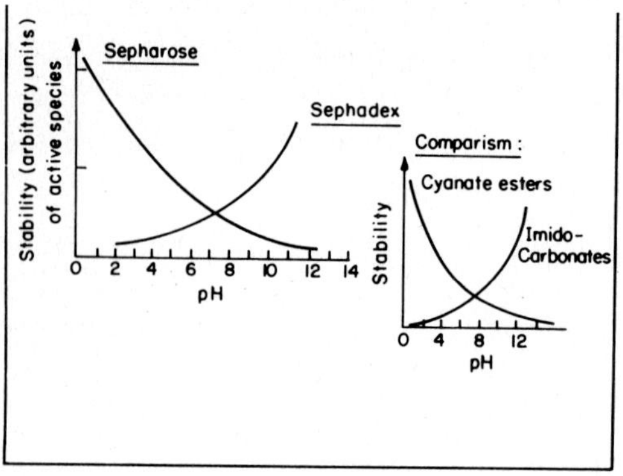

Fig. 3. Stability of coupling capacity as function of pH.

Prediction of coupling capacity of an activated resin.

Based on the assumption that coupling of ligand to cyanate esters and to imid carbonates occurs simultaneously and independently, the total coupling capacity of activated resin should be related to the amounts of cyanate esters and imidocarbor present on the resin according to the following equation:

Total coupling capacity = A [cyanate esters]+B [imidocarbonates]

In this equation A and B represent the respective coupling yields under specific experimental conditions.

When a sample of activated Sepharose, hydrolysed in 0.1N HCl at 20°C for 60 m was analysed, it was found to contain 25 μmol/100 mg resin of cyanate esters and r imidocarbonates. When Met was coupled to this sample, 19.0 μmol/100 mg resin of M bound covalently to the resin, as determined by the H_2O_2 oxidation procedure. Thi corresponds to a coupling yield of 82%. The experiment was represated several time with different samples of resin containing only cyanate esters. It was always fou that the amount of coupled Met corresponds to 80% (±5%) of the amount of cyanate esters present on the resin prior to coupling (fig. 4).

The same kind of experiment was repeated with different samples of activated Sepharose, which were hydrolysed in 0.1N NaOH at 20°C for 10 min and which therefc contained only imidocarbonates and no cyanate esters, as verified by analysis. Af coupling, the amount of bound Met always corresponded to 15% (±3%) of the amount c imidocarbonates initially present on the sample (fig. 4).

Assuming the value of A and B to be 0.80 and 0.15 respectively, the given equation was used to predict the coupling capacities of various samples of activated Sepharose. The results collected showed that it is possible to predict the coupling capacity of activated resins towards small model ligands with satisfactory accuracy.

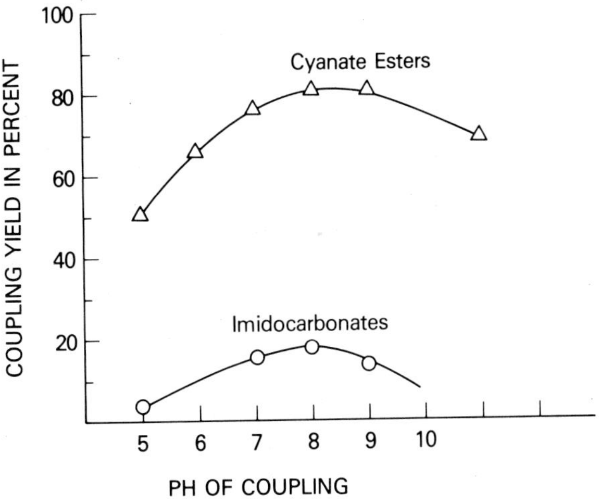

Fig. 4. Coupling of methionine to Sepharose containing only cyanate esters or imidocarbonates.

FINAL COMMENTS

The fortunate circumstance that cyanate esters and imidocarbonates behave in a directly opposite fashion during hydrolysis in acid or base makes it possible to prepare and use resins containing exclusively cyanate esters or imidocarbonates. The preparation of such resins facilitated the determination of the upper limit of the resin's coupling capacity. A stoichiometrical relation between the amount of active groups and the amount of coupled ligand was demonstrated for small monovalent model ligand. The exact amount of coupling of macromolecules, which contain many possible points of attachment, cannot be predicted easily.

Since 70-85% of the total coupling capacity of activated Sepharose is due to cyanate esters, this study can be considered as direct evidence for cyanate esters as the predominantly active species on Sepharose as suggested by us earlier.

REFERENCES

1 W.B. Jakoby, and M. Wilchek (ed.), Methods in Enzymol. 34 (1974) 1-755.

2 R. Axen, J. Porath, and S. Ernback, Nature, 214, (1967) 1302-1304.

3 T.H. Finlay, V. Troll, M. Levy, A.J. Johnson, and L.T. Hodgins, Anal. Biochem., 87 (1978) 77-90.

5 L. Ahrgren, L Kagedal, and S. Akerstrom, Acta Chem. Scand., 26 (1972) 285-288.

6 M. Wilchek, T. Oka, and Y.J. Topper, Proc. Nat. Acad. Sci. USA, 72 (1975) 1055-1058.

7 J. Kohn and M. Wilchek, Biochem. Biophys Res. Commun., 84 (1978) 7-14.

8 M. Wilchek and C.S. Hexter, Meth. Biochem. Anal., 23 (1976) 347-385.

9 J. Kohn and M. Wilchek, Anal. Biochem. (in press).

10 R. Axen and P. Vretblat, Acta Chem. Scand., 25 (1971) 2711-2716.

11 J. Houben and E. Schmidt, Ber., 46 (1913) 2447-2460.

12 K.A. Jensen, M. Due, A. Holm, and C. Wentrup, Acta Chem. Scand., 20 (1966) 2091-2106.

13 A.A. Albanese, J.E. Frankston, V. Irby, J. Biol. Chem., 156 (1944) 293-302.

14 A.L. Vogel, Textbook of Quantitative Inorganic Analysis, 3rd edn., Wiley and Son New York, 1961, pp. 343-362.

T.C.J. Gribnau, J. Visser and R.J.F. Nivard (Editors),
Affinity Chromatography and Related Techniques

FINAL AFFINITY SUPPORT CHARACTERIZATION : (IN)HOMOGENEITY, STABILITY OF LIGAND-MATRIX LINKAGE

J.LASCH, R. KOELSCH, S. WEIGEL, K. BLAHA[+] and J. TURKOVA[+]
Physiologisch-chemisches Institut der Martin-Luther-Universität, Halle/Saale (G.D.R.)
[+] Institute of Organic Chemistry and Biochemistry, Czechoslovak Academy of Science, Prague (Czechoslovakia)

ABSTRACT

The homogeneity of the distribution of protein ligands coupled to Sepharose and Sephadex has been studied by electron microscopy and microfluorometry. Depending on molecular size, matrix porosity and reactivity an even distribution as well as radial gradients of ligand density have been found.

A second type of heterogeneity resides in the diversity of ligand-matrix interactions. Circumstantial evidence from ligand leakage, electrophoretic desorption and thermal inactivation studies points to the existence of at least three distinct populations of immobilized proteins.

INTRODUCTION

The usual characterization of bioaffinity supports with respect to the amount of attached ligand, carrier geometry, binding capacity or specific biological activity will as a rule not reveal inhomogeneities of the population of fixed ligands. Working with immobilized proteins, mainly enzymes, we have taken advantage of a number of techniques specific for proteins to address this problem. This includes visualization of proteins on the microscopic level, thermal inactivation, leakage studies and electrophoretic desorption experiments.

Electron microscopic and microfluorometric techniques (ref. 1) were employed to study the distribution of bovine eye lens leucine aminopeptidase (ref. 2) within polysaccharide carrier beads. The immune ferritin technique and labelling with the fluorescent dye

FITC+ (molar FITC/protein ratio = 5.4) were used to visualize the enzyme. Scanning of carrier beads grafted with fluorescein-tagged enzyme was done with the FLUOVAL 1 photometrie (VEB Carl Zeiss, Jena equipped with a high pressure mercury lamp HBO 200, excitation filte 2xKP 490, BG 23/4, dichroic mirror TS 500, barrier filters GG 5, OG Zeiss objectives apo 16/0.40 and the Tesla photomultiplier 650 PK 41 The amplified photomultiplier signal was either recorded or fed to t A/D converter TEC 1 and printed.

Electrophoretic desorption studies with Sepharose-bound leucine aminopeptidase was done by a slightly modified PAGE+ technique. Smal samples of the enzyme gel suspension corresponding to 140 or 350 μg protein were applied on the top of polyacrylamide gels, run at 3 mA per tube for 2 or 3 hours and stained with Coomassie Brilliant Blue G 250.

Earlier leakage studies with proteins bound to Sepharose with the CNBr+ method (ref. 3) were extended to a number of model peptides, amino acid derivatives and albumin covalently attached to hydroxy-alkyl methacrylate gels derivatized with epichlorohydrin (Separon H 300 E_{max}), (ref. 4).

Leakage experiments were performed with 200 or 400 mg portions of the dry gel-peptide or gel-protein conjugates which were mixed with 200 or 400 μl of the appropriate sterilized buffer and left at room temperature with occasional shaking. After preselected times the gel was spun down and the supernatant analyzed for its nitrogen content by a modified micromethod (ref. 5).

The leakage data were fitted to the leakage function derived earlier (refs. 3,6) :

$$c_o/a = F(1 - exp(-kt))^n \qquad (1)$$

c_o = concentration of free (released) ligand, a = total ligand conce tration, n = number of anchoring bonds, k = pseudo first order rate constant of ligand detachment, F = fraction of ligand released and t denotes time. The stability of different bioaffinity supports towards leakage is most conveniently compared by their half-lives. Putting in eq.(1) $c_o/a = 1/2$ leaves us with:

$$T_{1/2} = - \ln(1 - \sqrt[n]{0.5})/k \qquad (2)$$

+Abbreviations used: FITC = fluorescein isothiocyanate, CNBr = cyan-ogen bromide, PAGE = polyacrylamide electrophoresis, LAP = leucine aminopeptidase, LpNA = leucine p-nitranilide.

All calculations and curve fitting were performed on a Hewlett-Packard desk calculator HP-85.

RESULTS AND DISCUSSION

Distribution of immobilized proteins

With sufficiently low concentrations of CNBr ($c < 50$ mg/ml activation mixture) a homogeneous distribution of the protein ligand within Sepharose beads is obtained (Fig. 1).

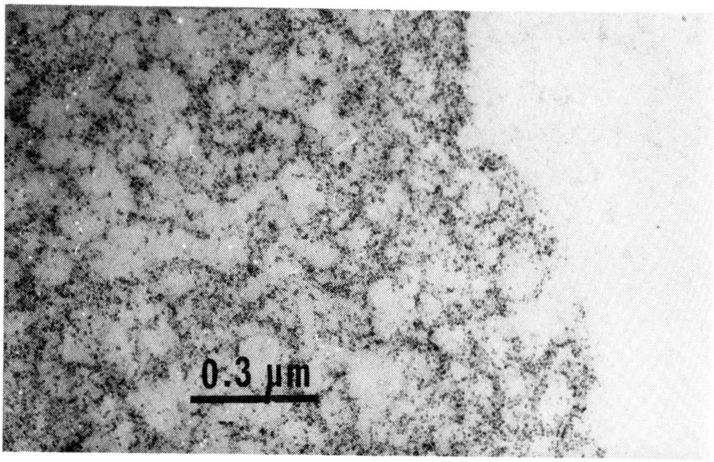

Fig. 1. Leucine aminopeptidase bound to Sepharose 6B with CNBr ($c = 20$ mg/ml activation mixture) and visualized with the immune ferritin technique.

Depending on molecular size, ratio between coupling capacity and amount of ligand available, matrix porosity and reactivity a radial gradient of the density of protein ligands might be obtained (Fig. 2). To a good approximation this type of distribution can be described by power functions of the particle radius: $f(r) = Ar^n$; A = constant, $n = 2,...,m$ (ref. 7). Radial matrix density gradients of the carrier beads can be excluded as source of the non-uniform enzyme distribution (ref. 8).

A third type of protein ligand morphology is the formation by choice or chance of a protein shell on the bead surface. An example is shown in Fig. 3.

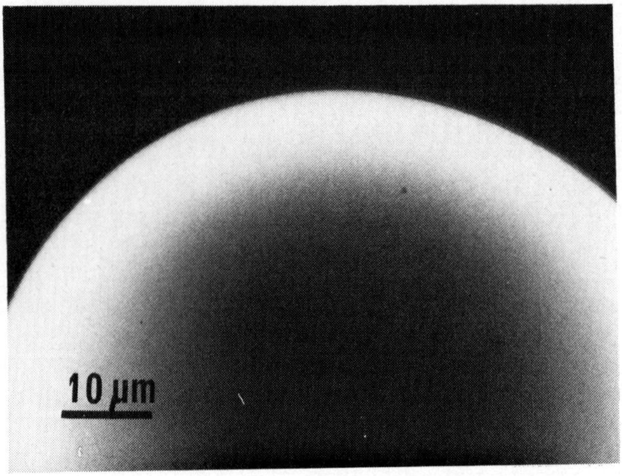

Fig. 2. Microfluorograph of a ligand density gradient produced by coupling FITC-labelled aminopeptidase to strongly CNBr-activated Sepharose 6B (c = 80 mg/ml activation mixture).

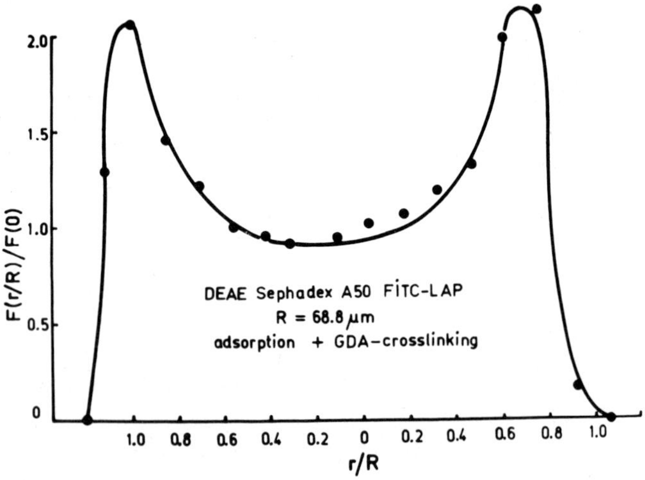

Fig. 3. Fluorescence intensity scan of a carrier bead covered with FITC-labelled leucine aminopeptidase.

The measured fluorescence profiles of Figs. 2 and 3 can be compared with calculated profiles and thus traced down to enzyme distributio f(r/R) within the beads (Figs. 4 and 5).

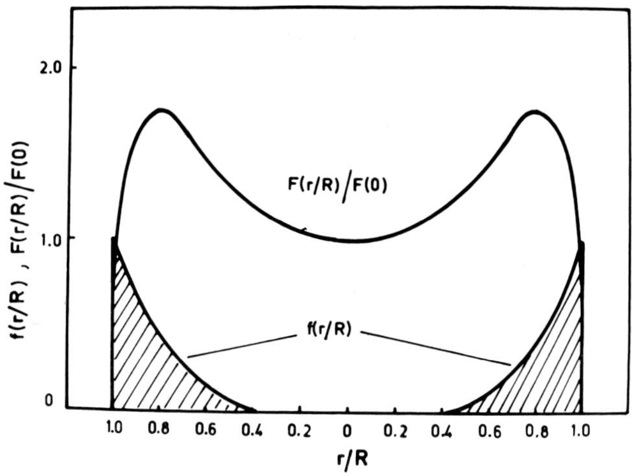

Fig. 4. Enzyme distribution according to the function $f(r) = Ar^4$ (hatched area) and corresponding normalized fluorescence profile F(r/R)/F(0).

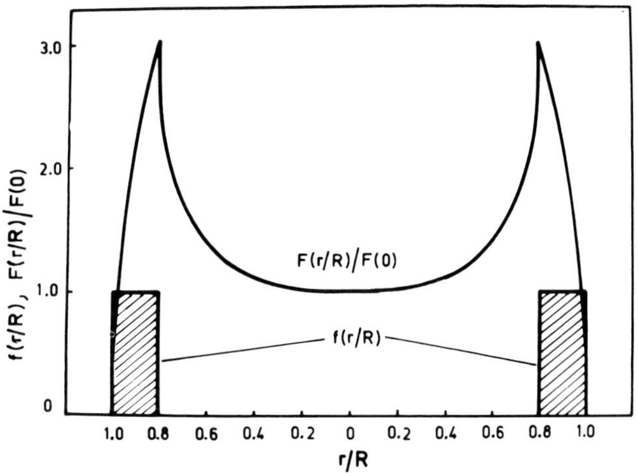

Fig. 5. Shell-like enzyme distribution at the bead surface (hatched area) and corresponding normalized fluorescence profile F(r/R)/F(0).

Thermal inactivation

Biphasic thermal inactivation of immobilized enzymes has been reported as early as 1973 by Zaborsky (ref. 9) and confirmed by many others later (ref. 10).Even with enzymes for which autolysis or hydrolysis by contaminating endoproteases after temperature-induced release can be absolutely excluded, a biphasic inactivation is observed (Fig. 6). As with many other enzymes (ref. 10) an exponential polynomial with 2 terms gives an excellent fit to the inactivation data. Thus the immobilized enzyme molecules fall into 2 classes, a labile and a stable one. The amount of the stabile fraction is very low at 70 °C (cf.Fig. 6) but increases to 60% at 55 °C.

The most conspicuous result of the evaluation of our data as well as numerous data from the literature is that the inactivation rate constant of the labile fraction differs only slightly from this same constant of the soluble enzyme. We interprete this to mean that the labile fraction, either preexisting or produced by the heating, represents enzyme molecules which are not bound covalently.

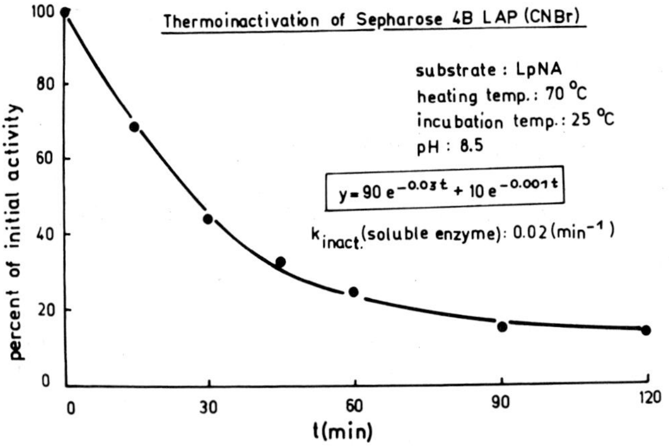

Fig. 6. Thermal inactivation of leucine aminopeptidase bound to Sepharose CL-4B, labile fraction = 90%, stable fraction = 10%, k_1 = 0.03 (1/min), k_2 = 0.001 (1/min), $k_{soluble}$ = 0.02 (1/min).

Leakage studies

It is well established now that the CNBr coupling method yields ligand support conjugates with relatively high leakage rates (refs. 11, 12, 13). Therefore much effort has been put into the developmen of reactive groups on the carrier which yield leakage-resistant ligand-matrix bonds (ref.13). Sepharose bearing oxirane groups intr duced by Sundberg and Porath proved to give very stable conjugates with nucleophilic ligands the stability of which seems to be limite only by the stability of the carrier matrix (ref. 14). Coupling various model peptides to hydroxyalkyl methacrylate gels derivatized with epichlorohydrin and studying the release of nitrogen at various pH we obtained the results shown in Fig. 7 and Table 1. Leakage data of the model protein albumin are shown in Fig. 8.

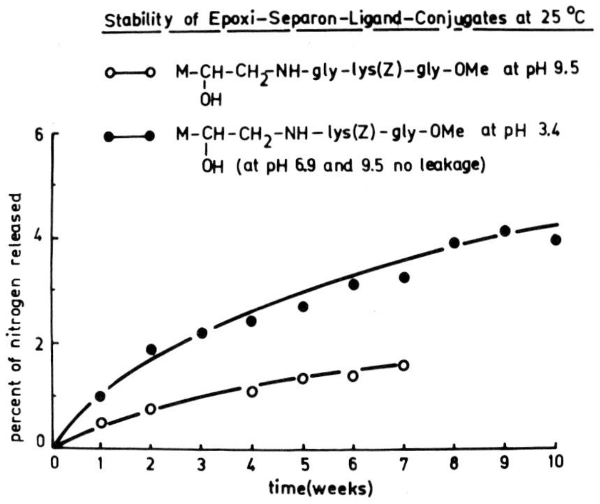

Fig. 7. Release of nitrogen from peptide-Separon conjugates, upper curve: dipeptide derivative at pH 3.4 ; lower curve: tripeptide derivative at pH 9.5

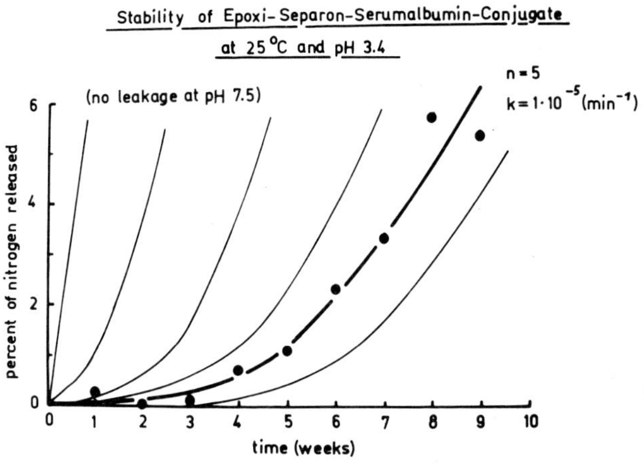

Fig. 8. Release of nitrogen from albumin attached to Separon H 300 via oxirane groups. The set of curves corresponds to n = 1, 2, ..., 6 of the leakage function eq.(1).

No leakage was observed with gly-OMe and $(lys\text{-}ala\text{-}ala)_x$ at acid pH (3.4) and with lys(Z)-gly-OMe and cyclo-gly-L-tyr at alkaline pH (9.5) even after 10 weeks. The release of nitrogen from the di- and tripeptide derivatives (cf.Fig. 7) levels off at a few percent the total amount coupled (at 5% if the data are fitted to the leaka function, cf.Table 1). These findings (pH dependence, very long hal lives and leakage stability of amino acid esters) point to the possibility that the nitrogen release is caused by peptide bond splitting.

TABLE 1

Leakage of nitrogen from ligands coupled to hydroxyalkyl methacryla gels derivatized with epichlorohydrin (Separon H 300 E_{max})

Ligand	Amount bound[a] (µmole/g)	pH	F (%)	10^5k (min^{-1})	n	$T_{1/2}$ (days)
Lys(Z)-gly-OMe	298	3.4	5	1.7	1	28.7
Gly-lys(Z)-gly-OMe	347	9.5	5	6.3	1	77.0
Gly-OMe	878	3.4	NL[b]			
$(Lys\text{-}ala\text{-}ala)_x$	84	3.4	NL			
Albumin	19[c]	3.4	100	1.0	5	142.8
		7.5	NL			

[a]determined by amino acid analysis, [b]NL = no leakage, [c]mg protein/g dry carrier

Electrophoretic desorption

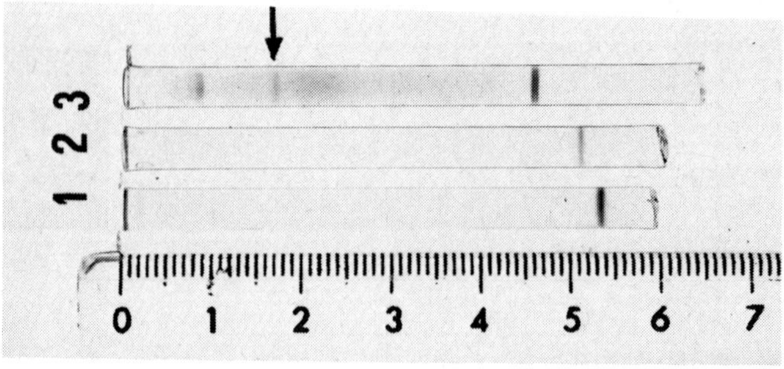

Fig. 9. Disc gel electrophoresis of aminopeptidase-Sepharose gel, 7.5% separation gel pH 8.3; gel 1 and 2: unsubstituted Sepharose blanks, 200 and 500 µl gel suspension (1:2), respectively; gel 3: 200 µl of aminopeptidase-Sepharose gel suspension (1:2), the arrow indicates the position of the unmodified soluble enzyme.

When aminopeptidase-Sepharose which was prepared by CNBr-coupling is desorbed electrophoretically substantial amounts of protein migrated into the polyacrylamide gel. Furthermore, when prior to electrophoretic desorption the immobilized enzyme was kept at room temperature until no further leakage could be detected, i.e. until about 30% of the initial amount of protein had detached, then washed and subjected to disc gel electrophoresis, an additional fraction of (leakage-proof) protein (10 - 20% of the initial amount) could be removed (Fig. 9, gel number 3). This protein was resolved into several bands one of which was the native enzyme the others more slowly migrating bands represented modified enzyme of an increased size and/or reduced charge (Fig. 10). This can be seen more clearly in Fig. 10 where a larger sample and a more sensitive staining technique has been employed.

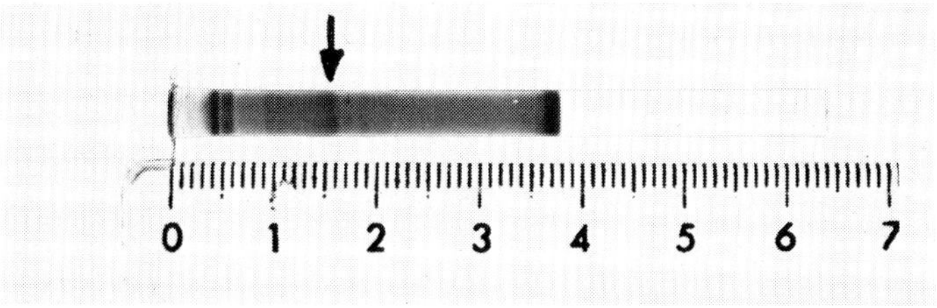

Fig. 10. Disc gel electrophoresis of aminopeptidase-Sepharose gel, 7.5% separation gel pH 8.3, 500 μl enzyme gel suspension (1:2), the arrow indicates the position of the soluble enzyme.

Summarizing all data the following picture emerges: there are 3 populations of protein ligands, one released slowly into solution by solvolytic cleavage of anchoring bonds and/or desorption from the polymeric network, a second one adsorbed unusually strongly to the matrix which can be detached only by the application of an electric field - it is known that extremely stable chelates can be dissociated in this way - and a third population which is stable. This is, of course, a methodological classification. The 3 populations comprise chemically different species (cf.Fig. 10).

REFERENCES

1 M. Sernetz and H. Puchinger, in K.Mosbach (Ed.), Methods in Enzymology, Vol.44, Immobilized Enzymes, Academic Press, New York, 1976, pp. 373-379.

2 H. Hanson and M. Frohne, in L. Lorand (Ed.), Methods in Enzymolog Vol.45B, Proteolytic Enzymes, Academic Press, New York, 1976, pp. 504-521.
3 J. Lasch and R. Koelsch, Eur.J.Biochem., 82 (1978) 181-186.
4 R. Koelsch, J. Lasch, K. Blaha and J. Turkova, in preparation.
5 P. Bohley, Hoppe-Seyler's Z.physiol.Chem., 348 (1967) 100-110.
6 J. Lasch, Experientia, 31 (1975) 1125-1126.
7 J. Lasch, in E. Hofmann, W.Pfeil and H. Aurich (Eds.), Proc.12th FEBS Meeting, Dresden, July 2-8, 1978, Pergamon Press, Oxford, 1979, pp.495-505.
8 M. Sernetz, O. Hannibal-Friedrich and M. Chun, Microscopica Acta, 81 (1979) 393-406.
9 O.R. Zaborsky, in A.C. Olson and C.L. Conney (Eds.), Immobilized Enzymes in Food and Microbial Processes, Plenum Publishing Corpor tion, New York, 1973, pp.187-203.
10 J. Fischer, R. Ulbrich, R. Ziemann, S. Flatau, P. Wolna,M. Schle V. Pluschke and A. Schellenberger, J.Solid-Phase Biochem., 5 (1980) 79-96.
11 G.I. Tesser, H.-U. Fisch and R. Schwyzer, Helv.chim.Acta, 57 (1974) 1718-1730.
12 M. Wilchek, T. Oka and Y.J. Topper, Proc.Natl.Acad.Sci.USA, 72 (1975) 1055-1058.
13 M. Wilchek, in E.K. Pye and H.H. Weetall (Eds.), Enzyme Engineer Vol.3, Plenum Publishing Corporation, New York, 1978,pp.283-289.
14 L. Sundberg and J. Porath, J.Chromatogr., 90 (1974) 87-98.

T.C.J. Gribnau, J. Visser and R.J.F. Nivard (Editors),
Affinity Chromatography and Related Techniques

PURIFICATION OF VARIOUS PECTIC ENZYMES ON CROSSLINKED POLYURONIDES

F.M. ROMBOUTS*, C.C.J.M. GERAEDS*, J. VISSER** and W. PILNIK*
Department of Food Science* and Department of Genetics**,
Agricultural University, P.O. Box 9101, 6700 HB Wageningen, The Netherlands

ABSTRACT

A convenient and reproducible method is described for the preparation of crosslinked polyuronides with different degrees of crosslinking. The preparations are very rigid, do not swell or shrink with changes in ionic strength or pH and, depending on the degree of crosslinking, may flow according to Darcy's Law, the specific permeability being 26 cm per h. Pectinesterases which are inhibited by pectate bind at pH 6 to crosslinked pectate and those of onion are eluted in at least a 100-fold purified form. Also, their multiple forms are separated. Endopolygalacturonases bind similarly to crosslinked pectate and crosslinked alginate, but the latter matrix is not degraded. The capacity of both matrices exceeds 5,000 units of yeast polygalacturonase per g. The use of crosslinked polyuronides for the isolation on a large scale of endopolygalacturonases from commercial "pectinases" requires further study.

INTRODUCTION

Pectic substances are structural polysaccharides occurring in plant cell walls. They consist primarily of a main chain of partially methyl-esterified (1,4)-α-D--galacturonan, but their fine structure is often more complicated. Pectic enzymes occur in plants and are also produced by plant-pathogenic and saprophytic microorganisms. Various types of enzyme exist: pectinesterases, polygalacturonases, pectate lyases and pectin lyases (cf. ref.1).

An obvious approach to purify these enzymes, and to separate their multiple forms, is by affinity chromatography on insolubilized forms of the substrate and structurally related polysaccharides. A method to crosslink sodiumpectate with epichlorohydrin was described by Tibenský and Kuniak (ref. 2). Endopolygalacturonase was selectively purified from the culture filtrate of *Aspergillus niger* by Rexová-Benková and Tibenský (ref. 3) and it was shown that the enzyme binds with its catalytic centre to the insoluble matrix (ref. 4).

We found a convenient and reproducible method of crosslinking in suspending powdered polyuronide preparations in a mixture of ethanol (3 volumes) and water

(1 volume) to which sodiumhydroxide and epichlorohydrin are added (ref. 5). In this system a suspension of only slightly swollen polyuronide granules is obtained and the solubility of epichlorohydrin is greatly increased. Both crosslinked pectate and crosslinked alginate proved to be very useful for the purification of pectate lyase and the separation of multiple forms of this enzyme (ref. 6). This paper summarises the results that we have obtained using crosslinked pectates and alginates in the purification of pectinesterases and some polygalacturonases, amongst others from commercial "pectinase".

RESULTS

Crosslinking of polyuronides

Sodiumpectate of sufficiently high molecular weight (> 30,000) is obtained by alkaline de-esterification of citrus pectin (Sunkist Growers, Ontario, California, USA) at 4°C as described by Rombouts *et al.* (ref. 5). Sodium alginate (Kelco Gel HV) is from Kelco Co., Chicago, Illinois, USA). Rapidase, a commercial "pectinase" is from Gist-Brocades, Delft, The Netherlands.

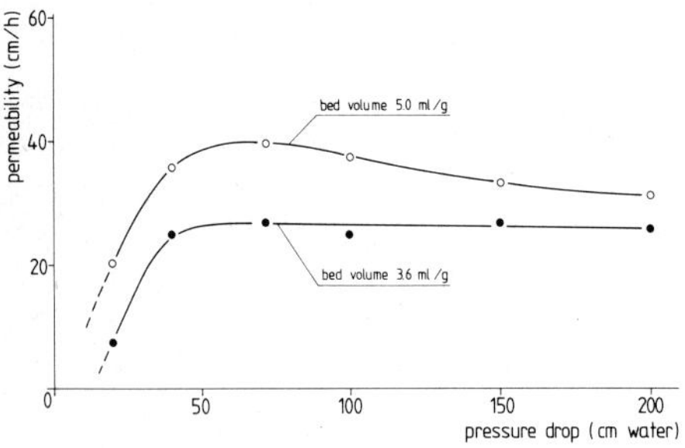

Fig. 1. Flow properties of columns (1.0x80 cm) with crosslinked pectates.

For crosslinking, 5-gram portions of sodiumpolyuronide are mixed with 1.1 ml of epichlorohydrin (epichlorohydrin to anhydrouronide monomers ratio= = 0.82) or 3.1 ml of epichlorohydrin (ratio = 2.34) in 15 ml of 95% ethanol. Then 5.0 ml of a 5 M sodiumhydroxide solution is added. The reaction mixtures are incubated on a rotary shaker (200 rpm) at 40°C for 4 h. The reaction is stopped by adding 1 M acetic acid to pH 5.0. After filtration on a medium fine (G3) sintered-glass filter, the crosslinked material is washed with 95% ethanol and dried in air. The bed volumes are 5.0 and 3.6 ml per g for preparations crosslinked with molar ratios of epichlorohydrin to anhydrouronide monomers of 0.82 and 2.34, respectively.

The flow properties of crosslinked pectates are shown in Fig. 1. The material with a high degree of crosslinking (bed volume 3.6 ml per g) obeys

Darcy's Law, the permeability being 26 cm per h. The other preparation shows a slight decrease of permeability with increased pressure. The preparations are very rigid, and do not swell or shrink with changes in pH and ionic strength.

Pectinesterase

The usefulness of crosslinked pectates for the purification and separation of multiple forms of plant pectinesterases is demonstrated in Figs 2 and 3. Crosslinked pectate with a bed volume of 3.6 ml per g binds two forms of pectinesterase from onion, which are separately eluted in about a 120-fold (pectinesterase II) and 100-fold (pectinesterase III) purified form. The isoelectric points of these enzymes are 9.4 (pectinesterase II) and 10.5 (pectinesterase III). A third enzyme (pectinesterase I, isoelectric point 5.75) can be separated (150-fold purification) from the material that does not bind to the column of Fig. 2 by chromatography under the same conditions on a column of pectate with a lower degree of crosslinking (Fig. 3).

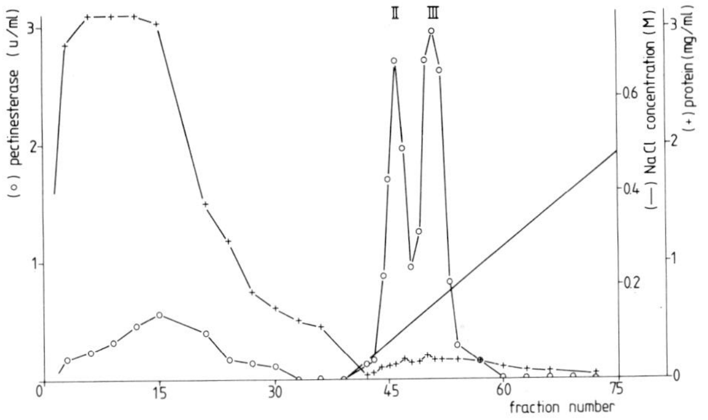

Fig. 2. Chromatography of onion pectinesterase on crosslinked pectate (bed volume 3.6 ml per g). Crude onion pectinesterase (182 ml, 1.82 units per ml), dialysed against 0.02 M sodium succinate buffer, pH 6.0, was applied to a column (2.2 x 4.0 cm) of crosslinked pectate. Elution was done with 0.02 M sodium succinate up to fraction 38 and then with a gradient of sodium chloride in this buffer. The flow rate was 10 ml per h and fractions of 10 ml were collected.

Pectate is a competitive inhibitor for plant pectinesterases and for certain fungal enzymes but not for that of *A. niger*. It is obvious from Table 1 that those enzymes which are inhibited by pectate, bind to crosslinked pectate, but there is not a good correlation between the value for Ki, or any other parameter, and the binding behaviour. In general, lowering the degree of crosslinking and the pH favour binding, but conditions under which *A.niger* pectinesterase would bind have not been found.

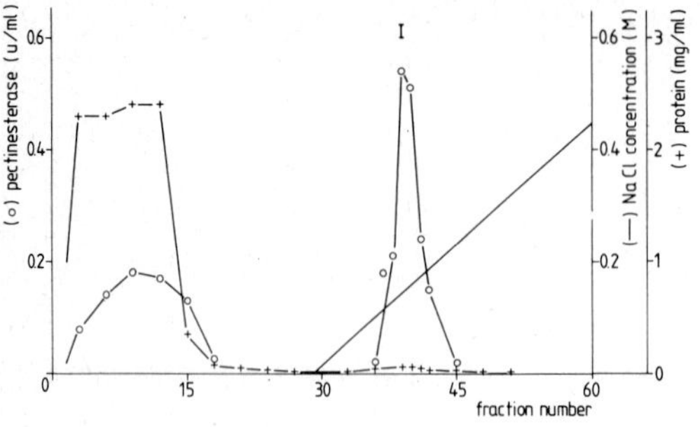

Fig. 3. Chromatography of onion pectinesterase on crosslinked pectate (bed volume 5 ml per g). Fractions 9 to 24 from Fig. 2 were applied to a column (2.2 x 5.7 cm) and chromatographed under the same conditions as in Fig. 2.

TABLE 1

Molecular and kinetic properties of pectinesterases from various sources and their binding behaviour at pH 6.0 on crosslinked pectate (bed volume 3.6 ml per g.)

Source	Enzyme	M.W.	I.E.P.	pH opt	K_m (mg/ml)	V_{max} ($\frac{\mu moles}{u.min}$)	K_i [a] (mg/ml)	Salt conc. required for elution (M)
orange	PE I	36,200	10.0	7.6	0.083	0.81	0.42	0.08 [b]
orange	PE II	36,200	11	8.0	0.0046	0.74	0.0016	0.24
grape-fruit	PE	54,000	10.5		0.133	0.48	0.065	0.07
onion	PE I		5.75					0.15 [b]
	PE II		9.4	7.0	1.96	1.09	0.33	0.09
	PE III		10.5					0.16
A.niger	PE	39,000	4.0	4.5	3.0	1.7	– [c]	0 [d]
A.niger	PE	30,000	4.0	4.5	3.6	1.96	–	0
Botrytis	PE I	28,400		7.0	0.06	0.98		0
cinerae	PE II	27,800		6.5	0.004	0.60		0.5 [e]

a) for pectate, a competitive inhibitor
b) bind to pectate with a lower degree of crosslinking (bed volume 5 ml/g) only
c) no inhibition; d) no binding; e) binding and elution at pH 4.2 (ref. 7)

We tried to determine whether the enzymes bind to the surface of the matrix only, or whether they can enter into the pores of the granules. Blue dextran 2000, orange pectinesterase II (molecular weight 36,200) and glucose elute in that order from the columns mentioned in Fig. 1, under conditions that prohibite binding (0.2 M sodium chloride in 0.05 M sodium citrate buffer, pH 3.0). Stronger evidence that most of the matrix is accessible for pectinesterase is the fact that completely methyl-esterified pectate with a high degree of crosslinking (bed volume 3.6 ml per g) may be enzymically de-esterified to a large extent (over 50%, ref. 5).

Polygalacturonase

Endopolygalacturonases bind to crosslinked pectate at pH 4.0 - 4.2 (0.1 M sodium acetate) and elute at pH 6.0 or higher, with or without additional salt. The multiple forms of endopolygalacturonase from *Kluyveromyces marxianus* with isoelectric points ranging from 2.3 to 6.5 bind to crosslinked pectate (ref. 8), showing that the isoelectric point is not a decisive factor in binding.

A technical application of affinity chromatography on crosslinked pectate could be the separation of endopolygalacturonase from commercial "pectinase" preparations. These preparations, derived from *A.niger* have a low content of pectic enzymes (less than 1%), of which pectin lyase, polygalacturonase and pectinesterase are usually present. For certain applications (e.g. the production of cloudifying agents from citrus wastes for application in soft drinks) a pure endopolygalacturonase preparation is desirable. The disadvantage of the use of crosslinked pectate for this purpose is that some enzymic degradation occurs during intensive use with endopolygalacturonase. Our experiments with endopolygalacturonases from *K.marxianus* show that the enzymes bind similarly to crosslinked pectate and to crosslinked alginate, but the latter matrix is not degraded by the enzymes. One g of each of these matrices binds more than 5,000 units of *K.marxianus* enzymes, directly from the culture filtrate adjusted at pH 4.2.

As is shown in Fig. 4 the endopolygalacturonase from a crude "pectinase" also binds to crosslinked alginate but, as with our crosslinked pectate, considerable amounts of pectinesterase and pectin lyase are also retained and elute with the polygalacturonase. It should be pointed out that the retention of pectin lyase and pectinesterase from *A.niger* "pectinase" does not involve true binding: in fact we have not been able to find conditions under which these enzymes (in purified form) would bind to crosslinked polyuronides. On columns with a higher degree of crosslinking less pectin lyase and pectinesterase are retained, but part of the endopolygalacturonase does not bind either. Increasing the amount of enzyme loaded onto the column has the same effect. In conclusion, application of crosslinked polyuronides for large scale fractionation of "pectinases" requires further study.

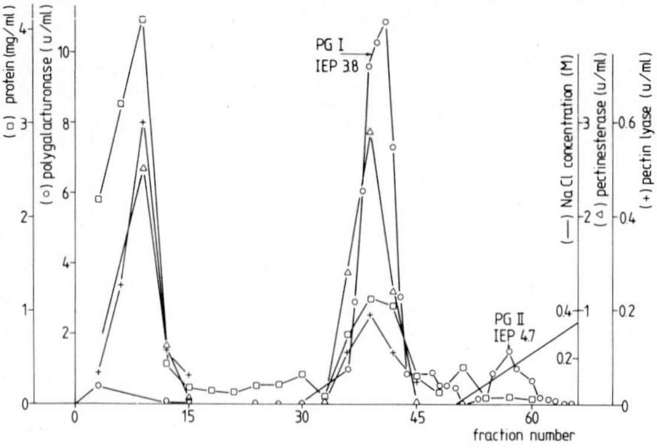

Fig. 4. Chromatography of "pectinase" (Rapidase) on crosslinked alginate (bed volume 3.6 ml per g). One g of enzyme, dissolved in 100 ml of distilled water was applied to a column (2.2 x 4.0 cm) equilibrated against 0.1 M sodium acetate buffer, pH 4.0. After washing with this buffer up to fraction 30, elution was done with 0.1 M sodium acetate buffer, pH 6.0 (up to fraction 50) and then with a linear gradient of sodium chloride in this buffer. The flow rate was 10 ml per h and fractions of 10 ml were collected.

REFERENCES

1 F.M. Rombouts and W. Pilnik, in A.H. Rose (Ed.), Economic Microbiology, Vol. 5, Microbial Enzymes and Bioconversions, Academic Press, London, 1980, pp. 227-282.
2 V. Tibenský and L'. Kuniak, Czech. Pat., 140,713 (1970).
3 L. Rexová-Benková and V. Tibenský, Biochim. Biophys. Acta, 268 (1972) 187-193
4 L. Rexová-Benková, Biochim. Biophys. Acta, 276 (1972) 215-220.
5 F.M. Rombouts, A.K. Wissenburg and W. Pilnik, J. Chromatogr. 168 (1979) 151-161.
6 J. Visser, R. Maeyer, R. Topp and F. Rombouts, in Les Colloques de l'INSERM, vol. 86, Affinity Chromatography, 1979, pp. 51-62.
7 L. Marcus and A. Schejter, privileged communication.
8 H.-P. Call, Optimierung, Reinigung und Charakterisierung einer "Endo-Polygalakturonase" der Hefe *Kluyveromyces marxianus*, Dissertation, R.-W. Technische Hochschule, Aachen, W.-Germany, 1980.

CHAPTER III

APPLICATIONS

- *ISOLATION*
- *PURIFICATION*

INTERNATIONAL SYMPOSIUM
ON AFFINITY CHROMATOGRAPHY
AND RELATED TECHNIQUES

T.C.J. Gribnau, J. Visser and R.J.F. Nivard (Editors),
Affinity Chromatography and Related Techniques

AFFINITY CHROMATOGRAPHY IN INDUSTRIAL BLOOD PLASMA FRACTIONATION

R. EKETORP
Research Department, Biochemistry, KabiVitrum AB, Stockholm, Sweden

INTRODUCTION

During the past twenty years very few new products derived from plasma have been introduced on the market. The products of major commercial importance are albumin, immunoglobulins, factor VIII and factor IX concentrates. In addition to these a limited extent of fibrinogen and a few other varied products have been produced.

The main reason why new products have not been manufactured is that their clinical importance is not known. However, it is worth mentioning that almost all plasma fractionators use the method described by Cohn et al., (ref. 1).

This particular method is based on ethanol fractionation in the cold. It is of utmost importance that all of the parameters, ethanol concentration, temperature, ionic strength, protein concentration and pH are most carefully controlled. The procedure is most sensitive to any of these parameters. (Illustrated in Fig. 1).

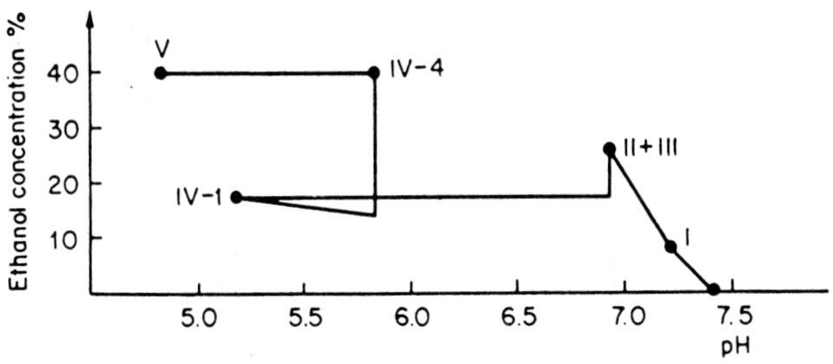

Fig. 1. Variation of pH and alcohol concentration when isolating plasma protein fractions according to Cohn method 6.

The risk of interfering with this process has made many fractionators hesitate to introduce other methods. Development work has been focused mainly on the increasing yields of albumin and factor VIII, due to the fact that the price of plasma is very high and that albumin is the major product.

Another way to improve the economy would be to isolate other plasma proteins which are of clinical interest and this of course is a field where affinity chromatography could play an important role. As plasma does contain at least fifty identified proteins, we can presume that at least a number of them play an important physiological role.

AFFINITY CHROMATOGRAPHY COMBINED WITH COHN FRACTIONATION

Many methods based on affinity chromatography in order to purify plasma proteins, i.e. plasminogen (ref. 2), antithrombin (ref. 3) and factor VIII (ref. 4) have been published during the last few years.

The main explanation why almost none of these methods have been applied to an industrial size scale until now is:

a) The lack of knowledge of how certain proteins function physiologically.
b) The risk of hepatitis.
c) The risk of contamination - i.e. ligands and carbohydrates.

Due to hereditary reasons, many persons are lacking certain proteins, and as various diseases now can be connected to congenital deficiency, the need for clinical preparations is increasing.

In giving an example, the frequency of thrombosis is considerably higher in humans with reduced antithrombin levels (ref. 5) and could perhaps be reduced by giving antithrombin. Another example of this is the importance of factor VIII and factor IX concentrates for haemophiliacs.

The risk of hepatitis is always present while preparing products from human plasma. Albumin preparations can be heat treated and will thereby lose their infectiability. Heat treatment, however, of other proteins, in the presence of stabilizers, has not yet been proven to be 100% effective. Tests performed in chimpanzees are, until the present time, the only way to prove the non-infectiability of a process.

It would be of great value if the heating procedures were tested in this way. Results have also been published recently, claiming that hepatitis B virus can be removed by affinity chromatography (ref. 6).

The risk of leakage from a chromatography matrix, with the possible effect of contaminating other products, must be carefully investigated.

In our laboratory we have carried out investigations in two ways. Firstly, we have investigated the nature and amount of leakage from the gels used. Secondly, we have followed the distribution of leakage products in our ordinary plasma fractionation.

In using radio-labelled glycine coupled to Sepharose Cl-6B® by the cyanogen-bromide method (ref. 7) and measuring the leakage it was found that there was quite a large amount of ligand leaking during the first 700 hours. The total loss of ligand was approximately 10% of coupled ligand and the leakage was greatly increased using buffers containing nucleophilic groups, i.e. ammonia.

The nature of the leakage products was also investigated and was found to be glycine or derivatives of glycine up to 99%. Only 0.1% high molecular weight material was found indicating glycine coupled to soluble agarose.

The high molecular weight material was also run through an ordinary plasma fractionation scheme. The amount of radioactivity found in different fractions is specified in Table 1.

TABLE 1

Distribution of radio-labelled leakage products from cross-linked agarose

Sample	% Radioactivity
Plasma	100
Supernatant I + II	99
γ-globulin solution	0.2
Supernatant IV	84
Supernatant V	80
Albumin	4

If plasma is believed to be in contact with the gel for 1 hour, we have calculated the maximum amount of contamination in albumin to be in the order of 0.2 ng/g albumin.

In preparing antibodies against leakage products attempts have not been successful, indicating a low immunogenicity of the carbohydrates.

The figure values for leakage of a ligand however, are so high that only ligands that are known to be harmless to man should be used.

This will therefore mean that methods where concanavalin A (purification of glucoproteins) and dyes (purification of albumin) have been used as ligands may be dangerous for preparation of proteins intended for clinical use.

What other aspects could be of interest in industrial plasma affinity chromatography?

The choice of matrix is extremely important and as have been stated by Egly and Porath (ref. 8) the following factors should be thoroughly considered when choosing the insoluble phase:

- Insolubility
- Rigidity
- Permeability
- Macroporosity
- Possibility to introduce ligands by chemical reactions
- Cost
- Minimal non-specific adsorption

Regarding the plasma fractionation, the macroporosity is extremely important. Thus agarose has become the matrix of choice in most applications. By cross-linking agarose, very satisfactory physical properties have been obtained in Sepharose Cl-6B®.

The excellent flow properties of these gels, allow extensive washing in a short period of time, thereby reducing the risk of bacterial contamination. The gels are also autoclavable, which in certain cases could be of interest.

RAW MATERIALS

In referring to what has been found regarding leakage, (if toxic ligands are not used) there is no reason not to use plasma directly for adsorption. Due to the risk of bacterial contamination, processes should be run in the cold and cold insoluble globulins should be removed prior to the introduction of affinity gels. A convenient source is fraction I supernatant after isolation of factor IX. One drawback for using this raw material is that the parameters for adsorption will be dependent on those used in the Cohn fractionation, as the plasma is further processed in order to produce albumin and immunoglobulins.

This can, in some cases, be overcome by using waste fractions such as fraction III and fraction IV. It should be emphasized however, that proteins may have been damaged by high ethanol concentrations or denaturing pH. The yield will also often be lower when starting with waste fractions, due to the wide distribution of a protein in the ethanol fractionation process. Usually the waste fractions must be processed prior to the affinity chromatography step as large amounts of denatured lipoproteins are contaminants. Filtration or centrifugation can be used after extraction, in order to remove undesired proteins. If plasma as a raw material is used, a system must be set up to continuously handle the large volumes. The system is illustrated in Fig. 2.

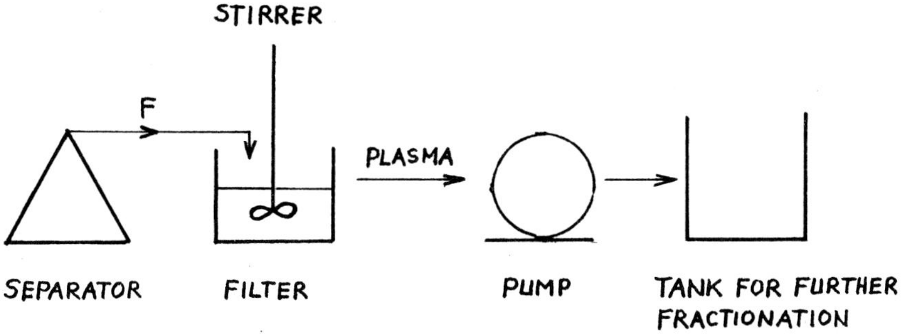

Fig. 2. Flow of plasma when using affinity gels in the Cohn fractionation.

Centrifuged plasma is continuously passed through a filter, containing the affinity gel. The filter is drained with the same flow rate as the flow rate of the separator. The plasma which has passed through the filter is brought back to the ordinary fractionation.

The volume of the filter is calculated according to the desired adsorption time. The content of the filter is continuously stirred, allowing a good flow rate and maximum contact area.

INDUSTRIALLY PREPARED PROTEINS

Only a few plasma proteins have been industrially prepared by affinity chromatography until now. One of them being plasminogen, the proenzyme of plasmin. Plasmin is the protease responsible for the fibrinolytic activity in plasma and it has therefore been of interest to use plasmin, (ref. 9) plasminogen (ref. 10) and streptokinase-plasminogen complex (ref. 11) in thrombolytic therapy.

Lysine coupled to agarose can be regarded as ε-aminocaproic acid (EACA) connected to a matrix. EACA is known to be a strong inhibitor of the activation of plasminogen to plasmin. The structure of lysine-Sepharose is illustrated in Fig. 3.

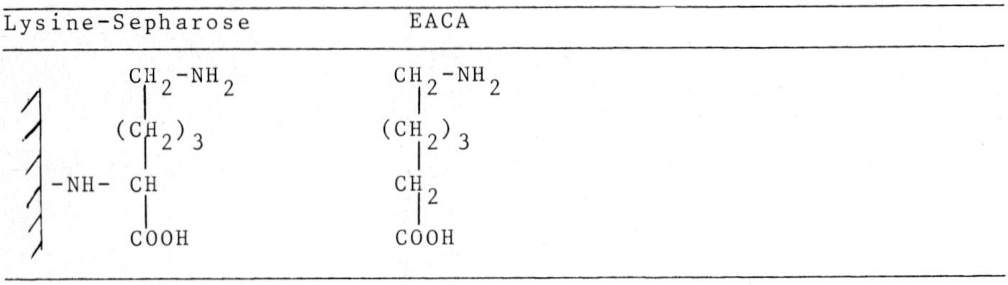

Fig. 3. Structure of Lysine-Sepharose compared with EACA.

Coupling of lysine should be carefully controlled as otherwise the incorrect aminogroup may be coupled to the matrix.

Plasminogen can be isolated, either from plasma or from Cohn fraction III (ref. 2), using lysine-Sepharose. If plasma is used, the preparations can follow the procedure described above. After adsorption, the gel can be transferred to a column washed with 1 M sodium chloride after which plasminogen is eluted with 0.1 M ε-aminocaproic acid.

Removal of EACA can easily be accomplished by either gel-filtration on Sephadex G-25 or by ultrafiltration.

In order to purify the potent inhibitor α_2-antiplasmin, plasminogen can be coupled to agarose (ref. 12). This inhibitor could be of great interest to use as an anti-dot during thrombolytic therapy.

In the native form plasminogen is a single chain protein with glutamic acid as N-terminal amino acid. When converted to plasmin the molecule is cleaved as illustrated in Fig. 4.

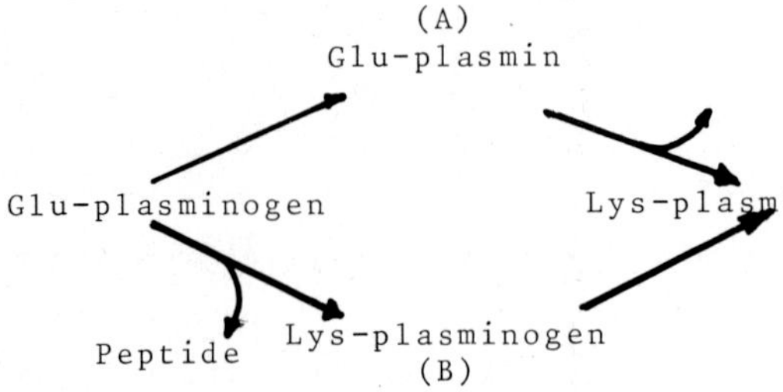

Fig. 4. Two possible pathways for activation of glu-plasminogen to lys-plasmin Pathway B seems to be the most probable (ref. 18).

The intermediate form is called lys-plasminogen (referring to the N-terminal amino acid. Kakkar (ref. 10) has reported good clinical results by using this intermediate form. It has also been reported that lys-plasminogen has a good deal higher affinity to fibrin than the native form (ref. 13).

Lys-plasminogen can be selectively prepared by using matrix-bound plasmin to convert native plasminogen to lys-plasminogen. This can be done in the cold under controlled circumstances. The product is free from plasmin and the degree of conversion is 100%. Lys-plasminogen manufactured in this way has been prepared in 50 g scale (corresponding to approximately 500 liter plasma) as a pyrogen-free stable product.

The conversion of glu-plasminogen to lys-plasminogen by plasmin-Sepharose is illustrated in Fig. 5.

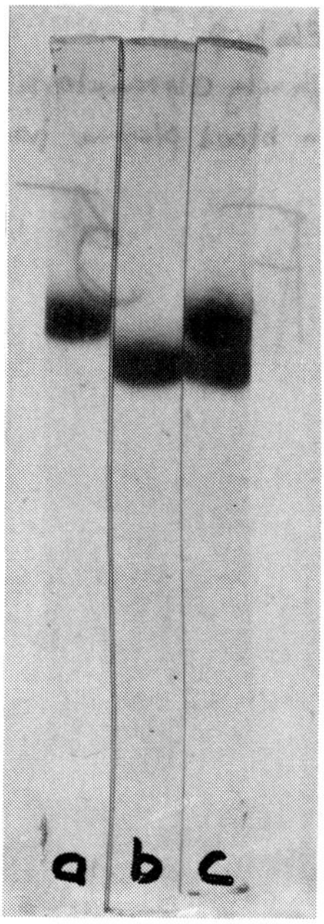

Fig. 5. SDS electrophoresis in 8 M urea in the presence of 0.1% mercaptoethanol a) glu-plasminogen b) lys-plasminogen and c) a and b mixed.

As can be seen from the SDS electrophoresis, in the presence of mercaptoethanol no plasmin can be detected. This has also been verified by using chromogenic substrates.

The purification of plasminogen is an example of how an extremely specific ligand can lead to a very simple procedure for isolation of a protein.

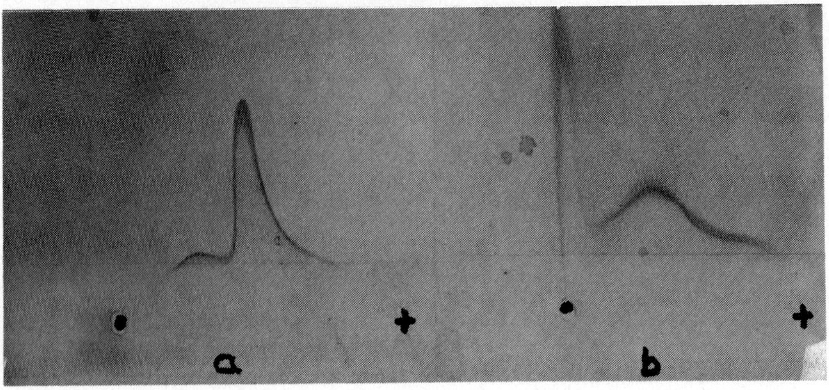

Fig. 7. Crossed immunoelectrophoresis against antithrombin according to Andersson, L.-O. et al., (ref. 17). Heparin added in the first dimension a) normal plasma and b) dissolved fraction IV-I.

When breaking the complex heparin-antithrombin with high salt concentrations, it should be noticed that the cofactor activity of antithrombin to a certain extent is effected.

Heparin-Sepharose has also been used in purifying factor VIII. The main contaminant in factor VIII preparations is fibrinogen. By treating dissolved cryoprecipitate with heparin-Sepharose, at least 95% of the fibrinogen can be removed and after concentration a highly purified factor VIII concentrate is achieved (ref. 4).

ECONOMY

From an economic point of view, some conclusions can be drawn regarding affinity chromatography in a large scale plasma fractionation.

a) The investments are fairly small as only small volumes of gel are handled if the gel-plasma ratio is kept low (1:15 - 1:50). Handling of plasma does not cause any problems as large filters (with an area of 0.5-1 m^2) are capable of handling several hundred liters of plasma within one hour, due to the excellent peoperties of the gels. Investments will mainly be in columns and laboratory equipment.

b) The used gels are expensive. It is therefore necessary that they can be reused several times. We have used the same heparin-Sepharose® for more than fifty large scale (1000 liter) preparations of antithrombin.

One example where the ligand is less specific is the purification of antithrombin. Antithrombin is also known as the cofactor of heparin. Heparin is therefore used as the ligand coupled to Sepharose (ref. 3). Heparin, a negatively charged polysaccharide, has affinity for lipoproteins, fibrinogen and antithrombin and other proteins due to the negative charge.

This has caused quite a few problems. First of all, the raw material should be free of fibrinogen. In using the supernatant I in the Cohn fractionation procedure this problem is solved.

By using polyethyleneglycol as a precipitating agent lipoproteins will be removed (ref. 14), but the plasma can not be further fractionated according to Cohn.

In looking at the elution pattern from heparin-Sepharose as described by Miller-Andersson et al., removal of contaminating material could be achieved by extensive washing with a carefully selected buffer.

In industrial scale, however, the separation has proved to be rather poor. By using a washing buffer containing 0.5 M/liter sodium chloride, approximately 30% of the antithrombin is lost (ref. 15).

The problem can be solved in a completely different way.

In looking at heparin-Sepharose as a cation exchanger with affinity properties, we can divide the proteins with affinity for the gel into two groups. One group being mainly based on affinity interaction (antithrombin and lipoprotein) and another group based on affinity for cation exchangers. As antithrombin and lipoproteins differ considerably in molecular weight, these proteins can be separated by polyethyleneglycol precipitation according to Thaler and Schmer (ref. 14) after elution from heparin-Sepharose.

Proteins adsorbed due to the cation exchanger character can be removed by introducing a cation exchanger later in the manufacturing procedure.

In order to remove traces of polyethyleneglycol and to concentrate the antithrombin solution, a DEAE-exchanger may be introduced.

The manufacturing process will then be as described in Fig. 6.

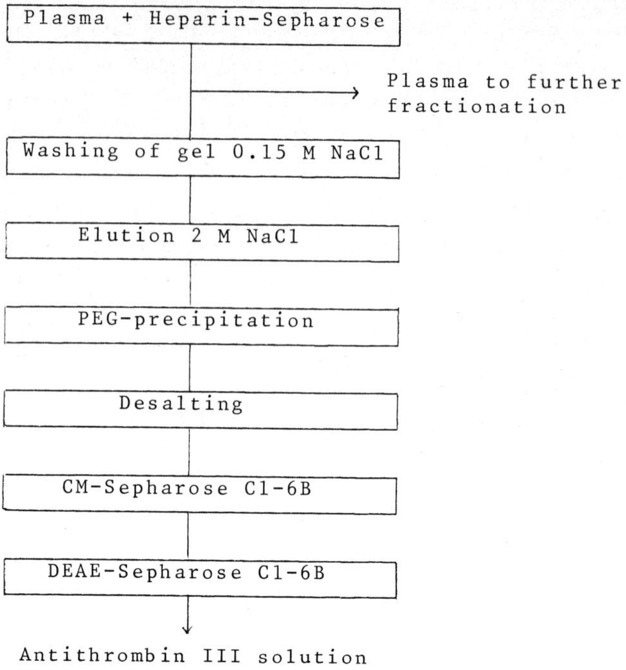

The introduction of the Sepharose-ion exchangers has opened up new possibilities as the flow rate can be held at approximately 25 cm/hour at $5^{o}C$. The high flow rates also reduce the risk of bacterial contamination. We have used the same gel for three years now without any problems. The same applies to heparin-Sepharose as the matrix is the same. Due to higher cross-linkage, flow rates of about 40 cm/hour can be achieved at $5^{o}C$.

Many papers have reported the use of waste fraction IV-I as starting material for antithrombin preparation, but as have been pointed out by Wickerhauser and Eketorp (ref. 15 and 16), a lot of the cofactor activity is lost in this fraction. Antithrombin is denatured due to the exposure of low pH and high ethanol concentration. This can be shown with crossed immunoelectrophoresis where heparin has been added in the first dimension run (ref. 17). See Fig. 7.

I feel quite strongly that there is a future for affinity chromatography in plasma fractionation. Several proteins which have been practically impossible to isolate in large quantities can now be purified several hundred times in one single step.

As the physiological tolerance, and not the purity, is the crucial point, I feel that during the next ten years we will see several new products on the market. It is also of great importance to get manufacturing methods. The knowledge - how to purify certain proteins - will probably be applicable even if plasma is not used as raw material. Gene-technology will probably be used for preparing, for example, albumin and cost benefit ratio of handling plasma may turn out to be negative.

Affinity chromatography could offer us ideas of how to isolate proteins applicable even to other sources than human plasma.

REFERENCES

1 E.J. Cohn, L.E. Strong, W.L. Hughes Jr., P.J. Mulford, J.N. Ashworth, M. Melin and H.L. Taylor, Journal of the Americal Chemical Society, 68, 1946, 459-475.
2 D.G. Deutsch and E.T. Mertz, Science, 170, 1970, 1095-1096.
3 M. Miller-Andersson, H. Borg and L.-O. Andersson, Thrombosis Research, 5, 1974, 439-452.
4 British Patent no 1460 607.
5 B. Blauhut, S. Necck, H. Kramar, R.H. Vinazzer and H. Bergmann, Thrombosis Research, 19, 1980, 775-782.
6 M. Einarsson, L. Kaplan, E. Nordenfeldt and E. Miller, Journal of Virological Methods, 1981 (submitted for publication).
7 J. Porath, R. Axén and S. Ernback, Nature, 215, 1967, 1491-1492.
8 J.M. Egly and J. Porath in O. Hoffmann-Ostenhof, M. Breteinbach, F. Koller, D. Kraft and O. Scheiner (Eds.), Affinity Chromatography, Pergamon Press, Oxford, 1978 pp 5-22.
9 R.A.G. Smith, R.J. Dupe, P.D. English and J. Green, Nature, 290, 1981, 505-508.
10 V.V. Kakkar, M. Lewis, S. Sagar, The Lancet, October 11, 1975, 674-676.
11 K.C. Robbins and L. Summaria, U.S. Patent no. 4,082,612.
12 M. Moroi and N. Aoki, Biochemical Journal, 251, 1976, 545-553.
13 S. Thorsen, Biochimica et Biophysica Acta, 393, 1975, 55-65.
14 E. Thaler and G. Schmer, British Journal of Haematology, 31, 1975, 223-253.
15 M. Wickerhauser, C. Williams and J. Mercer, Vox Sang, 36, 1979, 281-293.
16 R. Eketorp in J. Curling (Ed.), Methods of plasma protein fractionation, Academic Press, London, U.K., 1980, 177-188.
17 L.-O. Andersson, L. Engman and E. Henningsson, Journal of Immunological Methods, 14, 1977, 271-281.
18 P. Wallén in B. Blombäck and L.Å. Hansson (Eds.), Plasma proteins, J. Wiley and Sons, Chichester, 1979, 288-304.

T.C.J. Gribnau, J. Visser and R.J.F. Nivard (Editors),
Affinity Chromatography and Related Techniques

ISOLATION OF ANTITHROMBIN, HIGH AFFINITY HEPARIN, OR ANTITHROMBIN: HIGH AFFINITY HEPARIN COMPLEX BY SEQUENTIAL AFFINITY CHROMATOGRAPHY

R.E. Jordan, T. Zuffi and D.D. Schroeder
Cutter Laboratories, Inc., Berkeley, CA, USA

ABSTRACT

Most methods for the purification of plasma antithrombin employ an affinity step on immobilized heparin. However, this affinity support also binds a large number of other plasma proteins, making it difficult to achieve complete purification of antithrombin with differential salt elution. Alternatively, we have found that a very selective elution of antithrombin results when soluble heparin is used as the eluting agent.

This heparin can be removed from antithrombin by a second affinity step on Concanavalin A-Sepharose. The immobilized lectin binds tightly to the inhibitor, thereby allowing the elution of the heparin at high ionic strength. This second affinity step can also serve to separate a high affinity species of heparin or to provide the intact complex between antithrombin and high affinity heparin.

Thus, the coupled affinity system we describe provides a simple and rapid procedure for the simultaneous preparation of several components of interest in the anti-hemostatic system.

INTRODUCTION

Antithrombin, also known as antithrombin III, or heparin cofactor, is the primary plasma inhibitor of coagulation enzymes. It is a single chain glycoprotein containing approximately 10% carbohydrate and possessing separate functional sites for binding heparin and the target enzyme [1]. The interaction of antithrombin with heparin is electrostatic and reversible while that with an enzyme is an irreversible covalent attachment.

The purification of antithrombin from plasma has been greatly facilitated by the introduction of heparin affinity supports [1]. This methodology takes advantage of the particularly tight affinity between the protein inhibitor and the sulfated mucopolysaccharide. Affinity chromatography methods based on immobilized heparin allow purification of antithrombin of high purity and good yield on a laboratory scale and show considerable potential for scale-up.

The use of heparin-Sepharose, however, proves to be more complicated for this purpose than it first appears. Recent work on the structure/function relationships of heparin shows that only a limited sequence within the mucopolysaccharide chain is responsible for binding antithrombin [2, 3]. Only about 1 in 3 heparin molecules possesses this sequence which results in the anticoagulant effects by accelerating the rate at which antithrombin inhibits coagulation enzymes [4]. Since antithrombin interacts with only a small proportion of the total heparin, affinity supports employing the mucopolysaccharide are thus, predominantly composed of species which do not bind the inhibitor. This problem may be further compounded by inactivation of some of the heparin during the linkage reaction and/or steric inaccessibility of bound species. Experience shows, in fact, that only a few percent of the immobilized heparin serve to bind antithrombin during affinity chromatographic procedures (unpublished observations).

Many other plasma proteins display an appreciable affinity for heparin although in almost every case the interaction is relatively non-specific and independent of the sequence structure of heparin itself. Some of these interactions are probably due to simple ionic binding to the highly anionic heparin, while others, such as those of lipoproteins, are more specific for sulfated polymers. It is to be expected that most of the immobilized heparin chains are available for such interactions and this may explain the difficulty in obtaining pure AT III in one step from heparin-Sepharose chromatography.

MATERIALS AND METHODS

The heparin used in these studies was obtained from Sigma Chemical Co. and possessed an average anticoagulant activity of approximately 150 units/mg. Heparin-Sepharose-CL 4B was prepared by the cyanogen bromide activation method of Cuatrecasas [5]. Concanavalin A-Sepharose and dextran sulfate ($\bar{M}_w \sim 500,000$) were obtained from Pharmacia Fine Chemicals. All chromatography was done at ambient temperature.

Antithrombin concentrations were determined by absorbance at 280 nm, assuming an extinction coefficient ($E^{1\%}_{cm}$) of 6.5 [6].

Heparin concentrations of column eluates were monitored by the Azure A method of Jaques et al [7] as modified by Lam et al [4]. Quantitative determinations of heparin concentrations were performed by the carbazole method [8] assuming a uronic acid content for heparin of 30% [9].

Assays of heparin anticoagulant activity were carried out by modification of a two-stage amidolytic method [4]. In the present case, residual bovine thrombin (Miles Laboratories) was measured by its hydrolysis of the chromogenic substrate H-D-Phe-Pip-Arg-pNA (S-2238, Kabi Diagnostica). A single heparin preparation was used as a standard in all of these studies and its activity was taken to be its U.S.P. label value.

The assay of antithrombin activity also employed the above enzyme and chromogenic substrate, and was a modification of the method of Odegard et al. [10]. A unit of antithrombin activity corresponds to the amount present in 1 ml of pooled, normal plasma.

RESULTS AND DISCUSSION

The elution of protein from a heparin-Sepharose column using successive washes of 0.4 M NaCl and 1 M NaCl is shown in Fig. 1A. Human cryoprecipitate-poor plasma was applied and the column washed with 0.25 M NaCl (not shown) prior to elution. Assays for antithrombin demonstrate that the majority of the inhibitor elutes at 1 M NaCl although its purity depends on the removal of a large number of contaminating proteins at the intermediate ionic strength. In our experience, more extensive washing of the column at intermediate ionic strengths usually results in a lowered yield of antithrombin without complete removal of contaminating species.

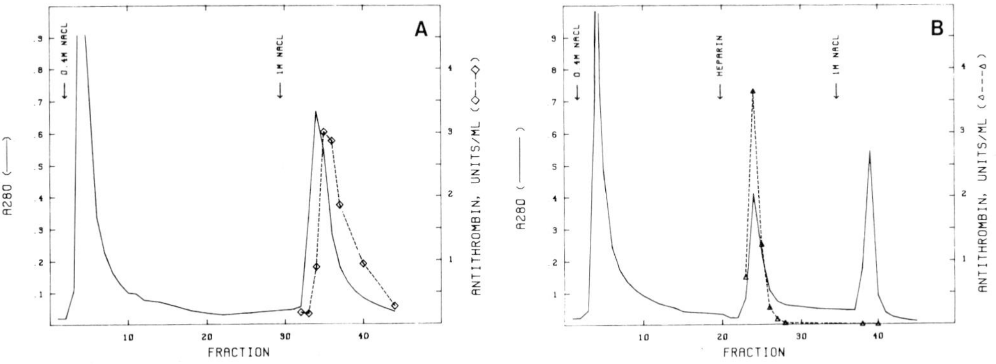

Fig. 1. Heparin-Sepharose chromatography of plasma. The sample (100 ml) was applied and the column (1.6 cm x 12.5 cm) then washed with the buffer (0.01 M Tris-HCl containing 0.25 M NaCl) to reduce the absorbance of the eluate to a constant baseline (not shown). A. Elution of bound proteins with successive buffer washes containing 0.4 M NaCl and 1.0 M NaCl respectively. B. Selective elution of antithrombin with heparin (2 mg/ml) at a low ionic strength (0.25 M NaCl).

An alternative approach for the elution of antithrombin from heparin-Sepharose is shown in Fig. 1B. In this case, after application and a wash with 0.4 M NaCl, the column was eluted with a lower ionic strength buffer containing heparin (2 mg/ml).

The antithrombin elutes in a sharp peak of considerable purity as evidenced by its high specific activity (6.3 units/mg), which is significantly greater than that obtained in the previous example (2.8 units/mg). The reasons for this increased purity are not completely clear since the heparin used as the eluting agent might be expected to compete for all protein:heparin-Sepharose interactions and thereby offer no advantage for the purification of antithrombin. The rather specific elution of antithrombin by heparin under the present conditions may relate to the unique mode of interaction between inhibitor and mucopolysaccharide.

Further support for the specific nature of this elution is provided by the observation that an additional protein peak is removed from the above column in a subsequent wash with 1 M NaCl. This peak is found to contain several protein species by SDS gel analysis (not shown), but contains very little antithrombin. These proteins are the same species which often contaminate antithrombin preparations eluted from heparin-Sepharose under high ionic strength conditions.

The apparently non-specific nature with which contaminating species bind to heparin-Sepharose led us to examine the effect of a heparin analogue, dextran sulfate. The results of this experiment are shown in Fig. 2. An elution step

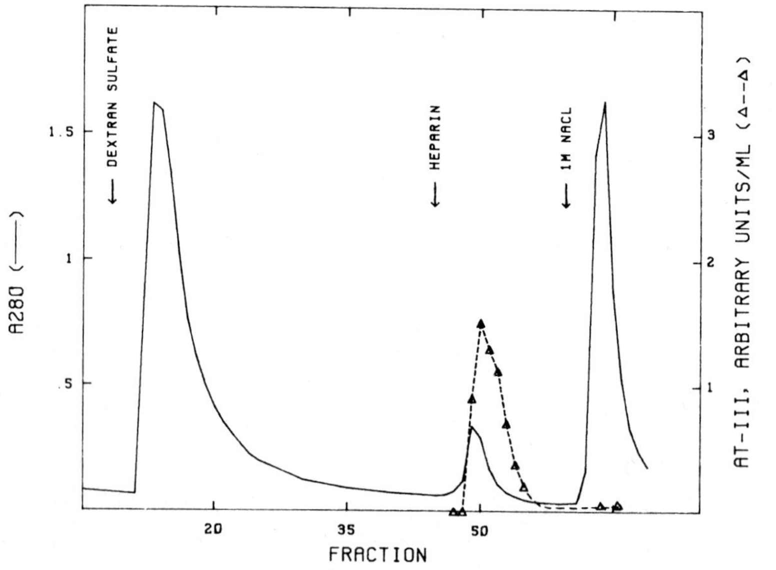

Fig. 2. Heparin-Sepharose chromatography of plasma. The sample was applied as described (legend to Fig. 1). Elution of bound proteins was accomplished with successive washes at low ionic strength (0.25 M NaCl) containing dextran sulfate (5 mg/ml) and heparin (2 mg/ml) respectively, and a final wash at 1.0 M NaCl. A short buffer wash was also done prior to the heparin elution step to remove dextran sulfate from the column.

with a buffer containing dextran sulfate (5 mg/ml) removes a large peak of protein from the column. This peak does not contain detectable antithrombin which, rather, elutes in a subsequent wash with a buffer containing heparin. This result demonstrates the specific nature of the antithrombin-heparin interaction in contrast to many of the other plasma components which bind to heparin-Sepharose.

Dextran sulfate added to the sample during application to the column was also found to prevent many of the undesired protein:heparin-Sepharose interactions without affecting the binding of antithrombin and thereby improving the purification of the inhibitor. Further use can be made of dextran sulfate in this system by the addition of calcium ions to the plasma sample. This results in a selective precipitation of the lipoprotein fraction [11] resulting in greatly improved chromatographic properties of the plasma on heparin-Sepharose. Antithrombin preparations of very high purity can be obtained in this way. A purification system based on this principle has recently been reported by McKay [12].

The antithrombin eluted with heparin from heparin-Sepharose is essentially free of other proteins but is mixed with a large molar excess of heparin (~20-fold). Antithrombin can be readily separated from this heparin, however, by a second affinity chromatographic step on Concanavalin A-Sepharose (Con A). This step takes advantage of the glycoprotein nature of antithrombin [13] and its tight interaction with the lectin. Heparin, in contrast, is not retained on a Con A-Sepharose column since it does not contain the appropriate carbohydrate groups [14]. The fraction of heparin which is tightly bound to antithrombin remains complexed with the inhibitor at low ionic strength on the Con A matrix. This high affinity heparin species can be displaced from the antithrombin in the presence of 1 M NaCl and eluted from the column without disrupting the tight interaction between Con A and antithrombin (Fig. 3). The antithrombin is subsequently recovered in a separate elution with a low ionic strength buffer containing an appropriate sugar (e.g., 0.2 M α-methyl-D-glucoside). Thus, in addition to providing a means of removing heparin from antithrombin, the Con A-Sepharose step can also serve to fractionate the heparin. The high affinity heparin which is obtained by the above sequence exhibits a specific anticoagulant activity several-fold higher than the commercial starting material (450-700 vs. 150 units/mg, respectively). The use of this method for the affinity fractionation of heparins of specified molecular weights into both high and low affinity species has recently been reported [15].

For the purpose of heparin fractionation, this Con A-Sepharose method is similar, in principle, to previously reported methods employing covalently immobilized antithrombin [16]. The present method also permits repetitive fractionation of heparin without displacing the antithrombin from the Con A-Sepharose column (unpub. res.). The advantage of the present method is that antithrombin can be recovered in a functionally active state.

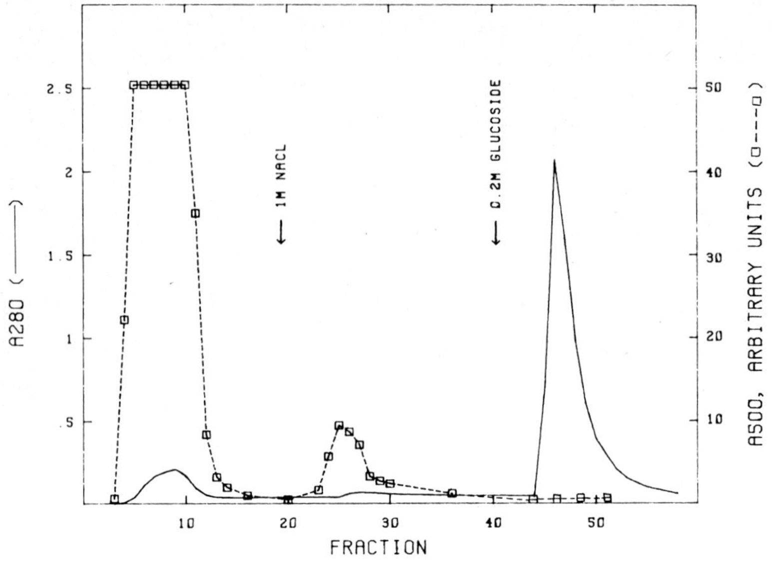

Fig. 3. Con A-Sepharose chromatography of an antithrombin/heparin mixture. The sample was applied to a column (2.5 cm x 7 cm) previously equilibrated with 0.25 M NaCl in 0.01 M Tris-HCl, pH 7.5. The column was washed with this buffer until heparin was not detectable in the eluate by the Azure A assay (expressed as arbitrary units of A_{500}). The bound heparin was eluted with a 1 M NaCl wash until no further heparin could be detected. Antithrombin was eluted with 0.2 M α-methyl D-glucoside in 0.25 M NaCl. All column buffers contained $CaCl_2$, $MgCl_2$ and $MnCl_2$ at 1 mM.

The most novel use of this coupled affinity chromatographic system, however, is the isolation of the complex formed between antithrombin and the high affinity heparin species. This is accomplished, as shown in Figure 4, by the simultaneous elution of both components from Con A-Sepharose with buffers containing appropriate sugars. Such elution conditions disrupt only the linkage between antithrombin and the immobilized lectin without interfering with the antithrombin-heparin interaction. The isolated antithrombin:high affinity heparin complex is stable at low ionic strength and can be dialyzed to remove the sugar.

We have confirmed that the heparin contained in the isolated complex possesses comparable activity (450-600 units/mg) to heparin eluted separately with 1 M NaCl. For this determination it was necessary to separate the heparin from the antithrombin. The separation was done by re-applying a small portion of the complex to Con A-Sepharose and eluting the heparin with high salt as used in Figure 3.

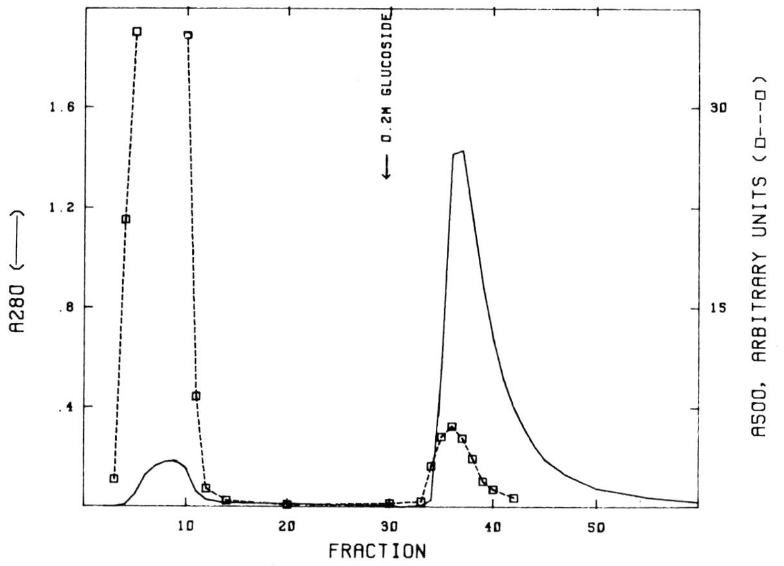

Fig. 4. Con A-Sepharose chromatography. An antithrombin/heparin mixture was applied as described (legend to Fig. 3). The antithrombin and bound heparin were eluted together with a buffer (0.25 M NaCl, 0.01 M Tris-HCl, pH 7.5) containing α-methyl D-glucoside (0.2M).

In preliminary experiments to characterize the nature of the isolated antithrombin complex we have found an approximately 2:1 molar excess of antithrombin to heparin. This is apparently not due to the presence of uncomplexed inhibitor but appears to result from the association of more than one antithrombin with each heparin chain. The results suggest that such higher-order complexes may be favored under the conditions employed in these fractionation experiments. Although 2:1 complexes have been previously observed between antithrombin and specific high molecular weight heparin fractions [17], the present results point to the preferential and stable formation of such complexes in the presence of a great excess of uncomplexed heparin.

The present method thus provides a means of obtaining the most active anticoagulant complexes between antithrombin and heparin. The detailed examination of the physico-chemical properties and in vivo activities of such complexes may provide considerable information toward understanding the hemostatic mechanism.

REFERENCES

1 M. Miller-Andersson, H. Borg and L.-O. Andersson, Thrombosis Res., 5(1974) 439-452.
2 R.D. Rosenberg and L. Lam, Proc. Nat. Acad. Sci., 76(1979)1218-1222.
3 U. Lindahl, G. Backstrom, M. Hook, L. Thunberg, L.-A. Fransson and A. Linker, Proc. Nat. Acad. Sci., 76(1979)3198-3202.
4 L. Lam, J.E. Silbert and R.D. Rosenberg, Biochem. Biophys. Res. Comm., 69(1976) 570-577.
5 P. Cuatrecasas, J. Biol. Chem., 245(1970)3059-3065.
6 B. Nordenman, C. Nystrom and I. Bjork, Eur. J. Biochemistry, 78(1977)195-203.
7 L.B. Jaques, F.C. Monkhouse and M.J. Stuart, J. Physiology 109(1949)41-48.
8 T. Bitter and H.M. Muir, Anal. Biochemistry, 4(1962)330-334.
9 M.Hook, I. Bjork, T. Hopwood and Lindahl U., FEBS Lett., 66(1976)90-93.
10 O.R. Odegard, M.Lie and U. Abeldgaard, Thrombosis Res., 6(1975)287-294.
11 M. Burstein and H.R. Scholnick, Adv. Lipid Res. 11(1973)67-108.
12 E.J. McKay, Thrombosis Res., 21(1981)375-382.
13 I. Danishfsky, A. Zweben and B.I. Slomiany, J. Biol. Chem., 253(1978)32-37.
14 I.J. Goldstein and C.E. Hayes, Advances in Carbohydrate Chem. and Biochem., 35(1978)127-340.
15 R. Jordan, L. Favreau, E. Braswell and R.D. Rosenberg, submitted for publication, 1981.
16 L.-O. Andersson, T.W. Barrowcliffe, E. Holmer, E.A. Johnson and G.E.C. Sims, Thrombosis Res. 9(1976)575-583.
17 R.D. Rosenberg, R.E. Jordan, L.V. Favreau and L.H. Lam, Biochem. Biophys. Res. Comm., 86(1979)1319-1324.

T.C.J. Gribnau, J. Visser and R.J.F. Nivard (Editors),
Affinity Chromatography and Related Techniques

THE APPLICATION OF IMMUNO-ADSORPTION ON IMMOBILIZED ANTIBODIES FOR LARGE SCALE CONCENTRATION AND PURIFICATION OF VACCINES

A.L. VAN WEZEL and P. VAN DER MAREL
Rijks Instituut voor de Volksgezondheid, P.O.Box 1, Bilthoven, The Netherlands

INTRODUCTION

Mass vaccination programmes are routinely applied in many countries in the world for protection of the population against bacterial and viral infectious diseases. In view of safety and in order to achieve and maintain high vaccination coverages, it is very important that no adverse reactions due to impurities in the vaccines occur after vaccination. Therefore, the use of highly purified antigens for the preparation of the vaccines must be considered as a prerequisite for Good Manufacture Practice in vaccine production. Sofar relatively good results are obtained with standard purification procedures based on the difference in physical and chemical properties of the antigens and impurities. However, due to forthcoming developments in vaccine production such as the application of non-tumorigenic continuous cell lines as substrate for virus multiplication and the use of genetic engineering for the production of the desired antigen, more specific procedures will be needed. Since affinity chromatography on immobilized antibodies might be considered as the most specific procedure for purification of protein antigens and in view of its successful application in this and other fields (ref. 1, 2, 3, 4), this system was studied by us for the purification of antigens in the field of vaccine production. As test model purification of polio virus with immobilized antibodies was evaluated in comparison to the routine procedures used at the large scale production of inactivated polio vaccine (scheme 1), (ref. 5, 6).

RESULTS AND DISCUSSION

Antibodies

Initially only polyclonal antibodies from animals hyper-immunized with highly purified antigens could be used. Recently also monoclonal antibodies prepared according to the hybridoma technique originally described by Köhler and Milstein (ref. 7) became available (ref. 8). The use of monoclonal antibodies has certain advantages above polyclonal antibodies from animals. The antigens used for immunization of the animals should be highly purified in order to obtain specific antibodies against the desired antigen. In the case of monoclonal antibodies this

SCHEME 1

Standard purification process for the removal of bovine serum proteins and cellular products from polio virus suspensions at the production of inactivated polio vaccin

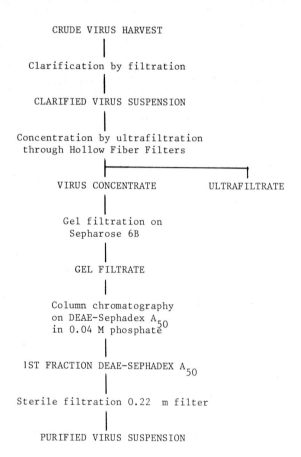

CRUDE VIRUS HARVEST

|

Clarification by filtration

|

CLARIFIED VIRUS SUSPENSION

|

Concentration by ultrafiltration
through Hollow Fiber Filters

|

VIRUS CONCENTRATE — ULTRAFILTRATE

|

Gel filtration on
Sepharose 6B

|

GEL FILTRATE

|

Column chromatography
on DEAE-Sephadex A_{50}
in 0.04 M phosphate

|

1ST FRACTION DEAE-SEPHADEX A_{50}

|

Sterile filtration 0.22 m filter

|

PURIFIED VIRUS SUSPENSION

specifity is achieved by cloning of the hybridoma cells. Monoclonal antibodies are only directed against one of several antigen determinants which may be present on the entire antigen. The advantage of this is that the antigen will be adsorbed to the immobilized antibodies by this single antigen determinant and therefore, will be less denatured and more easily removed. Furthermore as soon as the antibody producing hybridoma cells have been isolated, they can be stored in liquid nitrogen and reactivated at any time antibodies are required. This guarantees that always the same antibodies are available for the preparation of new immune adsorbents which will increase the reproducibility of the procedure.

However, monoclonal antibodies have one great disadvantage, e.g. that they are produced by a tumorigenic continuous cell line. This certainly will have consequences for their application in the purification of biological products for human use as potential oncogenic cell material, particularly nucleic acids, may turn up in the final product. Several precautions can be taken to overcome this problem. First the antibodies can be thoroughly purified, e.g. by affinity chromatography on protein-A columns (ref. 9). Elimination of the cellular nucleic acids might be demonstrated by hybridization techniques (ref. 10). The last traces of nucleic acids left may be removed by washing the immune adsorbent with nuclease or inactivated by a chemical treatment with glutaraldehyde or formalin. If such precautions are taken, most objections against the application of monoclonal antibodies for purification of biologicals will be met. For the experiments described below sofar only caprylated polyclonal sera from animals hyper-immunized with highly purified polio virus type 1 (Mahoney), type 2 (MEF_1) and type 3 (Saukett) were used (ref. 4).

Immobilization of antibodies

For the immobilization of the antibodies to support materials various methods have been described in literature (ref. 11, 12, 13). Four methods were compared by us:

a. Coupling to CNBr-activated Sepharose 4B according to the method of March et.al. (ref. 11).
b. Coupling to glutaraldehyde activated aminohexyl derivative of Sepharose 4B (ref. 12).
c. Adsorption of IgG on Spherosil XOC 005 at low ionic strength and fixation of the adsorbed IgG with glutaraldehyde or paraformaldehyde (ref. 14).
d. Coupling to Oxirane Acrylic Beads (ref. 13).

The binding capacity of the immune adsorbents for polio virus type 1 was determined in relation to the amount of coupled IgG as shown in figure 1 for CNBr-activated Sepharose 4B. For the other three supports the same pattern was found. The amount of coupled IgG was proportional to the amount of IgG added up to at least 8 to 12 mg IgG per ml packed support. However, the maximum binding capacity for the antigen was already achieved at a lower IgG concentration (fig. 1).

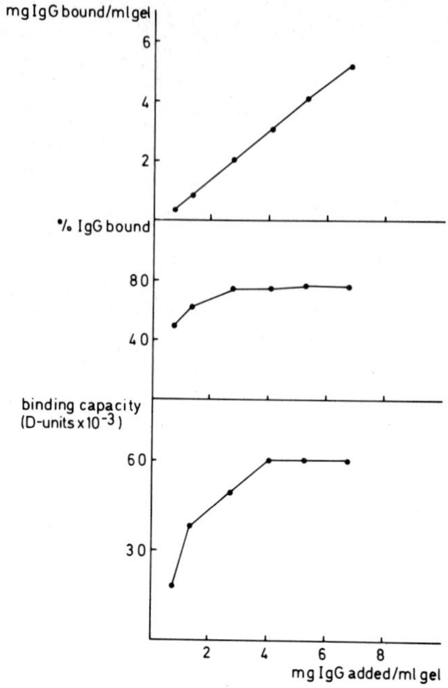

Fig. 1. The binding capacity of anti-polio type 1 Sepharose 4B immune adsorbent in relation to the amount of IgG bound.

The CNBr-activated Sepharose 4B appeared to give the highest maximum binding capacity at the lowest IgG concentration (table 1). It could be demonstrated by electron microscopical examination that as a relatively large antigenic particle was used, the pore size of the support material is playing a role in this. It was found that the polio virus particle could penetrate much further into the matrix of Sepharose 4B (fig. 2) than in Oxirane to which the virus was only adsorbed at the surface.

Comparable results were obtained for polio type 2 and 3. For all three types the virus from about 750 ml crude virus suspension could be bound to 1 ml of packed immune adsorbent.

Elution

The most crucial point was the elution of the antigen from the immune adsorbents as most eluents such as low and high pH solutions are destroying the virus antigen. The best results were obtained with the chaotropic salt NH_4SCN (fig. 3).

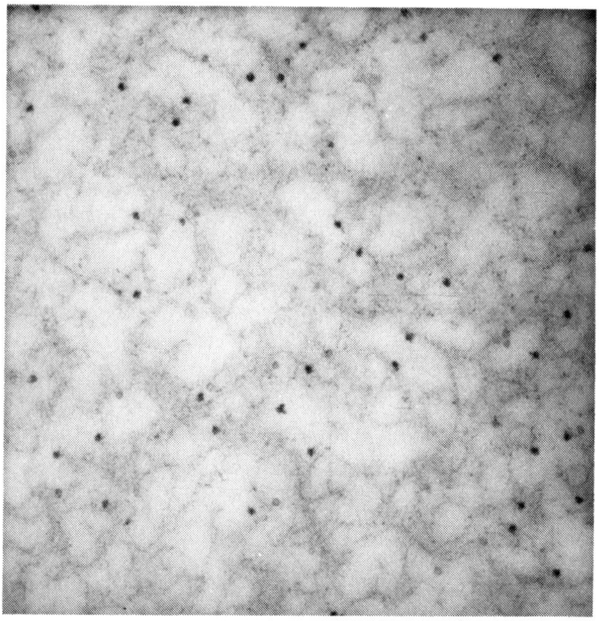

Fig. 2. Electron microscopical examination of the penetration of polio virus type 1 in the matrix of Sepharose 4B immune adsorbent.

TABLE 1

Comparison of the binding capacity of different immune adsorbents for polio virus type 1 per ml of packed immune adsorbent

	Max.binding cap. D-antigen U/ml	IgG bound mg/ml
CNBr-activated Seph. 4B	60×10^3	4.0
Glutaraldehyde act.Seph. AH 4B	50×10^3	6.6
Spherosil	40×10^3	9.0
Oxirane	30×10^3	5.0

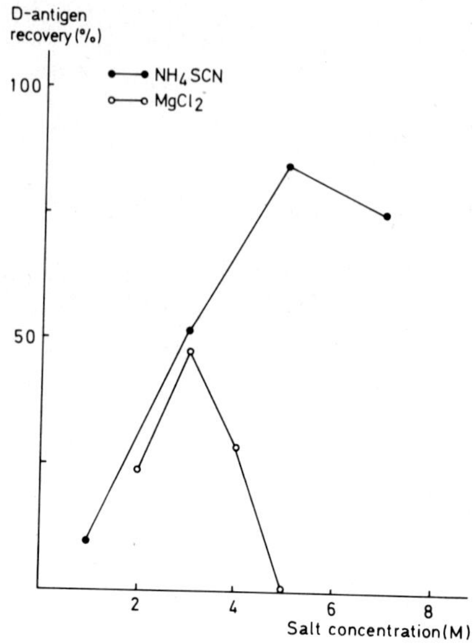

Fig. 3. Recovery of D-antigen polio type 1 from immune adsorbents at increasing chaotropic salt concentrations.

Concentrations up to 5 to 7 M NH_4SCN were needed to elute the virus antigen from the polyclonal anti-polio Sepharose 4B immune adsorbents with maximum binding capacity. However, at these concentrations of NH_4SCN the virus appeared to be not stable. Therefore, it was separated from the chaotropic salt solution by gel-filtration on Sephadex G50 straight on after elution. In later experiments it was demonstrated that lower concentrations of the chaotropic salt could be used for elution of the virus antigen, if less antibodies were coupled to the support. Although this will result in a decrease of binding capacity, this might have practical advantages. Also preliminary experiments indicate that with monoclonal antibody immune adsorbents a lower concentration of chaotropic salts may be applied for the elution of the virus antigen.

Large scale experiments

Large scale experiments were performed either by batch wise adsorption of the virus from a previously concentrated suspension or by percolating the clarified

TABLE 2

Purification of polio virus type 1 by affinity chromatography on immune adsorbents

	Batch wise					In columns				
	vol. ml	D-antigen U/ml	Bov. serum proteins µg/ml		Rec. %	vol. ml	D-antigen U/ml	Bov. serum proteins µg/ml		Rec. %
			Alb.	Imm.glob.				Alb.	Imm.glob.	
Virus suspension (filtrate)	600	1675	6600	1400	100	18600	89	30	37.5	100
Virus suspension (after adsorption)	600	n.d.[1]	n.d.	n.d.	n.d.	18600	1.7	30	37.5	2
Immune adsorbent eluate	190[2]	3311	< 0.06	3.75	63	250	3915	< 0.015	2.0	66
DEAE-Sephadex fraction	194[3]	2926	< 0.06	< 0.5	60	250	3010	< 0.015	< 0.2	63

[1] n.d. : not done

[2] PN content: 11 µg/ml

[3] PN content: 8 µg/ml

virus suspension through a column with immune adsorbent. For the batch wise adsorption about 50% excess of immune adsorbent was mixed with the virus suspension by means of a vibromixer for 2 to 4 hours. The beads were transferred to a small column and thoroughly washed with a phosphate buffered salt solution containing 0.5 M NaCl. The virus was eluted by a 7 M NH_4SCN solution in medium 199 and directly transferred to a Sephadex G50 desalting column in 0.04 M phosphate (ref. 4). The immune adsorbent eluate was further purified by column chromatography on DEAE-Sephadex. For adsorption experiments in columns the clarified virus suspension was percolated through a column packed with a 4-fold excess of immune adsorbent in order to achieve 100% adsorption at a reasonable flow rate. The actual flow rate used was 40 ml per cm^2 per hour for a column with a diameter of 5 cm and 5 cm high. After adsorption the column was washed and eluted according to the method described for the batch wise method. The results on control and recovery tests are summarized in table 2.

In comparison with the standard purification system (scheme 1, table 3) a good purification and reasonable recoveries were obtained by immune adsorbent purification.

TABLE 3

Purification of polio virus type 1 according to the procedure used at the routine production of inactivated polio vaccine

Virus suspension	Vol. l.	D-antigen U/ml	Recovery %	Serum content Albumin µg/ml	Serum content Imm.glob. µg/ml
Starting suspension	240	76	100	n.d.[1)]	n.d.
Filtrate (Pall-filters)	248	64.2	87	1,000	300
Concentrate	1	17,530	96	> 30,000	>30,000
Gel filtrate	4.5	3,465	85	0.23	2.0
DEAE-Sephadex fraction 1	4.5	3,465	85	0.03	< 0.23
Sterile filtrate	7.5	1,964	81	< 0.03	< 0.23
Monovalent vaccine	7.4	1,753	71	< 0.03	< 0.23

[1)] n.d.: not done

PN-content: gel filtrate 40 µg/ml
DEAE-Seph.fr.1 8 µg/ml

Only the removal of the bovine serum immunoglobulins was not as good as expected. In small scale experiments it was found that this was due to leakage of antibodies from the immune adsorbent column at the elution with 7 M NH_4SCN. There are two reasons for this leakage of antibodies. First of all it was found that normal Sepharose 4B is not stable in 7 M NH_4SCN. It totally gelates under these conditions. Although the immune adsorbent appears to be more stable, most probably because of extra cross links through the coupled IgG, it might be expected that still some complexes of IgG-agarose are leaking from the column at the elution step. This problem can be solved by using the extra cross linked Sepharose CL-4B, which appeared to be more stable in 7 M NH_4SCN than normal Sepharose 4B. The second factor which might cause leakage of IgG is breakages of the moderately stable isourea bonds between IgG and support material (ref. 15). This can be prevented by cross linking of the coupled IgG with low concentrations of glutaraldehyde as described by Köwal and Parsons (ref. 16). They found that glutaraldehyde cross linking decreases leakage of IgG from Sepharose 4B immune adsorbents to undetectable levels without much loss of antigen binding capacity. This could be confirmed by us (table 4). By taking these two precautions, leakage of IgG might be overcome which will not only have a positive effect on the purity of the product, but also on the recovery of the antigen and the stability of the immune adsorbent.

TABLE 4
The effect of the cross linking of immune adsorbents by glutaraldehyde on leakage of IgG and binding capacity

Glutaraldehyde added in vol/vol %	Leakage of IgG	Binding capacity D-antigen U/ml
0	++	28×10^3
0.01	+	23×10^3
0.02	-	24×10^3
0.04	-	24×10^3
0.10	-	22×10^3
0.20	-	24×10^3
0.40	-	21×10^3

+: positive in countercurrent immunoelectrophoresis
-: negative

According to our experience even the less stable anti-polio Sepharose 4B immune adsorbent could be used repeatedly without much loss of binding capacity. As it was very important that the immune adsorbent was not contaminated with common micro-organisms during storage, it was stored in 0.5% phenol which could be easily removed by thoroughly washing before use. The binding capacity of the immune adsorbent was not affected by the 0.5% phenol solution even after four weeks at 37°C or four days at 56°C. In this way desinfection and sterilization of the immune adsorbent will be possible which opens the possibility to perform the whole purification process with immune adsorbents under aseptical conditions.

CONCLUSIONS

From our results on the purification of polio virus by means of affinity chromatography on immune adsorbents it might be concluded that application of this system for the concentration and purification of vaccines on a large scale is feasible. Attention should be paid to the leakage of antibodies from the immune adsorbent. By using Sepharose CL-4B as support material and extra cross linking of antibodies with glutaraldehyde this problem may be overcome. Under these conditions the present purification process consisting of ultrafiltration, gel filtration on Sepharose 6B and column chromatography on DEAE-Sephadex may be replaced by affinity chromatography on immune adsorbents followed by gel filtration on Sephadex G50 in order to obtain virus suspensions of the same purity. If monoclonal antibodies are applied as ligands, special attention should be paid to the removal and/or inactivation of the potential oncogenic cell material originating from the hybridoma cells.

ACKNOWLEDGEMENTS

The authors wish to thank Dr.J.S. Teppema for the electron microscopical examination of the immune adsorbents and Dr.G. van Steenis for the preparation of hyper-immune anti-polio sera. They acknowledge the excellent technical assistance of Mr.A. Kooistra and Mr.A. Emmelot.

REFERENCES

1. S.T.Nakajima, J.Clin.Microbiol., 5 (1977) 635-639.
2. F.Brown, B.O.Underwood and K.H.Fantes, J.Med.Virol., 4 (1979) 315-319.
3. D.S.Secher and D.C.Burke, Nature, 285 (1980) 446-450.
4. P.van der Marel, A.L.van Wezel, A.G.Hazendonk and A.Kooistra, Develop.biol. Standard., 46 (1980) 267-273.
5. A.L.van Wezel, G.van Steenis, Ch.A.Hannik and H.Cohen, Develop.biol.Standard., 41 (1978) 159-168.
6. A.L.van Wezel, J.A.M.van Herwaarden and E.W.van den Heuvel-de Rijk, Develop. biol.Standard., 42 (1979) 65-69.
7. G.Köhler and C.Milstein, Eur.J.Immunol., 6 (1976) 511.
8. A.D.M.E.Osterhaus, A.L.van Wezel, G.van Steenis, A.G.Hazendonk and G.Drost, Proceedings joint IABS-ESACT meeting "On the use of Heteroploid and other Cell Substrates for the Production of Biologicals", Heidelberg, May 18-22, 1981, Develop.biol.Standard., in press.
9. P.L.Ey, S.J.Prowse and C.R.Jenkins, Immunochemistry, 15 (1978) 429-436.
10. P.S.Thomas, Proc.Natl.Acad.Sci. USA, 77 (1980) 5201-5205.
11. S.D.March, I.Parikh and P.Cuatrecasas, Anal.Biochem., 60 (1974) 149-152.
12. C.L.Cambiaso, A.Geffinet, J.P.Vaerman and J.F.Heremans, Immunochemistry, 12 (1975) 273-278.
13. O.Hannibal-Friedrich, M.Chun and M.Sernetz, Biotechnol.Bioeng., 22 (1980) 157-17
14. J.L.Tayot, Institut Mérieux, Lyon, personal communication.
15. V.Sica, E.Nola, I.Parikh, G.A.Puca and P.Cuatrecasas, Nature New Biol., 244 (1973) 36-37.
16. R.Köwal and R.G.Parsons, Anal.Biochem., 102 (1980) 72-76.

T.C.J. Gribnau, J. Visser and R.J.F. Nivard (Editors),
Affinity Chromatography and Related Techniques

PRACTICAL CONSIDERATIONS IN THE USE OF IMMUNOSORBENTS AND ASSOCIATED INSTRUMENTATION

J. W. EVELEIGH

E. I. du Pont de Nemours and Company, Wilmington, Delaware, U.S.A.

ABSTRACT

The properties of immunosorbents and the influence of system operating conditions on the separation of specific antibodies and the purification of antigens by immunoaffinity chromatography are examined in practical terms. Strategies are developed to accommodate the diversity of resources and the variable requirements of product quality and yield demanded of the method. Design criteria of dedicated instrumentation to automatically perform immunoaffinity separations are presented and assessed from the specific requirements of potential users.

INTRODUCTION

It is now generally recognised that the method of choice for the separation and purification of specific antibodies is immunoaffinity chromatography. Acceptance and use of this method as a routine procedure is however, yet to be fully realised in many laboratories. An increasing number of authors have exemplified acceptance by relegation of procedural details to a few sentences within the 'Materials and Methods' section of their contemporary publications. However, there still remains an apparent apprehension on the part of many others that the method is only predictable and reliable in the hands of experienced practitioners of the "immunoaffinity art". The purpose of this presentation is to reassure the uncertain among us, to encourage the unwilling and hopefully generate a wider acceptance of this elegant laboratory technique.

The immunoaffinity method of separation is basically very simple, being a direct application of adsorption-desorption chromatography. The crude antisera, or a derived immunoglobulin fraction, is passed over an immunosorbent consisting of the antigen chemically immobilised on an inert support. Specific antibodies are selectively retained on the immunosorbent by virtue of their avidity for the antigen, other components being eluted unhindered from the column. After washing the immunosorbent, the adsorbed antibodies are recovered by elution with a dissociating buffer. With suitable choice of dissociant the immobilised antigen will retain its immunochemical reactivity and specificity thus allowing repetition of the cycle. The converse, in which immobilised antibody is employed to purify antigen, is equally feasible; however, with alternative more established methods available, this immunoaffinity approach has been much less frequently used.

In common with all brief descriptions, the last paragraph states it all but tells us little. Reduction of principles to successful practice is invariably either a combination of insight and luck or implementation of the critical procedur details that result from previous empirical findings. Our objective therefore reduces to discussion of these critical factors essential to practical success of the method and the means to accomplish it.

INITIAL STRATEGY

A well established doctrine in experimental science is that success comes to those who are prepared for it; hindsight being a luxury that most can ill afford. It follows therefore, that an initial appreciation of externally imposed limitatior or constraints and of the overall purpose of the experiment is of paramount importance. In immunoaffinity chromatography this essentially reduces to what limitations are imposed by the resources available and what is the required product. How the separation or purification is effected in the most efficient manner, the strategy to be used, is largely determined by these two criteria.

Resources, in the context of strategy formulation, are primarily the supply of antigen and the amount of induced antisera to be processed. Essentially from this aspect, four differing circumstances can be discerned:

1) Unrestricted antigen supply and a large volume of antiserum.
2) Unrestricted antigen supply and a small volume of antiserum.
3) Restricted supply of antigen and a large volume of antiserum.
4) Restricted supply of both antigen and antiserum.

These possible alternatives are schematically depicted to the left of Figure 1.

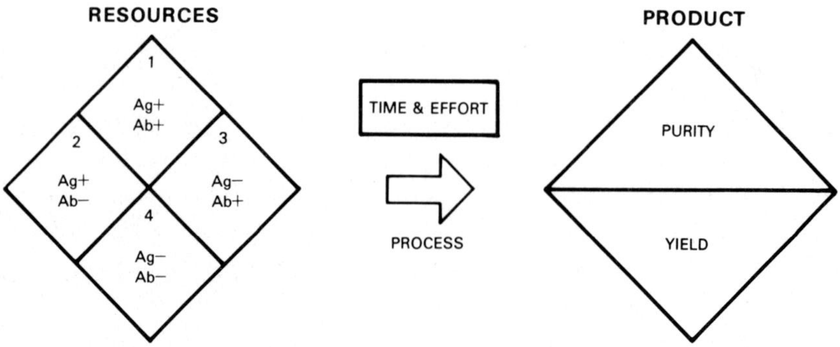

Fig. 1. Initial Strategy

The first two categories, in which readily obtainable antigens are being used, are relatively easy to accommodate. Immunosorbent preparation can be scaled to suit the volume of antiserum to be processed. Conventional immobilisation methods can be adopted and coupling efficiency becomes a secondary consideration. Furthermore, in the first category, optimisation trials can be undertaken as required to assure the desired product specifications. Where the supply of antiserum is limited however, the number of trial separations is curtailed demanding a greater emphasis on the chemical and physical characteristics of the immunosorbent and operational variables.

Categories 3 and 4, where a rare antigen is being used, essentially dictates a small volume separation column and the means for repetitive operation. The antigen must be immobilised efficiently to give maximal immunochemical reactivity and the separation process must be as optimal as is possible under the circumstances. One must be prepared to sacrifice expediency for efficiency. The last circumstance, in which both resources are limited, offers the greatest challenge and optimisation of both immunosorbent and operating efficiency becomes paramount.

The other side of this process strategy pertains to the required product specifications and is depicted on the right hand side of Figure 1. Purity of product is often the prime and sole reason for selecting the immunoaffinity chromatography method. The monospecific antibody should ideally be uncontaminated with any other serum proteins including immunochemically unreactive or cross-reactive immunoglobulins. There is also the natural desire to obtain high yields of product particularly if the antiserum supply is limited. Practically however, these two requirements become somewhat incompatible unless dedicated automatic instrumentation is available to obviate the extensive time and manual effort needed.

The connection between what we have and what we want is the chromatographic process. From these brief descriptions of differing circumstances, it should be apparent that associated strategies and operating protocols can be quite diverse. We can now begin to examine practical considerations of the process and components of the system in the light of these imposed conditions.

THE IMMUNOSORBENT

The first major decision following acceptance that immunoaffinity is the method of choice is that of selecting the type and means of preparing the immunosorbent. Since the first report by Lerman in 1953 on a specific immunosorbent based on cellulose (ref. 1), the literature contains a confusing spectrum of alternative supports and linking chemistries. With an unlimited supply of antigen and time, it is tempting to try an empirical approach to selection; with a rare antigen, however, it is more prudent to accept what has become the more commonly used materials. Supports having the desired physical and chemical characteristics compatible with extended column operation, reasonable capacity and low non-specific interaction are now commercially available. As molecular extenders are not necessary using protein antigens, the use of direct pre-activated supports is both convenient and generally adequate. When pure haptens are to be immobilised, considerable effort can be saved by using the derivatised supports commercially available for other affinity applications.

It is evident from even a cursory survey of the relevant literature that agarose is the most widely used support matrix. Activation of this material with cyanogen bromide was first reported by Porath (ref. 2) and the original method subsequently improved by the same author and his colleagues at Uppsala (ref. 3). Pharmacia, for example, offers many stabilised CNBr activated supports based on Sepharoses that allow direct coupling of protein antigens. An alternative method of linking through a N-hydroxysuccinimide active ester group was reported by Cuatrecasas and Parikh (ref. 4) and activated supports containing spacers are available from severa manufacturers including Pharmacia (Activated CH-Sepharose 4B) and Bio-Rad (Affi-Ge 10). Carbonyldiimidazole activation was described by Bethell et al (ref. 5) and Pierce Chemical Company offer similarly activated supports based on agarose, dextr and controlled pore glass.

It would be beyond the scope of this review to cover all of the many derivatise supports and linking chemistries that can be used to prepare immunosorbents. Othe than where specialised linking applications dictate their use, the various attributes and characteristics claimed for a particular support and chemistry usually relate to non-specific interactions and binding capacities. The absence of charge groups in the bridging linkage reduces the potential of ionic interactions resulti in improved purity of product. Other non-specific effects, related to hydrophobic and the amphoteric nature of immobilised protein, are however of equal significanc

Superficially, a support matrix with a high capacity for immobilised protein should have a definite advantage. The column will be smaller and the yield per cycle should be greater. For strategies in which resources are unrestricted, particularly of the immobilised species, and high yield is of primary concern this attitude is justified. Where material for immunosorbent preparation is limited however, efficiency in use is paramount. We have demonstrated (ref. 6) that immun chemical reactivity of an immunosorbent is significantly affected by the surface density of immobilised protein. The greater the degree of substitution, the lower becomes the specific reactivity of the immobilised protein. Thus, by attempting to increase capacity, one can in effect actually waste what is often valuable reagent. To obviate this effect, the strategy to adopt when resources are limited is to restrict the degree of substitution and compensate with a larger volume of immunosorbent.

Finally, mention must be made of the "reversed immunosorbent" technique that can be used with advantage in many applications. This method converts an immobili antigen immunosorbent to an antibody immunosorbent, or vice versa, by simple chemi cal cross-linking in situ (ref. 7). The approach avoids a separate isolation ste and can be very effective when the overall objective is to increase the supply of the limiting component. Furthermore, retention of multicomponent stoichiometry is largely retained allowing facile preparation of "balanced" antibody immunosorbents. For example, immobilised whole serum, after incubation with excess antisera and cross-linking, becomes an effective subtractive immunosorbent for the removal of serum components from culture media containing hybridoma antibodies. This immunosubtractive approach can be a viable alternative to the conventional adsorption-elution method when particularly labile components are involved.

SAMPLE PREPARATION AND COLUMN OPERATION

To ensure optimal and reproducible operation of an immunoaffinity system, the nature of the sample to be processed must be considered. When time is at a premium and the overriding requirement is product yield, it is common practice to process raw antisera. This course can give rise to subsequent problems involving not only purity of product but actual long term operation of the immunosorbent column. Firstly, it is imperative that clarified serum rather than plasma be used to avoid the real possibility that fibrin will be formed within the system. Secondly, it has also been found that the presence of lipid and denatured lipoprotein in the sample can dramatically reduce efficiency by coating the immunosorbent. Prior removal of the lipid using ether or dextran sulphate/calcium chloride precipitation (ref. 8) completely obviates this potential problem.

Where product is antibody and purity is of primary importance, the use of an IgG fraction rather than complete antisera is to be recommended. Ammonium or sodium sulphate precipitation is usually adequate although ion exchange chromatography is preferable if the amount of antisera is limited. This extra effort with the slight loss of antibody is justified by the avoidance of non-specific adsorption of serum proteins, particularly albumin, and subsequent contamination of product. Adsorption of non-specific IgG can be suppressed by the addition of 1M NaCl and 0.5% Tween 80 to the wash buffer. We have shown (ref. 6) that the inclusion of detergent alone results in an increase of at least 25% in the specific activity of eluted antibodies.

Determination of an optimal sample volume inevitably reduces to the empirical exercise of performing a limited series of trial separations. The many unpredictable and individualistic factors involved include active component concentration, specific reactivity of the immunosorbent and kinetic parameters related to column dimensions and application flow rate. With regard to the latter, the transport of reactive species to their complementary sites on a porous support is by bulk mixing within the fluid stream with the outer surface of the support and by diffusion within the pores. Where maximal utilisation of the immunosorbent is required, the application rate must therefore be low to accommodate the slow diffusion of macromolecules and in extreme cases, the flow can be temporarily stopped after sample application. However, where the major objective is to obtain a high yield in a minimum of time, with efficient utilisation of immobilised species being secondary, a very high flow rate is desirable with reaction restricted to the oute layers of the support (ref. 6). In most applications a compromise is acceptable and "reasonable" flow rates of between 1 and 10 ml per minute, for column bed volumes of 5 to 50 ml, are routinely adopted.

With respect to these trial separations, a phenomenon that has been consistenly observed with both antigen and antibody immunosorbents must be noted. Loosely termed the "first cycle effect", it is found that the first separation to be performed on a virgin column invariably gives atypical results. The effect is appare from Table 1 which reports the recovery of antigen from a 5 ml antibody column with repetitive addition of excess identical samples. The first cycle gives a lower recovery of product, originally specifically bound to the immunosorbent, than is obtained in succeeding cycles. It appears, in this example, that about 25% of the first sample becomes irreversibly bound and can be considered unrecover able. The effect persists to a lesser extent through subsequent cycles but this is largely now due to a less than optimal volume of chaotropic eluting buffer. This annoying loss of capacity is most apparent in highly substituted immunosorber and probably involves physical or steric entrapment within the support matrix.

TABLE 1

Effect of Cycling a 5-ml Column with Excess Antigen and Eluting with 15 ml of 2.5 M NH_4SCN

Cycle	Amount Applied (mg)	Amount Bound (mg)	Amount Eluted (mg)	Eluted/Bound (%)	Amount Retained (mg)
1	5.87	4.192	3.204	76.4	0.988
2	5.97	3.316	3.019	91.0	0.297
3	5.89	3.239	2.988	92.2	0.251
4	5.88	3.179	2.964	93.2	0.215
5	5.86	3.135	2.995	95.5	0.140
6	5.91	3.089	2.965	96.0	0.124

The volume of buffer wash following sample addition is very important from a product purity aspect. The porous immunosorbent column acts as a gel filtration system with elution of unreacted species being essentially diffusion controlled. Continuous U.V. monitoring of the column output reveals the exponential nature of this washing process and serves as an indicator of adequacy. The decision of when to add eluting agent is part of the overall strategy; high product purity demands that it be delayed until the absorbance indicates only the presence of low avidity dissociating product in the eluant. Maximum yield, however, dictates a minimum of delay to avoid losses and reduce overall cycle time. The compromise that in the majority of applications inevitably has to be accepted, is perhaps one of the more critical decisions in applying the immunoaffinity method.

The choice of eluting agent to displace the immunoreactive species is another operational decision. Chaotropic agents such as thiocyanate ion, iodosalicylates, and guanidinium salts are very effective and appear not to drastically affect gamma globulins, provided they are promptly removed. Several hundred cycles of protein antigen immunosorbent columns using phosphate buffered 2.5M ammonium thiocyanate have been obtained on automated systems with little or no loss of capacity. The use of acidic or basic eluants or concentrated urea solutions should be adopted only after prior testing of the antigen for stability to these potential denaturants. Some antigen-antibody complexes are notoriously difficult to dissociate and may require innovative approaches. Carbohydrate determinants and many hapten-conjugates give particular problems and often the only course is to use a lower avidity analogue rather than the original antigen. The volume of specific eluant to be used is again a matter of experimentation, however, a general rule of one to three column volumes is usually adequate. In repetitive cycling systems it is often more efficient from a yield per time basis to forego total displacement of the reacted component with large volumes of chaotrope in favor of a lesser but more rapid recovery. The undisplaced material remaining can then be recovered in a final cycle using a large volume of eluant.

INSTRUMENTATION

Application of the immunoaffinity method is considerably facilitated by the use of dedicated instrumentation. The system can take many forms depending upon the anticipated use, the degree of sophistication and fabrication facilities that are available. A simple manually operated assembly of chromatographic hardware will suffice for a few separations or purifications but if use of the method becomes routine then some form of automated system becomes essential. Not only does this reduce the tedium of manual operation and caretaking, but the improvement in quality and yield of product will completely justify the initial investment. Several years experience in automating immunoaffinity chromatography has produced a wide variety of dedicated instrument systems and has taught some valuable lessons in the process. The purpose of this section then is to critically examine the requirements and further offer some suggestions on the design and construction of these unique instruments.

An immunoaffinity system is basically very simple being essentially a collection of valves to control fluidic inputs to the column and to divert eluted components into collection vessels. A positive displacement pump is necessary to maintain fluidic supplies and an absorbance monitor on the column outlet is particularly essential to follow the chromatographic process. The first degree of sophistication is to control the operation of the fluidic valves with timers or a simple programmer and thus allow repetitive operation of the system. Higher degrees of elegance are provided by safety devices to allow unattended operation, integral desalting devices such as a gel filtration column to remove chaotropic ions, and fully automated control of the overall process by interrogative circuitry and related devices. The major components of an automated system having these features is schematically presented in Figure 2. Before attempting to describe this relatively complex instrument and its capabilities, it is necessary to briefly consider the required characteristics of the individual components.

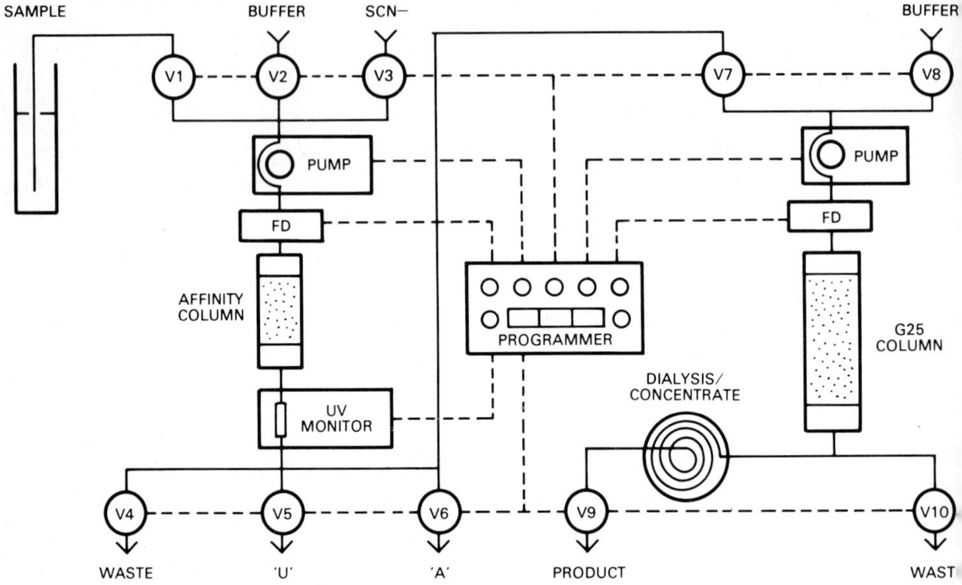

Fig. 2. Immunoaffinity System Schematic

FLUIDICS

Conventional chromatography columns with a high cross-sectional area to length ratio and preferably adaptors to reduce dead space, are adequate for immunoaffinity separations. Although refrigeration of the column is rarely necessary, provided wash buffer contains bactericide, it is prudent to retain sample and collected product in cooled containers. All fluid lines, connectors, and valve bodies should be of an inert plastic, as thiocyanate and acid are notoriously corrosive. The number of solenoid valves can be reduced by the use of multi-way valves although it must be recognized that in the event of power loss such configurations are not inherently "fail-safe". To prevent disasterous loss of sample by syphoning in this circumstance, a normally closed two-way sample valve can be considered essential. Latching solenoid fluid valves are to be recommended in obviating local heating and its associated problems. As an alternative to electrically operated valves, a mechanical, programmable, multi-way device designed for the purpose form the basis of an automated instrument built by the author (ref. 9). In the selection of a pump the corrosive nature of chaotropes limit the choice to those of the peristaltic type. The inherent deterioration of pumping rate is their major disadvantage in repetitive separations and may warrant the adoption of peak sensing facilities as opposed to simple time based operation.

PROTECTIVE DEVICES

A conventional chromatographic U.V. monitor with its associated recorder is essential in optimising operating conditions and for retrospective monitoring of unattended repetitive separations. Of equal importance is the incorporation of active fluid sensors at critical points within the fluid circuitry to detect line breakages and unanticipated loss of buffers, etc. These devices can be inexpensively fabricated from miniature photo and light emitting diodes and are electrically sensitive to the continued absence of fluid within an integral glass tube. Activation of one of these sensors initiates shut-down of the system and prevents further loss of sample or product. They are also useful in a positive sense for detecting completion of sample processing and monitoring buffer supplies and collection facilities. Other protective devices desirable for long term unattended operation are pressure sensors to indicate impending column blockage or valve malfunctions and power failure interrupts to avoid potential loss of synchronisation.

PROGRAMMER

The advent of the microprocessor based programmer has virtually satisfied the requirements for a versatile and reliable control system for this and other automated instrumentation. Early models of preparative immunoaffinity systems described by Anderson et al. (ref. 10). were programmed by mechanical resetable clocks or by cam-operated switches. These systems although adequate for routine separations proved to be somewhat inflexible and cumbersome to fine-tune for optimal performance. Process programming based on time sequencing is however the simplest technique to implement and is quite adequate for applications in which reasonable purity and yield is acceptable. Several microprocessor based timers are now commercially available and an immunoaffinity instrument can literally be constructed around them. The "UP-Timer" manufactured by Xanadu Controls, Springfield, New Jersey 07081, for example, provides highly versatile time sequencing of up to ten outputs using a penciled program card. An instrument based on this timer and incorporating all of the features previously discussed is illustrated in Figure 3 and has proven very reliable for routine purifications. Programmable fraction collectors such as the Siemens ES1 from ES Industries, Marlton, New Jersey 08053, are also readily adaptable to automatic repetitive operation.

A higher degree of sophistication and versatility is provided by the "programmable sequencers" recently becoming commercially available. Specifically designed for decision based sequential processing, they allow for interrogative operation

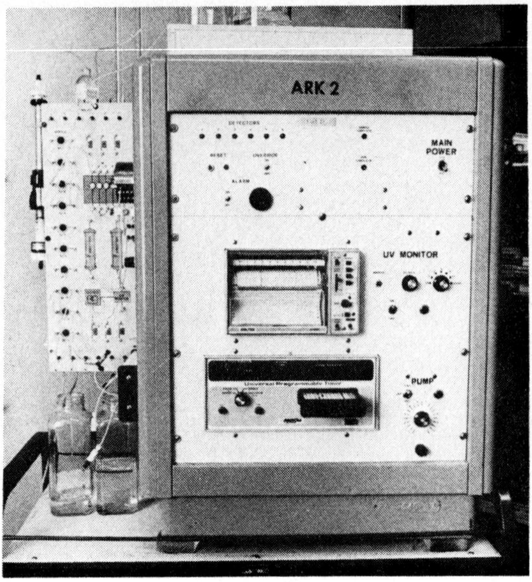

Fig. 3. Automatic Time-Based Instrument

and can be programmed to respond to feed-back information from the U.V. monitor for peak collection in conjunction with elapsed time control. Examples of these sequencers are the EPC-7101 from Encoder Products, Sandpoint, Idaho, and the Model 816 controller from CCI, Sunnyvale, California. The ultimate in versatility is provided by the use of a desk-top minicomputer such as the Commodore PET, the Apple series, and the Hewlett-Packard HP-85. Their large memory and basic programming facilities allow for extremely elegant software based routines that can automatically optimise and operate several separation modules simultaneously. To be recognised however, is one danger to adopting a minicomputer as the controll There will inevitable arise an overwhelming tendency to devise ambitious subroutines and to explore its elegant software potential - this leaves, as well I know, little or no time to do the mundane task of purifying antibodies!

FUTURE PROSPECTS

With the sophistication of contemporary microprocessor technology, the achievement of a versatile, completely automated and reliable immunoaffinity system is assured. By adopting the systematic practical approaches described earlier, consistent separations and purifications can become routine procedures rather than an art. It is certain that recognition of the reliability and relative ease of the method will result in a wider acceptability and more general use.

The growing interest in hybridoma technology for the production of monoclonal antibodies is an area in which immunoaffinity purifications can be of value. The specific isolation of the antibodies from culture media and ascites fluid is as simple and certainly more effective than non-specific Protein A adsorption. The required products of genetic engineering can potentially be isolated from bacterial or yeast homogenates and culture fluids in a single step using monoclonal antibodies or even traditional heterogeneous animal antisera. Radioisotopically labelled affinity purified antibodies for the location and therapy of certain tumors is of growing interest (ref. 11) and will undoubtedly necessitate automated production facilities for more general use. Finally, the increased use of immunoassays in clinical diagnostic testing, with their demand for consistent high purity reagents, is already taking the immunoaffinity method out of research laboratories into a manufacturing environment. What was once a highly specialised technique is now routine procedure and with it the protective "art", hopefully, history.

REFERENCES

1 L. S. Lerman, Nature, 172(1953)635.
2 J. Porath, Nature, 218(1968)834.
3 J. Porath, K. Asberg, H. Drvein and R. Axen, J. Chromatogr., 86(1973)53.
4 P. Cuatrecasas and I Parikh, Biochemistry, 11(1972)2291.
5 G. S. Bethell, J. S. Ayers, W. S. Hancock and M. T. W. Hearn, J. Biol. Chem., 254(1979)2572.
6 J. W. Eveleigh, J. Solid-Phase Biochem., 2(1977)45.
7 D. E. Levy and J. W. Eveleigh, J. Imm. Methods, 22(1978)131.
8 A. Van Dalen, H. G. Seijen and M. Gruber, Biochim. Biophys. Acta., 147(1967)421.
9 J. W. Eveleigh, J. Chromatogr., 159(1978)129.
10 N. G. Anderson, D. D. Willis, D. W. Holladay, J. E. Caton, J. W. Holleman, J. W. Eveleigh, J. E. Atrill, F. L. Ball and N. L. Anderson, Anal. Biochem., 66(1975)159; Anal. Biochem., 68(1975)371.
11 J. P. Mach, M. Forni, J. Ritschard, F. Buchegger, S. Carrel, S. Widgren, A. Donath and P. Alberto, Oncodevel. Biol. Med., 1(1980)49.

T.C.J. Gribnau, J. Visser and R.J.F. Nivard (Editors),
Affinity Chromatography and Related Techniques

A SIMPLE AND RAPID PROCEDURE FOR LARGE SCALE PREPARATION OF IgGS & ALBUMIN FROM HUMAN PLASMA BY ION EXCHANGE AND AFFINITY CHROMATOGRAPHY

J. SAINT-BLANCARD, J.M. KIRZIN, P. RIBERON and F. PETIT

Centre de Transfusion Sanguine des Armées "Jean Julliard", Clamart (France)

J. FOURCART, P. GIROT and E. BOSCHETTI

Réactifs IBF, Villeneuve-la-Garenne (France)

ABSTRACT

We describe here a new method for a large scale purification of IgGs and plasma albumin,and present the results obtained from a pilot plant. The process involves only three chromatographic columns and only three buffer solutions, the starting material is the human plasma cryosupernatant or Fraction-I supernatant.

The first column, filled with 50 litres of a new gel filtration medium, Trisacryl for desalting, realizes the equilibration of 17 litres of plasma in the right buffer. The second column, filled with 50 litres of DEAE-Trisacryl M, separates the totality of IgGs from the other proteins. The third column, filled with 50 litres of Blue-Trisacryl, performs the quantitative separation of the albumin from the other proteins.

This process is continuous. As a result of the high chromatographic efficiency of the supports utilized and of the simplicity and the continuity of the process, several cycles of fractionation can be performed per 24 hours.

The yield of the process is at least equivalent to the best known chromatographic procedures.

The quality of the IgGs and albumin is remarkable and meets the requirements of the Pharmacopeias.

INTRODUCTION

In recent years, one of the most significative progress in plasma proteins fractionation has been the application of chromatographic processes.

Using weak ion exchangers (DEAE-Sepharose, SP-Sephadex or CM-Sepharose) Curling and al. (ref. 1,2,3) have prepared a pure albumin from crude extract obtained by polyethylene glycol (PEG 4000) precipitation. We have applied this process in our laboratory (ref. 4). Also, with the same chromatographic supports, various combinations have been used for isolation of IgGs and/or albumin from different starting materials (plasma serum, cryosupernatant, Fractions II + III supernatant, euglobulin -free solutions, etc.) (ref. 5,6).

Recently, new synthetic matrix (Trisacryl) and derived exchangers (DEAE- and CM-Trisacryl M) have been developed. These materials show exceptional chromatographic properties and have been used by us in the separation of IgGs and albumin from cryosupernatant (ref. 7,8).

We have attempted to modify this process,replacing CM-Trisacryl by Blue-Trisacryl which is an immobilized form of Cibacron-Blue F3GA, a dye known to interact with albumin. So, the modified process involves the three following steps :

. Desalting on Trisacryl,
. Ion exchange on DEAE-Trisacryl,
. Affinity on Blue-Trisacryl.

MATERIALS AND METHODS

Chemicals

All reagents (analytical grade) were obtained from Fluka, Merck or Prolabo. Cel strips for electrophoresis, immunofilms for immunoelectrophoresis and Ouchterlony munofilms were purchased from Sebia.

The starting materials for chromatographic separations were plasma cryosupernat or Fraction I supernatant prepared in our laboratory.

The chromatographic supports, obtained from Réactifs IBF, are :

a) Trisacryl for desalting, a highly cross-linked copolymer of N-[tris(hydroxym thyl)-methyl]-acrylamide with exclusion limit of 4,000 daltons,

b) DEAE-Trisacryl M, a macroporous weakly cationic ion exchanger having a high solution power and a very low level of non-specific adsorption (ref.9).

c) Blue-Trisacryl, a derivative obtained by coupling Cibacron-Blue F3GA on a ch cally modified macroporous and non-ionic Trisacryl.

Chromatographic separation process

The fractionation was performed in a pilot plant using a set-up of the type sho in Fig. 1 and according to the schematic sequences shown in Fig. 2.

The sample was first equilibrated in a 0.025 M tris-HCl buffer, pH 8.8 contain 0.035 M NaCl (B1) on a column of Trisacryl for desalting. The column volume was ab 50 litres of gel (35 x 50 cm) and the sample volume was of 17 litres. The flow-ra was 35 litres per hour. During this step, the plasma proteins were equilibrated i above buffer and the salts and low molecular weight components were eliminated. Th volume of the desalted proteins sample obtained after this column was about 18 li the dilution factor is usually between 1.05 and 1.1.

The fraction containing the proteins was applied to a 50 litre DEAE-Trisacryl M column (35 x 50 cm) previously equilibrated in the above mentioned buffer (B1). In these conditions, all the proteins were adsorbed, excepted for the IgGs which were eluted in a sharp peak. The dilution factor of the IgG fraction was 1.1-1.2. The o plasma proteins were desorbed by elution with 0.05 M tris-HCl buffer, pH 8 contai 0.75 M NaCl (B2). The flow-rate was 30 litres per hour. The column of DEAE-Trisac was then reequilibrated in B1 buffer.

The latter fraction was passed through a 50 litre column of Blue-Trisacryl(35x5 previously equilibrated in B2 buffer. In these conditions, albumin and lipoproteins were totally adsorbed whereas the other proteins were collected in a first fraction. Albumin was eluted by 0.05 M tris-HCl , pH 8 containing 3.5 M NaCl (B3). The lipoproteins were successively desorbed by washing with distilled water.

Fig. 1. Pilot plant equipment for IgGs and albumin purification

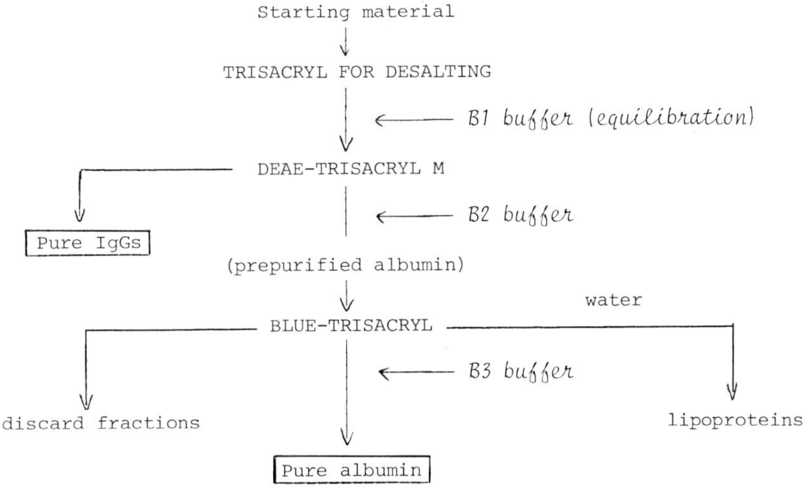

Fig. 2. Chromatographic sequences (see the formulations of buffers in the text).

Other operations and controls

The solution containing IgGs and albumin was rapidly desalted and concentrated b ultrafiltration on flat membrane in a Pellicon cell (Millipore). The final preparati were performed according to classical methods for the production of therapeutic form Thereafter, they were controled as required by the Pharmacopeias, particularly throu electrophoresis, immunoelectrophoresis and immunodiffusion (degrees of purity) and b gel filtration (detection of IgGs fragments or IgGs and albumin polymers). If neces- sary, the absence of bilirubin in albumin was controled.

Tris-barbital buffer, pH 9.2 was used for the electrophoresis which was carried c for 35 minutes at a voltage of about 160 volts. The immunochemical tests were perfo med on agarose strips against human plasma proteins antiserum (ref. 10). Fragments a polymers were detected on Ultrogel AcA 34 column according to Suomela's method (ref. Bilirubin was measured by Nosslin's method (ref. 11).

RESULTS

Desalting and reequilibration of starting material on Trisacryl for desalting are illustrated in Fig. 3A which shows the separation of high and low molecular weight components.

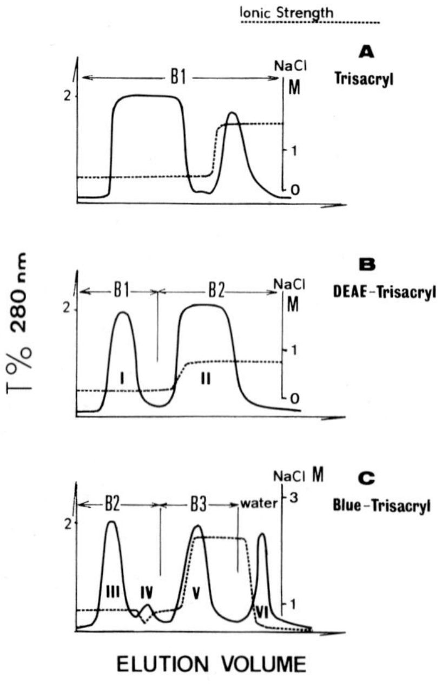

Fig. 3. Elution profiles obtained du- ring fractionation (starting material cryosupernatant)

It should not be out of place to mention here that in the experimental conditions (pH, ionic strength), there was no precipitation of euglobulins inside the column.

Ion exchange chromatography (Fig. 3B) results in separation of two fractions. The first fraction (I) contains 99.5% pure IgGs as attested by electrophoresis analysis (Fig. 4). It should be noted that IgGs never interact with chromatographic material, avoiding thus the risk of an eventual denaturation. Furthermore, in the IgG fraction, only traces of IgEs were visible. The second fraction (II) is composed of all the other plasma proteins.

Three main peaks are eluted during chromatography on immobilized dye (Fig. 3C). The protein desorbed in the first peak is mainly transferrin (fraction III). The second fraction (IV) contains α- and β-globulins in low quantities as well as traces of albumin. (Fig. 4).

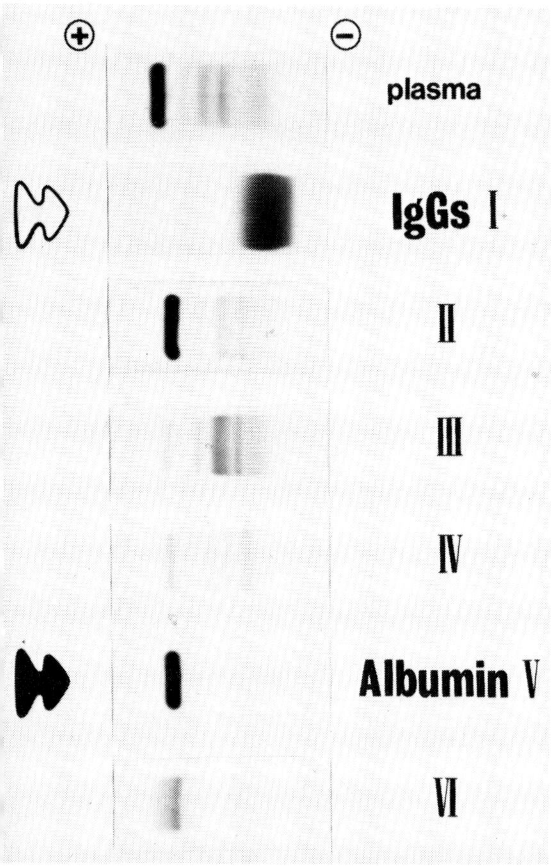

Fig. 4. Electrophoretic patterns of plasma protein fractions obtained according to the discussed procedure (starting material : cryosupernatant)

The greater part of albumin is in the following peak (fraction V) with a purity superior to 99% (Fig. 4 and 5).

The last fraction (VI) contains α- and β-lipoproteins (Fig. 6) and traces of a bumin desorbed by elution with water.

The yield of albumin as well as that of IgGs is higher than 80%.

The quality of the final products fulfils the Pharmacopeias requirements.

Especially, the albumin obtained by our affinity process was, after stabilizatio verified to be resistant at a heating of 60°C during 10 hours, in order to inactiva eventual Hepatitis B antigen. No polymers of albumin or bilirubin were detected. It should be noted the total absence of immunoglobulins or transferrin. On the other ha the colours of the albumin solutions meet the Pharmacopeias specifications. The use of hemolyzed plasma as starting material does not change the colours of final forms

With this process, it is perfectly feasible to produce pyrogen-free protein solu tions as demonstrated by our experiments.

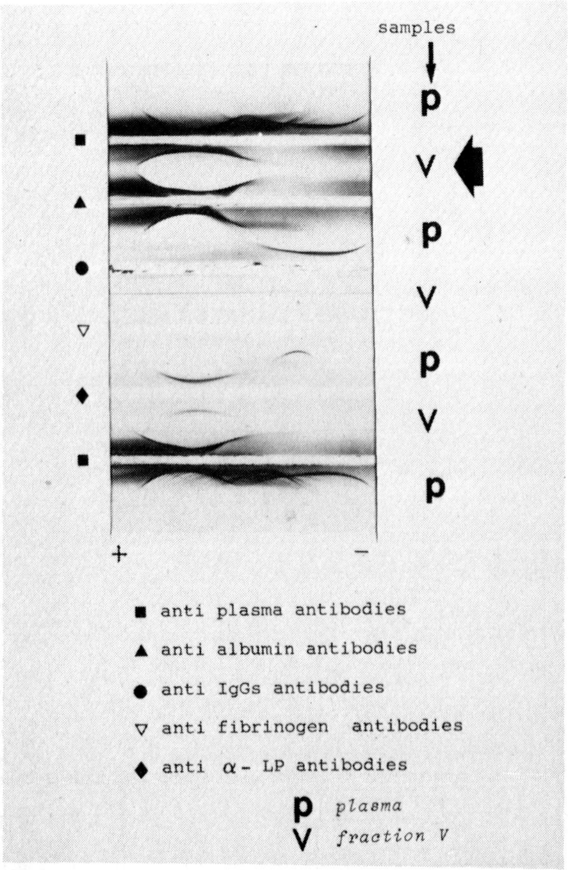

Fig. 5. Immunoelectrophoresis of frac tion V (V) and control of normal huma plasma (P) (Starting material : cryo-supernatant)

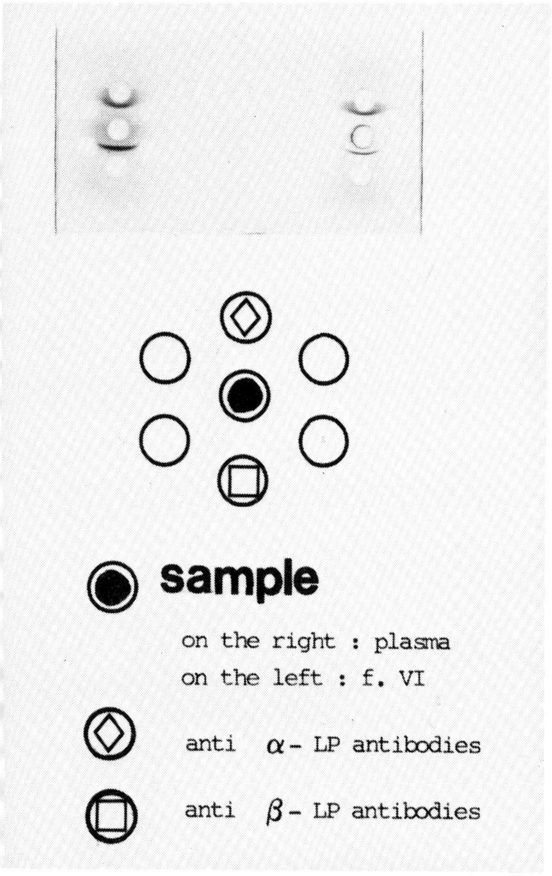

Fig. 6. Ouchterlony immunodiffusion of fraction VI and control of normal human plasma against anti-LP antibodies.

DISCUSSION

The major interest of this method is the continuity of the entirely chromatographic process, permitting to keep the products sterile and apyrogenic. Indeed, the gels are autoclavable and before the columns, sterilizing filters can be placed.

Since crude plasma cryosupernatant can be used as starting material, it is possible to isolate certain coagulation factors and/or fibronectin from cryoprecipitate (ref.12) Other sub-fractionations could also be performed on minor fractions.

The results we have got (quality, purity, yield) are widely dependent on the utilization of the new Trisacryl supports. Indeed, these supports are synthetic, non-biodegradable, resistant to heating as far as 120°C, stable in large scale of pH. They have a good power of resolution and high capacity. These properties guarantee a high productivity.

Up to now, total stability of the existing dye-gel adsorbents seems to have been limited as a consequence of ligand leakage (ref. 13). Our development on Blue-Trisacryl has been focused on the creation of stable covalent links between Cibacron Blue and the matrix, by using particular reaction conditions which eliminate any parasite and not stable fixation of the dye. The first tests of stability performed on this product have shown very encouraging results.

The high flow-rates accepted by the gels enable the entire operation to be performed in less than 8 hours. The use of this chromatographic procedure allows to avoid ethanol precipitations. Furthermore, since all the stages are performed at 20°C, without any final lyophilization step, this method permits a consequent economy of energy.In addition, the process can be totally automatized.

CONCLUSION

As it has been previously demonstrated, the advantages of this method combined with the quality of the gels utilized and the total reproductibility of the process guarantee high reliability.

So, large scale application of this new plasma protein fractionation method is quite feasible.

REFERENCES

1 J.M. Curling, L.O. Lindquist and S. Eriksson, Process Biochem., 12(1977),22-25.

2 J.M. Curling, J.H. Berglöf, L.O. Lindquist and S. Eriksson, Vox Sanguinis, 33,2 (1977), 97-107

3 J.M. Curling and J.H. Berglöf, International Workshop on Technology for Protein Separation and Improvement of Blood Plasma Fractionation, Reston, Virginia, September 7-9, 1977, reprint Pharmacia Fine Chemicals, Uppsala, pp. 1-7.

4 J. Saint-Blancard, M. Allary, F. Limonne, B. Desneux and J.M. Curling, XIe Congr National de Transfusion Sanguine, Strasbourg, Juillet 1979, Abstract N° 53.

5 J. M. Curling, Albumin Purification by Ion Exchange Chromatography, in J. M. Curling (Ed.), Methods of Plasma Protein Fractionation, Academic Press, London, 198 pp. 77-91.

6 H. Suomela, An Ion Exchange Method for Immunoglobulin G Production, in J. M. Curling (Ed.), Methods of Plasma Protein Fractionation, Academic Press, London, 198 pp. 107-116.

7 J. Saint-Blancard, J. Fourcart, F. Limonne and E. Boschetti, 18th Congress of th International Society of Blood Transfusion, Montreal, August 16-22, 1980, Abstra N° 813.

8 J. Saint-Blancard, J. Fourcart, F. Limonne, P. Girot and E. Boschetti, Annales Pharmaceutiques Françaises (1981), in press.

9 DEAE- / CM-TRISACRYL M, Manuel Pratique pour Chromatographie d'Echange d'Ions , Réactifs IBF, 1980.

10 Fiche technique " Immunofilms Sebia ".

11 L. Hartman, in Techniques Modernes de Laboratoire et Explorations Fonctionnelles tome I, L'Expansion, Paris 1977, pp. 513-515.

12 M. F. Roulleau, E. Boschetti, T. Burnouf, J.M. Kirzin and J. Saint-Blancard, Pro Aff. Chromat., Nijmegen, 1981.

13 M. J. Harvey, The Application of Affinity Chromatography and Hydrophobic Chromatography to the Purification of Serum Albumin, in J.M. Curling (Ed.), Methods of Plasma Protein Fractionation, Academic Press, London, 1980, pp. 189-200.

T.C.J. Gribnau, J. Visser and R.J.F. Nivard (Editors),
Affinity Chromatography and Related Techniques

SURFACE TOPOGRAPHY OF INTERFERONS: A PROBE BY METAL CHELATE CHROMATOGRAPHY

E. SULKOWSKI, K. VASTOLA, D. OLESZEK and W. VON MUENCHHAUSEN
Roswell Park Memorial Institute, Buffalo, New York 14263 (U.S.A.)

ABSTRACT

Mammalian interferons display significant diversity in their affinities for immobilized metal chelates of Co^{++}, Ni^{++}, Cu^{++} and Zn^{++}. Among interferons of human origin, HuIFN-β binds at neutral pH to all tested metal chelates. By contrast, HuIFN-αs display an affinity for Cu^{++} chelate only. MuIFNs bind only to Cu^{++} chelate and can be separated into two distinct populations with a falling pH-gradient. MuIFNs prepared in the presence of tunicamycin, a glycosylation inhibitor, bind stronger to a Cu^{++} column than glycosylated counterparts suggesting a modified surface topography. HaIFNs display an affinity for Cu^{++} chelate only, and their strength of binding is similar to that of MuIFNs.

The affinities of several "model" proteins and oligopeptides for metal chelates were also explored.

INTRODUCTION

Since Porath et al. (ref. 1) introduced metal chelate affinity chromatography, several proteins have been purified by its use. The application of this novel chromatographic principle was also rapidly extended to the purification of interferons: human fibroblast interferon (ref. 2,3,4), human leukocyte interferon (ref. 3,5,6) and hamster interferon (ref. 7,8).

Mammalian interferons share many biologic properties and, therefore, are expected to display a measure of similarity in their surface topography, the diverse antigenicity notwithstanding. Metal chelate affinity chromatography may serve as a facile surface topography probe: for example, the classification of human amnion interferon as a fibroblast interferon was aided by its display of affinity for Zn^{++} chelate (ref. 9). Recently, the production of ample amounts of many human interferons for therapeutic use has been made feasible by the insertion of their structural genes into E. coli and yeast. The "authenticity" of microbial products may be probed, *inter alia*, by metal chelate affinity chromatography.

The affinity of proteins for metal chelates is due to amino acid side chains of histidine, cysteine and, perhaps, tryptophan (ref. 1). The amino acid composition of human leukocyte interferons (HuIFN-α) reveals the presence of three histidines (his 7, his 34 and his 58) in all eight HuIFN-α (ref. 10,11). Chou-Fasman analysis, reported by Hayes (ref. 12), places his 7 in random coil while his 34 and his 58 are in β-turns of the interferon polypeptide; presumably then all these histidines are accessible to the solvent. Human fibroblast interferon (HuIFN-β) contains five histidines: his 93, his 97, his 121, his 131 and his 140 (ref. 13). His 93 and his 131 are assigned to β-pleated sheet and his 121 to α-helix. His 97 and his 140 are found in β-turns (ref. 12) and, presumably, are accessible. Thus, HuIFN-α and HuIFN-β might have, each, at least two histidyl side chains localized on their surfaces. However, their microenvironment may be quite different: all histidines (three) in HuIFN-α are in N-fragment whereas those in HuIFN-β (five) are in C-fragment of the interferon polypeptide. As a result, they may assume different topography upon folding of their respective interferon polypeptides.

The majority of HuIFN-α (six out of eight) has four cysteine residues: cys 1, cys 29, cys 99 and cys 139 (ref. 10,11). Those cysteines are involved in disulfide bonds (ref. 14). Two interferons-α have, in addition, cys 86, which is assigned to α-helix (ref. 12) and may not be accessible. Therefore, we assume that cysteinyl side chains of HuIFN-α may not serve as electron donor groupings for coordination to metal chelates. HuIFN-β has three cysteines: cys 17, cys 31 and cys 141. Two of them, cys 31 and cys 141, form a disulfide bond (ref. 14). One can expect that cys 17 will contribute to affinity of HuIFN-β for metal chelates.

Most of the HuIFN-α (seven out of eight) have two tryptophan residues, trp 77 and trp 141 (ref. 10,11). Both tryptophans are placed in α-helix (ref. 12). HuIFN-β has three tryptophans: trp 22, trp 79 and trp 143 (ref. 13). Trp 22 and trp 143 are assigned to β-pleated sheet while trp 79 is located in β-turn and is adjacent to unique glycosylation point, Asn 80 (ref. 12). Conceivably, trp 79 should be accessible to the solvent, although carbohydrate moiety may interfere with its coordination to immobilized metal chelates. The solvent accessibility of other tryptophans is not known. However, it is tempting to advance the following speculation. It is well established that cholera toxin inhibits the establishment of the antiviral state by interferon (ref. 15). Recently, it has been reported that tryptophan residues are involved in the binding and action of cholera toxin (ref. 16). The participation of aromatic residues (tryptophans?) in the interaction between interferons and their cell membrane receptors was postulated previously (ref. 17,18). If the tryptophan residues of interferons are indeed involved, then the most likely candidates would be trp 141 for HuIFN-α (present when unique as in LeIF B, ref. 11) and corresponding to trp 143 for HuIFN-β. If this were the case, then those tryptophan residues would be expected to be exposed to the solvent and could contribute toward the affinity of interferons for metal chelates.

The information pertaining to possible electron donor groupings on the surface of other mammalian interferons is not available at present.

RESULTS

Materials

HuIFN-β and mouse L-cell interferon (MuIFN-α,β) were prepared and assayed as described (ref. 19,20,21). Non-glycosylated MuIFNs were prepared in the presence of tunicamycin, 2 μg/ml. Pancreatic trypsin inhibitor was a gift from Dr. L. Kress. Micrococcal nuclease ("Foggi") was prepared as described previously (ref. 22). Bovine pancreatic ribonuclease A, lysozyme (egg white), angiotensins I and II and somatostatin were purchased from Sigma. Epoxy-activated Sepharose 6B was obtained from Pharmacia. Iminodiacetic acid (disodium salt) from Aldrich, was immobilized to agarose according to Porath et al. (ref. 1). The same batch of sorbent was utilized in all experiments reported here.

Chromatographic procedure

The columns (0.9 x 8 cm) were charged with Zn^{++} and Cu^{++} at pH 4.0 (ref. 3), equilibrated and developed as described previously (ref. 3,7): charging with Co^{++} and Ni^{++} was done at pH 6.0. All columns were prepared and developed at 4°. After applying a protein sample, 4 ml, a column was rinsed with 20 ml of equilibrating solvent, 0.02 M sodium phosphate, 1 M NaCl, pH 7.4 (↓) and initially developed with 25 ml of 0.1 M sodium acetate (1.0 M NaCl), pH 6.0 (↓). A pH gradient (↓) was then developed by mixing (a) 25 ml of 0.1 M sodium acetate (1.0 M NaCl), pH 6.0 and (b) 25 ml of 0.1 M sodium acetate (1.0 M NaCl), pH 4.0. Some columns were further rinsed with terminal solvent (↓). Any modifications of this protocol are given in figure legends. Protein concentration was measured by fluorometric assay (ref. 23).

Affinity of mammalian interferons for metal chelates

Table 1 gives the summary of the data on the binding of selected mammalian interferons to metal chelates.

TABLE 1

Affinity of interferons for Me^{++} chelates at neutral pH

IFN / Me^{++}	Co^{++}	Ni^{++}	Cu^{++}	Zn^{++}
HuIFN-β	+	+	+++	+
HuIFN-α	–	–	++	–
HaIFNs	–	–	+	–
MuIFNs	–	–	+	–

–, no binding ++, partially reversible binding
+, reversible binding +++, irreversible binding

All interferon preparations were chromatographed on the same batch of the sorbent in order to keep the ligand density constant. The affinity of interferons for Me^{++} chelates varies considerably. HuIFN-β binds to all metal chelates tested whereas HaIFN and MuIFN bind to Cu^{++} chelate only. Moreover, the binding of human interferons to Cu^{++} chelate is practically irreversible, under experimental conditions employed, whereas HaIFN and MuIFN can be completely recovered.

Human fibroblast interferon (HuIFN-β)

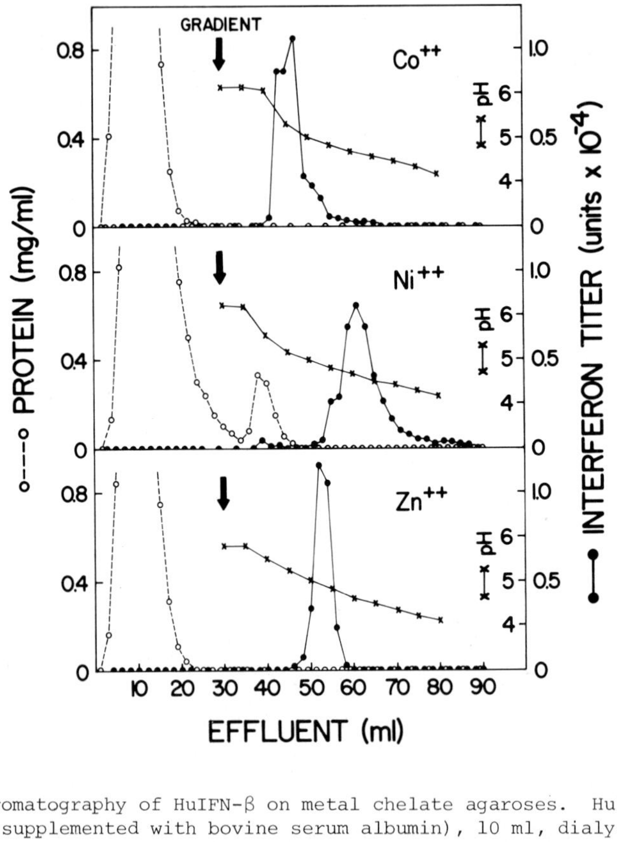

Fig. 1. Chromatography of HuIFN-β on metal chelate agaroses. HuIFN-β (purified preparation supplemented with bovine serum albumin), 10 ml, dialyzed against 0.1 M sodium acetate (1 M NaCl), pH 6.0 at 4°, was applied on a metal chelate column equilibrated with dialysis buffer. The column was first rinsed with the equilibrating solvent and then developed with a pH-gradient obtained by mixing: (a) 25 ml of 0.1 M sodium acetate (1 M NaCl), pH 6.0 and (b) 25 ml of 0.1 M sodium acetate (1 M NaCl), pH 4.0.

The affinity of HuIFN-β for Me^{++} chelates is quite strong as it remains bound to all of them even at pH 6.0. The binding to Cu^{++} chelate (not shown) is irreversible and only a partial recovery is possible upon decreasing the Cu^{++} density as reported previously (ref. 3). The binding of HuIFN-β to Co^{++},

Zn^{++} and Ni^{++} may be also shared by human amnion interferon which has been reported to bind to Zn^{++} chelate (ref. 9). It has been also observed that monkey interferon (ref. 24) binds to all listed Me^{++} chelates (D. Gurari-Rotman and E. Sulkowski, unpublished observations). Thus, the affinity for Co^{++}, Ni^{++} and Zn^{++} may be a distinguishing property of IFN-β produced by primates.

Mouse L-cell interferon (MuIFN)

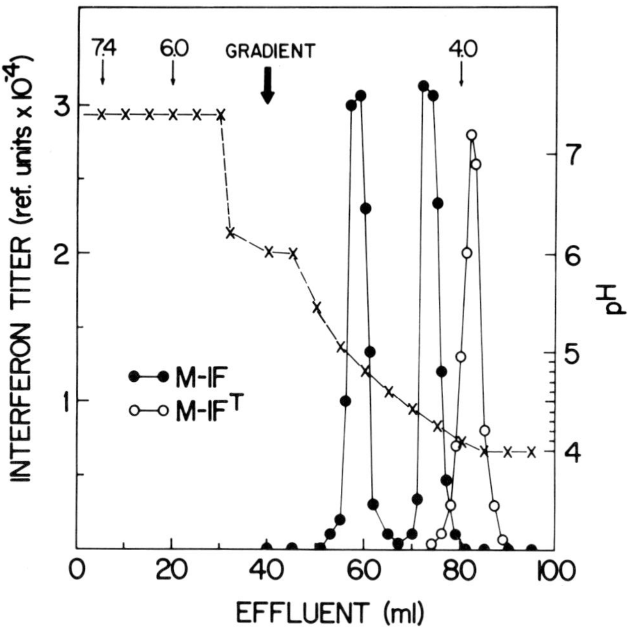

Fig. 2. Chromatography of MuIFNs on Cu^{++} chelate agarose. A preparation of interferon was dialyzed against 0.02 M sodium phosphate (1 M NaCl), pH 7.4 at 4°. An interferon sample, 5 ml, was applied on a column. The column was rinsed with 15 ml of equilibrating solvent (dialysis buffer) and developed with 20 ml of 0.1 M sodium acetate (1 M NaCl), pH 6.0. The column was further developed with a pH-gradient formed by mixing: (a) 20 ml of 0.1 M sodium acetate (1 M NaCl), pH 6.0 and (b) 20 ml of 0.1 M sodium acetate (1 M NaCl), pH 4.0. Finally, the column was washed with 20 ml of 0.1 M sodium acetate (1 M NaCl), pH 4.0. The protein elution profile is omitted for the sake of graphic clarity. ●—●, M-IF (glycosylated MuIFN); ○—○, $M\text{-}IF^T$ (non-glycosylated MuIFN prepared in the presence of tunicamycin).

Mouse L-cell interferon is a mixture of α and β types. When chromatographed on Me^{++} chelates, it does not show any affinity for Co^{++} chelate although it is retained on Ni^{++} chelate at pH above neutral. Interestingly, the L-cell interferon can be resolved into two subpopulations. Whether these represent α and β types remains to be established. Non-glycosylated mouse L-cell interferon is retained stronger than glycosylated counterpart and is not resolved into two components. It is quite possible that one subcomponent, when produced in the presence of tunicamycin, is not secreted into the culture medium. The stronger retention may indicate a better accessibility of electron donor grouping when the carbohydrate moiety is not attached to an adjacent site on the surface of the interferon molecule.

Affinity of model oligopeptides and proteins for Me^{++} chelates

In order to assess the affinity of a particular amino acid side chain for a particular metal chelate (Co^{++}, Ni^{++}, Cu^{++}, Zn^{++}), we have studied some readily available oligopeptides and proteins.

Angiotensin I and II

Human angiotensins I and II are oligopeptides (deca- and octa-) which contain two and one histidines, respectively. Cysteine and tryptophan are absent. There is no consensus on the conformation of angiotensins in solution (ref. 25). We assume that his 6 and his 9 (angiotensin I) and his 9 (angiotensin II) are accessible. If so, then one should observe stronger retention of angiotensin I on Me^{++} chelates. Figure 3 shows that this is the case. In fact, they can be readily separated on Zn^{++} and Co^{++} chelates.

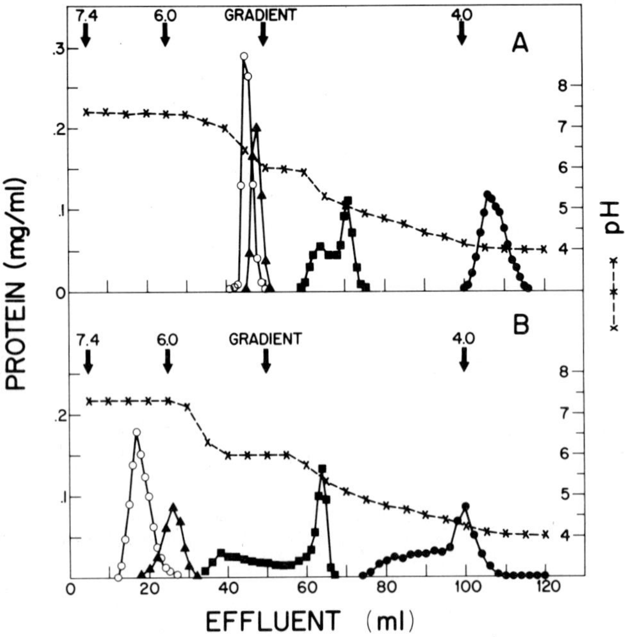

Fig. 3. Chromatography of angiotensin I (panel A) and angiotensin II (panel B) on Me^{++} chelates. ○—○, Zn^{++}; ▲—▲, Co^{++}; ■—■, Ni^{++}; ●—●, Cu^{++}.

Somatostatin

Somatostatin is a cyclic tetradecapeptide which contains one tryptophan residue (trp 8). Histidine is absent. Somatostatin has a stable conformation in solution. In the model advanced by Holladay and Puett (ref. 26), tryptophan residue is located at the β bend of the hairpin-like structure and is accessible to the solvent. Its chromatography on metal chelates revealed no retention at neutral pH on any metal chelates. There was, however, some delay in the elution from Ni^{++} chelate column and retention on Cu^{++} chelate at pH 7.4.

Pancreatic trypsin inhibitor

Pancreatic trypsin inhibitor is devoid of histidine, tryptophan and cysteine (disulfide bonds). The inhibitor, our "control" model protein, is not retained at pH 7 on any metal chelate tested (Co^{++}, Ni^{++}, Cu^{++}, Zn^{++}). It is retained on Cu^{++} chelate column at pH 7.4 presumably due to coordination *via* amino group(s).

Pancreatic ribonuclease A

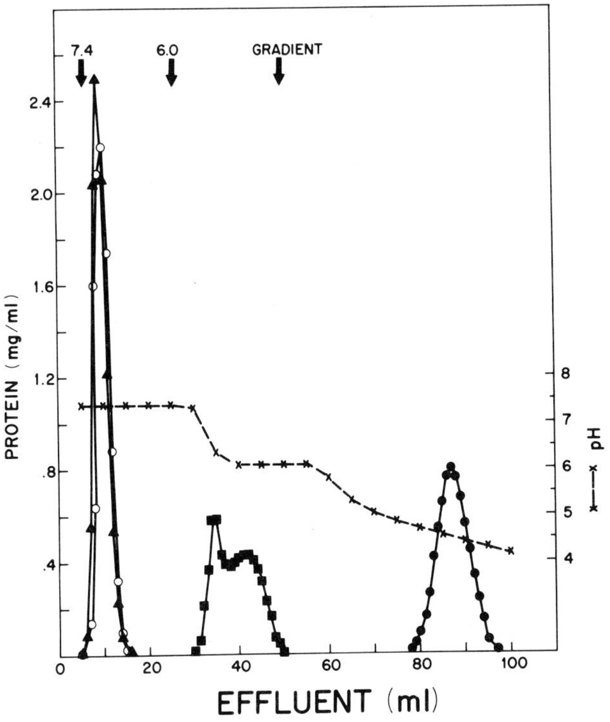

Fig. 4. Chromatography of ribonuclease A. ○—○, Zn^{++}; ▲—▲, Co^{++}; ■—■, Ni^{++}; ●—●, Cu^{++}.

Pancreatic ribonuclease A does not contain a tryptophan residue and all its cysteine residues are locked into disulfide bridges. It has four histidines: his 12, his 48, his 105 and his 119 which should be considered as metal ligands (ref. 27). However, his 12 and his 119 are too close to each other (ref. 28) to serve both as ligands for free metal ions and even less so for immobilized metal chelates. His 48 is buried. Therefore, we postulate that there are two histidyl side chains available for coordination: his 12 (or his 119) and his 105 (ref. 29).

Figure 4 illustrates the lack of affinity of ribonuclease A for Zn^{++} and Co^{++} chelates. The retention on Ni^{++} chelate is reversed at pH 6.0 and on Cu^{++} chelate at pH 4.5.

Micrococcal nuclease ("Foggi")

This protein is totally devoid of cysteine residues. The only tryptophan residue (trp 140) is buried. There are four histidines present: his 7, his 46, his 121 and his 124 (ref. 30). All these histidines seem to be accessible to the solvent. The chromatographic behavior of micrococcal nuclease on Me^{++} chelates is quite like that of ribonuclease A except that this protein is somewhat stronger retained on Cu^{++} chelate.

Lysozyme

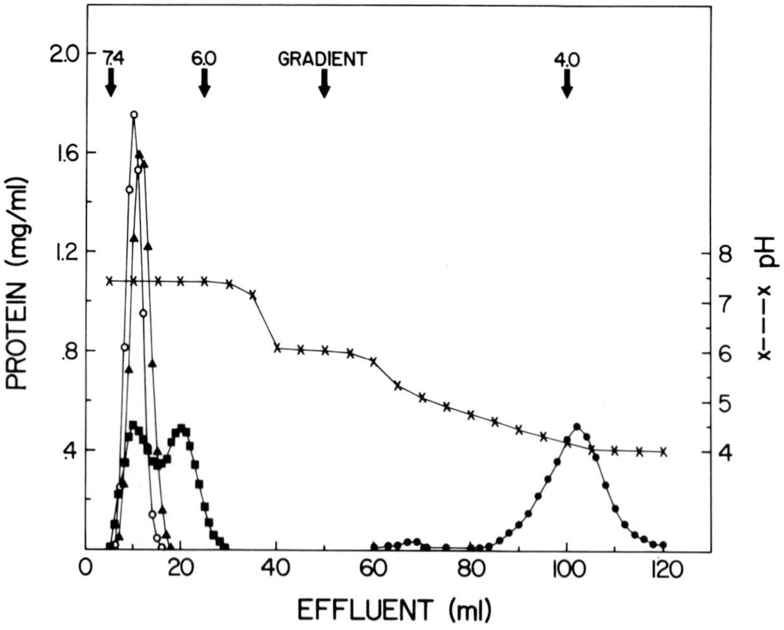

Fig. 5. Chromatography of lysozyme. ○—○, Zn^{++}; ▲—▲, Co^{++}; ■—■, Ni^{++}; ●—●, Cu^{++}.

Hen egg-white lysozyme contains one histidine (his 15) and six tryptophans (trp 28, trp 62, trp 63, trp 108, trp 111 and trp 123). All cysteines are involved in disulfide bridges. Trp 62, trp 63 and trp 123 are localized on the surface of the lysozyme (ref. 31,32).

Lysozyme, as shown in Figure 5, is strongly retained on Cu^{++} chelate only. It seems that the affinity for Cu^{++} chelate is too strong to be accounted for by coordination to his 15, especially that affinity for Ni^{++} chelate is rather weak too. It is difficult at this juncture to identify other possible ligands: (asp-52 and glu-35?) or, perhaps, one of the tryptophans.

DISCUSSION

Metal chelate chromatography has been used very successfully for the purification of some proteins, including interferons. It seems to us, however, that this technique can be taken advantage of as a probe into surface topography of proteins as well.

The affinity of interferons for immobilized metal chelates signals the presence on their surfaces of electron donor groupings. This in itself is of value as some specific questions of their involvement, if corroborated by results of chemical modification, in the biologic activity of interferons can be asked. The affinity for metal chelates may prove also to be of analytical value in assessing the surface topography of multiple members of IFN-α family before crystallographic data become available. There is also a possibility that metal chelate affinity chromatography may be useful in assessing the posttranslational modification of interferon (Fig. 2). Recently, two subpopulations of HuIFN-β have been observed on Zn^{++} chelate chromatography (ref. 33). It has been suggested, although far from being proven, that these may represent glycosylation variants.

There is a vast amount of data on the binding of *free metal ions* to proteins. However, these observations are not directly applicable to the understanding of the affinity of *immobilized metal chelates* for proteins. It will suffice to observe, without discussing the steric considerations of the coordination process, that human serum albumin binds Zn^{++} ions to almost all of its 16 imidazole groups (ref. 34) but itself is not retained on Zn^{++} chelate agarose (ref. 1). Therefore, there is a need to study the affinity of "model" proteins, i.e., of known tertiary structures, for immobilized metal chelates. The actual involvement of the potential donor groupings in the coordination has to be buttressed but specific chemical modifications of the former. Our observations on the affinity of some model proteins for metal chelates signal the need for more information.

REFERENCES

1 J. Porath, J. Carlsson, I. Olsson and G. Belfrage, Nature, 258(1975)598-599.
2 V.G. Edy, A. Billiau and P. De Somer, J. Biol. Chem., 252(1977)5934-5935.
3 K.C. Chadha, P.M. Grob, A.J. Mikulski, L.R. Davis, Jr. and E. Sulkowski, J. Gen. Virol., 43(1979)701-706.
4 J.W. Heine, M. De Ley, J. Van Damme, A. Billiau and P. De Somer, Ann. N.Y. Acad. Sci., 350(1980)364-373.
5 K. Berg and I. Heron, Scand. J. Immunol., 11(1980)489-502.
6 S. Yonehara, Y. Yanase, T. Sano, M. Imai, S. Nakasawa and H. Mori, J. Biol. Chem., 256(1981)3770-3775.
7 E. Bollin, Jr. and E. Sulkowski, Arch. Virol., 58(1978)149-152.
8 E. Bollin, Jr. and E. Sulkowski, J. Gen. Virol., 52(1981)227-233.
9 P.C.P. Ferreira, M. Paucker, R.R. Golgher and K. Paucker, Arch. Virol., 68(1981)27-33.
10 M. Streuli, S. Nagata and C. Weissmann, Science, 209(1980)1343-1347.
11 D.V. Goeddel, D.W. Leung, T.J. Dull, M. Gross, R.M. Lawn, R. McCandliss, P.H. Seeburg, A. Ullrich, E. Yelverton and P.W. Gray, Nature, 290(1981)20-26.
12 T.G. Hayes, Biochem. Biophys. Res. Comm., 95(1980)872-879.
13 T. Taniguchi, S. Ohno, Y. Fujii-Kuriyama and M. Muramatsu, Gene, 10(1980)11-15.
14 R. Wetzel, Nature, 289(1981)606-607.
15 E.F. Grollman, G. Lee, S. Ramos, D.S. Lazo, H.R. Kaback, R.M. Friedman and L.D. Kohn, Cancer Res., 38(1978)4172-4185.
16 M.J.S. De Wolf, M. Fridkin, M. Epstein and L.D. Kohn, J. Biol. Chem., 256 (1981)5481-5488.
17 E. Sulkowski, M. Davey and W.A. Carter, J. Biol. Chem., 251(1976)5381-5385.
18 E. Sulkowski, K. Vastola and H.V. Le, Ann. N.Y. Acad. Sci., 350(1980)339-346.
19 E.A. Havell and J. Vilček, Antimicrob. Agents Chemother., 2(1972)476-484.
20 K.C. Chadha, M.W. Davey, D.M. Byrd and W.A. Carter, Infect. Immun., 10(1974) 1057-1061.
21 M.W. Davey, E. Sulkowski and W.A. Carter, J. Biol. Chem., 251(1976)7620-7625.
22 E. Sulkowski and M. Laskowski, Sr., J. Biol. Chem., 241(1966)4386-4388.
23 P. Böhlen, S. Stein, W. Dairman and S. Udenfriend, Arch. Biochem. Biophys., 155(1973)213-220.
24 E. Yakobson, M. Revel and D. Gurari-Rotman, Arch. Virol., 59(1979)251-255.
25 R.E. Lenkinski, R.L. Stephens and N.R. Krishna, Biochemistry, 20(1981)3122-3126
26 L.A. Holladay and D. Puett, Proc. Nat. Acad. Sci. USA, 73(1976)1199-1202.
27 E. Breslow and A.W. Girotti, J. Biol. Chem., 241(1966)5651-5660.
28 A.M. Crestfield, W.H. Stein and S. Moore, J. Biol. Chem., 238(1963)2421-2428.
29 C.R. Mattews, J. Recchia and C.L. Froebe, Anal. Biochem., 112(1981)329-337.
30 A. Arnone, C.J. Bier, F.A. Cotton, V.W. Day, E.E. Hazen, Jr., D.C. Richardson, J.C. Richardson and A. Yonath, J. Biol. Chem., 246(1971)2302-2316.
31 D.C. Philips, Proc. Nat. Acad. Sci. USA, 57(1967)484-495.
32 R. Cassels, C.M. Dobson, F.M. Poulsen and R.J.P. Williams, Eur. J. Biochem., 92(1978)81-97.
33 J.W. Heine, J. Van Damme, M. De Ley, A. Billiau and P. De Somer, J. Gen. Virol., (1981) in press.
34 F.R.N. Gurd and D.S. Goodman, J. Am. Chem. Soc., 74(1952)670-675.

T.C.J. Gribnau, J. Visser and R.J.F. Nivard (Editors),
Affinity Chromatography and Related Techniques

BOVINE AND HUMAN FIBRONECTINS: LARGE SCALE PREPARATION BY AFFINITY CHROMATOGRAPHY

M.F. ROULLEAU and E. BOSCHETTI

Réactifs IBF, Villeneuve-la-Garenne, (France)

T. BURNOUF, J.M. KIRZIN and J. SAINT-BLANCARD

C.T.S.A. "Jean Julliard", Clamart (France)

ABSTRACT

The large scale purification of fibronectin is connected with several problems : the stability of the chromatographic support ; the need of sterilization, the flow-rate of the column ; the cost of the adsorbent and its productivity.

We present here a simple procedure involving a new type of adsorbent prepared by mixing and polymerizing agarose with gelatin. By this way, we guide the substitution ratio (quantity of gelatin immobilized per ml of gel) on which depends the adsorption power.

The cross-linking by bifunctional reagent as glutaraldehyde assures a high chemical stability of the adsorbent and its sterility. The gel thus obtained has an excellent mechanical resistance and is particularly adapted to large scale applications.

With this gelatin-agarose copolymer (Gelatin-Ultrogel), we have performed numerous cycles of production of bovine and human fibronectinswithout any loss of the adsorption capacity.

The fibronectins were identified by cross-reaction against specific antibodies, and their homogeneity controled by SDS-polyacrylamide gel electrophoresis. Their biological activity was demonstrated by a spreading test performed on BHK cells.

INTRODUCTION

Fibronectin is a large glycoprotein present in plasma and synthetized by fibroblasts in culture (1,2). It interacts with several extracellular matrix components like collagen (3,4), glycosaminoglycans (5-7) and gangliosides (8).

As a consequence of these multiple interactions, the functions of the fibronectin are in relation with the cell-cell and cell-substratum adhesion processes.

Recent reports show that the role of fibronectin in the final step of the coagulation (9) and its interest in therapeutic applications such as von Willebrand's disease, thrombasthenia (10), trauma and burn (11,12).However, one of the major factors limiting the therapeutical use of fibronectin is its unavailability in sufficiently large and sterile quantities.

The problems of its industrial production are connected with the preparation of a well adapted affinity chromatography adsorbent having high adsorption capacity and selectivity,and permitting, furthermore, repeated uses without loss of the capacity and without leakage of the ligand. The isolation of plasma fibronectin by affinity chro tography is possible by three techniques : by using either immobilized antibodies (14) or immobilized heparin (7) or immobilized gelatin (15-17).

The high cost of the specific antibodies and the loss of their biochemical activ when used repeatedly represent serious limitations of this first technique.

In spite of the good stability and availability of heparin as a ligand, immobili heparin forms with fibronectin very stable complexes which can be dissociated only highly denaturating conditions (8 M urea) (6).

The use of the gelatin presents several advantages such as its low cost and the mation of a reversible complex with fibronectin. In addition, gelatin leads to the formation of specific complexes with a low level of non-specific adsorption.

Gelatin has already been immobilized on cyanogen bromide activated agarose beads and used as an affinity adsorbent on the analytical scale (15). However, this adsor bent is not ideal in consideration of the gelatin leakage and of the high toxicity the cyanogen bromide, which represent limitations for the scaling up in view of the peutic uses.

For these above-mentioned reasons, we have studied a new chromatographic support avoiding the majority of the above-described limitations.

In this paper, we describe the preparation and the evaluation of a new affinity sorbent called Gelatin-Ultrogel and its use for the large scale preparation of fibronectir

MATERIAL AND METHODS

Chemicals

Gelatin-agarose gels were prepared according to the method described below or ob tained from Réactifs IBF, Villeneuve-la-Garenne, France (Gelatin-Ultrogel). Pooled human plasma and plasma cryoprecipitate were a gift from C.T.S.A., Clamart, France. lulose acetate strips (Cellogel) for electrophoresis as well as dried agarose plate for immunoelectrophoresis were purchased from Sebia, Issy-les-Moulineaux, France. T other ragents were of analytical grade.

Preparation of gelatin-agarose gels

A hot aqueous solution of agarose (5% w/v at 85°C) was mixed with a well known q tity of high molecular weight gelatin (from 0.5% to 5% w/v, see Table I). After sol bilization, the solution was emulsified in an oil phase according to classical meth of preparation of agarose beads (18).

After gelification, the gel particles were stabilized by treatment with a 1% aque solution of glutaraldehyde for one night at 4°C.

The gel thus obtained was washed repeatedly in sterile conditions with a 0.1 M t HCl buffer pH 7.8, a 1 M sodium chloride solution and finally with demineralized water

The gelatin-agarose gel was filled in a chromatographic column according to the method described below.

Determination of the fibronectin adsorption capacity of gelatin-agarose

This parameter was determined by measuring the quantity of human fibronectin adsorbed and desorbed on the gels using purified fibronectin.

The volume of the gel was 5 ml, the dimensions of the column were 1.6 x 2.5 cm, the adsorption buffer was 0.05 M tris-HCl pH 7.5, the elution solution was 4 M urea in tris-HCl buffer.

Preparation of bovine fibronectin

5 litres of whole bovine plasma containing 500 mg/litre of benzamidine were injected into a chromatographic column containing 4 litres of Gelatin-Ultrogel (column dimensions : 12.6 x 25 cm). The linear flow-rate was of 10-15 $cm.h^{-1}$. The gel was then washed with a 0.05 M tris-HCl buffer pH 7.5, with a 4 M sodium chloride solution in the same buffer, and finally with 0.1 M urea solution in the same buffer.

The adsorbed fibronectin was eluted with a 0.05 M tris-HCl buffer pH 7.5 containing 4 M urea at a linear low-rate of 2.5-4 $cm.h^{-1}$.

The fibronectin obtained was immediately dialyzed against a 0.05 M tris-HCl buffer pH 9 containing 0.15 M sodium chloride and 1 mM calcium chloride at 4°C and stored at - 20°C.

Preparation of human fibronectin

The preparation of human fibronectin was performed by two different methods : affinity chromatograpy on column and selective adsorption in batch using freeze-dried Gelatin-Ultrogel. The starting materials from which the fibronectin was extracted were whole human plasma or human plasma cryoprecipitate.

1) Column affinity chromatography method using whole plasma as a starting material: The procedure was the same as described for bovine plasma.

2) Column affinity chromatography method using plasma cryoprecipitate as a starting material : The cryoprecipitate from 8 litres of human plasma was solubilized in 6 litres of 0.05 M tris-HCl buffer 7.5 containing 0.15 M sodium chloride under stirring. The solution was then centrifuged, the supernatant injected into 4 litres Gelatin-Ultrogel column and the purification of the fibronectin was performed as described above.

3) Removal of fibronectin from whole human plasma by batch selective adsorption using freeze-dried Gelatin-Ultrogel : The swollen gel was first lyophilized after extensive washing in order to eliminate the salts. Then 2.5 g of lyophilized powder corresponding to about 50 ml of swollen gel was added to 300 ml of human plasma and the suspension stirred for 18 hours at 20°C. The gel was separated by filtration and the fibronectin-depleted plasma collected without dilution. The fibronectin was recovered in the same way as indicated on the chromatographic process.

Determination of the purity of the fibronectin

1) Electrophoresis : Two types of electrophoresis were used : the first was performed on cellulose acetate strips according to the manufacturer's instructions. The con-

ditions of migration and staining are indicated on figure legends. The second elect phoresis method was performed on 5% cross-linked polyacrylamide containing 0.1% sod dodecyl sulphate and 0.01% 2-mercaptoethanol as described earlier (19). The migratic conditions are indicated on the figure legend.

2) Immunoelectrophoresis : These analytical determinations were applied to all t fractions obtained from the chromatographic column in order to check the presence c foreign proteins in the fibronectin preparations and to determine the protein compc sition of the intermediate fractions.

The immunoelectrophoreses were performed according to the manufacturer's instruc tions ; the conditions of work are indicated on figure legends.

3) Determination of the biological activity of fibronectin : It was determined b using a spreading test on BHK-21-C-13 cells according to a modification of the Gri nel's technique (20). Briefly, the cells were washed with Eagle's MEM medium withou serum and then placed in Petri dishes either with 5 µg/ml of fibronectin or withou fibronectin as a control, and incubated at 37°C. During the incubation, the extent spreaded cells was determined visually with an inverted microscope.

RESULTS AND DISCUSSION

Characteristics of the chromatographic adsorbent

The gelatin-agarose gels were prepared in view of their application to preparati purification of fibronectin. They should correspond to the following characteristic

- . rigid and porous particles in order to obtain high flow-rates,
- . high adsorption capacity,
- . absence of gelatin leakage,
- . easy protection towards bacterial contaminations,
- . level of non-specific adsorptions as low as possible.

As a matrix, we chose 5% agarose gel possessing a high porosity and a good rigic The porosity is in fact an important point because of the high molecular weight of fibronectin (440,000).

We studied the influence of the quantity of gelatin immobilized on the adsorptic capacity of the gel. The results of this study are represented on Table I.

Table I - Adsorption capacity of Gelatin-agarose gels

Composition of gel		Immobilized gelatin (mg/ml)	Exclusion limit *	Adsorption capacity (mg FN/ml gel)
Agarose (%)	Gelatin (%)			
5	0.5	3.3	6-8 x 10^6	0.1
5	1	6.5	4.5 x 10^6	0.65
5	2	13	1.7 x 10^6	1.35
5	3	20	0.7 x 10^6	1.05
5	4	26	0.25 x 10^6	0.9
5	5	32	0.12 x 10^6	0.8

* for globular proteins

Table I shows that initially the capacity of the gel increases with the concentration of the ligand. Simultaneously, an increase of the ligand concentration leads to a progressive diminution of the gel porosity which limits the accessibility of the fibro nectin to the ligand. From these experiments, it can be seen that maximal capacity of the gel towards fibronectin corresponds to a gelatin concentration of about 13 mg/ml.

The risks of ligand leakage were avoided by using glutaraldehyde as a strong cross-linking reagent of gelatin. In the conditions of work described in the Material and Method section, we never found any resolubilization of gelatin : control of the absence of gelatin was made with a detection method having a sensitivity level of 100 ng (Bradford method (21) applied to the 10-20 fold concentrated chromatographic fractions obtained in absence of the sample).

Another interest of the use of glutaraldehyde is the sterilization effect of this reagent : since the final step of preparation of the support is the glutaraldehyde cross-linking, Gelatin-Ultrogel can be used in sterile conditions without autoclaving. Furthermore, in case of bacterial contamination, the gel can be resterilized by washing with a 0.5% glutaraldehyde solution.

Finally, it must be noted that the preparation of the gelatin-agarose gel does not involve any activation step of the matrix : as a consequence, the non-specific adsorption sites of the gel are minimized and due essentially to the ligand itself.

Fibronectin preparation procedure

A representative elution profile is given on Fig. 1. During the adsorption step, several proteins bind to gelatin and are successively removed by washing with 4 M sodium chloride and 1 M urea. In these conditions however, the fibronectin is practically not eluted. Fibronectin elution is obtained by washing with 4 M urea buffered solution.

The analysis of the intermediary protein fractions by electrophoresis and immunoelectrophresis (Fig. 2 and 3) indicates that the 4 M sodium chloride fraction contains small quantities of several serum proteins like albumin, β-globulins and γ-globulins. The 1 M urea fraction seems to be comparable to the preceding fraction but contains in addition a small amount of fibronectin. The last protein fraction does not contain any other plasma protein than fibronectin.

Recent reports indicate that it is possible to desorb fibronectin in non-denaturating conditions either with arginine or with spermidine and putresceine (17,22), but because of the high cost of those products, they cannot be used in preparative scale. Furthermore, when the elution is done by the arginine for example, the gel has to be regenerated with a 8 M urea solution while in the method presented here, the gel is totally regenerated after elution with a 4 M urea solution.

The use of plasma cryoprecipitate instead of whole plasma renders the purification easier in consideration of the composition of the starting material.

In the batch procedure, the selective removal of fibronectin from human plasma using lyophilized Gelatin-Ultrogel gives a yield lower than the chromatographic procedure.

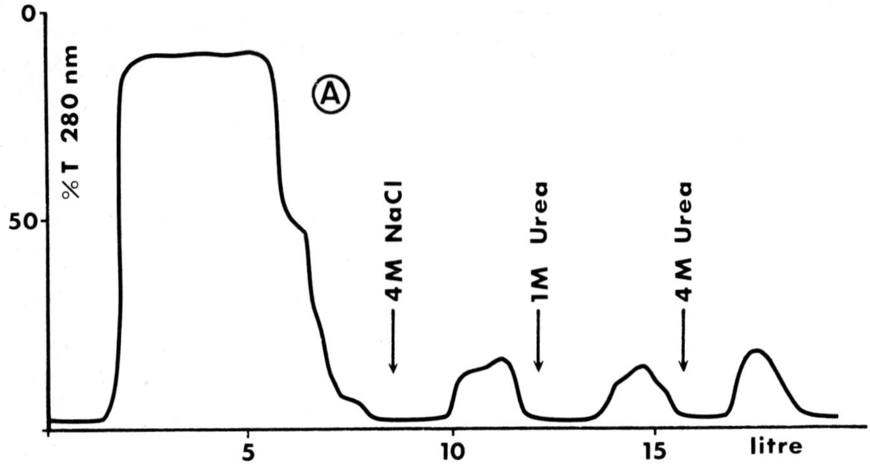

Fig. 1. Preparation of human fibronectin by affinity chromatography on Gelatin-Ultrog
Sample volume : 5 litres of plasma ; column : 12.6 x 25 cm ; initial buffer : 0.05
HCl, pH 7.5; washing solutions and elution solutions were prepared in the initial
buffer ; flow-rates : adsorption and washing linear flow-rates were 10-15 cm.h^{-1},
elution flow-rate was 2.5-4 cm.h^{-1} ; temperature : 20°C.

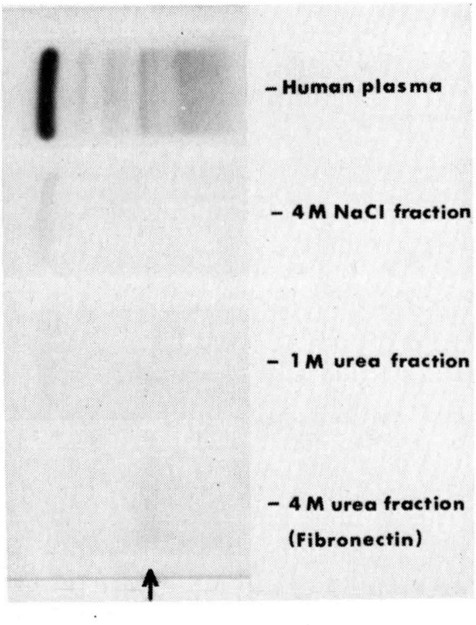

Fig. 2. Cellulose acetate electrophoresi
the fractions obtained from affinity ch
tography column.
Buffer : Barbital-tris pH 9.2 (10.3 g of
dium barbital, 7.2 g of tris, 1.84 g of
bital acid for 1 litre of buffer) ; migr
time : 35 minutes at 160 volts ; migrati
temperature : 20°C, staining : ponceau S
lution (5 g in 1 litre of 5% trichloroac
acid).

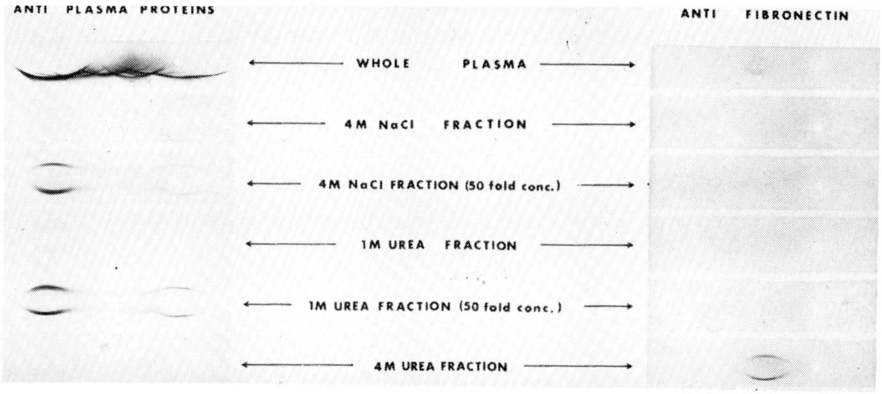

Fig. 3. Agarose immunoelectrophoresis of the fractions obtained from affinity chromatography column.
Buffer : Barbital-tris pH 9.2 (see Fig. 2); migration time : 70 minutes at 150 volts and at 20°C ; sample volume : 1 µl ; antiserum volume : 50 µl ; diffusion time : 48 h; staining : amido black solution (5 g in 1 litre of a mixture of methanol/acetic acid/ water 5:1:4 , v/v/v).

In fact, after the adsorption step, a certain amount of fibronectin remains in the plasma even after three sucessive adsorptions. Nevetheless, this procedure is rapid and allows an easy recovery of the partially fibronectin-depleted plasma without dilution or addition of salts and buffer. The recovery of the fibronectin is still performed following the method described above. Although this batch procedure was studied only on a small scale, it seems very promising for large scale applications.

Results obtained with column chromatography column procedure permit the recovery of 150-200 mg of purified fibronectin per litre of plasma. This method is reproducible and the gel can be used at least 10-fold without any significative loss of its adsorption capacity.

Analytical results

SDS-polyacrylamide gel electrophoresis of the fibronectin obtained from the column chromatography process (Fig. 4) shows a main band with a characteristic mobility (14).

The small amount of foreign proteins does not correspond to the principal constituents of the plasma ; in fact, the detection assays effected in order to check albumin, trans ferrin, immunoglobulin G and fibrinogen by immunoelectrophoresis gave us negative results. Results of the determination of the fibronectin biological activity as regards its spreading activity for BHK cells are represented on Table II.

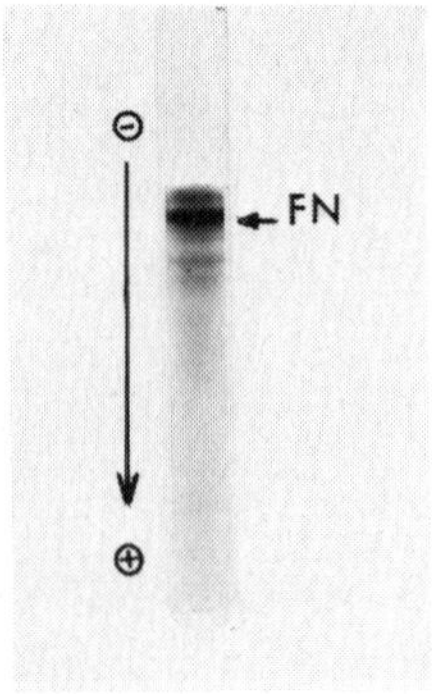

Fig. 4. SDS-polyacrylamide gel electrophoresi
of a human fibronectin preparation.
Amount of fibronectin : 30 µg (previously tre
at 37°C for 15 h with the migration buffer);mi
gration buffer : phosphate 0.1 M, pH 7 contai
0.1 SDS and 0.01 2-mercaptoethanol ; migr
time : 4.5 h at 10°C; staining : coomassie R-
solution (2.5 g in 1 litre of a mixture of me
nol:acetic acid:water 45:5:50 , v/v/v).

The presence of the fibronectin favorizes considerably the percentage of sprea cells. In the first 45 minutes of incubation, the number of spreaded cells placed presence of fibronectin was about 5-fold superior to the control ; beyond this ti the difference diminishes. This phenomenon has to be assigned to the self-product of fibronectin by the BHK cells themselves which progressively induce a spreadin effect even in absence of exogeneous fibronectin.

TABLE II

Cell spreading activity of human fibronectin (BHK 21-C-13)

Incubation time (min.)	% Spread cells	
	Without fibronectin	With fibronectin (5 µg/ml)
15	2	18
30	7	33
45	16	45
60	21	50
90	29	62
120	32	70

CONCLUSION

The purification procedure we present here as well as the analytical results sh it is possible to set up fibronectin production on a preparative scale by a simpl rapid and economic affinity chromatography method. The first step to consider was preparation of a chromatographic support defined as a function of several objecti of productivity and stability ; this gel had also to be capable to produce a pure logically active and sterile fibronectin in consideration of further therapeutic biological applications. Although our experiments have been limited to the treatm of 8 litres of plasma, this procedure can easily be scaled up.

ACKNOWLEDGEMENTS

We wish to thank Prof. M. Monsigny for his help in the determination of the cell spreading activity of the fibronectin and for his useful comments during the development of this work.

REFERENCES

1 E. Ruoslahti and A. Vaheri, J. Exp. Med., 141 (1975) 497-501.
2 E. Ruoslahti, A. Vaheri, P. Kuusela and E. Linder, Biochim. Biophys. Acta, 322 (1973) 352-358.
3 E. Engvall, E. Ruoslahti and E. J. Miller, J. Exp. Med., 147 (1978) 1584-1595.
4 F. Jilek and H. Hormann, Hoppe-Seyler's Z. Physiol. Chem., 359 (1978) 247-250.
5 J. Laterra, R. Ansbacher and L. Culp, Proc. Natl. Acad. Sci. (USA), 77 (1980) 6662-6666
6 E. Ruoslahti and E. Engvall, Biochim. Biophys. Acta, 631 (1980) 350-358.
7 K.M. Yamada, D.W. Kennedy, K. Kimata and R.M. Pratt, J. Biol. Chem., 255 (1980) 6055-6063.
8 H.K. Keinman, G.R. Martin and P.H. Fishman, Proc. Natl. Acad. Sci. (USA), 76 (1979) 3367-3371.
9 D.F. Mosher and E.M. Williams, J. Lab. Clin. Med., 91 (1978) 729-735.
10 I. Cohen, E.V. Potter, T. Glaser, R. Entwistle, L. Davis, J. Chediak and B. Anderson, J. Lab. Clin. Med., 97 (1981) 134-140.
11 M.A. Scovill, T.M. Saba, F.A. Blumenstock, H. Bernard and S.R. Powers, Ann. Surg. 188 (1978) 521-529.
12 T.M. Saba and E. Jaffe, Am. J. Med., 68 (1980) 577-584.
13 M. Vuento, M. Wrann and E. Ruoslahti, FEBS Lett., 82 (1977) 227-231.
14 E. Ruoslahti and E. Engvall, Ann. N.Y. Acad., 312 (1978) 178-191.
15 E. Engvall and E. Ruoslahti, Int. J. Cancer, 20 (1977) 1-5.
16 W. Dessau, F. Jilek, B.C. Adelmann and H. Hormann, Biochim. Biophys. Acta, 523 (1978) 227-237.
17 M. Vuento and A. Vaheri, Biochem. J., 183 (1979) 331-337.
18 J. Berges, E. Boschetti and R. Tixier, C.R. Acad. Sci., Paris, 273 (1971) 2358-2360.
19 K. Weber, J.R. Pringle and M. Osborn, Meth. Enzymol., 26 (1972) p. 3.
20 F. Grinnell, D. Hays and D. Minter, Exp. Cell Res., 110 (1977) 175-181.
21 M.M. Bradford, Anal. Biochem., 72 (1976) 248-254.
22 M. Vuento and A. Vaheri, Biochem. J., 175 (1978) 333-336.

T.C.J. Gribnau, J. Visser and R.J.F. Nivard (Editors),
Affinity Chromatography and Related Techniques

AFFINITY ELUTION FROM ION-EXCHANGERS - PRINCIPLES, PROBLEMS AND PRACTICE

R.K. SCOPES
Department of Biochemistry, La Trobe University, Bundoora, Victoria 3083, Australia

ABSTRACT

The principles of affinity elution chromatography using ion-exchange materials as adsorbents are discussed, comparing the procedure with conventional affinity (adsorption) chromatography. For certain classes of proteins it offers a cheap and simple method for purification, but is not applicable to proteins with low isoelectric points, unless they have positively-charged ligands. A simple, rapid procedure for purifying yeast pyruvate kinase using both a substrate and an effector ligand to elute the enzyme is described.

INTRODUCTION

Conventional affinity chromatography allows for two stages of biospecific selection: during adsorption and in the elution procedure. Elution with free ligand is sometimes less successful than using a non-specific procedure such as increasing salt concentration, so many methods only make use of selectivity during adsorption. It is also possible to carry out affinity chromatography in which the biospecific selectivity is only at the elution stage; in this case a non-specific adsorbent such as an ion-exchanger can be used. Generally known as affinity elution chromatography, the latter method has a number of advantages and disadvantages compared with the conventional affinity (adsorption) technique, and is probably more widely applicable than has been generally realised [1-7]. This paper will describe some of the limitations and uses of affinity elution chromatography, with particular reference to the purification of pyruvate kinases from both muscle tissue and yeast.

PRINCIPLES

A protein should adsorb to an anion exchanger at pH's above its isoelectric point (pI) and to cation exchangers at pH's below pI. For reasons which will become clear, affinity elution is mainly used with cation exchangers; the principles will be illustrated with examples of proteins adsorbing to CM-cellulose, at pH values below their pI. The isoelectric point of a protein depends on the buffer composition, as binding of charged molecules causes a shift from the isoionic point as defined at zero ionic strength. Even low concentrations of specific ligands will bind to proteins, and if the ligand is charged, a shift in isoelectric point occurs. Consequently in the presence of charged ligand the strength of adsorption to an ion exchanger changes, in particular at pH's close to the isoelectric point, as illustrated in Fig. 1. The objective of a successful affinity elution procedure

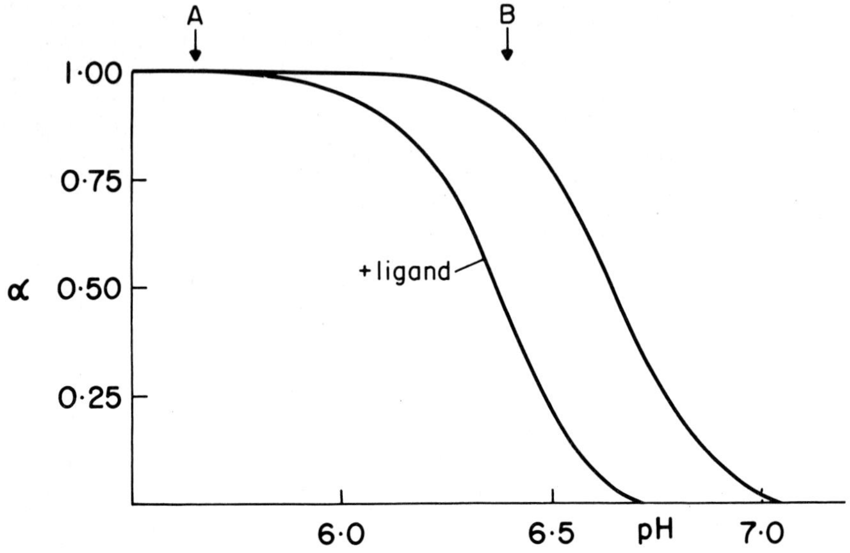

Fig. 1. Effect of negatively-charged ligand on adsorption of enzyme to cation exchanger. α = partition coefficient of adsorption.

is to adsorb the enzyme totally, at a pH where the partition coefficient is close to 1.0 (e.g. A in Fig. 1), then to raise the pH until the binding is weakened, so that inclusion of ligand causes a large decrease in partition coefficient (B in Fig. 1). It will be noted that proteins that do not bind that particular ligan should not be eluted, provided that the ligand concentration does not markedly increase the ionic strength of the buffer, nor should proteins that bind the ligan but which are more strongly adsorbed, i.e. those that have much higher isoelectric points. Ligand concentrations of 5-10 times the dissociation constant are usually sufficient, typically from 0.1 to 1mM. The more charges on the ligand, and the more binding sites for it, the better.

The following restrictions apply:

(1) The enzyme must be stable when adsorbed to the ion-exchanger at below its isoelectric point (for cation exchangers).

(2) At the point equivalent to B in Fig. 1, the enzyme must bind the ligand; it need not be catalytically active, indeed for a single substrate enzyme it may be desirable not to react.

(3) The ligand must have the same charge as the adsorbent. This is (a) so that adsorption is decreased, not increased, and (b) the ligand does not itself bind to the adsorbent.

Because most enzyme ligands are negatively charged, the technique is largely limited to cation exchangers. However we have observed rather non-specific affinity elution from DEAE-cellulose using magnesium ions, and a purification procedure for ornithine transcarbamoylase has been developed using ornithine affinity elution, also from DEAE-cellulose [8]. Since restrictions (1) and (2) above limit the pH range of operation, the proteins being purified need to have fairly high isoelectric points (>6). The restriction of stability at the isoelectric point also applies to isoelectric focussing and chromatofocussing as enzyme purification methods [9].

We have purified over twenty different enzymes, several of them from more than one species, using affinity elution chromatography [4,6,7]. Unlike affinity adsorption, species variation in behaviour can be considerable because of the dependence of adsorption on isoelectric point. Thus a successful method might be developed by switching source material to a species in which the enzyme has a higher isoelectric point. In many cases enzymes from microorganisms and plants tend to have lower isoelectric points than corresponding animal enzymes, though there are many exceptions to this rule.

RESULTS WITH PYRUVATE KINASE

To illustrate the principles and problems associated with affinity elution chromatography I have chosen the example of pyruvate kinase, from both muscle and yeast. Rabbit muscle pyruvate kinase is isoelectric at around pH 8 [10]; at $I = 0.01$ it adsorbed to CM-cellulose below pH 7.5, and the partition coefficient for adsorption increased to close to 1.0 by pH 6.5 (Fig. 2). The presence of magnesium ions in the buffer did not alter the adsorption, but each of the negatively-charged substrates ADP and phosphoenolpyruvate decreased the adsorption in accordance with expectation. Somewhat unexpected was the effect of fructose 1,6-bisphosphate which also decreased the adsorption, despite the fact that it is not an allosteric effector of the muscle enzyme. Phosphoenolpyruvate plus fructose 1,6-bisphosphate effects were additive, and this combination, at about pH 6.8, appears to be ideal for affinity elution of pyruvate kinase from CM-cellulose. Indeed, on a column it worked well. Unfortunately crude preparations of muscle pyruvate kinase are certain to be contaminated with large amounts of fructose 1,6-bisphosphate aldolase, which also adsorbs to CM-cellulose. Inclusion of fructose 1,6-bisphosphate elutes the aldolase as well [2]. Consequently it is better to use only phosphoenol pyruvate, or a combination with ADP (but lacking magnesium so that reaction does not occur).

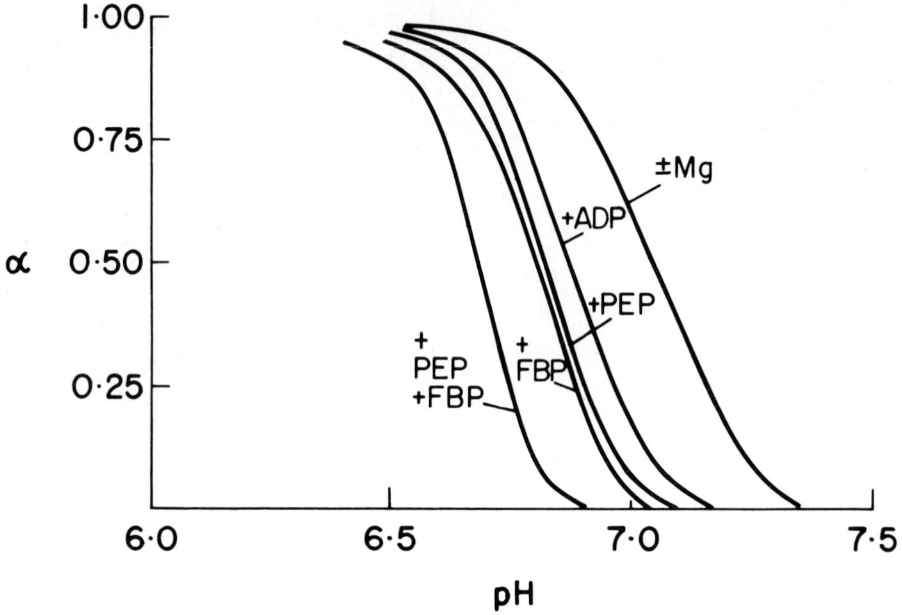

Fig. 2. Partition coefficient/pH curves for rabbit muscle pyruvate kinase. Mg^{2+} (1mM) did not affect curve. ADP, phosphoenolpyruvate (PEP) and fructose 1,6-bisphosphate (FBP) were added as indicated, at a concentration of 0.1mM.

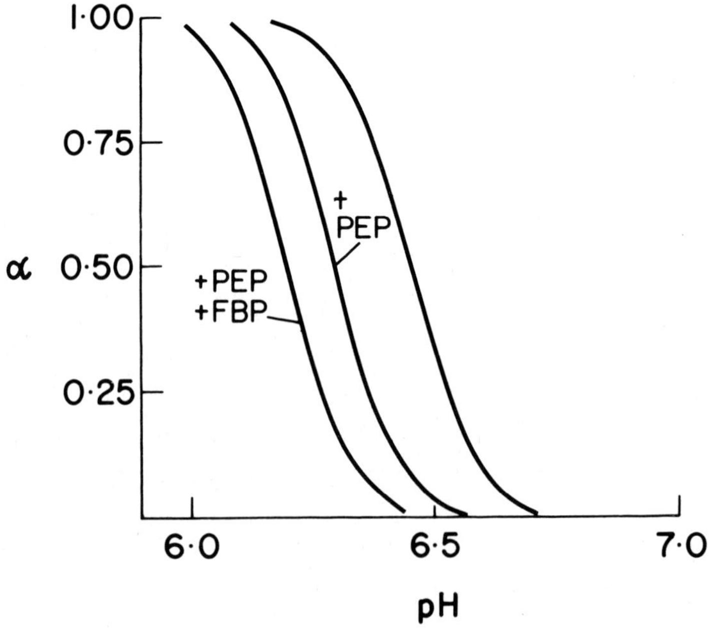

Fig. 3. Partition coefficient/pH curves for yeast pyruvate kinase. Phosphoenolpyruvate (PEP) and fructose 1,6-bisphosphate (FBP) were added as indicated, at a concentration of 0.1mM.

The yeast enzyme has a lower isoelectric point, 6.7 [11], and adsorbs to CM-cellulose at pH's below this, the partition coefficient rising rapidly to nearly 1.0 at pH 6.0 (Fig. 3). In this case phosphoenol pyruvate plus fructose 1,6-bisphosphate has proved to be very successful for affinity elution. Being an acidic protein, yeast aldolase does not adsorb to CM-cellulose. However, after early successful purifications we had problems in adsorbing the pyruvate kinase even at pH 6.0. This was due to an omission of a step to remove nucleic acids using protamine sulphate. In the presence of a negatively-charged polymer like DNA or RNA, positively-charged proteins can bind preferentially to the soluble polymer rather than to the ion-exchanger, resulting in non-adsorption. Pre-treatment with protamine avoids this; any experiment to see if an enzyme will bind to a cation exchanger should be preceded by some step which removes nucleic acids. For example yeast phosphoglycerate kinase which adosrbs to CM-cellulose even above pH 7, will not adsorb reliably at pH 6 if nucleic acids are present [12]. The elution profile of yeast pyruvate kinase from CM-cellulose is illustrated in Fig. 4. The overall procedure is summarised in Table 1; this whole preparation can be completed in

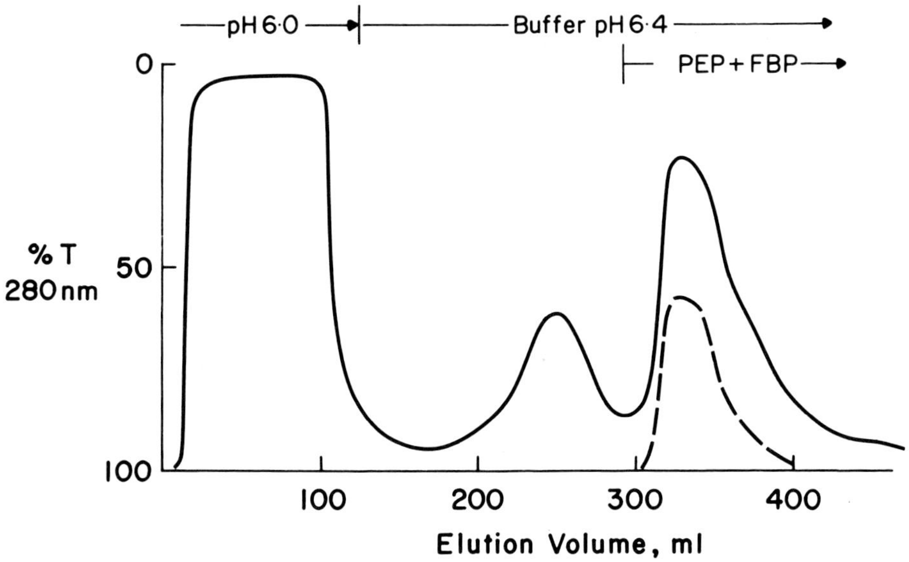

Fig. 4. Elution of pyruvate kinase from CM-cellulose. Sample was applied at pH 6.0, washed with 3 column volumes of pH 6.4 buffer, then 0.2mM each of phosphoenol-pyruvate and fructose 1,6-bisphosphate was added.

6-7 hours. The final specific activity (680) is considerably higher than the best previously reported (400), though this is mainly due to using more optimal assay conditions.

TABLE 1

Purification of yeast pyruvate kinase from 100g "Active Dried Yeast", extracted with a Vibrogen Cell Mill

	Volume	mg protein	k units	specific activity
Extract	450	16600	139	8.4
Acetone pH 6.0, 28-38% v/v	105	4100	106	26
$(NH_4)_2SO_4$ pH 6.0, 0-60% saturation	45	730	104	120
After de-salting and protamine treatment	95	680	96	142
Affinity eluted from CM-cellulose	75	112	76	680

DISCUSSION

Affinity elution chromatography has several advantages compared with affinity adsorption techniques. Firstly, the labour of preparing a suitable adsorbent (or the expense of purchasing one) is eliminated. Secondly, the adsorption capacity of ion-exchangers can be up to 100 times greater than affinity adsorbents, which usually more than compensates for the non-specificity of adsorption. Thirdly, elution of the desired protein occurs in a narrow pH range, so that other proteins with the same ligand but different pI's may remain behind, or be washed off before the specific elution step. The principle disadvantage is that the necessary opera pH, i.e. below the pI on cation exchangers, may be below the pH stability range - only relatively high pI proteins are suited to the method. If the protein should have a positively charged ligand, the converse arguments can be made concerning anion exchangers. It is possible that synthesis of a substrate analogue with posi charges could extend the range of operation; however this would increase the developmental work to a level comparable with developing affinity adsorbents.

For successful operation the ligand should have at least two (negative) charges or it should at least add one charge to the protein per 30,000 Daltons [4]. The shift in isoelectric point is the principle reason for the decrease in adsorption in the presence of ligand. However there are two other factors to be considered which can influence the strength of adsorption to an ion exchanger. These are (i) localised clusters of like-charged residues on the protein molecule, particula at the ligand binding site, which can interact strongly with the adsorbent; in so cases this causes adsorption on the wrong side of the isoelectric point [13]. In these cases, addition of the ligand displaces the strong interaction, and the ove weakening is more than expected from simply the charge addition alone. (ii) Bindi of ligand to the protein may result in a conformational change, which can cause certain areas of the protein surface to move closer or further away from the adsorbent. The strength of the interaction with adsorbent may thus be influenced by shape change as well as charge addition. In most cases that we have investigated the shift in isoelectric point seems to be the dominant factor [13,14].

The principles of affinity elution described here can be extended to the use of other adsorbents, including affinity adsorbents and mixed function columns [15]. The procedure of (a) adsorb strongly, (b) weaken adsorption by change of buffer composition and (c) elute with low concentration of ligand should be followed always. However, step (b) is often omitted, so that high (expensive) concentrations of ligand are needed, and an increase in ionic strength due to the ligand can cause non-specific elution of other proteins.

REFERENCES

1 B.M. Pogell, Biochem. Biophys. Res. Comm., 7(1962)225-230.
2 E.E. Penhoet and W.J. Rutter, Methods Enzymol., 42(1975)240-249.
3 F. Von der Haar, Methods Enzymol., 34(1974)163-171.
4 R.K. Scopes, Biochem. J., 161(1977)253-263.
5 R.K. Scopes, Biochem. J., 161(1977)265-277.
6 T. Fifis and R.K. Scopes, Biochem. J. 175(1978)311-319.
7 J.R. Davies and R.K. Scopes, Analyt. Biochem., 112(1981), in press.
8 N.J. Hoogenraad and S. Edwards, (unpublished results).
9 "Chromatofocussing", Pharmacia Fine Chemicals, Uppsala, Sweden.
10 W.A. Susor, M. Kochman and W.J. Rutter, Science, 165(1969)1260-1262.
11 A.E. Aust and C.H. Suelter, J. Biol. Chem., 253(1978)7508-7512.
12 R.K. Scopes, Biochem. J., 122(1971)89-92.
13 R.K. Scopes and E. Algar, FEBS Letts., 106(1979)239-242.
14 R.K. Scopes, Analyt. Biochem., 112(1981), in press.
15 R.J. Yon, Trans. Biochem. Soc., 9(1981).

CHAPTER IV

APPLICATIONS

- DIAGNOSTIC
- BIOMEDICAL

INTERNATIONAL SYMPOSIUM
ON AFFINITY CHROMATOGRAPHY
AND RELATED TECHNIQUES

T.C.J. Gribnau, J. Visser and R.J.F. Nivard (Editors),
Affinity Chromatography and Related Techniques

USE OF IMMOBILIZED REAGENTS IN IMMUNOASSAY

A.H.W.M. Schuurs, T.C.J. Gribnau, J.H.W. Leuvering

Organon SDG, P.O. Box 20, 5340 BH OSS, The Netherlands

INTRODUCTION

Immobilized reagents for use in immunoassay may be defined as immune reagents (usually antigens, haptens or antibodies) bound to a solid phase.

The solid phase can serve as label, and have the form of particles of sizes varying between 0,05 and 10 μm. Immune reagents bound to the particle surface react with their counterparts as they do in the fluid phase. By this immune reaction the physico-chemical properties of the particles change. This change, called flocculation or agglutination, or its inhibition is used to detect the immune reaction; the particles as such serve as visually/spectrophotometrically detectable label and are indicated as "sensitized particles". In other applications the particle material itself is determined in the final stage of the assay (similar to the measurement of radioactivity in radio-immunoassay). Both applications of solid particles will be discussed in the first part.

The solid phase can also take the shape of larger particles or objects, or of the inner surface of reaction vessels. In these cases the immune reagent bound to the solid phase is meant to remove its counterpart in the reaction from the reaction fluid, e.g. an antigen bound to the inner wall of a test tube binds its specific antibody and removes it from the fluid phase. This type of process is usually coined bound/free or B/F separation. This will be the subject of discussion in the second part.

In both parts, some information will be given as to the means of binding of immune reagents to the solid phase.

SENSITIZED PARTICLES

In this survey three types of particles used in immune reactions will be discussed: (i) latex particles; (ii) gold sol particles; (iii) erythrocytes.
Elsewhere in these proceedings Gribnau et al. describe the use of colloidal dye particles for this purpose.

Latex particles

In 1956 Singer and Plotz (1) were the first to use (polystyrene) latex particles coated with immunoglobulin for the detection of the rheumatoid factor (RF) by means of the agglutination or flocculation of these particles. During the early sixties, several investigators (2-5) described latex coated with human chorionic gonadotrophin (HCG) which agglutinated with a properly diluted anti-HCG serum. This agglutination could be inhibited by free HCG present in a test sample, e.g. urine from a woman who is pregnant. A simple pregnancy test was born: by rocking a small volume of test urine with an HCG-coated latex suspension + anti-HCG serum for 2 - 3 minutes on a glass slide, the presence or absence of HCG could be easily detected.

Another procedure to monitor agglutination (inhibition) is direct analysis of the reaction mixture by nephelometry or light scattering, reviewed by Deverill and Reeves (6). Of the more sophisticated methods the particle counting immunoassay (PACIA) in which the non-agglutinated particles are counted in a highly mechanized and automated equipment should be specifically mentioned (7).

Apart from RF and pregnancy testing many other applications have been described in the course of time. In these instances, the particles were coated with either antigen or antibody and used in agglutination or agglutination inhibition tests (e.g. hepatitis B surface antigen, anti-tetanus, anti-measles).
If antibody-coated particles are used in combination with serum or plasma as test fluid for purposes other than the detection of RF, the latter may interfere and cause false positive reactions. This can be avoided by using $F(ab')_2$ fragments of the antibody rather than the whole Ig molecules.

Latex agglutination can also be used for the determination of small molecules (haptens), e.g. thyroxine, by a system explained in Fig. 1 (8).

Latex particles are often coated with the desired protein by simple physical adsorption. In aqueous solutions (as used in immunochemistry) this adsorption is fast (9) and, to a large extent, irreversible (10, 11). Nevertheless, several groups have investigated chemical binding. In principle, this may give a better control of the amount and the orientation of the bound molecules. It is also possible to insert spacer molecules between the surface of the particle and the immunoreactive molecules. A number of possibilities are presented in Table 1. A combination of physical adsorption and covalent binding has been described by Limet et al. (7).

The practical advantages in agglutination (inhibition) tests of chemical binding over physical adsorption are very poorly documented in the literature. To our knowledge on

Limet et al. (7) present experimental evidence that in PACIA (see above) the use of particles sensitized by covalently bound immunoreactant may give more sensitive tests than particles to which antibody has been physically adsorbed.

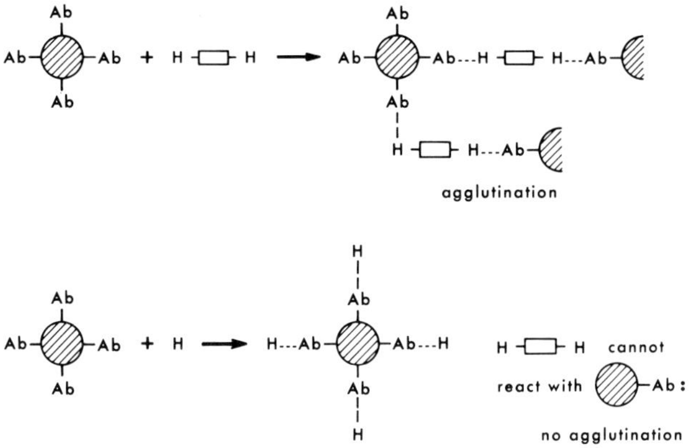

Fig. 1. Reaction of "univalent" hapten (H) with antibody-coated particles does not cause agglutination. A hapten made polyvalent, e.g. by coupling to protein or dextran (H ─[ ]─ H), does cause agglutination (top part of figure). This agglutination can be inhibited with free H (lower part) (cf. ref. 8).

Gold sol particles

Gold sol particles prepared according to procedures developed long ago in colloid-chemistry (cf. 12) and coated with antibody adsorptively (13 - 15) can be used in several ways: (i) as electron-dense markers in immune electron microscopy (13, 14); (ii) as sensitized particles in agglutination (inhibition) tests (16 - 18); (iii) as labels in non-isotopic immunoassays (15). The latter two applications named SPIA or sol particle immunoassay will be discussed in some more detail.

Sol particle agglutination (inhibition) assays. A stable gold sol has a deep red colour. Flocculation of this hydrophobic sol, e.g. by electrolytes, first causes a colour change to purple and blue. After some time the flocculated particles sediment leaving a clear supernatant. The flocculation rate is, amongst others, dependent on the electrolyte concentration. The same phenomenon occurs when gold sol particles to which antibodies have been adsorbed are mixed with a solution of the corresponding antigen. The concentration of the antigen in the test solution is a determinant in the flocculation rate. Flocculation as a function of time by a constant amount of human placental lactogen

TABLE 1

Immobilization of immune reagents

a. Survey of methods

- adsorption: ⁎ to plain (native) polymer
  ⁎ to surface-modified polymer
- entrapment
- covalent coupling: ⁎ reactive ligand to (functionalized) polymer ++)
  ⁎ ligand to polymer with reactive+) groups ++)
  ⁎ ligand to (functionalized) polymer, applying cross-linking reagents

+) present as such, or after activation of functional groups
++) with/without insertion of spacer

b. Current functional groups in ligands/polymers

C-OH	(aliphatic, aromatic)	Si-OH
$C\text{-}NH_2$	(aliphatic, aromatic)	
C-COOH	$C\text{-}CONH_2$	$C\text{-}CONHNH_2$
C-SH		
$C\text{-}SO_3H$		

c. Current reactive groups in polymers (P) and corresponding appropriate functional groups in ligands (L)

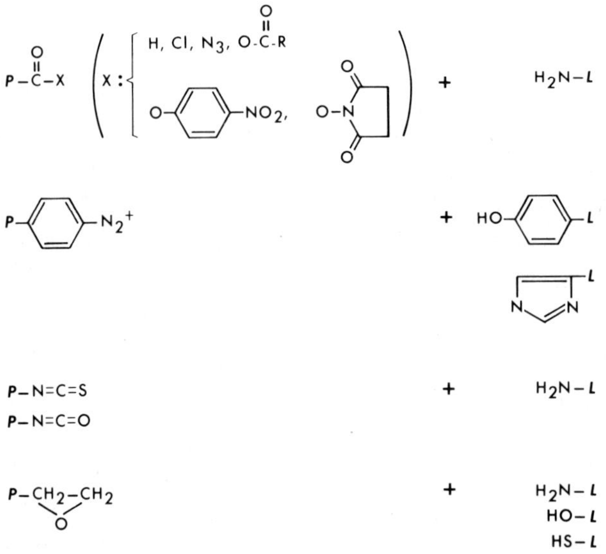

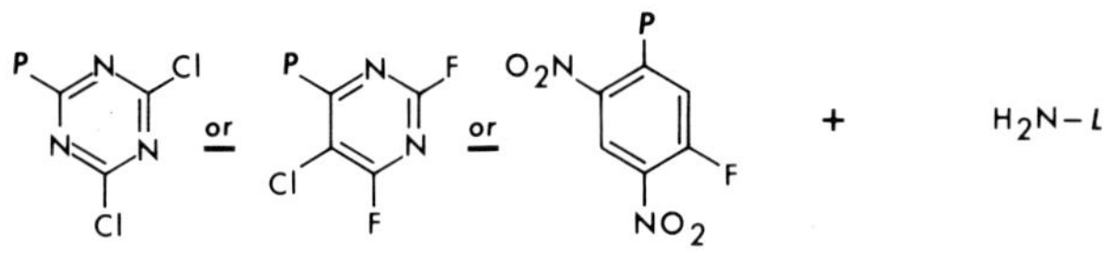

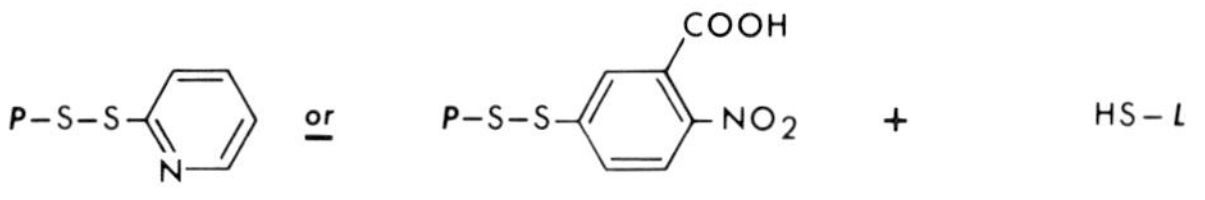

d. Examples of cross-linking reagents

* homobifunctional

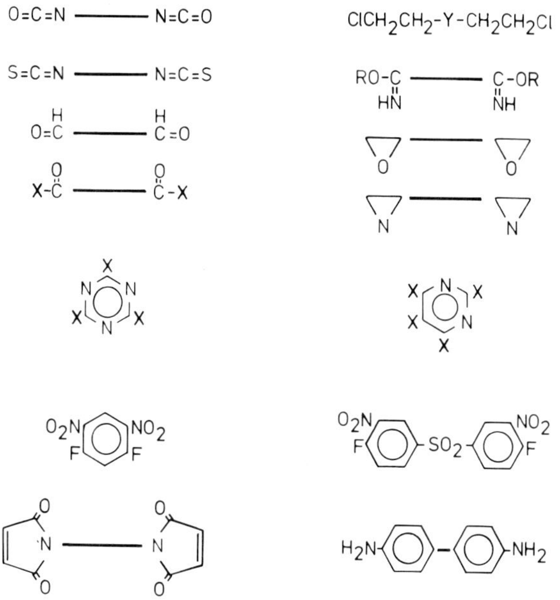

* heterobifunctional

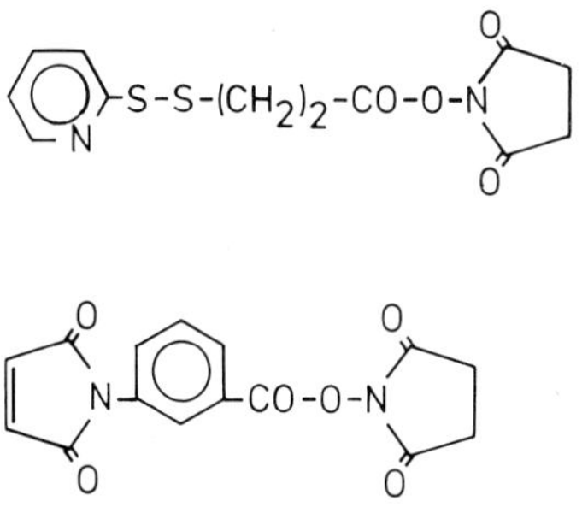

e. Examples of spacers

$H_2N-(CH_2)_n-COOH$	oligo- or polypeptides
$H_2N-(CH_2)_n-NH_2$	oligo- or polysaccharides

(HPL) is shown in Fig. 2a while a dose-response curve at a chosen time interval is presented in Fig. 2b. Flocculation was monitored spectrophotometrically by measuring the absorbance at 540 nm (A_{540}). An example presented elsewhere (16, 18) is a test for human chorionic gonadotrophin (HCG). The flocculation process is accelerated whe it is allowed to take place in a centrifugal analyser. In this way a test result similar to the one of Fig. 2b can be obtained in less than 15 min (Fig. 3). The flocculation can also be followed nephelometrically (Leuvering et al., unpublished data) or by means of sedimentation patterns in microtitration plates (17).

A variation of this test system and suitable for testing haptens (8) is shown in Fig. 1. In our laboratory, this approach has been shown to be also feasible for gold so particles (Leuvering et al., to be published).

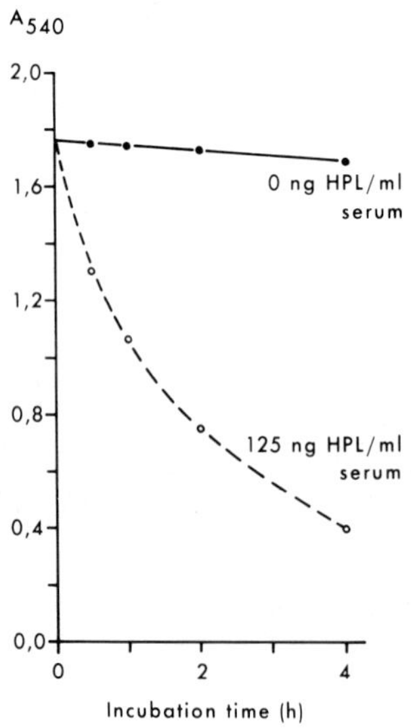

Fig. 2a. Flocculation monitored spectrophotometrically by measuring the absorbance at 540 nm (A_{540}) as a function of time in the presence or absence of antigen (HPL).

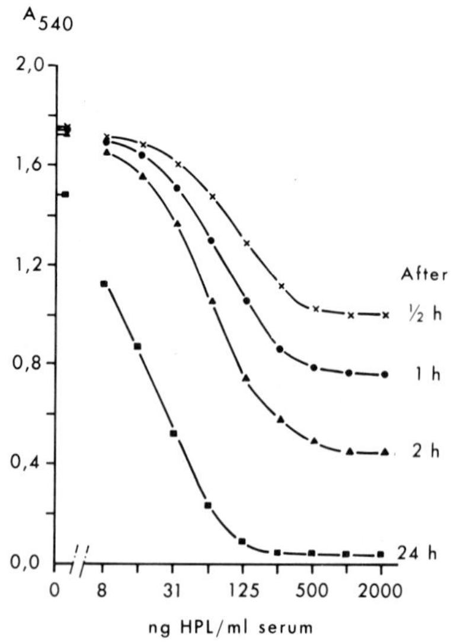

Fig. 2b. Dose-response curves for HPL determined after four different reaction times.

Sol particle immunoassays. The principle involved in this type of assay is the same as used with a radioactive isotope or an enzyme as label; it is discussed in the second part. The application of this set-up for the determination of HCG and HPL has been described recently (15).

Erythrocytes

Agglutination, i.e. clumping, of erythrocytes has been known since Landsteiner's discovery of blood groups in 1901. For example, antibodies present in the serum of an individual with blood group A react with certain antigens present on the erythrocytes of a person having blood group B. This reaction becomes visible by the clumping of the erythrocytes. In this case the antigen is originally present on the surface of erythrocyte. However, it is also possible to coat erythrocytes with a desired antigen artificially. Polysaccharide antigens, e.g. from bacteria, can be passively adsorbed onto erthrocytes but for proteins certain tricks are needed. In 1951 Boyden (19) described the treatment of erythrocytes with tannic acid allowing the adsorption of protein antigens on to their surface. Erythrocytes coated ("sensitized") with antigen, polysaccharides or proteins, agglutinate by the addition of corresponding antiserum in a process, usually called passive haemagglutination.

Real cell clumping is caused by appropriate antibody in sufficient concentration. However, a more sensitive test is possible by interpreting the patterns with which erythrocytes upon standing settle in tubes with hemispherical bottoms. Agglutinated cells produce a "carpet" on the bottom while non-agglutinated cells form a ring or a button in the centre of the hemisphere (20; Fig. 4).

In the test models described above, fresh erythrocytes are used (cells treated with tannic acid and subsequently coated with antigen can only be used for one or two days. To avoid this disadvantage erythrocytes are treated with formalin prior to tannic acid treatment and sensitization (21). In this way, reagents can be prepared with long shelf lives (e.g. 3 year at 2 - 8°C). On the basis of the pioneering work of Strausser (22) and Wide (23) this approach is widely used in commercial pregnancy tests.

Experimental details of these and other procedures to prepare sensitized erythrocytes have been reviewed (24).

Antibody-coated particles

A difficulty peculiar to the coating of whatever type of particles with antibody is the following. Since antibodies with a required specificity are chemically indistinguishab from those with other specificities, a preparation made from an antiserum by a classi cal biochemical procedure contains a high percentage of antibodies irrelevant for the particular test. These antibodies "dilute" the relevant antibody and can even affect th specificity of the test. Yet, using hyperimmune sera it is possible to arrive at useful test systems. It is also possible to isolate the required antibodies by immuno-affinit chromatography. However, the resulting antibodies have often suffered from the deso tion procedures. The most elegant solution is the use of a combination of monoclonal antibodies.

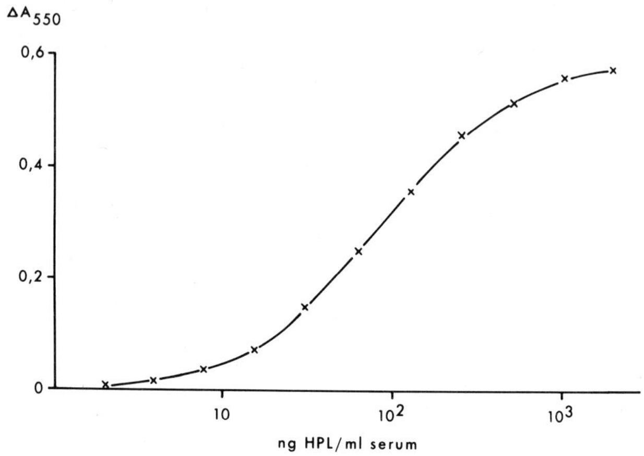

Fig. 3. Dose-response curve for HPL dissolved in human serum measured with CentrifiChem (10 min).

Fig. 4. Erythrocytes sedimentation patterns in round-bottom tubes. Ring pattern: no agglutination: erythrocytes sediment independently from each other and form a ring where the slope of the bottom has become negligible. "Even carpet" pattern: agglutination; erythrocytes form a sort of network and cover after sedimentation the bottom evenly.

IMMOBILIZED REAGENTS FOR BOUND/FREE SEPARATION

Many types of immunoassays make use of labelled reagents. Radioactive isotopes are used in radio-immunoassay (RIA), enzymes in enzyme-immunoassay (EIA), fluorescent molecules in fluoro-immunoassay (FIA), gold sol particles in SPIA, dye sol particles in DIA (cf. Gribnau et al., these Proceedings). In most cases a label is not changed by the antigen-antibody reaction. It is essential to separate labelled reagent which has reacted with its counterpart from reagent which has not. The ways to accomplish this separation will be described for various test principles.

Competitive immunoassays

In the competitive RIA as originally described by Yalow and Berson (25) radioactively labelled antigen or hapten competes with unlabelled antigen (hapten) in a test fluid for a limited amount of antibody. The amount of labelled antigen (hapten) bound to antibody is then a measure for the amount of antigen (hapten) in the test fluid. To determine the amount of antibody-bound labelled antigen it must be separated from unreacted or free labelled antigen, hence the term B/F separation.

In early work, complicated B/F separation methods, e.g. chromato-electrophoresis, were used. These procedures were unsatisfactory when large number of samples had to be assayed. An important improvement was the introduction of solid phase RIA whereby the antibody is covalently coupled to insoluble polymer particles (26 - 28). The particles can be removed from the reaction mixture by simple centrifugation and washed free from contaminants. Another approach was the use of antibodies physically adsorbed to plastic discs or to the inner wall of test tubes (29, 30). In this way the B/F separation became quite easy and the solid phase antibody could be easily washed.

Immunoassays based on antibody-antigen reactions in the solid liquid interface tend to be rather imprecise and insensitive. By using a solid phase bound "second antibody" (i.e. an antibody directed against the first antibody which, in its turn, is directed against the antigen under test) the reaction between this antigen and the first antibody can first proceed in the fluid phase whereafter the antibody-bound labelled antigen can be easily removed. An additional advantage is that the solid phase second antibody reagent can be used in all assays where first antibody raised in the same species is used. Originally, second antibody was coupled to diazotized aminoarylcellulose (31). Recently, we developed a more efficient solid phase reagent by using 2,4,6-trifluoro-5-chloropyrimidine (FCP), or FCP-based reactive azo-dyes, to couple antibodies to cellulose (32, 33).

Sandwich-type immunoassays

Sandwich assays have been described in many different variations comprising usually three but also four or more "layers" of reactants (cf. 34). This is exemplified in Fig. 5.

The solid phase usually consists of discs, rings, balls or the inner wall of the reaction tube or of the well of a microtitration plate.

This type of assays can only be used for (bi- or multivalent) antigens. For the assay of haptens a sandwich-type inhibition test can be used. An example of such an assay is presented in Fig. 6.

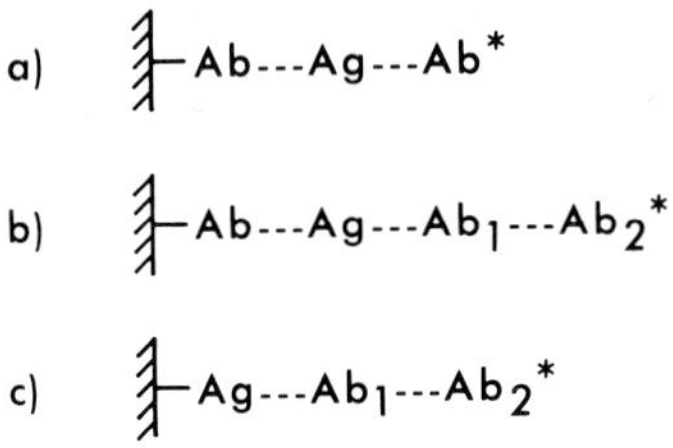

Fig. 5. Examples of sandwich-type immunoassay. symbolises the solid phase --- the antigen-antibody binding, Ab, Ab_1 = (first) antibody, Ab_2 = second antibody (directed against Ab_1), Ag = antigen, * = label (radioactive isotope, enzyme gold sol particle, etc.). In a) Ag is determined, in b) Ag or Ab (= Ab_1) and in c) Ab (= Ab_1)

Some preparative procedures for solid-phase reagents

Physical adsorption and some covalent coupling procedures have already been mentioned above. We shall now discuss them in some more detail.

Tubes, and microtitration plate wells. In solid phase immunoassay test procedures the immunosorbent is usually the inner surface of polystyrene tubes or of wells of polystyrene microtitration plates. Apart from polystyrene, polyvinyl chloride, polypropylene and polycarbonate are also used. By far the most frequently used coating method is physical adsorption (cf. first part "latex particles"). Its great advantage is simplicity The antigen or antibody to be coated is allowed to be adsorbed from a solution containi 1 - 25 µg/ml at pH 7 - 8 or 9,6, at 4°C, room temperature or 37°C for a period varyin from a few hours to overnight. After this incubation the tubes or wells are washed, e.g with phosphate-buffered saline with or without Tween. Sometimes, a second coating

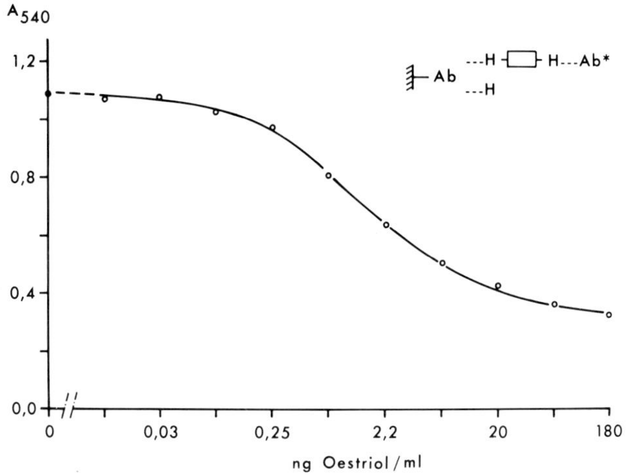

Fig. 6. Sandwich inhibition assay for oestriol. The principle is explained in the right top corner: for explanation of symbols see Fig. 5. A hapten (H) present in a test fluid competes with a molecule in which the same hapten is "polymerized", e.g. by binding of several haptens to a protein (H —[ ]— H). The amount of Ab* bound to the solid phase is inversely related to the amount of hapten present in the test fluid. Gold sol particles were used as label. Both the wells of a microtitration plate and the gold sol particles were coated, by physical adsorption, with the immunoglobulin fraction of rabbit antiserum against oestriol. "Polymerized hapten" was prepared by coupling oestriol-16/17-hemisuccinate to bovine serum albumin.

with an "inert" protein is applied to avoid adsorptions of other proteins from the reaction mixture. The solid phase reagent is normally used immediately but, under dry conditions, it can also be kept for long periods of time at 2 - 8°C.

Depending on the nature of the immune reagent to be bound the adsorption conditions are important. This is exemplified by the adsorption of monoclonal antibodies to polyvinyl chloride whereby ionic strength and pH are important variables (35).

Polysaccharide antigens which attach poorly to polystyrene are bound indirectly through specific antibody leading to test principle b) in Fig. 4 for measuring antibody levels (e.g. 36) or via Concanavalin A or other sugar-specific lectins.

An interesting approach is the coupling of reagent to polymer pretreated with 1% glutardialdehyde. This method is reported to increase the affinity of the surface for proteins. After the incubation with the reagent to be coated on the surface the unused reactive groups of glutardialdehyde can be blocked with 4-aminobutyric acid (37 - 39). Cells can also be fixed on to vessel walls and serve as solid phase antigen (40 - 42). Related to this procedure is the preparation of solid phase virus antigen: a monolayer of appropriate cells infected with virus is fixed with an ethanol-acetone mixture; non-infected cells serve as control (38). A most interesting application is the incubation of a cell monolayer with throat washings of a patient. The growth of virus is detected by enzyme labelled specific antibody. In this test procedure results can be obtained within 24 - 28 hours instead of the usual 1 - 2 weeks needed for virus isolation (38).

Particles and other objects. If, instead of tubes or plate wells, other objects such as beads, balls, rings, pieces of variable shape are chosen, there is a greater variety of methods of preparing immobilized immune reagents. Nylon can be coated adsorptively but also covalently by using partially hydrolyzed polymer in combination with glutardialdehyde or a carbodiimide (43, 44). Porous glass beads activated with reactiv groups containing silanes, e.g. 3-aminopropyltriethoxysilane (in combination with glutardialdehyde) or epoxypropyltriethoxysilane, can covalently bind immune reagents (45, 46). Cellulose (e.g. disks of filter paper) and agarose (e.g. beads) activated by cyanogen bromide can bind proteins (47 - 50). The use of FCP or FCP-based reacti dyes (32, 33) leads to stronger bonds between polysaccharides and proteins (51, 52). Partially hydrolyzed polyacrylamide beads can be used in combination with a carbodiimide (53). Immunoglobulin has been coupled to enzacryl polythiolactone (54). Antibodies against a hapten, e.g. thyroxine, have been microencapsulated in nylon membranes which are permeable for thyroxine but not for the antibodies (55). Magnetic particles coated with polymer to which immune reagent is adsorptively or covalently bound can be easily separated from the reaction medium (56 - 59).

Physical adsorption versus covalent binding

However easy the adsorptive procedure, it has important disadvantages. Exposure of the solid phase reagent to successive incubation and washing steps causes detachment of the solid adsorbed reagent from the surface giving a loss in sensitivity and precision of the test system (44). An uncertain factor is also the possibility of denatu ration of the protein in the adsorption process. It appears that solid phase reagents pre pared by covalent binding generally lead to assays with steeper dose-response curves. This usually implies higher sensitivity and precision and, for qualitative tests, a highe positive/negative response ratio. Furthermore, the ratio $\frac{\text{non-specific binding}}{\text{specific binding}}$ (nois to signal ratio) is less favourable for adsorptively prepared reagents. Such differences in properties between solid phase reagents prepared by adsorption or by covalent bindir have been found by various investigators (36, 44, 60). Finally, there is sometimes a great variation in adsorption properties between different batches of polymers even to t extent that some batches are unsuitable (44).

CONCLUSION

This short survey gives an impression of the many applications of immobilized reagents in immunoassay. Practically nothing has been said about their use in histology and cytology. A recent paper aptly describes such application (61). It is clear that immobilized reagents are occupying an important position amongst immune reagents in general.

REFERENCES

1 J.M. Singer and C.M. Plotz, Amer. J. Med. 21 (1956) 888 - 892.
2 W. Pollack (Ortho Pharmaceutical Corp.): US Patent 3,234,096 (March 7, 1961).
3 J.L. Robbins, G.A. Hill, B.M. Carle, J.H. Carlquist and S. Marcus, Proc. Soc. exp. Biol. (N.Y.) 109 (1962) 321 - 325.
4 M. Treacy (Ortho Pharmaceutical Corp.): US Patent 3,309,275 (August 23, 1963).
5 A.H.W.M. Schuurs (N.V. Organon): US Patent 3,551,555 (priority April 15, 1965).
6 J. Deverill and W.G. Reeves, J. Immunol. Meth. 38 (1980) 191 - 204.
7 J.M. Limet, C.H. Moussebois, C.L. Cambiaso, J.P. Vaerman and P.L. Masson, J. Immunol. Meth. 28 (1979) 25 - 34.
8 C.L. Cambiaso, A.E. Leek, F. de Steenwinkel, J. Billen and P.L. Masson, J. Immunol. Meth. 18 (1977) 33 - 44.
9 J.M. Singer, I. Oreskes, F. Hutterer and J. Ernst, Ann. Rheum. Dis. 22 (1963) 424 - 428.
10 G.C. Bernhard, W. Cheng and D.W. Talmage, J. Immunol. 88 (1962) 740 - 762.
11 H.G. de Bruin, C.J. van Oss and D.R. Absolom, J. Coll. Interface Sci. 76 (1980) 254 - 255.
12 G. Frens, Nature, Phys. Sci. 241 (1973) 20 - 22.
13 W.P. Faulk and G.M. Taylor, Immunochemistry 8 (1971) 1081 - 1083.
14 M. Horisberger, J. Rosset and H. Bauer, Experientia 31 (1975) 1147 - 1149.
15 J.H.W. Leuvering, P.J.H.M. Thal, M. van der Waart and A.H.W.M. Schuurs, J. Immunoassay 1 (1980) 77 - 91.
16 J.H.W. Leuvering, P.J.H.M. Thal, M. van der Waart and A.H.W.M. Schuurs, Fresenius Z. Anal. Chem. 301 (1980) 132.
17 W.D. Geoghegan, S. Ambegaonkar and N.J. Calvanico. J. Immunol. Meth. 34 (1980) 11 - 21.
18 J.H.W. Leuvering, P.J.H.M. Thal, M. van der Waart and A.H.W.M. Schuurs, J. Immunol. Meth., accepted for publication.
19 S.V. Boyden, J. exp. Med. 93 (1951) 107 - 120.
20 J.E. Salk, J. Immunol. 49 (1944) 87 - 98.
21 L.R. Cole and V.R. Farrell, J. exp. Med. 102 (1955) 631 - 645.
22 H. Strausser, Thesis Rutgers University, New Brunswide, NJ, Dissert. Abstr. 20 (1959) 430.
23 L. Wide, Acta endocr. (Kbh.) Suppl. 70 (1962).
24 W.J. Herbert, in D.W. Weir (Ed.), Handbook of experimental Immunology, Vol. 1 Immunochemistry, 3rd edn. Blackwell Scientific Publications, Oxford, London, Edinburgh, Melbourne, 1978, Ch. 20.
25 R.S. Yalow and S.A. Berson, J. Clin. Invest. 39 (1960) 1157 - 1175.
26 K. Catt, H.D. Niall and G.W. Tregear, Biochem. J. 100 (1966) 31c - 33c.

27 L. Wide and J. Porath, Biochim. Biophys. Acta 130 (1966) 257 - 260.
28 L. Wide, Acta endocr. Suppl. 142 (1969) 207 - 218.
29 K.J. Catt and G.W. Tregear, Science 158 (1967) 1570 - 1572.
30 K.J. Catt, Acta endocr. Suppl. 142 (1969) 222 - 243.
31 F.C. den Hollander, A.H.W.M. Schuurs and H. van Hell, J. Immunol. Meth. 1 (1972) 247 - 262.
32 T.C.J. Gribnau, Th. v. Lith, F. Roeles, C.J. van Wijngaarden, H. van Hell and A.H.W.M. Schuurs, in H. Peeters (Ed.), Protides of Biological Fluids, 27th Colloquium, Pergamon Press Oxford and New York (1979) 793 - 796.
33 T.C.J. Gribnau, Th. v. Lith, A. van Sommeren, F. Roeles, H. van Hell and A. Schuurs, INSERM 86 (1979) 175 - 186.
34 A.H.W.M. Schuurs and B.K. van Weemen, J. Immunoassay 2 (1980) 229 - 249
35 T.J. McKearn, in R.H. Kennett and K.B. Bechtol (Eds.), Monoclonal Antibodie Plenum Press, New York and London (1980) 388 - 390.
36 D.J. Barrett, A.J. Ammann, S. Stenmark and D.W. Wara, Infect. Immun. 27 (1980) 411 - 417.
37 T. Boenisch, in H. Peeters (Ed.), Protides of the Biological Fluids, Pergamon Press, Oxford, 24 (1976) 743 - 749.
38 F.R. Bishai, in J. Langan and J.J. Clapp (Eds.), Ligand assay, Masson Publis USA Inc, New York, 1981, Ch. 8.
39 Y. Carlier, A. Colle, P. Tachon, D. Bout and A. Capron, J. Immunol. Meth. 40 (2982) 231 - 238.
40 J.W. Stocker and C.H. Heusser, J. Immunol. Meth. 26 (1979) 87 - 95.
41 R.A. Polin and R. Kennett, J. Pediatr. 97 (1980) 540 - 544.
42 R.H. Kennett, in R.H. Kennett and K.B. Bechtol (Eds.), Monoclonal Antibodie Plenum Press, New York and London,(1980) 376 - 377.
43 R.M. Hendry and J.E. Herrmann, J. Immunol. Meth. 35 (1980) 285 - 296.
44 O.-P. Lehtonen and M.K. Viljanen, J. Immunol. Meth. 34 (1980) 61 - 70.
45 P.J. Robinson, P. Dunnill and M.D. Lilly, Biochim. Biophys. Acta 242 (1971) 659 - 661.
46 H.H. Weetall, Separation and Purification Methods, Vol. 2, Marcel Dekker, (19 199.
47 R. Axén, J. Porath and S. Ernback, Nature 214 (1967) 1302 - 1304.
48 J. Porath, K. Aspberg, H. Drevin and R. Axén, J. Chromatogr. 86 (1973) 53 - 5
49 S.C. March, I. Parikh and P. Cuatrecasas, Anal. Biochem. 60 (1974) 149 - 152
50 A.H. Nishikawa and P. Bailon, Anal. Biochem. 64 (1975) 268 - 275.
51 T.C.J. Gribnau, Thesis, Nijmegen (1977)
52 T.C.J. Gribnau, G.I. Tesser and R.J.F. Nivard, J. Solid-Phase Biochem. 3 (1978) 1 - 33.
53 Bio Rad Labs, Bulletin LO45E (1977), Catalogue F (1980).
54 A. Kardana, M. Squires, T. Adam, G.K. Goka and K.P. Bagshawe, J. Immuno Meth. 30 (1979) 47 - 53.
55 E.P. Halpern and R.W. Bordens, Clin. Chem. 25 (1979) 860 - 862, 1561 - 1563
56 J.L. Guesdon and S. Avrameas, Immunochemistry 14 (1977) 443 - 447.
57 K.O. Smith and W.D. Gehle, J. Infect. Dis. 136 Suppl. (1977) S329 - S336.
58 R.D. Margessi, J. Ackland, M. Hassan, G.C. Forrest, D.S. Smith and J. Land Clin. Chem. 26 (1980) 1701 - 1703.
59 K.O. Smith and W.D. Gehle in H. v. Vunakis, J.L. Langone (Eds.), Methods i Enzymology, Academic Press 1980, Vol. 70, Part A, 388 - 416.
60 B. Ziola and H. Tuokko, Acta path. Microbiol. Scand. Sect. C 88 (1980) 127 -
61 A. Rembaum and W.J. Dreyer, Science 208 (1980) 364 - 368.

T.C.J. Gribnau, J. Visser and R.J.F. Nivard (Editors),
Affinity Chromatography and Related Techniques

BLOOD COMPATIBLE ADSORBENT HEMOPERFUSION FOR EXTRACORPOREAL BLOOD TREATMENT

T.M.S. CHANG
Artificial Cells and Organs Research Centre, Faculty of Medicine, McGill University, 3655 Drummond Street, Montreal, Canada, H3G 1Y6

ABSTRACT

Potentially a large number of adsorbents could be used for extracorporeal blood treatment. However, most adsorbents are not blood compatible and as a result methods have to be developed to make these blood compatible before they can be used clinically. One of the most commonly used methods to make adsorbent blood compatible is the principle of artificial cells. This way, adsorbents are either microencapsulated within a spherical membrane or microencapsulated by ultrathin coating. This approach has made it possible for the use of adsorbents like activated charcoal, resins, and immunosorbents for extracorporeal treatment of patients.

INTRODUCTION

In theory the principle of chromatography should be very useful for the selective removal of unwanted material from the blood of patients. However, a number of requirements have to be met in order to adopt chromatography systems for the treatment of extracorporeal blood. Thus, as a minimal, it is essential to make sure that the system is acceptable both in the way of blood compatibility and in the way of lack of release of harmful material to the patients. Initial attempts to use chromatography type of columns in the form of ion-exchange resin [1], activated charcoal [2] and others for possible extracorporeal blood treatment have failed because in all cases the adsorbent system adversely affected blood cells especially platelets. Furthermore some adsorbent systems also released unwanted materials, for instance particulates or pyrogens into the perfusing blood. The principle of artificial cells [3,4] has been applied to prepare suitable extracorporeal hemoperfusion systems [5]. With this beginning a rapidly increasing amount of work has been carried out on the use of different hemoperfusion systems for extracorporeal blood treatment. These studies involve an increasing number of centres and the results have been reviewed in detail in a number of publications [5-11]. Since this field is now very extensive, only a few typical examples are used here to illustrate the principle.

BASIC PRINCIPLE

In the majority of cases the systems are made blood compatible and prevented from releasing unwanted material by the principle of artificial cells. In principle, the approach is to prevent the direct contact of the adsorbent with blood cells by surrounding each granule of adsorbent with a blood compatible membrane. Different types of membrane material can be used. However some are more blood compatible than others. The most blood compatible approach was the approach whereby adsorbents are coated with albumin-cellulose nitrate [12-14]. Other materials include the use of hydrogel, cellulose nitrate, cellulose acetate, cellulose heparin complexed membrane and others [5-11]. All these have various degrees of blood compatibility but the albumin coating approach at present still results in the best blood compatible systems. The exact method of applying the membrane depends on the particular application. The principles may be summarized as follows:

1. Suspensions or solutions of adsorbent, enzyme or protein can be enclosed within spherical ultrathin membrane artificial cells. The membrane thickness is usually about 200 Å [3,4]. The membranes retain protein or macromolecules inside the artificial cell but allow peptides and smaller molecules to equilibrate rapidly across. This way, molecules smaller than macromolecules can effectively cross the membrane. The ultrathin membrane (200 Å) and the large surface-to-volume relationship (2 M^2/10 ml) is such that the rate of equilibration is at least 200 times faster than a standard hemodialysis machine.

2. In the case of solid granules, the easiest way is to apply the membrane directly as a coating onto the surface of each granule. Being supported by the granule this way ultrathin membranes can still have very good mechanical strength. With this type of coating the adsorbent can be separated from circulating blood cells and the blood compatibility would depend on the blood compatibility of the coating material. The coating membrane, if properly applied, would also be effective in preventing the elution of fine powders from the adsorbent into the circulating blood. In the coating procedure it's important to have as thin a membrane as possible. In the original coating approach, ultrathin coating of less than 500 Å was used [12]. However in some other systems coatings as thick as 3 micron have been used. The ultrathin coating would result in a much more permeable system than the much thicker coating, as a result, most recent improved commercial systems follow the original ultrathin coating technique [12].

3. In cases where it is desirable to remove macromolecules the coating can also be applied but in this case it has to be applied as an ultrathin coating. By applying an ultrathin coating the coating material only covers the surface of the granule without covering over the surface entrance pores of the granule [14]. This way the porosity of the granule is not affected and as a result, macromolecules from blood can still enter the pores readily. On the other hand, the adsorbent surface can now be made blood compatible.

With these types of general principles a number of systems have been treated this way to result in blood compatible system for use in extracorporeal blood treatment [3-14]. Coated activated charcoal is one of the most commonly used for extracorporeal blood treatment of patients. It is now used very extensively in clinical applications and therefore will be given as a detailed example. Resins, immunosorbents, synthetic immunosorbents, and others will be more briefly discussed.

Activated charcoal coated by blood compatible material for hemoperfusion

Activated charcoal is a very powerful adsorbent which can remove many chemicals, drugs and other exogenous and endogenous toxins. It has been used for many years for detoxification in industry, for removal of toxic gases, for purification of water supply and others. In medicine it has been used at one time or another for gastrointestinal administration to remove ingested toxic materials and drugs. Since in many cases of acute drug intoxication and in metabolic diseases like renal failure and liver failure, it is desirable to remove the unwanted material directly from blood, the possibility of using activated charcoal for direct blood hemoperfusion was tested [2]. A number of drugs and some uremic metabolites can be effectively removed from blood by charcoal [2]. Unfortunately this type of treatment in patients was discontinued very shortly because of the findings that charcoal hemoperfusion rapidly deplete platelets. Furthermore charcoal hemoperfusion also resulted in the release of harmful charcoal powder into the circulation with emboli to lung, kidney, liver and other vital organs. These observations had resulted in the termination of further application of charcoal hemoperfusion in patients. In 1966 it was demonstrated here that the principle of artificial cells [3,4] could be applied to microencapsulate charcoal to prevent the powder from embolizing and at the same time to separate the charcoal from the blood cells of circulating blood [5]. With this finding, further development was carried out to test a number of blood compatible materials for coating activated charcoal. Having investigated a number of systems, the albumin-cellulose nitrate coating of activated charcoal (ACAC) [12,15] was developed for extensive clinical trial here. Compared to control the coating has virtually eliminated any lowering of platelets or white blood cells during hemoperfusion treatment [12-14]. Scanning electronmicroscopic study of the surface demonstrated that those coated with albumin-cellulose nitrate do not have any deposits of platelets or fibrin or other cellular material, those not coated with albumin approach have platelets and fibrin deposits [14]. Clinical study was carried out to analyze the effectiveness of the ACAC approach for the treatment of acute drug intoxication [16-18], for chronic renal failure [13,19-23], and for liver failure [24-29]. The result shows that in patients the albumin coating prevents platelet removal and the whole system was blood compatible with no removal of white cells or red blood cells or platelets. There was also no adverse effect in patients and there was no release of particulates or pyrogens. In acute intoxicated patients this approach was much more effective in removing drugs from the blood of patients than the standard hemodialysis machine. Furthermore even though the hemodialysis machine cannot remove any significant amount of protein-bound drugs, the ACAC system, because of the ultrathin membrane and the albumin coating, was effective in removing those drugs which are loosely protein-bound. This adds further to the effectiveness of this approach as compared to standard hemodialysis machine. In study on patients with chronic renal failure it was found that ACAC hemoperfusion was much more effective in removing middle molecules, creatinine, uric acid, phenol, and other waste metabolites from the renal failure patients than the standard hemodialysis machine. Patients treated by hemoperfusion alone become symptom-free with much shorter treatment time than when using hemodialysis machine. However hemoperfusion by itself cannot remove water, salt, electrolytes or urea. To solve this problem it is used alternatingly with hemodialysis. Another approach is to use it in series with hemodialysis machine. A third approach is to use this in series with a small ultrafiltrator to remove water, salt and with oral adsorbent to remove urea and other electrolytes. Study in liver failure patients demonstrated that in patients with Grade IV liver failure, hemoperfusion resulted in the complete recovery of the consciousness of patients most likely due to removal of some toxic substances.

Extensive study has since then been carried out in many centers and many industrial hemoperfusion systems have been developed in many countries [6-11, 29-43]. The results of clinical applications using the different hemoperfusion systems varied tremendously. In the earlier large-scale production by the industries, membranes used for coating were much thicker, up to 3 micron. As a result the permeability and the rate of removal of material were not as effective. Furthermore, some industrial systems did not use very blood compatible membranes. With further development and the use of ultrathin membrane and more blood compatible membranes the available commerical systems are improving and as a result our earlier study using laboratory prepared system is now being supported by many centers [6-11,29,30,35,37-39,41]. The conclusion at present is that using the best type of hemoperfusion systems there are no problems with blood compatibility. Furthermore there are no problems with the release of particulates. Insofar as the clinical results are concerned, the approach of extracorporeal treatment of blood using coated charcoal is now accepted universally as a treatment for severe acute drug poisoning if the drug has a large volume distribution. In the treatment of renal failure patients it has also been demonstrated that this approach is much more effective than hemodialysis in removing toxic materials like middle molecules and other waste metabolites. In fact patients on standard hemodialysis who still have complications like pericarditis, pruritis, peripheral neuropathy, nausea, vomiting, have been treated by hemoperfusion. By doing this the complications in many cases could be resolved. In the case of treatment of liver failure, it has been firmly demonstrated that hemoperfusion of Grade IV coma patients has resulted in the majority of them recovering complete consciousness temporarily. However the long-term effects and the time of initiation of treatment for complete recovery is still being investigated at present.

Other systems

Resin is another group of adsorbents which would be very effective for removing more selectively certain drugs and certain materials from the circulating blood [42]. However the use of resin for hemoperfusion did not receive wide acceptance because of the problem of blood compatibility since it tends to lower systemic platelets. With the demonstration here that microencapsulated resin can be prevented from contact with blood and at the same time continue to remove unwanted material from circulating blood [5], further studies were carried out. The approach of using albumin coating for making surfaces blood compatible [12,14] has been applied to the coating of resins [43]. This way they find that resin can be used in liver failure patients, who usually have very sensitive platelets, without adversely affecting the platelet levels. Further studies carried out by other groups [9,44] have also demonstrated that albumin coating on resins is very effective for making the surface compatible. As a result, there is now increasing investigations into the possible uses of different types of resins for hemoperfusion.

A number of synthetic immunosorbents could be made available to remove specific antigens and antibodies from blood. However most of these immunosorbents cannot be used for direct hemoperfusion because of the possible adverse effect on blood elements. Recent study carried out here includes the coating of a synthetic immunosorbent with albumin and cellulose nitrate [45]. This is a silicate material with synthetic antigenic group that can remove blood group A antibodies or blood group B antibodies. Study here demonstrated that without coating with albumin this system resulted in platelet removal from blood. However after coating with albumin and ultrathin cellulose nitrate membrane, this system no longer causes removal of platelets but at the same time can still remove anti B or anti A selectively from blood [45]. Another type of immunosorbent for hemoperfusion was developed from the observation that albumin on the albumin-cellulose nitrate coated activated charcoal system [12,15] can effectively remove antibodies to albumin from circulating blood [46]. With this observation this was extended further to substitute the albumin on the ACAC with other antigens or antibodies. This way this type of immunosorbent can be used for removing antigen or antibody from plasma [47].

Research has also been carried out to apply the principle of artificial cells to other forms of biologically active materials. For example microencapsulation of enzymes like urease, asparaginase, tyrosinase for use in hemoperfusion to convert specific substrates [4,48-52]. Multienzyme systems with cofactor recycling in the artificial cell has also been prepared for more complex action [53,54].

In conclusion many types of adsorbents or biologically active materials which cannot be used for extracorporeal blood treatment can now be treated by the principle of artificial cells to result in blood compatible systems. With the availability of a large number of specific adsorbents and other biologically active materials, different types of blood compatible hemoperfusion systems could be developed for extracorporeal blood treatment in patients.

ACKNOWLEDGEMENT

The support of the Medical Research Council of Canada since 1965 and the present special project grant (MRC-SP-4) is gratefully acknowledged.

REFERENCES

1 G.E. Schreiner, Arch. Int. Med., 102(1958)896-900.
2 D. Yatzidis, Proc. Eur. Dial. Transplant. Assoc., 1(1964)83.
3 T.M.S. Chang, Science, 146(1964)524-525.
4 T.M.S. Chang, Artificial Cells, Thomas, Springfield, Illinois, 1972.
5 T.M.S. Chang, Trans. Am. Soc. Artif. Intern. Organs, 12(1966)13-19.
6 T.M.S. Chang, Artificial Kidney, Artificial Liver, and Artificial Cells, Plenum, New York, 1978.
7 R.M. Kennedi, J.M. Courtney, J.D.S. Gaylor and T. Gilchrist, Artificial Organs, MacMillan, London, 1977.
8 V.G. Kikolaef and V.V. Strelko, Hemoperfusion, Naukova Dumka, Kiev, 1979.
9 H. Klinkmann, D. Falkenhagen and J.M. Courtney, Int. J. Artif. Organs, 2(1979)296.
10 R. Williams and I.M. Murray-Lyon, Artificial Liver Support, Pitman, London, 1975.
11 V. Bonomini and T.M.S. Chang, Hemoperfusion, S.Karger AG, Basel, Switzerland, in press.
12 T.M.S. Chang, Can. J. Physiol. Pharmacol., 47(1969)1043-1045.
13 T.M.S. Chang, A. Gonda, J.H. Dirks and N. Malave, Trans. Am. Soc. Artif. Intern. Organs, 17(1971)246-252.
14 T.M.S. Chang, Can. J. Physiol. Pharmacol., 52(2)(1974)275-285.
15 T.M.S. Chang, Kidney Int., 10(1976)S218-S224.
16 T.M.S. Chang, J.F. Coffey, C. Lister, E. Taroy and A. Stark, Trans. Am. Soc. Artif. Intern. Organs, 19(1973)87-91.
17 T.M.S. Chang, Kidney Int., 10(1976)S305-S311.
18 T.M.S. Chang, Clin. Toxicol., 17(1980)529-542.
19 T.M.S. Chang, A. Gonda, J.H. Dirks, J.F. Coffey and T. Burns, Trans. Am. Soc. Artif. Intern. Organs, 18(1972)465-472.
20 T.M.S. Chang, M. Migchelsen, J.F. Coffey and A. Stark, Trans. Am. Soc. Artif. Intern. Organs, 20(1974)364-371.
21 T.M.S. Chang, E. Chirito, P. Barre, C. Cole and M. Hewish, Trans. Am. Soc. Artif. Intern. Organs, 21(1975)502-508.

22 T.M.S. Chang, Clin. Nephrol., 11(1979)111-119.
23 T.M.S. Chang, E. Chirito, P. Barre, C. Cole, C. Lister and E. Resurreccion, Artif. Organs, 3(1979)127-131.
24 T.M.S. Chang, Lancet, 2(1972)1371-1372.
25 T.M.S. Chang and M. Migchelsen, Trans. Am. Soc. Artif. Intern. Organs, 19(1973)314-319.
26 E. Chirito, B. Reiter, C. Lister and T.M.S. Chang, Artif. Organs, 1(1)(1977) 76-83.
27 T.M.S. Chang, C. Lister, E. Chirito, P. O'Keefe and E. Resurreccion, Trans. Am. Soc. Artif. Intern. Organs, 24(1978)243-245.
28 Y. Tabata and T.M.S. Chang, Trans. Am. Soc. Artif. Intern. Organs, 26(1980) 394-399.
29 T. Agishi, N. Yamashita and K. Ota, in S. Sideman and T.M.S. Chang (Eds.), Hemoperfusion: Kidney and Liver Support and Detoxification, Part I, Hemisphere, Washington, D.C., 1980, p.255.
30 I. Amano, H. Kano, H. Takahira, Y. Yamamoto, K. Itoh, S. Iwatsuki, K. Maeda and K. Ohta, in T.M.S. Chang (Ed.), Artificial Kidney, Artificial Liver, and Artificial Cells, Plenum, New York, 1978, p.89.
31 J.D. Andrade, R. Van Wagenen, M. Ghavamian, J. Volder, R. Kirkham and W.J. Kolff, Trans. Am. Soc. Artif. Intern. Organs, 18(1972)235.
32 M.C. Gelfand, J.F. Winchester, J.H. Knepshield, S.L. Cohan and G.E. Schreiner, Trans. Am. Soc. Artif. Intern. Organs, 24(1978)239-242.
33 A.E.S. Gimson, R.D. Hughes, P.G. Langley, J. Canalese and R. Williams, in S. Sideman and T.M.S. Chang (Eds.), Hemoperfusion: Kidney and Liver Support and Detoxification, Part I, Hemisphere, Washington, D.C., 1980, p.285.
34 H.W. Leber, M. Neuhauser and G. Goubeaud, Eur. Soc. Artif. Organs, 3(1976) 202.
35 K. Maeda, A. Saito, S. Kawaguchi, T. Niwa, R. Sezaki, K. Kobayashi, H. Asada, Y. Yamamoto and K. Ohta, in S. Sideman and T.M.S. Chang (Eds.), Hemoperfusion: Kidney and Liver Support and Detoxification, Part I, Hemisphere, Washington, D.C., 1980, p.349.
36 Z. Niu, S.R. Jia, D.Y. Zhang, C.X. Xu, X.J. Tang, W.K. Fan, Y.P. Luo and Z.M. Li, Chongqing Medical College Bull., November(1980)1-6.
37 M. Odaka, Y. Tabata, H. Kobayashi, Y. Nomura, H. Soma, H. Hirasawa and H. Sato, in T.M.S. Chang (Ed.), Artificial Kidney, Artificial Liver, and Artificial Cells, Plenum, New York, 1978, p.79.
38 M. Odaka, H. Hirasawa, H. Kobayashi, M. Ohkawa, K. Soeda, Y. Tabata, M. Soma and H. Sato, in S. Sideman and T.M.S. Chang (Eds.), Hemoperfusion: Kidney and Liver Support and Detoxification, Part I, Hemisphere, Washington, D.C., 1980, p.45.
39 R. Oules, H. Asaba, M. Neuhauser, V. Yahiel, S. Baehrendtz, B. Gunnarsson, J. Bergstrom and P. Furst, in T.M.S. Chang (Ed.), Artificial Kidney, Artificial Liver, and Artificial Cells, Plenum, New York, 1978, p.153.
40 B.M. Chavers, C.M. Kjellstrand, C. Wiegand, J. Ebben and S.M. Mauer, Kidney Int., 18(1980)386.
41 S. Stefoni, L. Coli, G. Feliciangeli, L. Baldrati and V. Bonomini, Int. J. Artif. Organs, 3(1980)348.
42 J.L. Rosenbaum, in T.M.S. Chang (Ed.), Artificial Kidney, Artificial Liver, and Artificial Cells, Plenum, New York, 1978, p.217.
43 H.Y. Ton, R.D. Hughes, D.B.A. Silk and R. Williams, Artif. Organs, 3(1979)20.
44 J.L. Rosenbaum, personal communication, 1981.
45 T.M.S. Chang, Trans. Am. Soc. Artif. Intern. Organs, 26(1980)546-549.
46 D.S. Terman, T. Tavel, D. Petty, M.R. Racic and G. Buffaloe, Clin. Exp. Immunol., 28(1977)180.
47 D.S. Terman, in C. Giordano (Ed.), Sorbents and Their Clinical Applications, Academic, New York, 1980, p.470.
48 T.M.S. Chang, Biomedical Applications of Immobilized Enzymes and Proteins, Vols. 1 & 2, Plenum, New York, 1977.
49 T.M.S. Chang, Artif. Organs, 4(4)(1980)264-271.
50 C.D. Shu and T.M.S. Chang, Int. J. Artif. Organs, 4(1981)82-84.

51 D.L. Gardner, R.D. Falb, B.C. Kim and D.C. Emmerling, Trans. Am. Soc. Artif. Intern. Organs, 17(1971)239.
52 C. Kjellstrand, H. Borges, C. Pru, D. Gardner and D. Fink, Abstr. Am. Soc. Artif. Intern. Organs, 10(1981)47.
53 T.M.S. Chang, C. Malouf and E. Resurreccion, Artif. Organs, 3(1979)S284-S287.
54 Y.T. Yu and T.M.S. Chang, FEBS Letters, 125(1981)94-96.

T.C.J. Gribnau, J. Visser and R.J.F. Nivard (Editors),
Affinity Chromatography and Related Techniques

AFFINITY CHROMATOGRAPHY AND AFFINITY THERAPY

J. Kálal, J. Drobník and F. Rypáček
Institute of Macromolecular Chemistry of Czechoslovak Academy of Sciences, 162 06 Prague 6 (Czechoslovakia)

ABSTRACT

The principal of the affinity therapy is suggested as a controlled movement of a polymeric drug in the body. The controlling factors such as properties of polymer on the one hand and physiological interactions involved in the process on the other are discussed. Experimental approach including the variable polymer model system, the methods of tracing of polymer fate in the body is presented together with the results ilustrating the possibility of controlled interaction of polymers with cells.

The principle of affinity chromatography is based on the control of the movement of a molecule in a liquid phase by specifically interacting groups on a solid phase. These interactions are usually adjusted by the design of the immobilised ligands, to have a biological affinity for substances to be separated (ref. 1). The high efficiency obtained by affinity chromatography is also the main goal of the affinity therapy, the idea of which is based on a similar principle: to control the movement of a drug in the body by specific biological mechanisms or/and interactions with some groups on surfaces of cells and tissues. Since in this case the immobilised side - the ligands on the matrix - is designed by the nature, we have to modify the moving party. In this regard we are interested in the affinity of soluble polymers as drug carriers to various organs and tissues. And not only we. The idea of using soluble macromolecules as affinity-drug carriers has attracted attention of many scientists (refs. 2-5). The question may arise: why use soluble polymers or even synthetic polymers? There are two main reasons, first - soluble polymers can really serve as drug carriers, second - synthetic soluble polymers represent a powerful model system for the study of this special problem. We hope that this question will become clearer at the end of this lecture.

In this research we have to deal with some theoretical and methodical problems. In the theory we have to specify the nature of the factors controlling the movement of the polymer in the body.

We can imagine these factors as belonging to two groups. The first group is formed by polymer properties, while the different physiologi cal mechanisms involved form the second group. We should consider relations between these two partners.

Some features, such as high molecular weight (and consequently the degradability or nondegradability), polydispersity, uniformity of the main chain in connection with random conformation, are common properties of all synthetic polymers. We should always remember them when evaluating their role in general physiological mechanisms, such as membrane permeability, ways of excretion and mechanisms of nonspecifi retention. In addition to these factors, the polymers may acquire mor specific properties derived from a specific chemical modification. Th latter may be directed to specific places in physiological routes (specific cell interactions, immunological determinants, etc.).

The high molecular weight of polymers is recognized by the organism owing to the different permeability of membraneous barriers between various body compartments. Among all these size dependent transfers, the transport of substances across the glomerular membrane in the kidney, across the cell membrane and membrane of lysosomes, seems to us to be the most important from the point of view of affinity therap

Glomerular filtration is the main route by which the polymers are excreted from the body. The molecular weight, or better to say, the hydrodynamic volume of the molecule is the principal parameter in thi process, since it can be described mostly as ultrafiltration through the membrane with a maximum pore radius about 5.5 nm (refs. 6,7). The gradient of hydrodynamic pressure is the driving force in this proces (ref. 8). It should be stressed, that a soluble polymer (which is in general polydisperse) is always fractionated durir the glomerular excretion, which means that the size distribution of the sample which may be involved in other physiological processes is progressively shifted. The one of the greatest importance is cellular pinocytosis, i.e. mechanisms by which the macromolecules can cross th barrier between extra- and intracellular space and enter the cells. Pinocytosis seems to be a common feature of all mammalian cells, but some kinds of cells are specialized for this function. Pinocytosis starts with the invagination of the plasma membrane around the material present in the external medium or bound to the cell surface, i.e., formation of pinocytic vesicle -phagosome, which consequently fuses with lysosome and its content is exposed to the action of lysosomal hydrolases. Macromolecules of natural origin (i.e., proteins nucleic acids, polysaccharides etc.) are split to low molecular weight components which are used in the cell metabolism. A synthetic polymer-a drug carrier - present in the surrounding medium should be internalised in this way also. If it is nondegradable, it remains bound by the lysosomal membrane, due to its limited permeability for substances of molecular weight higher than approx. 300 daltons (refs. 9,10). Therefore,we should not be surprised finding synthetic polymers within the cells. The quantitative distribution of polymer in the organs and tissues mostly reflects the distribution of pinocytic activity of the cells, i.e., with the highest concentration of polymer in the liver, spleen, bone marrow, i.e., tissues of reticuloendothelial system (RES) (refs. 11-13).

To this pattern of polymer distribution we arrive without taking into account any other property of polymer except the high molecular weight. To speak about polymeric drug carriers in affinity therapy, we should change this natural tissue distribution of polymer, i.e., we should regulate the affinity of polymer for cellular uptake. This idea is not a new one: it was introduced by de Duve at the beginning of the seventies (ref. 14). However, lysosomotropic agents (nucleic acids carrying cytostatic drugs) used by de Duve and others (refs. 15,16), in spite of their lower toxicity, exhibited only that kind of tissue specifity which was based on the assumption of a higher pinocytic activity of malignant cells compared with the others. This discriminating factor seems to be unsufficient to protect the other cells, especially RES - with a high rate of pinocytosis, from these agents.

Therefore, what room is there left for selective regulation of pinocytosis? Let us look once again at the scheme of the process just mentioned (Fig. 1). We should understand it in general as one of the types of"adsorptive pinocytosis" in which the substances in the extracellular medium exhibit either "zero" or "non-zero" adsorptive interaction with the cell membrane. Some nonspecific factors, i,e, charge, hydrophobicity, size, conformation etc. may take part in the control of affinity of macromolecules for the cell surface, and we should imagine them as producing either positive or negative adsorption. The specific factors, on the other hand (all ligand- receptor associated interactions are mentioned now) manifest themselves only in the positive adsorption effect.

Ligand-receptor associated endocytosis (LRAE) is a mechanism through which the entry of many important metabolites, hormones, enzymes becomes very efficient. The nature of this process is known from the prominent work on the transport of cholesterol bound to the carrier lipoprotein and targetting the cell membrane receptors competent for the lipoprotein carrier (refs. 17,18), as well as on the transport of vitamin B 12 on transcobalamins (ref. 19) the growth factors (ref. 20) etc. In other cases, more simple low molecular weight compounds may be sufficient to provide the adsorptive uptake via specific cell receptors. This is the case of 6-phosphomannose as ligand for the uptake of lysosomal hydrolases in fibroblasts (refs. 22), while another one, viz. N-acetylglucosamine, directs the lysoso enzymes preferentially to Kupffer cells of liver (refs. 23,24). We should not skip another type of ligand-receptor pairs, formed by ant bodies and corresponding surface antigens. Immunospecific determinan were used several times for directing the cytostatic drugs or toxins to cancer cells. This subject has been recently considered by Olsnes (ref. 25). We should like only to mention the main hindrances on thi seemingly simplest way to the affinity drugs. They are of two types. Firstly, immunoglobulines, either with cytostatic drugs or as immuno toxins are foreign proteins, which should act as antigens and elicit the neutralization antibodies. Secondly, the properties of these com plexes are mostly given by the bulk of the protein component. Such chemically modified proteins are usually very efficiently cleared by the RES cells, and therefore these cells are "in vivo" at least the second target for this killing devices.

At the end of this theoretical and speculative part, let us draw some conclusions which serve also as premises for our further experimental approach to the subject of affinity therapy.

It seems that there will be no problem to find a sufficient amount of ligand-receptor pairs to achieve positive interaction with the ce membrane and to make the cell uptake more efficient. The problem remains in choosing such ligand-receptor systems which are specific on for the target tissue.

In addition to positive adsorption provided by specific ligand-receptor interactions, the whole complex should acquire a sufficient amount of property mediating the negative sorption effect, which should compensate or diminish the untargeted nonspecific uptake by t cells with the "inherited" high rate of pinocytosis, e.g., RES. We believe that this combination of nonspecific (-) effect with highly specific ligand-receptor associated endocytosis provides room for the introduction of synthetic polymers. Therefore, in our study we examine factors by which the synthetic macromolecules may alter the rate of its entry into the cells.

In the experimental approach to this study we met two categories of problems. Firstly, we should develop a variable polymer system, which makes possible an easy handling of polymer properties. It means to have polymers ready for chemical modification and obtainable in a wide spectrum of variants, including noninteracting physiologically inert polymers as well as polymers with specific tissue interactions. The routes for preparation of such polymers have been adequately developed in polymer chemistry (ref.26). Secondly, we need these polymers labeled to be able to trace them in the body and to follow their interactions. Labeling with radioactive isotopes and/or fluorescence labeling can be used for tracing these polymers (refs. 27,28). One family of polymers which offers a very useful model are represented by the derivatives of polyaspartic acid, viz., polyaspartamides. Polysuccinimide, obtained by the thermal polycondensation of aspartic acid, serves as a common reactive precursor, from which the various polyaspartamides may be prepared through stepwise aminolysis (ref. 29). In Fig.2 we demonstrate a set of modified polyaspartamides from which the study of nonspecific factors affecting the rate of adsorptive pinocytosis of polymers was started. We would like to correlate some chemical or structural properties of polymers with manifestation of either positive or negative interaction with cell surfaces in the way as discussed above. How these variations in chemical structure of polymers, all derived from the same type, affect the rate of their pinocytosis can be demonstrated by the rate of their accumulation by tubular cells of mouse kidney and by the Kupffer cells in liver.

The cells of resorption epithelium of kidney proximal tubules are very efficient in pinocytosis, because this is the main mechanism by which the low and middle molecular weight proteins are reabsorbed from primary urine and catabolised in lysosomes of tubular epithelium. The Kupffer cells on the other hand, are another type of cells having endocytosis as a characteristic function. The Kupffer cells in mice and rats represent more than 90% of the endocytic capacity of reticulo-endothelial system.

Quantitative relations in the intensity of cellular accumulation of polymers I through V are shown in Fig.3.

We see that polymers II and III with the aliphatic chains and the aromatic groups, respectively, exhibit a very strong positive effect on the rate of pinocytosis in the kidney cells, but on the other hand they are practically without effect on the Kupffer cells.

These results were obtained in experiments with the whole animal. I this case we should take into account their residental time in the plasma compartment, during which they are accessible to liver macrophages. As mentioned above, the polymers are at the same time filtere through the kidney glomerular membrane out of the plasma compartment according to their molecular weight. All polymers (I through V) used were of molecular weight about 10000 daltons in order to be filtered without great restrictions and accessible to tubular cells. Therefore the experiments were performed with an explanted liver to evaluate solely the role of the Kupffer cells. We see in Table 1 a comparison of the rate of accumulation in the liver for polymers I and III. No measurable differences in the rate of uptake between these two polyme were found in the isolated liver, which supports the results obtained with the whole animal.

The cellular uptakes of polymers I and III were also compared using visceral yolk sacs of rat embryos - an experimental model on which th pinocytosis is very extensively studied. These experiments were performed in collaboration with department of Professor J.B. Lloyd at th University of Keele. We can demonstrate that modification by aromatic side chains in our case greatly enhances the rate of cellular uptake in comparison with unmodified poly[(2-hydroxyethyl)aspartamide](Fig.4 The experiments on yolk sacs also proved that mechanism of pinocytosi is really involved in the cellular uptake, since it can be suppressed to zero level by addition of metabolic inhibitors, e.g. 2,4-dinitrophenol or iodoacetic acid.

The enhamcement of cellular uptake of polymers by aromatic phenol rings or hydrophobic aliphatic side chains should be related to the promoting effect on adsorptive pinocytosis, and we should like to stress that this effect is dependent not only on the type of the chemi cal modification of the polymer, but also on the type of the cells. I is important from the point of view of affinity therapy and of the possible role of synthetic polymers in it.

We would also like to mention the role of the charge of polymers. This problem needs special attention, since many polyions, namely, polyanions, such as DIVEMA, polyaminoacids etc., were found to exhibi some biological and also therapeutical effects themselves. We can present some preliminary results demonstrating that the charge will also play an important role in the regulation of adsorptive pinocytosis.

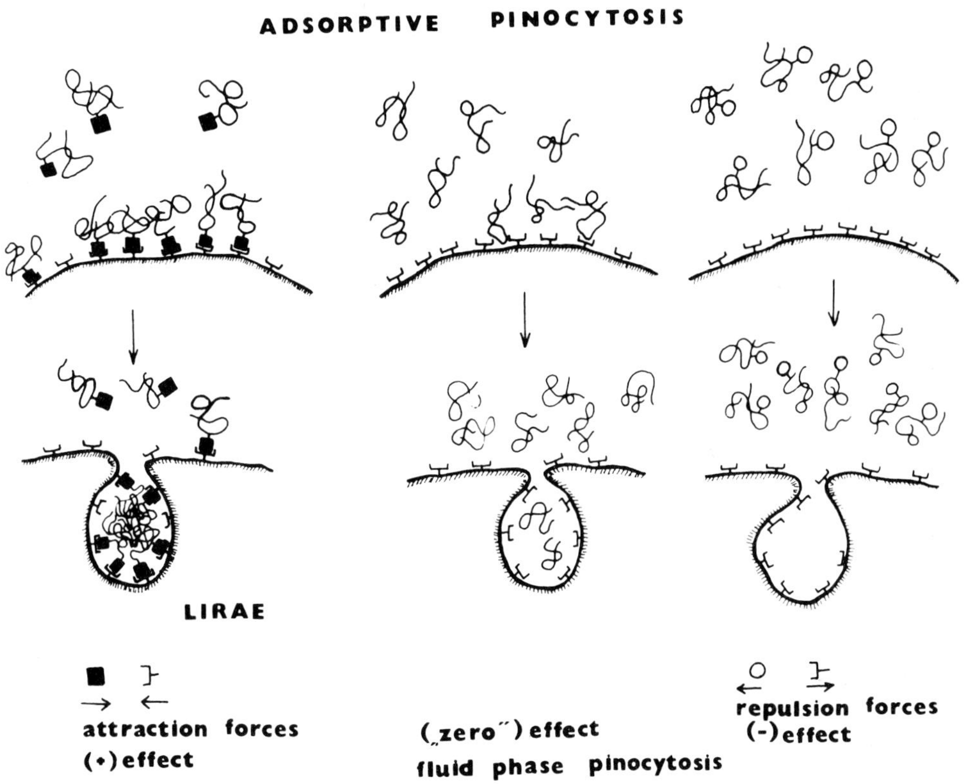

Fig. 1 Scheme of suggested types of interactions of polymer with the cell surface in the initial phase of pinocytosis.

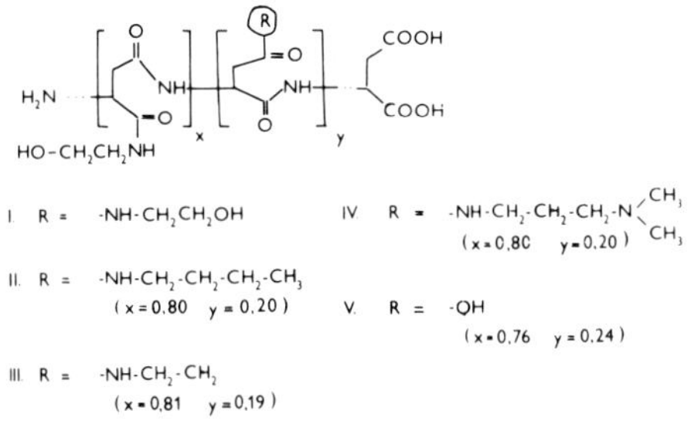

Fig. 2 The modified polyaspartamides - model polymers.

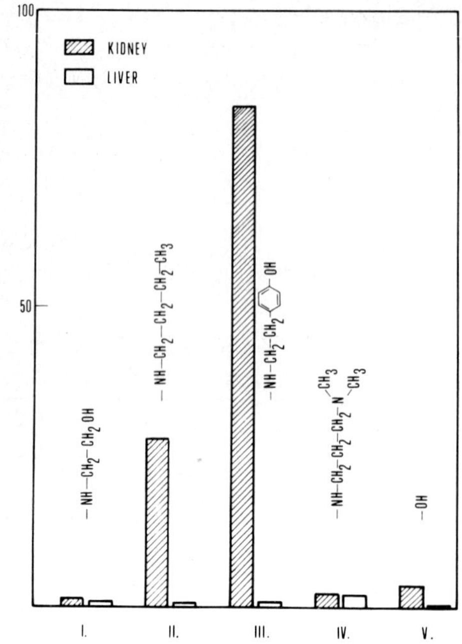

Fig. 3 Quantitative relation of intensity of cellular uptake of polymers I through V in the kidney and the liver of mice.

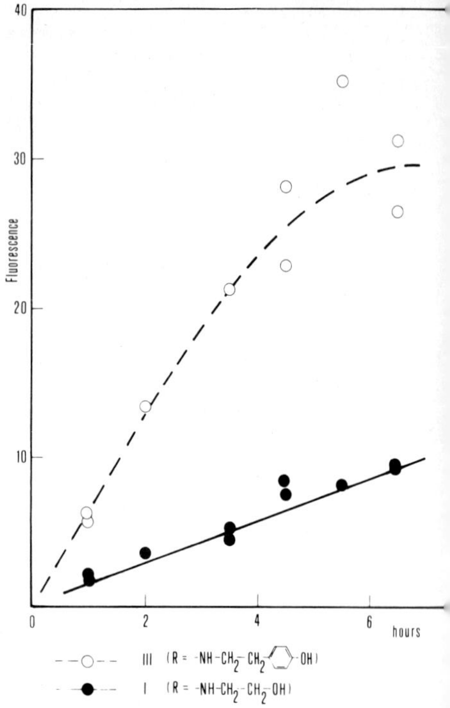

Fig. 4 The typical experiment showing the different rate of uptake of fluorescence labeled polymers I and III by the cells of rat yolk sac.

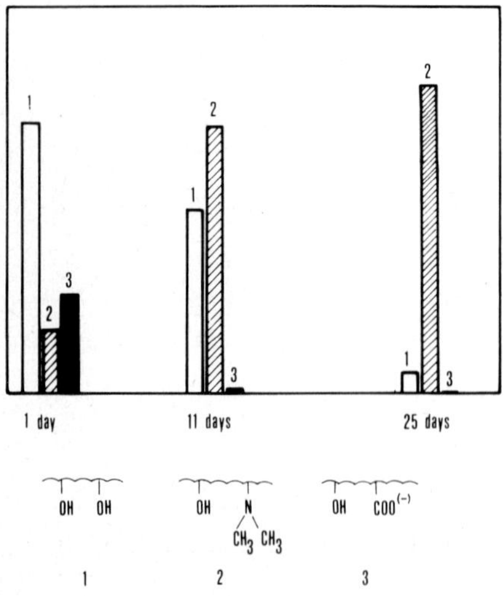

Fig. 5 Quantitative relation of amount of polymers, I (neutral), IV (positively charged) and V (negatively charged) in the splee of mice in different time interva after administration.

Affinity of RES cells to positively or negatively charged polymers is reflected in the rate of disappearance (or accumulation) in the spleen of mice during the redistribution of polymer in the body (Fig.5).

We hope that further progress in this field will bring some new data.

TABLE 1

Amount of polymer I and III accumulated in perfused rat liver during 150 minutes of perfusion, expressed as % of total perfused dose

Polymer		$\bar{M}_w$	%
I	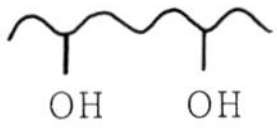	8 600 51 400 70 000	1.5 ± 0.5 2.1 ± 0.5 1.0 ± 0.4
III	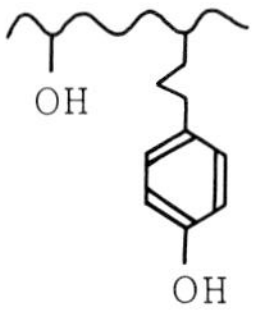	35 000	1.9 ± 0.7

REFERENCES

1 P. Cuatrecasas and C.B. Anfinsen, Ann. Rev. Biochem., 40(1971) 259-278.
2 H. Ringsdorf, J. Polym. Sci., Polym. Symp., 51(1975)135-153.
3 H. Ringsdorf, 17th Microsymp. on Macromolecules. Medical Polymers - Chemical Problems, Prague 1977.
4 A. Zaffaroni and P. Bonsen, in L.G. Donaruma and O. Vogl (Eds), Polymeric Drugs, Academic Press, New York, 1978.
5 E.P. Goldberg, in L.G. Donaruma and O. Vogl (Eds), Polymeric Drugs, Academic Press, New York, 1978.
6 K.E. Jørgensen and J.V. Møller, Am. J. Physiol., 236(1979)F103-F111.
7 J. Hardwicke, B. Hulme, J.H. Jones, and C.R. Rickets, Clin. Sci., 34(1968)505-514.
8 B.M. Brenner, W.M. Deen, and Ch.R. Robertson, Ann. Rev. Physiol., 38(1976)9-19.
9 S.C. Silverstein, R.M. Steinman, and Z.A. Cohn, Ann. Rev. Biochem., 46(1977)669-722.
10 B. Ehrenreich and Z.A. Cohn, J. Exp. Med., 129(1969)227-243.
11 W. Wessel, M. Schoog, and E. Winkler, Arzneim. Forsch. (Drug Res.), 21(1971)1468-1482.
12 H.A. Ravin, A.M. Seligman, and J. Fine, New Engl. J. Med., 247(1952) 921-929.
13 H. Marek, H. Koch, and K. Seige, Z. Ges. Exp. Med., 150(1969)213-222.
14 C. de Duve, T.de Barsy, B. Poole, A. Tronet, P. Tulkens, and F. Van Hoof, Biochem. Pharmacol., 23(1974)2495-2531.
15 A. Trouet, D.D. Campeneere, and C. de Duve, Nature, New Biol., 239(1972)110-112.
16 G. Barbanti-Brodano and L. Fiume, Nature, New Biol., 234(1973) 281-283.

17 M.S. Brown and J.L. Goldstein, Science, 191(1976)150-154.
18 J.L. Goldstein and M.S. Brown, Ann. Rev. Biochem., 46(1977)897-930
19 P. Youngdahl-Turner, L.E. Rosenberg, and R.H. Allen, J. Clin. Inve
61(1973)133-141.
20 G.Carpenter and S. Cohn, J. Cell. Biol., 71(1976)159-171.
21 E.F. Neufeld, G.N. Sando, A.J. Garvin, and L.H. Rome, J. Supramol. Struct., 6(1977)95-101.
22 A. Kaplan, D.T. Achord, and W.S. Sly, Proc. Nat. Acad. Sci., 74(1977)2026-2030.
23 D.T.Achord, E.E. Brot, and W.S. Sly, Fed. Proc., 36(1977)653.
24 P.D. Stahl, J.S. Rodman, M.J. Miller, and P.H. Schlesinger, Proc. Nat. Acad. Sci., 75(1978)1399-1403.
25 S. Olsnes, Nature, 290(1981)84.
26 J. Kálal, J. Drobník, J. Kopeček, and J. Exner, in L.G. Donaruma (Ed.), Polymeric Drugs: Synthetic Polymers in Chemotherapy, Academic Press, New York 1978.
27 L. Šprincl, J. Exner, D. Štěrba, and J. Kopeček, J. Biomed. Mater. Res., 10(1976)953-963.
28 F. Rypáček, J. Drobník, and J. Kálal, Anal. Biochem., 104(1980) 141-149.
29 J. Vlasák, F. Rypáček, J. Drobník, and V.Saudek, J. Polym. Sci., Polym. Symp., 66(1979)59-64.

T.C.J. Gribnau, J. Visser and R.J.F. Nivard (Editors),
Affinity Chromatography and Related Techniques

POLYMERIC AFFINITY DRUGS FOR TARGETED CHEMOTHERAPY: USE OF SPECIFIC AND NON-SPECIFIC CELL BINDING LIGANDS

E. P. GOLDBERG, H. IWATA, R. N. TERRY, W. E. LONGO, M. LEVY, T. A. LINDHEIMER AND J. L. CANTRELL

Department of Materials Science, MAE 217, University of Florida, Gainesville, Florida, 32611, U.S.A.

ABSTRACT

The conceptual framework for affinity chromatography, which achieves highly specific separation of biological molecules, is also applicable to the design of unique drugs for affinity therapy. This is a potentially important and rapidly developing new field of research. Recent studies involving antibodies, hormones, cationic polypeptides and lectins as biospecific ligands for targeting chemotherapy to tumor cells are briefly reviewed in this paper.

In parallel studies, simple physical targeting (i.e. direct injection of cytotoxic drugs into solid tumors) has been shown to cure hepatomas in guinea pigs even with metastasis. To enhance the efficacy of such intratumor (i.t.) immunochemotherapy, prolonged i.t. drug residence time has been sought by preparing both soluble and insoluble drug-macromolecule compositions possessing tissue-binding properties. Adriamycin (AD) and mitomycin C (MC) have been attached to various macromolecular carriers having cell surface binding ligands. These need not be highly specific but are used to prolong drug conjugate residence time in the injected tumor tissue.

Many types of such polymeric drug compositions have been prepared and studied for chemical-pharmacological behavior. These include soluble conjugates: AD-polyglutaradehyde, AD-aldehyde dextran, AD-succinylCon A and MC-succinylCon A. Insoluble compositions based upon functionalized albumin or cross-linked dextran microspheres as well as stable polymer-drug salts have also been prepared. Initial studies encompassing (a) multiple injection *in vivo* experiments with AD and MC, (b) tumor tissue analyses for AD retention vs. time, (c) drug release from polymer conjugates *in vitro*, and (d) dose response/cure rate/tumor staging studies in the guinea pig hepatoma point to therapeutic opportunities for i.t. immunochemotherapy using polymeric drugs having physical or covalent cell-binding properties.

INTRODUCTION

Conventional cancer chemotherapy is systemic and relatively non-specific. Serious toxic side-effects make aggressive therapy hazardous. As a consequence, solid tumors of the lung, breast and colon, which readily metastasize, often defy effective treatment. This has motivated the growing search for new drugs as well as new therapeutic methods which would confine cytotoxic activity to tumor cells.

The idea that chemotherapy might be targeted is to be found in the work of Ehrlich even as early as his 1878 PhD thesis which deals with specific cell stains (Ref. 1). During the past decade, advances in tumor immunology have suggested that unique antigenic characteristics of neoplastic cells (tumor associated antigens) might be utilized for therapeutic targeting. Some of the important attempts to utilize tumor-specific antibodies will be briefly reviewed in this paper. In addition, studies with less specific targeting agents, such as carbohydrate-binding lectins and electrostatic-binding polycations (polylysine) will also be noted.

Our own approach to tumor-specific drug delivery originally emphasized tumor-specific antibody (Ab) targeting with polymeric carriers to prepare polymeric affinity drugs (Ref. 2). However, beginning in 1976 work on BCG i.t. immunotherapy with C. A. McLaughlin, J. L. Cantrell and E. Ribi stimulated a parallel investigation of i.t. chemotherapy.

We have shown that direct i.t. injection of drugs such as adriamycin (AD) and mitomycin C (MC) can cure metastatic line 10 hepatomas in guinea pigs. Most important, cured animals were shown to have specific immune resistance to a subsequent line 10 tumor challenge (Ref. 3). This confirmed an earlier report by Borsos et al. (Ref. 4). However, because we have found very rapid disappearance of low molecular weight drugs from the tumor injection site as well as enhanced drug efficacy when given in sequenced multiple injections, more effective i.t. therapy might be achieved using polymeric drugs having tissue-binding functionality This paper therefore surveys some of our recent work on the preparation and propert of both soluble and insoluble polymer-drug compositions with emphasis on those whic contain ligands for cell surface association.

AFFINITY CHEMOTHERAPY

Tumor-Specific Antibodies as Tumor Affinity Ligands

During the past decade, Ghose and coworkers have played a leading role in explor ing possibilities for using tumor-specific Abs to localize antitumor activity. Early work with chlorambucil bound to EL4 mouse lymphoma or human melanoma Abs suggested that both drug activity and Ab binding could be retained for covalent conjugates (Ref. 5). Studies by Davies and O'Neill questioned *in vivo* targeting but suggested a beneficial synergism between unattached Ab and drug when given together (Ref. 6). Soon after, Rowland demonstrated preferential antitumor activity for a nitrogen mustard-polyglutamic acid-EL4 mouse Ab conjugate and for a similar drug-dextran-Ab conjugate (Ref. 7). Hurwitz and Arnon were also among early workers in this field reporting the preparation of daunomycin and adriamycin conjugates with dextran and mouse tumor Abs. They showed selective *in vitro* drug activity but little *in vivo* benefit (Ref. 8). However, their work did disclose a beneficial aspect to dextran binding of drugs; reduced toxicity and a higher therapeutic index. Ghose too has used dextran (periodate oxidized to aldehyde-dextran) as a carrier for Ab conjugates. He has reported that these conjugates lose activity if the unsaturated amine-aldehyde Schiff base coupling is reduced with borohydride; a practive often employed to eliminate the possibility of hydrolytic cleavage of the C=N bond. Rowland, Arnon and Hurwitz, and Ghose have recently reviewed various aspects of Ab conjugate drug targeting (Refs. 9, 10).

The advent of hybridoma techniques has spurred a tremendous increase in research on Ab targeting. Dozens of papers on tumor-specific monoclonal Ab localization and drug conjugates have already appeared in the literature (Ref. 11). Thorpe et al. have recently reviewed studies on monoclonal Ab conjugates with special emphasis on targeting and modifying abrin, ricin and diphtheria toxins (Ref. 12). These toxins possess a carbohydrate binding B-chain linked by disulfide groups to a toxic A-chain. B-chain binding to the cell surface may cap endocytic receptors and facilitate cell uptake of the A-chain. This is pertinent to the work reported here on drug-Con A conjugates. The lectin conjugates may be considered synthetic analogs of the toxins wherein the cytotoxic drug ligand performs the A-chain function and the lectin moeity binds to cell membrane receptors.

Although the emphasis of this paper is on our tissue-binding polymer drug studies, brief mention should be made of the line 10 hepatoma in strain 2 guinea pigs as an *in vivo* model for Ab targeting. We have used this animal model in both our Ab and i.t. injection studies. Magee et al. (Ref. 13) has recently reported *in vivo* localization of fractionated sheep Abs to this hepatoma in primary lesions as well as in metastases. Rather than raise antisera, we have chosen a route to autologous Abs isolated from the tumor-bearing guinea pig ascites fluid. Isolation of tumor-specific IgG and IgM fractions using protein A chromatography and the preparation of drug conjugates is in progress and will be reported elsewhere.

Although there have been many reports of success for tumor specific Ab localization *in vitro*, targeting with significant therapeutic benefit has been difficult to achieve *in vivo*. Important problems include:

(1) circulatory antigen and antigen-Ab complexes
(2) metabolic/biochemical changes in conjugates with loss of activity
(3) transport kinetics to tumor tissue vs. competitive binding and metabolism
(4) changing and cross-reactive antigenicity
(5) masking or interiorization of tumor cell specific antigens

To deal with such problems, it may be necessary to explore a number of new approaches (perhaps in combination) involving:

(1) complexing or removing (e.g. in *ex-vivo* shunt chromatography) circulatory antigen and immune complexes

(2) use of $(Fab')_2$ portion of IgG to avoid Fc complement binding and reduce molecular size

(3) use of Abs to tumor antigen associated Abs

(4) therapy with both i.t. and i.v. injections of Ab conjugates

(5) surgical or radiation reduction of primary lesion tumor burden coupled with systemic administration of Ab conjugate for elimination of metastasis.

Hormones, Lectins and Polycations as Tumor Affinity Ligands

These ligand types have shown some degree of specificity in provoking antitumor activity. Varga has evaluated the hormone, melanotropin, as a conjugate with daunomycin and shown good *in vitro* activity but *in vivo* results were disappointing (Ref. 10). Polylysine conjugates with methotrexate exhibit some tumor specificity and selectivity among different tumor lines (e.g. active against Ehrlich ascites but inactive against L1210) (Ref. 10, L. Arnold et al.). Tumor cell affinity and specificity may be related to electrostatic association and facilitated cell membrane transport by the polycation (Ref. 14). This is shown schematically for a poly-L-lysine conjugate with a carboxylated adriamycin derivative prepared in our laboratory (Fig. 1).

Lectins such as concanavalin A (Con A), abrin and ricin also give evidence of tumor specific association presumably by carbohydrate binding, receptor capping, preferential endocytosis and by agglutinating properties. Con A plus daunomycin (Ref. 15) and Con A conjugates with methotrexate and chlorambucil have reportedly shown enhanced antitumor activity (Ref. 16). As noted in the following discussion, both specific and non-specific tumor cell affinity and uptake of lectins make them uniquely interesting for our application to i.t. therapy.

INTRA-TUMOR AFFINITY IMMUNOCHEMOTHERAPY

Studies by Klein and coworkers (Ref. 17) on the treatment of physically accessible tumors, particularly skin cancers, by topical application or i.t. injection of drugs were conducted with considerable success almost 20 years ago. Clinically, a 5-fluorouricil ointment is today used effectively to eradicate many dermal tumors. However, direct i.t. injection of cytoxic agents remains a relatively unused and untried method for drug localization. The principal reasons seem to be:

(1) necrotic/inflammatory reactions

(2) rapid diffusion of coventional drugs out of the tumor

(3) the need to treat metastases as well as primary lesions (which can be removed surgically)

(4) uncertainties regarding the provokation of metastasis by i.t. injections

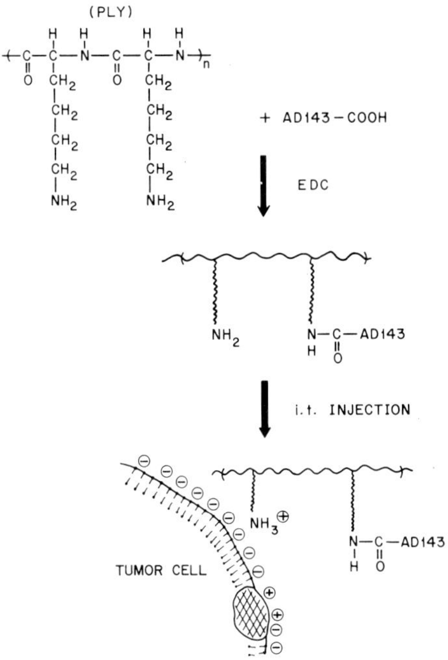

ELECTROSTATIC TISSUE BINDING OF
CATIONIC DRUG CONJUGATE (Polylysine – AD143)

Figure 1

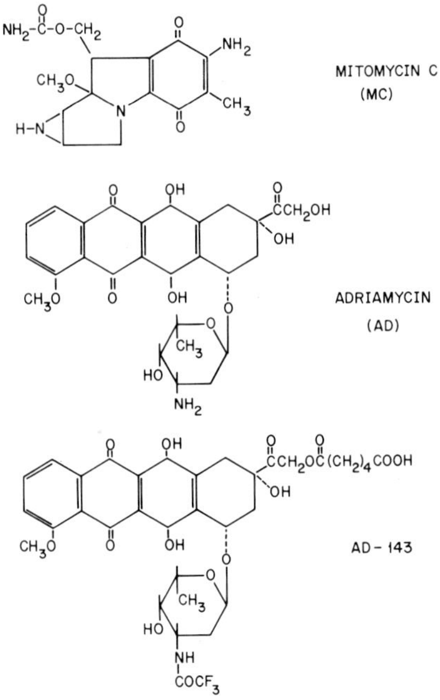

ANTITUMOR DRUG STRUCTURES

Figure 2

As noted earlier, it has been found by us (Ref. 3) and by Borsos et al. (Ref. 4) that a single i.t. injection of a cytotoxic drug in a guinea pig hepatoma achieves an immunological result. The injected tumor regresses, metastases are eliminated and the animal develops specific immune resistance to a subsequent lethal challenge with the same tumor line. Enhanced efficacy for i.t. chemotherapy followed by i.t. BCG immunotherapy has also been demonstrated (Ref. 3). Intratumor chemotherapy may therefore be regarded as a special type of immunotherapy (i.e. "immunochemotherapy"). The mechanistic reasons for this phenomenon are under investigation in several laboratories and will not be considered here.

An important potential limitation of i.t. injection may be the rapid clearance of drug from the tumor. We have worked principally with such clinically important drugs as adriamycin (AD), mitomycin C (MC) and AD-143. Structures for these are given in Fig. 2. Functional groups for polymer or protein binding are $-NH_2$ for AD, -COOH for AD143 and the aziridine ring NH for MC. In Fig. 3, the rapid rates of clearance of AD from guinea pig hepatomas and of bleomycin from rat hepatomas following i.t. injection are shown. Also noted is the longer residence time of bleomycin when given as an oil emulsion (Ref. 18) and the somewhat slower disappearance of AD (which may be related to its slightly basic properties). A major objective of our research has therefore been the development of tissue binding drug compositions of the following types to prolong high tumor drug concentrations and reduce systemic toxicity:

(A) Albumin microspheres containing AD
(B) Cross-linked dextran microspheres functionalized by carboxymethylation with AD associated as a stable salt (CMDEXS/AD)
(C) Cross-linked dextran microspheres with aldehyde functionality (by periodate oxidation) covalently coupled to AD via C=N bonds (see Fig. 4) with free -CHO groups for covalent tissue binding
(D) Soluble polyglutaraldehyde covalently coupled to AD via C=N bonds
(E) Soluble aldehyde-dextran covalently coupled to AD via C=N bonds
(F) Soluble covalent conjugates of succinylated Con A with AD or MC (AD-SCon A and MC-SCon A)

In view of their unique and especially interesting properties, only DMDEXS/AD (type B) and MC-SCon A (type F) will be discussed in more detail here.

Based upon our earlier studies (Ref. 19) showing that basic drugs such as AD readily form stoichiometric water-stable insoluble salts with anionic polypeptides (e.g. polyglutamic acid), CMDEXS/AD containing as much as 30-40 wt% AD was prepared as shown schematically in Fig. 5. Injection i.t. produces physical immobilization of the microspheres and release of AD by salt dissociation as the spheres are permeated by diffusing tissue electrolytes. This is a novel rapid release system (hours) with release kinetics controlled by microsphere crosslinking and porosity, electrolyte diffusion (in) and AD diffusion (out). This behavior can be demonstrated *in vitro* in a microsphere column release experiment (Fig. 6). The microspheres are quite stable in water but upon changing to physiological saline, AD comes off almost quantitatively in a few hours. CMDEXS/AD may be modified with surface tissue binding affinity ligands and with covalently bound drug to enhance immobilization and to prolong drug release as desired. Initial tests in i.t. therapy of guinea pig hepatomas indicate positive drug activity and reduced toxicity.

We regard the use of lectin conjugates for affinity binding of cytotoxic or immunostimulant ligands by i.t. injection as a most promising approach. There are difficult chemical attachment problems if one desires high drug loading with retention of good conjugate solubility. One successful approach is shown schematically in Fig. 7. Starting with the tetrameric Con A, succinylation produced a dimeric succinyl Con A (SCon A) with derivatized amino groups.

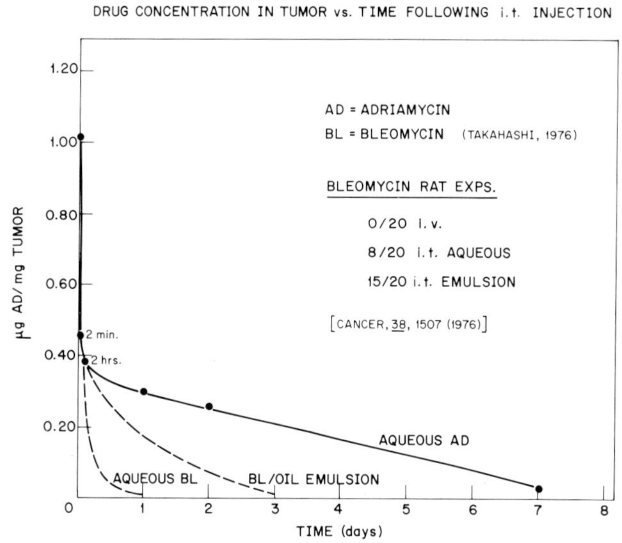

Figure 3

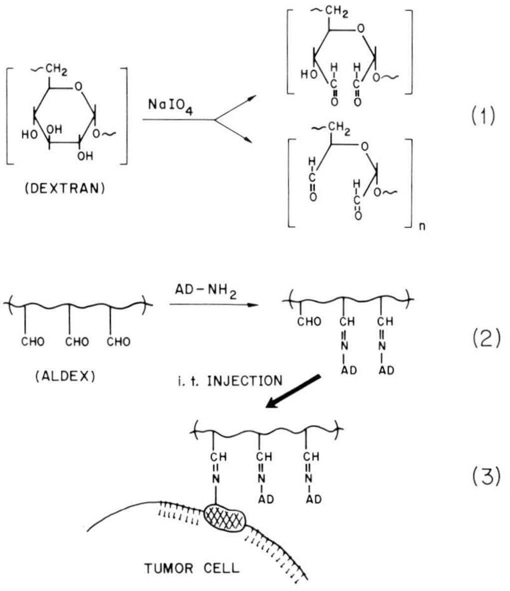

COVALENT TISSUE BINDING OF
ALDEX-DRUG CONJUGATE VIA IMINE COUPLING

Figure 4

Carbodiimide coupling with MC (or AD) then yielded soluble conjugates which retain carbohydrate affinity binding properties. This is shown for MC-SCon A on a Sephadex G-100 column (Fig. 8). The conjugate binds to the dextran until eluted with 0.1M glucose. Tumor tissue binding at the site of injection is shown schematically in Fig. 9. Mechanistically, several processes may explain the favorable i.t. therapy activity of these conjugates:

(1) Cell surface binding via lectin receptors
(2) Antimetabolic and cytotoxic activity of intact conjugate
(3) Capping of endocytic receptors favoring cell uptake and lysosomal cleavage
(4) Release of free MC locally

Fig. 10 shows evidence for *in vivo* tissue binding of MC-SCon A. MC concentrations in rabbit blood were monitored vs. time following i.v. and i.m. injections of free drug. High initial blood levels diminished rapidly as indicated. However, i.m. MC-SCon A remained fixed at the site of injection with no significant blood levels even after 4 hours.

Drug activity was demonstrated *in vitro* for MC-SCon A using a ^{3}H-thymidine uptake assay in EL4 culture (Fig. 11). Results are interesting in that the conjugate demonstrated an order of magnitude greater cytotoxicity than free MC. Lectin conjugation therefore significantly enhances *in vitro* antitumor activity. Initial experiments in the metastatic guinea pig hepatoma have also been very promising. Injections into 6 day tumors ($\sim$12 mm diam.) have produced $\sim$50% cures and 100% prolonged survival (2X untreated survival time) for dose levels of 1mg MC in 8%MC-SCon A per animal. This is a dose level at which free MC also exhibits antitumor activity but with appreciable systemic toxicity. Drug administration systemically (i.v.) is completely ineffective. Also interesting are i.t. results with Con A and SCon A in control experiments. SCon A showed no i.t. activity up to 24mg/animal whereas Con A did exhibit some activity; $\sim$25% prolonged survival and $\sim$10% cures. This difference in activity may be ascribed to the dimeric, more anionic and non-agglutinating properties of SCon A as compared with Con A. Further studies using various tissue binding lectins and covalent binding ligands with AD and MC are in progress for enhancing i.t. immunochemotherapy and will be reported in detail in the near future.

ACKNOWLEDGEMENTS

This work was supported in part by the State of Florida Biomedical Engineering Center and by National Institutes of Health Biomedical Research Grants. We are indebted to Dr. E. Ribi for continued encouragement, to Dr. C. A. McLaughlin for helpful discussions, to Dr. S. Shimizu for the ^{3}H-thymidine assays, and to

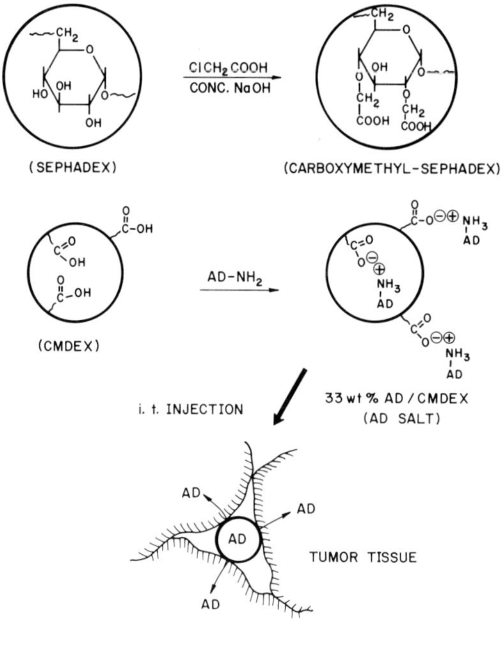

Figure 5

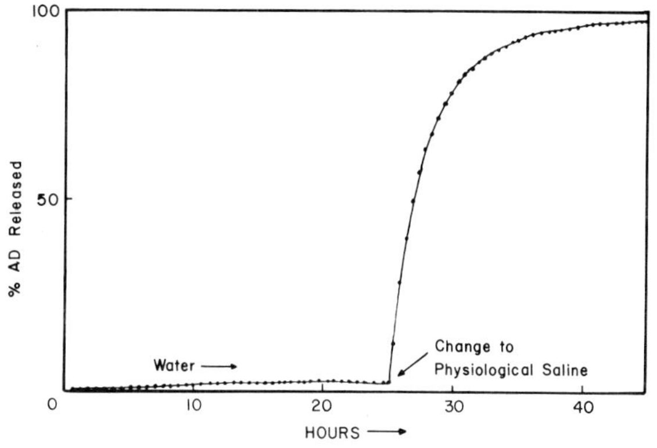

AD Release from 33% AD/CM-Sephadex (flow rate = 0.077ml/min)

Figure 6

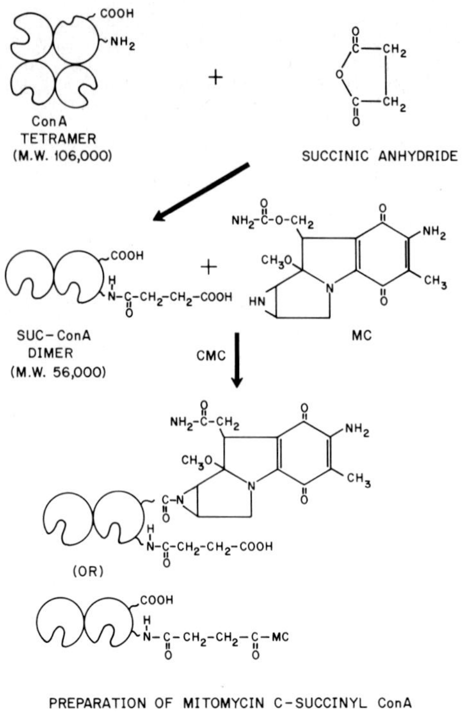

Figure 7

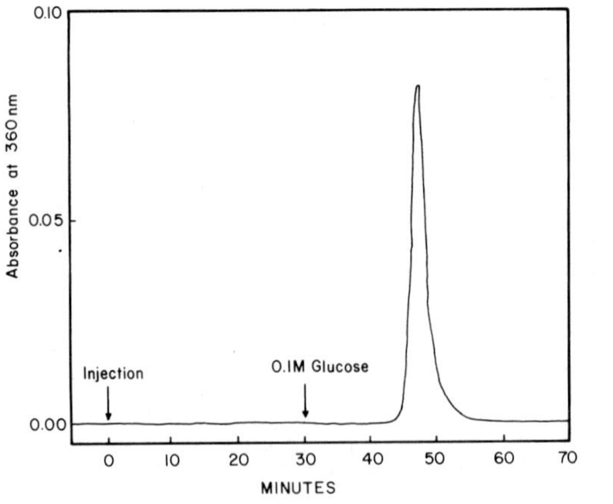

Affinity Chromatography of MC-SConA (MC=7.9 wt%) on Sephadex G-100.

Figure 8

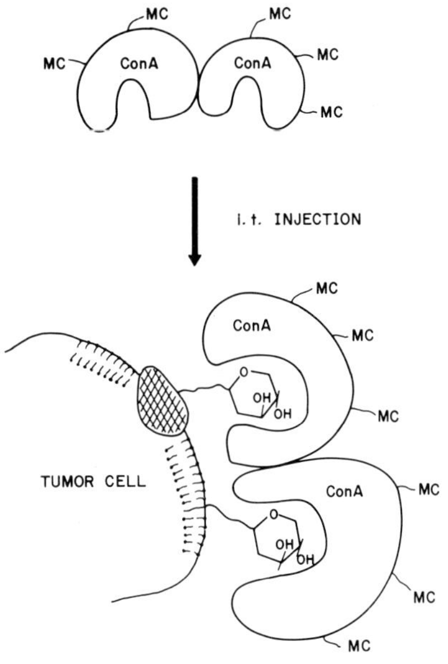

CELL BINDING OF MC-SConA CONJUGATE

Figure 9

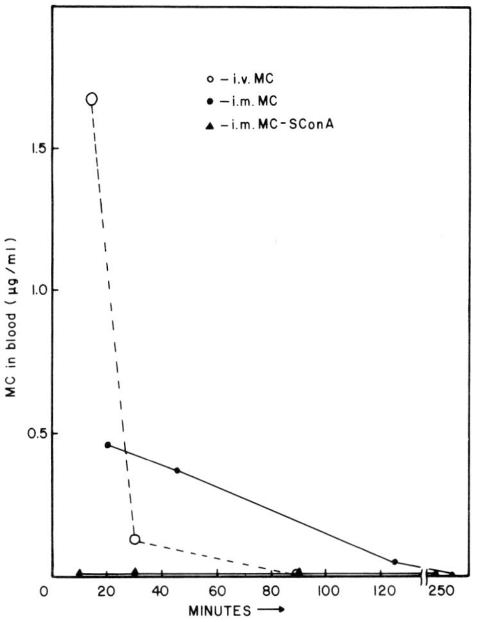

SYSTEMIC MC CONCENTRATION vs. TIME IN RABBITS.

Figure 10

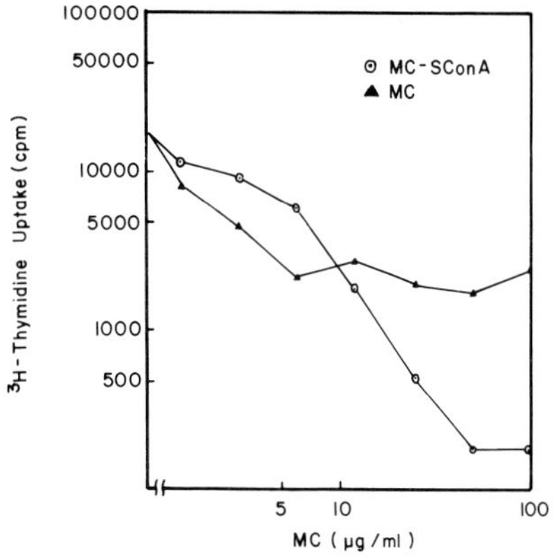

^{3}H-THYMIDINE Uptake Dose Response in EL4 Culture.

Figure 11

M. Smith for great help with this manuscript. Dr. Cantrell was affiliated with the NIH-Rocky Mountain Laboratory and is now at Ribi Immunochem Research (Hamilton, Montana), Dr. Levy is at the Weizmann Institute (Israel) and R. Terry is now at the Southern Research Institute (Birmingham, Alabama). We also wish to express our gratitude to Dr. Arcamone of Farmitalia (Italy) and Dr. Bradner of Bristol Laboratories for generous samples of AD and MC and to Dr. Israel of the Farber Cancer Institute at Harvard for collaboration on functionalized AD and the preparation of AD-143.

REFERENCES

1. P. Ehrlich, PhD Dissertation, University of Leipzig, June 1878.
2. E. P. Goldberg, in G. Donaruma and O. Vogl (Eds.), Polymeric Drugs, Academic Press, N.Y., 1978, pp. 239-261.
3. C. A. McLaughlin, J. L. Cantrell, E. Ribi and E. P. Goldberg, Cancer Res., 38 (1978) 1311-1316.
4. T. Borsos, R. C. Bast Jr., S. H. Ohanian, M. Segerling, B. Zbar and H. J. Rapp, Annals N.Y. Acad. Sci., 276 (1976) 565-572.
5. T. Ghose et al., British Med. J., 3 (1972) 495-499.
6. D. A. L. Davies and G. J. O'Neill, Br. J. Cancer, 28 Suppl. I (1973) 285-298.
7. G. Rowland, Europ. J. Cancer, 13 (1977) 593-596.
8. E. Hurwitz, et al., Cancer Res., 35 (1975) 1175-1181.
9. T. Ghose, et al., J. Natl. Cancer Inst., 61 (1978) 657-676.
10. Chapters in E. P. Goldberg (Ed.), Targeted Drugs, Wiley, N.Y., in press.
11. See (a) Proceedings of the 5th Congress of Immunology, Paris, July, 1980 and (b) Federation Proceedings, 40, No. 3 (1981).
12. P. E. Thorpe, et al., in Faber and McMichael (Eds.) Monoclonal Antibodies in Clinical Medicine, Academic Press, N.Y., in press.
13. W. E. Magee, et al., Federation Proceedings, 40, No. 3 (1981) p. 1042, Abstr. No. 4603.
14. W-C. Shen and H. J.-P. Ryser, Molecular Pharmacol., 16 (1979) 614-622.
15. S. G. Bradley, et al., in T. K. Chowdhury and A. K. Weiss (Eds.) Advances in Experimental Medicine and Biology, Plenum Press, N.Y., 1975, pp. 291-307.
16. J-Y Lin, et al., JNCI, 66 (1981) 523-528.
17. E. Klein, N.Y. State J. Med., 68 (1968) 877.
18. T. Takahash, et al., Cancer, 38 (1976) 1507-1514.
19. E. P. Goldberg, et al., Am. Chem. Soc. Preprints, Div. Org. Coatings Plastics Chem., 44 (1981) 132-136.

CHAPTER V

APPLICATIONS

•ORGANIC DYES
•DYE-LIGANDS

INTERNATIONAL SYMPOSIUM
ON AFFINITY CHROMATOGRAPHY
AND RELATED TECHNIQUES

T.C.J. Gribnau, J. Visser and R.J.F. Nivard (Editors),
Affinity Chromatography and Related Techniques

SOME PREPARATIVE AND ANALYTICAL APPLICATIONS OF TRIAZINE DYES

C.R. LOWE, Y.D. CLONIS, M.J. GOLDFINCH
Department of Biochemistry, University of Southampton, Southampton, SO9 3TU (U.K.)
and
D.A.P. SMALL and A. ATKINSON
Microbial Technology Laboratory, PHLS Centre for Applied Microbiology and Research, Porton Down, Wiltshire, SP4 OJG (U.K.)

ABSTRACT

The unique reactivity, spectral properties and ability of triazine dyes to imitate the binding of natural heterocycles to biological macromolecules has prompted their application in biotechnology. Thus, the chemically reactive triazine dyes have been exploited as active site directed affinity labels for nucleotide-dependent enzymes, as pseudo-affinity ligands for the purification of several enzymes by affinity chromatography, as ligands for high performance liquid affinity chromatography (HPLAC) and as potential ligands in a new type of direct electrochemical sensor, the affinity electrode. It is anticipated that the low cost and ready availability of these dyes will make their large scale application in affinity chromatography an attractive proposition.

INTRODUCTION

The introduction of the Procion range of chemically reactive dyes by ICI completely revolutionised large scale printing and dyeing technology in the 1950's. The commercially available dyes are derived mainly from azo, anthraquinone and phthalocyanine chromophores attached to reactive dichlorotriazinyl (Procion MX range) or monochlorotriazinyl (Procion H, HE or P range) functional groups by -NH-bridges (ref. 1). Anthraquinone and phthalocyanine chromophores produce bright blue and turquoise shades respectively, whilst mixed chromophores of the anthraquinon-stilbene, anthraquinone-azo and phthalocyanine-azo classes produce green shades. The majority of the yellow, orange and red dyes are derived from the azo class, whilst rubine, violet, navy, brown and black dyes are generally metal complexes of O,O'-dihydroxyazo or O-carboxy-O'-hydroxyazo dyes. The structures of typical dyes from each of the main classes of chromophore are shown in Fig. 1.

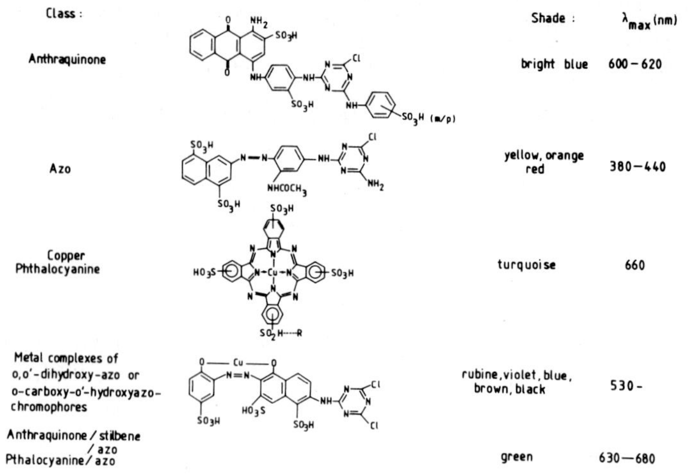

Class :
Shade :
λ_{max} (nm)
Anthraquinone
bright blue
600 – 620
Azo
yellow, orange red
380 – 440
Copper Phthalocyanine
turquoise
660
Metal complexes of o,o'-dihydroxy-azo or o-carboxy-o'-hydroxyazo-chromophores
rubine, violet, blue, brown, black
530 -
Anthraquinone / stilbene / azo
Pthalocyanine / azo
green
630 — 680

Fig. 1. The main types of triazine dye chromophore.

The triazine dyes have a number of characteristics which makes their application in biotechnology an attractive proposition (ref. 1): (i) they are readily available at low cost in large quantities and with a variety of chemically distinct chromophores; (ii) the dyes display characteristic spectral properties wit a wide range of λ_{max} values covering the entire spectral range and have high molar extinction coefficients (up to 60,000 $l.mol^{-1}.cm^{-1}$); (iii) the triazine group is reactive towards nucleophiles such as the hydroxyls of polysaccharides or metal oxides or the side chain functional groups of proteins, and (iv) the dyes exhibit a remarkable ability to bind biospecifically to a wide range of proteins and enzymes, particularly those binding nucleotides, coenzymes and other heterocyclic molecules (refs. 2-5). A large body of information has been accumulated over the last decade or so on the interaction of proteins with the anthraquinone dye Cibacron Blue F3G-A (Procion Blue H-B). This dye selectively interacts with pyridine-nucleotide dependent dehydrogenases, kinases, CoA-dependent enzymes, hydrolases, polynucleotide-dependent enzymes, restriction endonucleases, synthetases, and a number of blood proteins including serum albumin, clotting factors, lipoproteins, complement factors and interferon (ref. 1). Not surprisingly, agarose-immobilised Cibacron Blue F3G-A and, more recently, Procion Red HE-3B and other dyes, have been extensively exploited in the purification of their complementary proteins by affinity chromatography (refs. 2,5). The considerable interest in the application of triazine dyes in preparative biochemistry has prompted the initiation of a number of studies to establish the molecular basis for these selective interactions. As a result of these studies, it has been suggested that the polysulphonated aromatic chromophores of the triazine dyes mimic the naturally occurring biological heterocycles such as the nucleotide mono-, di- and tri-phosphates, NAD^+, $NADP^+$, flavins, coenzyme A and folic acid (refs. 1,6,7). Indeed, x-ray crystallographic studies on the binding of Cibacron Blue F3G-A to horse liver alcohol dehydrogenase have highlighted remarkable similarities in chromophore and coenzyme binding (ref. 8). This report examines potential applications of triazine dyes based on these highly selective interactions.

RESULTS AND DISCUSSION

Triazine dyes: A new class of affinity labels for nucleotide-dependent enzymes

Since triazine dyes are highly coloured, are chemically reactive and mimic the binding of natural biological heterocycles, an obvious application which exploits all these characteristics is as active site directed irreversible affinity labels. The reaction between an active site directed reactive dye (D) and an enzyme (E) may be formulated by (ref. 9):

$$E + D \underset{k_2}{\overset{k_1}{\rightleftharpoons}} E.D \xrightarrow{k_3} ED$$

where E.D is the enzyme-dye Michaelis complex, ED is the irreversibly inhibited enzyme and k_3 is the rate limiting step. A steady state treatment of the process yields the equation:

$$\frac{1}{k_{obs}} = \frac{1}{k_3} + \frac{K_D}{k_3} \cdot \frac{1}{[D]}$$

where k_{obs} is the observed rate of enzyme inactivation for a given concentration of dye, D, k_3 is the maximal rate of inactivation (min^{-1}) and K_D is the dissociation constant (k_2/k_1) of the enzyme-dye complex.

Pig heart lactate dehydrogenase may be inactivated by a number of triazine dyes at pH 8.5 and 35°C. Of the dyes tested (Table 1) significant inactivation rates were only obtained with the more reactive dichlorotriazinyl dyes of MX designation (ref. 9). In contrast, monochlorotriazinyl dyes of H, HE or P designation do not significantly inactivate lactate dehydrogenase even at dye concentrations as high as 200 μM. The inactivation of lactate dehydrogenase by dichlorotriazinyl dyes at pH 8.5 and 35°C follows approximately pseudo-first order kinetics at low dye concentrations. The double reciprocal plot of $^1/k_{obs}$ versus $^1/[D]$ for a number of dyes yields a straight line with a positive ordinate intercept and is indicative of saturation kinetics for the inactivation process. The maximum rate of inactivation (k_3) determined from the ordinate intercept for the dichloro-triazinyl (MX) dyes listed in Table 1 is remarkably constant at 0.09-0.22 min^{-1}

despite a wide divergence in K_D values for the same dyes. The dissociation constants for the H, HE and P type dyes were calculated from data inserted into the equation:

$$\frac{1}{k_{obs}} = \frac{1}{k_3} + \frac{K_D}{k_3} \cdot \frac{1}{[D]} \left(1 + \frac{[I]}{K_I}\right)$$

where I is a competitive inhibitor and K_I its dissociation constant and obtained by competition between Procion Blue MX-R and the monochlorotriazinyl dyes. Table 1 shows that the dissociation constants for triazine dyes encompass a 154-fold range from 0.8-122.9 μM and are lowest for blue-green dyes and highest for yellow-orange dyes.

TABLE 1

Maximal rates of inactivation (k_3) and dissociation constants (K_D) of Procion dyes for pig heart lactate dehydrogenase (ref. 9).

Procion Dye	k_3 (min^{-1})	K_D (μM)
Blue HE-RD	-	0.8
Green HE-4BD	-	1.8
Green H-4G	-	2.5
Blue H-B	-	5.6
Brown H-2G	-	6.1
Blue MX-R	0.12	7.2
Yellow H-5G	-	7.3
Red H-3B	-	10.1
Red HE-3B	-	10.1
Turquoise H-A	-	13.4
Violet H-3R	-	15.6
Red P-3BN	-	20.0
Rubine MX-B	0.22	31.9
Red MX-5B	0.09	32.2
Yellow H-A	-	36.0
Yellow MX-R	0.16	72.8
Scarlet MX-G	0.09	88.9
Orange MX-G	0.12	93.4
Yellow MX-8G	0.13	122.9

Yeast hexokinase is inactivated by both mono- and dichlorotriazinyl dyes at pH 8.5 and 33°C (ref. 4). The enzyme is speedily inactivated by 100 μM Procion Green H-4G, Blue H-B, Turquoise H-7G and Turquoise H-A, is more slowly inactivated by Procion Brown H-2G, Green HE-4BD, Red HE-3B and Yellow H-5G and is not inactivated at all by Procion Yellow H-A. Hexokinase quantitatively inhibited by Procion Green H-4G and lactate dehydrogenase by Procion Blue MX-R, contain approximately 1 mol dye/mol subunit. In each case, the inhibition is irreversible and cannot be recovered on incubation with high concentrations of substrates or coenzymes.

Both lactate dehydrogenase and hexokinase are progressively protected from inactivation with triazine dyes by increasing concentrations of their appropriate coenzymes. Partial protection from inactivation of lactate dehydrogenase by Procion Blue MX-R is observed with 1 mM AMP and 1 mM NAD^+ whilst complete protection is observed with 1 mM NADH (ref. 9). Fig. 2 demonstrates that increasing concentrations

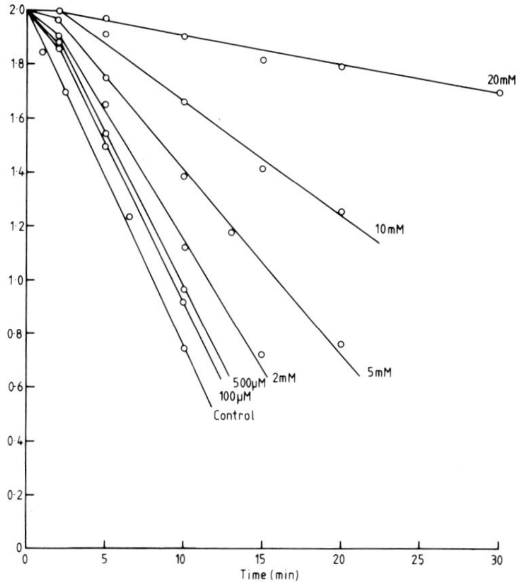

Fig. 2. The effect of increasing concentrations of ATP on the inactivation of yeast hexokinase by Procion Green H-4G.

of ATP progressively protect yeast hexokinase from inactivation by Procion Green H-4G. Similarly, ADP and AMP afford partial protection whilst sugar substrates such as D-glucose, D-mannose and D-fructose protect and non-substrates such as D-arabinose and D-galactose do not (ref. 4). These observations substantiate the contention that triazine dyes are active site-directed reagents.

Interestingly, 10 mM Mg^{2+} also reduces the rate of inactivation of yeast hexokinase by 100 μM Procion Green H-4G by approximately 80%. Of the dyes tested, the protective effect of Mg^{2+} appears to be unique to the inactivation of yeast hexokinase by Procion Green H-4G. The effect of Mg^{2+} is to increase the affinity of the dye for the enzyme. The enzyme-dye dissociation constant is reduced from 199.0 μM to 41.6 μM in the presence of 10 mM Mg^{2+}. Of a number of divalent metal ions tested, only Mg^{2+} and to a lesser extent Sr^{2+} and Ca^{2+} protected yeast hexokinase against inactivation by Procion Green H-4G. No significant protection against inactivation was observed with the transition metal ions, Mn^{2+}, Co^{2+}, Cu^{2+} or Zn^{2+}. Presumably, the binding of Mg^{2+}, and to a lesser extent Ca^{2+} and Sr^{2+}, to Procion Green H-4G induces a conformation which favours binding of the dye whilst at the same time mo the triazine ring away from the reactive nucleophile. In yeast hexokinase, the apparently essential thiol of cysteine-243 is the most likely candidate for alkylation by the reactive dye, since alkylation of this thiol causes complete inactivation of the enzyme with protection being afforded by substrates of both nucleotide and hexose classes. In lactate dehydrogenase, a potential candidate fo the nucleophile might be the ε-amino group of lysine-58 which lies close to the pyrophosphate bridge of NAD^+ and the $O^{3'}$ of the adenosine ribose moiety. In both cases, it seems likely that the topography of the coenzyme binding site allows som latitude in binding polysulphonated aromatic chromophores.

Affinity chromatography on immobilised triazine dyes

The exploitation of triazine dyes as active site directed affinity labels for several nucleotide-dependent enzymes (refs. 4,9) gives some foundation to a more rational approach to the application of such dyes in affinity chromatography. Thu Mg^{2+} promotes the binding of yeast hexokinase to agarose-immobilised Procion Green H-4G but not to any of the other dyes tested (ref. 4). Elution of bound hexokinas could be effected by omission of Mg^{2+} from the column irrigants or by inclusion of MgATP or D-glucose but not by non-substrates such as D-galactose. The Mg^{2+}-promoted adsorption of hexokinase to immobilised Procion Green H-4G and subsequent elution with 10 mM MgATP may be exploited to purify yeast hexokinase nearly 7-fold from a crude yeast extract.

Similarly, a systematic investigation into the interaction of several triazine dyes with two enzymes from purine metabolism, IMP dehydrogenase (EC 1.2.1.14) and adenylosuccinate synthetase (EC 6.3.4.4) has been conducted (ref. 3). Evidence from kinetic inhibition studies, enzyme inactivation with specific nucleotide affinity labels and specific elution techniques from agarose-immobilised dyes indicate that triazine dyes such as Procion Blue H-B, Red HE-3B and Red H-3B are a to differentiate between the nucleotide binding sites of these multi-nucleotide dependent enzymes. It appears that Cibacron Blue F3G-A (Procion Blue H-B) may be better analogue of monophosphates such as IMP than di- or tri-phosphate nucleotide such as NAD^+ or ATP. Thus, IMP is a particularly effective eluant for E.coli IMP dehydrogenase from immobilised Cibacron Blue F3G-A. On the other hand, Procion Re HE-3B appears to bind to the NAD^+-binding site of IMP dehydrogenase; in this case, AMP and NAD^+ are effective eluants of the enzyme from immobilised Procion Red HE-3 columns, whilst IMP, GMP and XMP are completely ineffectual (ref. 3). This information has been exploited to design specific purification protocols for these enzymes by affinity chromatography.

High performance liquid affinity chromatography (HPLAC)

The two techniques of high performance liquid chromatography (HPLC) and affinity chromatography (AC) have been combined to yield a new approach termed high performance liquid affinity chromatography (HPLAC) which includes the inherent speed and resolving power of HPLC with the biological specificity of affinity chromatography (refs. 10-12). The almost ubiquitous applicability of immobilised triazine dyes in affinity chromatography suggested a study of their value as ligands in HPLAC. Two procedures have been used to couple triazine dyes to microparticulate silica: in the first, which is more applicable to monochlorotriazinyl dyes, 6-aminohexyl substituted dye is coupled to microparticulate silica epoxysilylated with γ-glycidoxypropyl trimethoxysilane to generate ligand substitutions in the range 3-6 μmol dye/g dry weight silica (refs. 11,12). Alternatively, dye adsorbents may be prepared by coupling the reactive dyes, particularly dichlorotriazinyl dyes, directly to glycol-silylated silica (ref. 12). However, in the latter approach ligand substitutions in the range 1.5-3.0 μmol dye/g dry weight silica are more commonplace. Fig. 3 shows the structure of a typical HPLAC adsorbent comprising 6-aminohexyl-

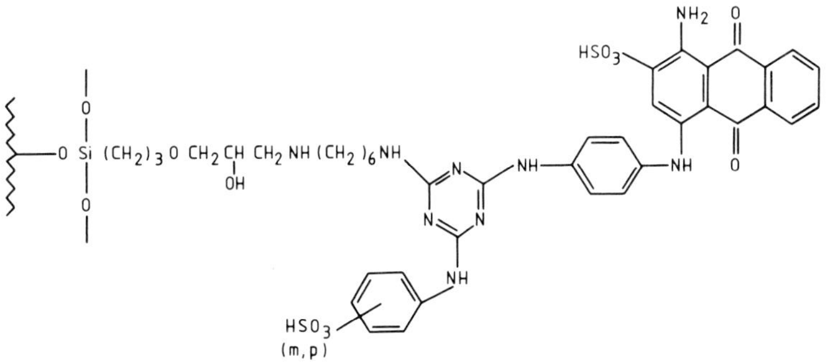

Fig. 3. The structure of the Cibacron Blue F3G-A adsorbent.

Cibacron Blue F3G-A immobilised to microparticulate (5 μm) silica. The Cibacron Blue F3G-A adsorbent equilibrated with 0.1M potassium phosphate pH 7.5 quantitatively adsorbs pig heart and rabbit muscle lactate dehydrogenase, pig heart malate dehydrogenase and horse liver alcohol dehydrogenase and retards L. mesenteroides glucose-6-phosphate dehydrogenase (ref. 11). Synthetic mixtures of serum albumin, pig heart lactate dehydrogenase and horse liver alcohol dehydrogenase may be resolved by sequential elution with 1 mM NAD^+, 1 mM NAD^+/0.1 mM pyrazole and 1 mM NAD^+/0.1M pyruvate with continuous on-line analysis of protein and enzyme activity (ref. 11).

The two principle isoenzymes of lactate dehydrogenase may also be resolved in 10 mi or less by NADH gradient elution from the Cibacron Blue F3G-A adsorbent. The general versatility of this HPLAC adsorbent is also demonstrated with the resolutic by ternary complex formation of yeast hexokinase and 3-phosphoglycerate kinase from crude yeast extracts and ribonuclease A from crude bovine pancreatic extracts.

One problem experienced with the Cibacron Blue F3G-A adsorbent is the relatively low recoveries of enzyme activity (~25%) and the fact that adsorption could only be promoted at ionic strengths $\gtrsim$0.1M phosphate. Such features are highly indicative that non-specific hydrophobic interactions play an important role in the adsorptior process. Lactate dehydrogenase can however be quantitatively desorbed from an adsorbent comprising Procion Blue MX-R linked to glycol-silylated silica with a 200 μl pulse of 200 mM KCl or 0.1 mM NAD^+/0.1M pyruvate. It seems therefore, that in certain cases, dyes other than Cibacron Blue F3G-A might prove beneficial. For example, the Mg^{2+}-promoted adsorption of yeast hexokinase to immobilised Procion Green H-4G may be exploited in HPLAC. Exclusion of Mg^{2+} from the irrigating buffer results in quantitative elution of the applied sample of hexokinase in the void volume of the column (ref. 12). Inclusion of 6 mM Mg^{2+} in the column irrigants results in quantitative adsorption of the enzyme with prompt elution bei effected with a pulse of 10 mM ATP/10 mM D-glucose. Yeast hexokinase adsorbed to silica-immobilised Procion Green H-4G in the presence of Mg^{2+} may be eluted with Mg^{2+}-free or EDTA-containing buffers. Interestingly, the binding of hexokinase to silica-immobilised Cibacron Blue F3G-A is independent of Mg^{2+} ion concentration.

The binding of enzymes to HPLAC adsorbents may also be promoted in the presence of transition metal ions. For example, carboxypeptidase G2 binding to silica-immobilised Procion Red H-8BN was increased from 40% to over 90% by including 0.2 Zn^{2+} in the enzyme sample (ref. 12). Subsequent elution of the bound enzyme with 0.5M KCl results in a 5-fold increase in specific activity. Similarly, calf intestinal alkaline phosphatase may be bound to an adsorbent comprising 6-aminohexyl-Procion Yellow H-A immobilised to epoxysilylated silica in the presence of 2 mM Zn^{2+}. Elution of the bound enzyme could be effected with approximately 80% recovery with a pulse of 10 mM EDTA. Substituting other metal ions such as Mn^{2+}, Co^{2+}, Ni^{2+}, Mg^{2+}, Ca^{2+}, K^+ and Na^+ for Zn^{2+} in the column irrigants had no apparent effect on the binding of alkaline phosphatase to silica-immobilised Proci Yellow H-A.

Analytical applications: the affinity electrode

The affinity electrode is a novel type of direct electrode comprising a biospecific ligand attached to an oxidised metal electrode (ref. 13). The concept has been realised by noting potential changes associated with the binding of proteins to the monochlorotriazinyl dyestuff, Cibacron Blue F3G-A, covalently attached to oxidised titanium electrodes. The electrode responds linearly to added human serum albumin within the range 0-15 μg/ml but shows a progressive saturation effect at >15 μg/ml albumin. The electrode is regenerated by immersion in 8M urea followed by washing to remove excess urea and can be used continuously over a period of several months without apparent loss in response.

The Cibacron Blue F3G-A electrode also responds to yeast alcohol dehydrogenase with a linear calibration curve in the range 0-5 μg/ml enzyme (Fig. 4). The biospecificity of the interaction between yeast alcohol dehydrogenase and the Cibacron Blue F3G-A electrode is demonstrated by the addition of competitive inhibitors to the solution bathing the electrodes. For example, 2 mM NAD^+ lowers the

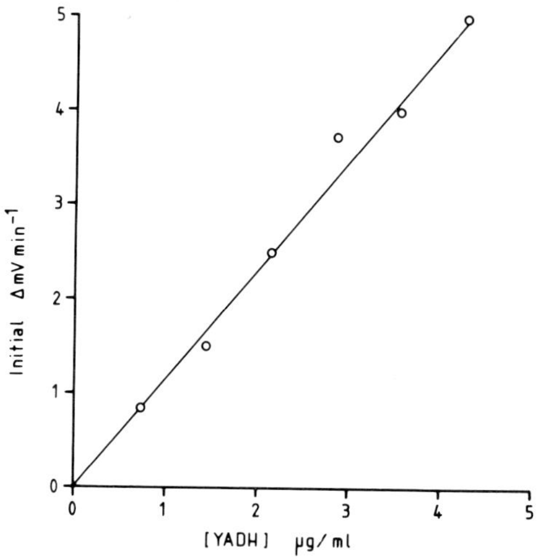

Fig. 4. Calibration curve for yeast alcohol dehydrogenase (YADH) for the Cibacron Blue F3G-A affinity electrode.

response by nearly 30% whilst formation of the ternary complex, enzyme-NAD^+-hydroxylamine, reduces it by over 60% and in the presence of 2 mM NADH the response is entirely abolished. Furthermore, the electrode response is progressively diminished as the concentration of competing NAD^+ is increased to 7.5 mM. Other dye electrodes comprising Procion Green HE-4BD, Red HE-3B and Yellow H-A immobilised to titanium produced different responses to albumin and yeast alcohol dehydrogenase. The sensitivity and selectivity of the green electrode for albumin was considerably reduced compared to the blue electrode. Titanium-immobilised Procion Red HE-3B proved the best electrode material for the assay of yeast alcohol.

CONCLUSIONS

The three characteristic features of triazine dyes, namely their ability to bind to biological macromolecules, their spectral properties and their chemical reactivity with nucleophiles ensures a wide range of applications for these dyes. Protein purification by affinity chromatography is probably the broadest and most well documented application to date (refs. 1,5). The low cost, ready availability and ease of coupling to matrix materials represents a major advantage of triazine dyes over other adsorbents. Not surprisingly, triazine dye adsorbents are finding application in the large scale purification of enzymes of diagnostic or therapeutic importance (ref. 1). A logical extension of such studies is the application of a number of triazine dyes as ligands in high performance liquid affinity chromatograp (refs. 11,12) in order to improve resolution and expedite the separation process.

The characteristic spectral properties and the functional reactivity of the triazine group have been exploited in the investigation of active site topography in their use as specific active site directed irreversible affinity labels for nucleotide-dependent enzymes (refs. 4,9). Finally, the ability of triazine dyes to selectively bind proteins may be exploited analytically in the design of specific electrochemical sensors (ref. 13).

REFERENCES

1 C.R. Lowe, D.A.P. Small and A. Atkinson, Int. J. Biochem., 13 (1980) 33-40.
2 C.R. Lowe, M. Hans, N. Spibey and W.T. Drabble, Analyt. Biochem., 104 (1980) 23-
3 Y.D. Clonis and C.R. Lowe, Biochim. Biophys. Acta, 659 (1981) 86-98.
4 Y.D. Clonis, M.J. Goldfinch and C.R. Lowe, Biochem. J., 197 (1981) 203-211.
5 P.D.G. Dean and D.H. Watson, J. Chromatogr., 165 (1979) 301-319.
6 J. Baird, R. Sherwood, R.J.G. Carr and A. Atkinson, FEBS Lett., 70 (1976) 61-66.
7 R.A. Edwards and R.W. Woody, Biochemistry, 18 (1979) 5197-5204.
8 J-F. Beillmann, J-P. Samama, C.I. Brändén and H. Eklund, Eur. J. Biochem., 102 (1979) 107-110.
9 Y.D. Clonis and C.R. Lowe, Biochem. J., 191 (1980) 247-252.
10 S. Ohlson, L. Hansson, P.O. Larsson and K. Mosbach, FEBS Lett., 93 (1978) 5-9.
11 C.R. Lowe, M. Glad, P.O. Larsson, S. Ohlson, D.A.P. Small, A. Atkinson and K. Mosbach, J. Chromatogr. (1981) in press.
12 D.A.P. Small, A. Atkinson and C.R. Lowe, J. Chromatogr. (1981) in press.
13 C.R. Lowe, FEBS Lett., 106 (1979) 405-408.

T.C.J. Gribnau, J. Visser and R.J.F. Nivard (Editors),
Affinity Chromatography and Related Techniques

THE POTENTIAL OF ORGANIC DYES AS AFFINITY LIGANDS IN PROTEIN STUDIES

A. ATKINSON, J.E. McARDELL, M.D. SCAWEN, R.F. SHERWOOD, D.A.P. SMALL,
Microbial Technology Laboratory, PHLS, Centre for Applied Microbiology & Research,
Porton, Wilts. SP4 0JG

C.R. LOWE,
Department of Biochemistry, University of Southampton, Southampton.

and

C.J. BRUTON
Department of Biochemistry, Imperial College, London.

ABSTRACT

Methods for the immobilisation of triazine dyes to agarose are discussed. The use of such matrices as the sole means of accomplishing the large scale purification of the enzymes 3-hydroxybutyrate dehydrogenase and malate dehydrogenase from *R. sphaeroides* is described. Tryptophanyl- and other aminoacyl- tRNA synthetases from *B. stearothermophilus* have also been purified on triazine-dye agarose.

Kinetic data on the application of *Procion* Brown MX-5BR as a probe for the tryptophan binding site of tryptophanyl-tRNA synthetase is presented. The synthesis of a high performance liquid affinity chromatography (HPLAC) matrix composed of *Procion* Brown MX-5BR linked to microparticulate silica is described. Information on the binding and elution characteristics of tryptophanyl-tRNA synthetase on this matrix is presented.

INTRODUCTION

Organic dyes have now found wide application in both protein purification and other areas of protein research (ref. 1-2). Although the majority of the effort has concentrated on the application of the monochlorotriazinyl dye, *Cibachron* Blue F3G-A, this is only one of a large family of readily available triazine dyes. One manufacturer (ICI) alone produce over 70 different triazine *(Procion)* dyes for the textile industry. Structurally these dyes fall into 2 main groups; the monochlorotriazinyl, *Procion* H, series and the dichlorotriazinyl, *Procion* MX, series both of which are derived principally from the anthroquinone, azo or phthalocyanine chromophores. The structure of two such dyes is shown in Fig. 1

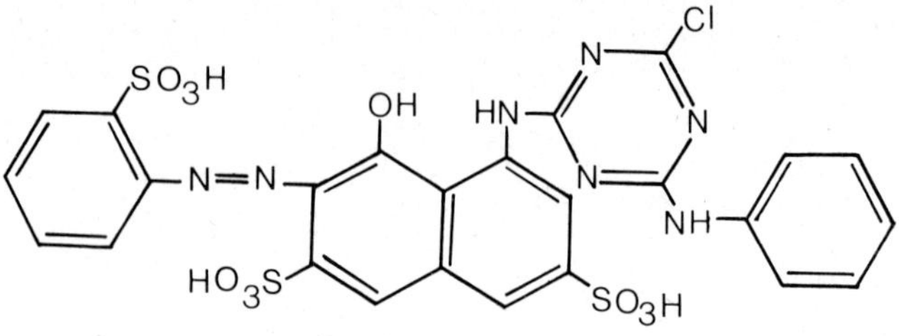

Procion Red H-3B

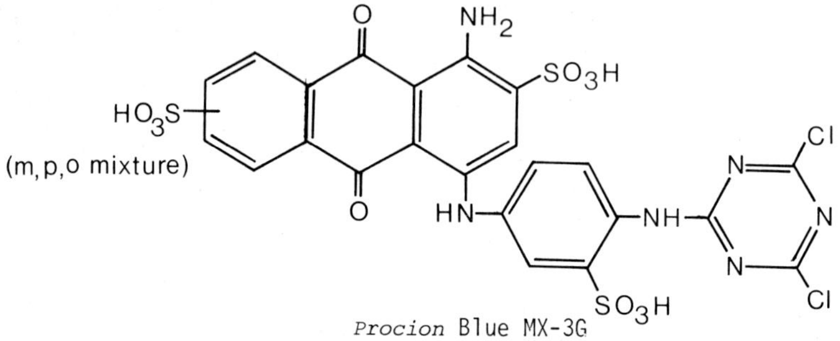

Procion Blue MX-3G

Fig. 1

Many enzymes which interact with nucleotides and other heterocyclic molecules have been shown to bind to a number of triazine dyes (ref. 3-4). Thus immobilised triazine dyes have been exploited as group specific adsorbents for the purification of a wide variety of enzymes including dehydrogenases, kinases, esterases, peptidases and nucleic acid binding proteins (ref. 5). In addition to their use in 'affinity' chromatography the triazine dyes have found other uses in protein research (ref. 1): for example as active site directed probes and affinity labels (ref. 6) and as affinity electrodes (ref. 7). Probably the major advantage of the application of triazine dyes to protein research is that the dyes are not biodegradable; that they will form a stable linkage with a matrix, such as agarose, or indeed any molecule with a free hydroxyl, amino or sulphydryl group; that the conditions required for coupling the dyes to another molecule are relatively mild and that the dyes themselves are inexpensive.

This paper discusses the parameters of the immobilisation of triazine dyes to agarose, the purification of enzymes on such matrices and the use of triazine dyes as affinity labels. In addition the paper describes the covalent attachment of a triazine dye to microparticulate porous silica and the use of this matrix in high performance liquid affinity chromatography (HPLAC).

RESULTS AND DISCUSSION

The synthesis of triazine dye agarose matrices

The direct formation of an ether link between the triazine ring of the dye and the agarose hydroxyl groups can be readily achieved (ref. 5) and is illustrated in Table 1. The salient features of this protocol are presented below:

1. The individual dyes are not homogeneous and contain as much as 50% by weight stabilisers (phosphate) and antidust agents. These can be removed by prior precipitation of the dye as the K^+ salt, with potassium acetate, from aqueous solution.
2. Increasing the dye concentration to appreciably over $4 mg.ml^{-1}$ does not significantly increase the amount of dye covalently linked to the agarose.
3. Following equilibration between the agarose and dye, NaCl is added to force the dye onto and into the matrix. NaCl also helps to disintegrate "stacked" dye molecules, which can present subsequent problems.
4. The alkaline conditions for promoting coupling of the MX-dyes is critical (0.008 to 0.014M) as higher concentrations apparently rapidly hydrolyse the second chlorine atom on the triazine ring, resulting in a lower degree of coupling.

TABLE 1

The synthesis of *Procion*-Agarose matrices		"H" Dyes	"MX" Dyes
(i)	Agarose	Sepharose 4B or 6B (20g) in H_2O (100ml)	
	add:		
(ii)	Dye	400-500mg in water (10ml)	
	after 10 min add:		
(iii)	NaCl solution	4M NaCl (10ml)	
	after 30 min add:		
(iv)	NaOH solution to give	0.1M (0.05 to 0.2M)	0.01M (0.008 to 0.014M)
(v)	Temp/Time swirl	72h at ambient	4h at ambient
	or:	48h at 30°C	2h at 30°C
(vi)	Wash	Copiously with (a) H_2O; (b) 1M NaCl, 25% ethanol; (c) H_2O; (d) 1M NaCl.	
(vii)	Store	0.1M phosphate buffer, pH 7.0	

5. The monochlorotriazinyl, H, dyes are much less reactive than the MX-dyes and require longer to attain a reasonable substitution level. The rate of couplin of the H dyes can be increased 4-fold by raising the temperature to 60°C and using a cross-linked agarose.
6. Following coupling, MX-dye conjugates should be either left at pH 8.5 for 3 days or treated with 2M NH_4Cl, pH 8.5 for 4 hours to convert the remaining chlorine residues on the triazine ring to hydroxyl or amino functions respecti
7. Matrices can be regenerated after use and washed free of protein with either high salt or 5M urea before re-equilibration.

This method of preparing triazine dye-Sepharose conjugates routinely gives dye concentrations in the region 2 to 4mg (ca. 2 to 8μmoles) ml^{-1} agarose. Higher or lower dye concentrations can be achieved by varying the conditions, or, can occur due to batch variation in the reactivity of the dyes. We have immobilised both triazine dyes and aminohexyl derivatised triazine dyes to many different matrices, including those based on cellulose, dextran, acrylamide, agarose and co-polymers of these. For all enzymes studied to date optimal enzyme binding and elution has been obtained utilising either the 4% or 6% agarose matrices or their cross-linked counterparts. It is worth noting that some dye molecules have 2 triazine rings, spacially well separated in the dye structure. Such dyes appear to increase the rigidity of the agarose beads, presumably by cross-linking within the agarose bead structure.

Occassional problems with dye leaching or dye-protein being eluted from these matrices have been found. These are usually due to non-covalently bound "stacked" dye molecules and can often be overcome by rigorous and extensive washing of the matrix. Currently, however, the method of synthesising these matrices is being re-examined with the specific aim of obtaining conditions in which over 60% of the added dye becomes covalently bound to the agarose (cf. 14% at present). Such binding has been achieved with cellulose and if achievable with agarose would make it simpler to obtain a specific dye substitution level and would also dramatically reduce the problems presented by "stacked" dye molecules.

The purification of 3-hydroxybutyrate dehydrogenase (HBDH) and malate dehydrogenase (MDH) from *R. sphaeroides*

Both the versatility and advantage of triazine dye chromatography are illustrate by the preparative scale purification, using both co-factor and salt elution, of HBDH and MDH from *Rhodopseudomonas sphaeroides*. The former enzyme is a valuable diagnostic tool for estimating the ketone bodies, acetoacetate and 3-hydroxybutyrat The conventional purification of HBDH is tedious however and involves 8 steps with an overall 9% yield of enzyme (ref. 8). In the protocol described below triazine dye 'affinity' chromatography is the only protein purification technique used to take both HBDH and MDH to homogeneity. Furthermore although each enzyme requires two columns to achieve this purification only two triazine matrices in total are used. All the stages are developed in 10mM potassium phosphate buffer, pH 7.5.

One kilogram of the cell free extract of *R. sphaeroides*, prepared by pressure disruption of the cells at 500kg/cm^2 followed by centrifugation, is adsorbed on a 1.8 litre *Procion* Red H-3B Sepharose 4B column. HBDH is eluted with a step of 1M KCl in 90% yield, followed by MDH elution with a step of 2mM NADH in 1M KCl. Following desalting the HBDH pool is adsorbed on an 800ml column of *Procion* Blue MX-4GD Sepharose 4B matrix and after washing the column with 1M KCl, approximately 300mg of homogeneous HBDH is eluted, in 70% over yield with 2mM NADH in 1M KCl. Partially purified MDH, obtained from the first column is rigorously desalted to remove NADH and then bound on the *Procion* Blue MX-4GD Sepharose 4B column. About 1g of pure MDH can be recovered by elution with a 6 litre linear gradient from 0 to 700mM KCl; MDH eluting at about 400mM KCl. A summary of the data is presented in Table 2.

TABLE 2
Purification of HBDH and MDH

Step	Enzyme	Units	Protein (mg)	Spec. Act. (u/mg)	Yield %
Cell Extract	HBDH	6,900	76,200	0.09	100
	MDH	414,000	76,200	5.4	100
Procion Red H-3B	HBDH	6,800	2,200	3.1	99
	MDH	327,000	3,100	105	
Procion Blue MX-4GD	HBDH	5,400	280	19.2	78
	MDH	265,000	910	292	64

With the *Procion* Blue MX-4GD matrix the capacity of the column for HBDH increases with dye substitution and reaches a plateau between 1 and 3mg dye ml^{-1} Sepharose. Above this dye substitution level the capacity of the matrix for HBDH slowly decreases; however the affinity of matrix for the enzyme increases since the ability of 2mM NADH to elute this enzyme decreases (and increasing NADH is required to elute HBDH as the dye substitution level increases).

One further interesting feature of this system is a comparison (Table 3) of the binding and elution characteristics of MDH from yeast (Y-MDH) and *R. sphaeroides* (RS-MDH) on a number of triazine dyes immobilised on Sepharose 4B.

TABLE 3

Behaviour of Y-MDH and RS-MDH on *Procion* matrices

Matrix	Y-MDH	RS-MDH
Sc. MX-G, Or. MX-G, Ye. MX-G, Ye. H-5G, Ye. MX-R Gr. H-4G, Bl. MX-R, Bl. MX-G, Bl. MX-3G, Bl. MX-4GD Bl. H-B, Bl. F3G-A, R. HE-7B, Gr. HE-4BD	k	k
Ye. MX-8G, Ye. H-A, Tu. H-A, Tu. H-7G Bl. MX-7RX	o	k
R. P-3BN, R. H-8BN, R. H-3B, R. HE-3B R. MX-2B, R. MX-5B, Bl. HE-R3D	k	N
Ru. MX-B	N	k

Matrices: *Procion*: Sc. - Scarlet; Or. - Orange; Ye. - Yellow; Gr. - Green; Bl. - Blue; R. - Red; Tu. - Turquoise; Ru. - Rubine

Binding and elution: k - eluted with KCl; N - eluted with NADH; O- no binding to matrix

This data indicates that the same enzyme from 2 different sources does not necessarily behave similarly on the same matrix.

The purification of aminoacyl-tRNA synthetases

The aminoacyl-tRNA synthetases are a group of enzymes, each of which possess an amino acid, ATP and tRNA binding site. Since they are a group of 20 enzymes with a similar function and are often difficult to purify, a study has been made of their binding to *Procion*-Sepharose (ref. 4). In addition to developing new purification regimes it was also hoped that this study might lead to the development of new affinity labels and probes for these enzymes since structural work on the methionyl- and tyrosyl- enzymes is almost complete and on others is advanced, including the tryptophanyl-enzyme (ref. 9).

The binding of 13 of the synthetases to 32 immobilised dyes indicated that the interactions were complex (ref. 4). However, the purification of 3 enzymes, the tyrosyl (YTS)-, tryptophanyl (WTS)- and methionyl (MTS)- enzymes, was developed on these matrices. The separation of YTS from the glyceraldehyde-3-phosphate dehydrogenase (GAPDH) of *B. stearothermophilus* is tedious and difficult. Two systems were developed to achieve this; in the first both proteins were adsorbed on *Cibachron* Blue F3G-A and GAPDH eluted with NAD followed by YTS with 2M KCl; in the second *Procion* Orange MX-G was used to selectively bind GAPDH only.

A preliminary screen of the dyes had shown that WTS could be bound to immobilised *Procion* Brown MX-5BR at pH 6.5, and eluted with 50mM tryptophan at this pH. This procedure was used to purify WTS to homogeneity from a crude extract (Table 4). Following preliminary purification on DEAE-cellulose and DEAE-Sephadex, 13mg of pure WTS was eluted from a 50ml *Procion* Brown MX-5BR column to which 2g of crude protein had been applied. Recovery of the enzyme was excellent and a 137-fold purification achieved in this step.

TABLE 4

The Purification of WTS from *B. stearothermophilus* (2Kg)

	Total Protein (mg)	Total Units	Sp. Activity (u/mg)	Yield %
Cell Extract	182,000	25,500	0.14	100
DE-23	60,500	22,950	0.38	90
DEAE-Sephadex	2,010	15,280	7.6	60 (100)
Brown MX-5BR	12.9	13,420	1040	53 (88)

The dyes *Procion* Green HE-4BD, Blue MX-4GD and Red H-8BN proved to be almost universal adsorbents for the synthetases studied (ref. 4). The polyanionic nature and sulphonic acid group distribution of certainly Green HE-4BD, and to some extent Blue MX-4GD, together with their remarkable susceptibility to phosphate, rather than chloride, for the elution of these enzymes could be significant. Green HE-4BD may mimic the phosphodiester backbone of tRNA and since it is a large linear molecule (40 A^{0}) it could occupy a substantial portion of the tRNA binding cleft or groove in the synthetases. MTS has been purified to homogeneity on *Procion* Green HE-4BD Sepharose by phosphate gradient elution. The enzyme was eluted at 400mM phosphate in a 70% yield with a 12-fold purification.

Although several kinases can be eluted from *Procion* matrices with ATP (ref. 1, 2, 5) none of the synthetases tested could be recovered in significant yield with this substrate (ref. 4). It should be pointed out however that in the study (ref. 4) ATP was only applied after amino acid had been applied to the matrices. Since the ATP and amino acid binding sites in the enzyme must be close in order to form the aminoacyladenylate intermediate; it is probable that those enzymes eluted from the *Procion* matrices by their cognate amino acid will also be sensitive to elution by ATP. Preliminary evidence for this can be obtained from the published data (ref. 4).

A *Procion* active site probe for the tryptophanyl-tRNA-synthetase

The tryptophan dependent elution of WTS from *Procion* Brown MX-5BR Sepharose and also the partial elution of YTS from the same matrix with tyrosine, suggests that Brown MX-5BR might mimic these two aromatic amino acids and bind to the enzymes at their respective cognate amino acid binding site.

Kinetic studies have therefore been carried out to ascertain the exact nature of the interaction between Brown MX-5BR and WTS, with the aim of using the dye as a probe for the tryptophan binding site of this enzyme. Since Brown MX-5BR is not homogeneous the dye was freed from additives by precipitation with acetone from methanol solution.

Early studies showed that, at pH 8.5 and $20^{o}C$, the WTS was 85% inactivated in 1 hour by a 6-fold molar excess of the reactive dye. Even at a 60-fold molar excess of dye over enzyme, protection against this inactivation was obtained in the presence of tryptophan, but not the non-cognate amino acids glycine or tyrosine indicating the specificity of the interaction. In nucleotide protection studies, effectiveness against inactivation by the dye decreased in order : ATP>ADP>PPi>GTP. AMP gave only a slight protection. Significantly the protective effect of the nucleotides was reduced by the presence of equimolar concentrations of $MgCl_2$, which is required for catalytic activity of the enzyme. The effect of $MgCl_2$ was proportionally greater at the higher ATP concentrations.

Since the Mg-ATP (Km 50μM) and tryptophan (Km 9μM) sites in WTS must be close in order to form tryptophanyl-adenylate, protection against Brown MX-5BR inactivation of WTS by both is not surprising. Since $MgCl_2$ has no effect alone on the dye inactivation of WTS, it must be assumed that the binding of Mg-ATP to WTS opens the enzyme structure, thus making it more susceptible to the dye, and less protected, than the enzyme with bound ATP alone. Predictably from the known reaction mechanism of these enzymes the protective effects of low and sub-optimal tryptophan and ATP concentrations on dye inactivation of WTS were found to be approximately additive.

Brown MX-5BR is a small dye containing 4 aromatic rings, one sulphonic acid group, one carboxyl group and a chromium-atom per monomeric unit. At least 3 different isomers around the chromium atom are possible and therefore for further work the dye was purified by preparative TLC on silica in isobutanol (20): isopropanol (40):ethylacetate (10):water (30). The band with the highest Rf was used for subsequent experiments. The reactive dye affinity constant (Kd) for Brown MX-5BR on WTS was determined at 7μM.

Replacement of the chlorine atoms of the dichlorotriazine moiety of Brown MX-5BR with amino (by NH_3), hydroxyl (by alkali) or methoxyl (by methanol/alkali) can be readily achieved. Amino Brown MX-5BR is a competitive inhibitor (Ki 60μM) of tryptophan for WTS. Both hydroxy Brown MX-5BR and methoxy Brown MX-5BR however are lower affinity mixed-competitive inhibitors of tryptophan (Ki 500μM and 1,250μM respectively) indicating that they can combine with the enzyme at a second site separate from the tryptophan binding site.

In the light of this data the behaviour of WTS on the *Procion* Brown MX-5BR Sepharose matrix is significant. The synthetis of the original matrix involved a final incubation in NH_4Cl to convert remaining chlorine atoms of the dichlorotriazine ring to amino functions. WTS bound well to this matrix and was eluted by tryptophan in high yield - as could be predicted from the subsequently derived kinetic data. Similarly, it can be predicted that identical matrices, but with hydroxyl residues in place of amino functions, would have a much reduced efficiency for WTS and would not be as sensitive to tryptophan elution. This was subsequently confirmed by binding and elution studies on a matrix of this type.

The kinetics of amino Brown MX-5BR inhibition of Mg-ATP in the WTS reaction are complex and are still being examined. Amino Brown MX-5BR is not however a competitive inhibitor of this substrate. Although other kinetic inhibition studies remain to be performed; current effort is being directed at identifying the specific amino acid in WTS to which reactive Brown MX-5BR covalent attaches by purifying and sequencing the dye-peptide from a proteolytic digest of Brown MX-5BR-WTS. Since the full sequence of this protein is known it should then be possible to identify the tryptophan binding site of WTS.

High performance liquid affinity chromatography (HPLAC) on silica immobilised triazine dyes

The reactive triazine dye, *Cibachron* Blue F3G-A has been covalently attached to microparticulate porous silica and used for the high speed resolution of dehydrogenases, lactate dehydrogenase isoenzymes, kinases and other proteins, some from complex mixtures (ref. 10). Subsequently a number of other reactive triazine dyes were immobilised on microparticulate silica and used to resolve other enzymes such as alkaline phosphate, carboxypeptidase G2 and tryptophanyl-tRNA synthetase (ref. 11). In this latter work, the effect of divalent metal ions such as Mg^{2+} and Zn^{2+} on promoting the adsorption of metalloenzymes to triazine dye adsorbents was also investigated.

To prepare the HPLAC matrix, microparticulate silica (5 or 10μm) was first silylated with γ-glycidoxypropyltrimethoxysilane. Triazine dyes were then covalently attached to the derivatised silica in one of two ways. By silylating silica at pH 5.5 to 7.0 the terminal epoxide group (oxirane) of the epoxysilylated silica is left unaltered and can be reacted directly at pH 8.6 with aminoalkyl-derivatised triazine dyes to yield dyes attached to silica via a long methylene spacer arm. Alternatively, following the silylation reaction, brief exposure of the matrix to pH 3.5 converts the epoxide of the epoxysilane to a glycol which may be reacted directly with the triazine dye at pH 8.6 to yield a triazine dye-silica matrix with a short spacer arm. An outline of this protocol is shown in Figure 2.

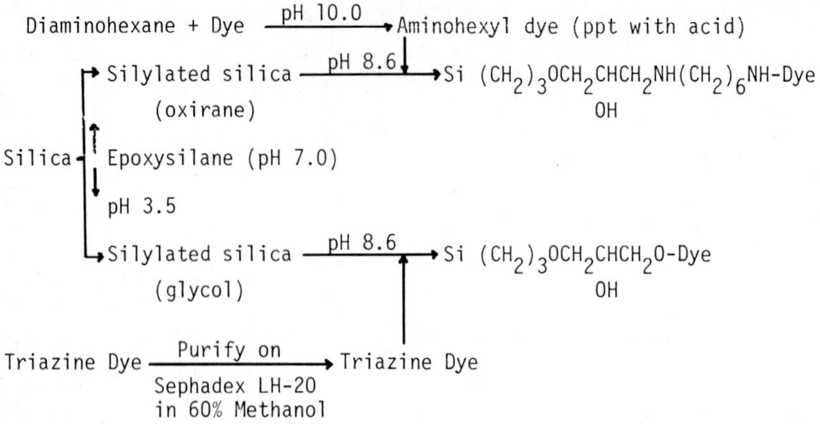

FIGURE 2
Synthesis of Triazine dye-silica for HPLAC

Triazine dye concentrations of 3 to 7μmole/g silica are usually obtained with the aminoalkyl/oxirane method. Substitution levels with the same dye but using the reactive dye/glycol method are usually about 60% of this. This lower substitution does not appear to adversly affect protein binding capacity or resolution however.

The requirements for binding of an enzyme to an HPLAC matrix differ from those for the same dye immobilised on Sepharose. The presence of phosphate appears to prevent the binding of several enzymes to the HPLAC matrices whereas it has been widely used for the same enzymes on the triazine agarose matrices. Thus tryptopha tRNA synthetase (WTS) will not effectively bind to *Procion* Brown MX-5BR silica in 10mM phosphate, pH 6.5, conditions under which it binds well to the same dye covalently attached to agarose. This enzyme however binds well to Brown MX-5BR silica in 10mM HEPES, pH 7.0.

Partial elution (40%) of WTS from this matrix can be achieved with a pulse of 40mM tryptophan, although 10mM tryptophan has relatively little effect. 40mM concentrations of the non-cognate amino acids tyrosine or glycine do not elute WTS from this column. Pulse nucleotide elution of WTS from Brown MX-5BR silica is unaffected by $MgCl_2$ and 10mM nucleotide concentrations are effective in the decreasing order: ATP (80% elution)>ADP = GTP (50% elution)>PPi (40% elution). No significant elution is obtained with AMP, although Pi does elute a very small amount of WTS. In view of the inhibition of binding of WTS to this matrix by Pi, the effects of PPi and Pi on elution are perhaps not surprising.

The *Procion* Brown MX-5BR silica matrix used in these studies was synthesised via the reactive dye/glycol route. Thus the triazine ring carries a free hydroxyl (in place of the second chlorine atom) and is linked to the short spacer arm via an ether link. The kinetic data on the nature of the hydroxy- and methoxy-Brown MX-5BR inhibition of tryptophan binding to WTS agrees closely with the elution characteristics of WTS from Brown MX-5BR silica.

Although the reduced binding of WTS to this column in phosphate is a characteristic of several enzymes in their interaction with the HPLAC matrices there are obviously exceptions. Lactate dehydrogenase (Porcine heart) binding to *Procion* Blue MX-R silica, synthesised by the reactive dye/glycol coupling system, exhibits the reverse in that it will not bind to this matrix in HEPES, pH 7.0 but will bind in phosphate at the same pH. The interaction of this enzyme with the matrix is interesting when compared with its interaction with the structurally related dye *Cibachron* Blue F3G-A immobilised on silica and synthesised by the aminohexyl-dye/oxirane coupling (ref. 10, 11). Both matrices bind tightly lactate dehydrogenase (LDH) which can be quantitatively recovered from both by biospecific ternary complex elution with 0.1mM NAD^+/0.1M pyruvate, whilst NAD^+ alone is ineffective. However the strength of adsorption of LDH to the *Cibachron* Blue F3G-A matrix appears to weaken at lower ionic strength suggesting a major contribution from hydrophobic forces, possibly due to the long spacer arm. In contrast the Blue MX-R matrix binds well at low ionic strength and displays none of these effects.

CONCLUSION

The data presented here is not ment to imply that the interaction between a specific enzyme and dye is mutually exclusive. Indeed, WTS has been shown to bind to 7 of the 32 dyes immobilised on agarose (ref. 4). In addition this enzyme will also bind to the HPLAC matrix, aminohexyl-*Cibachron* Blue F3G-A silica (ref. 11), even though it shows no binding to the same dye linked directly to agarose (ref. 4). The capacity of this HPLAC matrix for WTS is relatively low and although some enzyme can be pulse eluted by 40mM tryptophan it is likely that many of the forces involved in the interaction are hydrophobic; particularly since the enzyme is known to bind well to aminohexyl-Sepharose, a matrix formerly used in its purification (ref. 12).

Similarly although *Procion* Brown MX-5BR appears to be very specific for the active site(s) of WTS, this dye will also rapidly and differentially inactivate other enzymes. Thus it rapidly and strongly inactivates Carboxypeptidase G2 but has a lesser effect on alcohol dehydrogenase from Yeast.

The value of the triazine dye-agarose matrices for both laboratory- and large-scale enzyme purification has been recognised for some time. It is only recently, however, that the value of these dyes as active site directed probes has been recognised (ref. 6). Now the development of HPLAC matrices with their speed of resolution, high resolving power and value as tools for kinetic characterisation serves to further demonstrate the potential of triazine dyes in protein studies.

REFERENCES

1 C.R. Lowe, D.A.P. Small and A. Atkinson, Int. J. Biochem., 13 (1980) 33-40.
2 P.D.G. Dean and D.H. Watson, J. Chromatog. 165 (1979) 301-319.
3 J. Baird, R. Sherwood, R.J.G. Carr and A. Atkinson, FEBS Lett. 70 (1976) 61-66.
4 C.J. Bruton and A. Atkinson, Nucleic Acids Res. 7 (1979) 1579-1592.
5 T. Atkinson, P.M. Hammond, R.D. Hartwell, P. Hughes, M.D. Scawen, R.F. Sherwood, D.A.P. Small, C.J. Bruton, M.J. Harvey and C.R. Lowe, Biochem. Soc. Trans., (August 1981) in press.
6 Y.D. Clonis and C.R. Lowe, Biochem. J. 191 (1980) 247-251.
7 C.R. Lowe, FEBS Lett. 106 (1979(405-408.
8 H.U. Bergmeyer, K. Gawehn, H. Klotzsch, H.A. Krebbs and D.H. Williamson, Biochem. J. 102 (1967) 423-431.
9 C.J. Bruton, in Nonsense Mutations and tRNA Suppressors, Academic Press, New York, 1979, pp 47-68.
10 C.R. Lowe, M. Glad, P.O. Larsson, S. Ohlson, D.A.P. Small, T. Atkinson and K. Mosbach, J. Chromatog. (1981) in press.
11 D.A.P. Small, T. Atkinson and C.R. Lowe, J. Chromatog. (1981) in press.
12 T. Atkinson, G.T. Banks, C.J. Bruton, M.J. Comer, R. Jakes, T. Kamalagharan, A.R. Whitaker and G.P. Winter, J. Appl. Biochem. 1 (1979) 247-258.

T.C.J. Gribnau, J. Visser and R.J.F. Nivard (Editors),
Affinity Chromatography and Related Techniques

THE APPLICATION OF COLLOIDAL DYE PARTICLES AS LABEL IN IMMUNOASSAYS: DISPERSE(D) DYE IMMUNOASSAY ("DIA")

T. GRIBNAU, F. ROELES, J. v.d. BIEZEN, J. LEUVERING, A. SCHUURS
Organon Scientific Development Group, Oss, The Netherlands

ABSTRACT

The application of disperse dye sol particles as label in immunoassays was investigated as an alternative to current types of labeled immunoassays. Dye particles, coated with anti-(Human Chorionic Gonadotrophin), were used for an agglutination assay of this hormone ("DIA/agglutination"). In this case the particles serve mainly as support material, changing their physicochemical properties due to the immune reaction. The chromophoric properties of the dye particles were applied more particularly by using them as label in sandwich type immunoassays ("DIA/sandwich"). The advantage over labeling by covalent coupling of single dye molecules was clearly proven. Quantitative results of the determination of Human Placental Lactogen (HPL), Human Chorionic Gonadotrophin (HCG), Prolactin (PRL), testosterone and anti-Rubella are given. Finally, the possibility of the simultaneous determination of two antigens (HCG, HPL), by using differently coloured antibodies, was demonstrated.

INTRODUCTION

The development of clinical chemical assays has been highly advanced by utilizing the specificity of antigen-antibody interactions combined with the application of sensitive labeling and detection techniques.
Quite a number of these so-called "labeled immunoassays" has been developed during the past 30 years, based on similar immunochemical principles, but using a variety of different labels: erythrocytes, latex particles, bacteriophages, atoms/molecules (radioisotopes, free radicals, fluorophores, enzymes, electron-dense compounds), inorganic colloidal particles.

The radio-immunoassay has proven its value as a reliable, sensitive and, therefore widely applied technique (1, 2). The disadvantages of this method - high costs of reagents and equipment, short shelf life of the radioactive label, government regulations and waste disposal problems - stimulated the development of alternatives, leading to the introduction of the homogeneous and heterogeneous enzyme-immunoass (3-7). The application of inorganic colloidal particles as label in immunoassays was introduced quite recently (8).

Particulate labels are applied in two essentially different ways:

a) total particles as such ("physical" label):

formation of immunocomplex → visually/optically detectable aggregates

b) particle material itself ("chemical" label):

formation of immunocomplex → analysis of particle material in immunocomp or remaining non-bound fraction

This paper describes the use of colloidal organic dye particles as physical as well as chemical label. Readily commercially available "disperse dyes" (9) were taken as starting material for the preparation of the dye sols. These are defined, according to t Colour Index, as: "A class of substantially water-insoluble dyes originally introduced for dyeing cellulose acetate, and usually applied from fine aqueous suspensions; now also widely used for the colouration of all hydrophobic synthetic fibres" (10).

Dye sol particles can be applied as physical label in so-called agglutination assay similar to the use of erythrocytes, latex particles or gold sol particles (cf. review, ref. 11). The agglutination, for example, of antibody coated dye particles, caused by the addition of antigen, can be observed visually (qualitative) or the decreasing absorbance of the reaction mixture can be measured spectrophotometrically (quantitativ

The characteristic properties of colloidal dye particles are far more utilized, howe by applying them as chemical label. This can be illustrated using a sandwich immunc assay as model system (Fig. 1).

The advantage of dye sol labeling over molecular labeling lies in the difference between the values of Y and X. Approximative calculations can be made yielding the results given in Table 1. The molar absorbance and the molecular weight of the molecular dye label and of the dye sol particle label were supposed to be equal, and a density of 1 g/cm^3 was used for the latter. The particles were considered to be spherical, and to be coated with a closely packed monolayer of antibodies. The IgG molecules were also considered as spheres with a radius of ~6 nm (12), and with about 200 side chain functional groups/molecule, suitable for the covalent coupling of dye molecules: -OH (Ser), -NH_2 (ε-Lys, N-terminus) and imidazole (His).

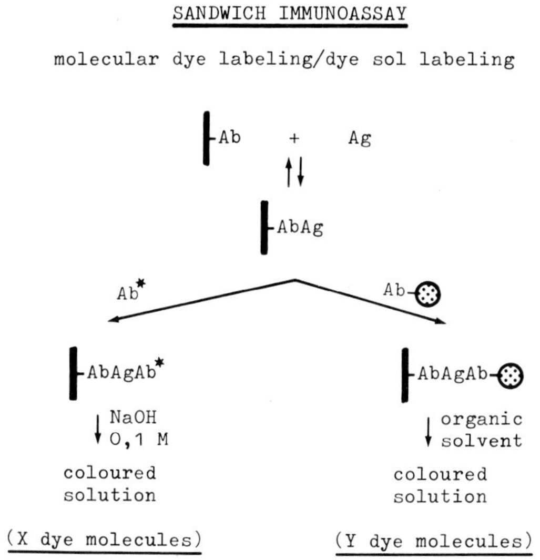

Fig. 1. Principle of a sandwich immunoassay for the determination of an antigen (Ag), using an antibody (Ab) labeled with covalently coupled dye molecules (*) or a dye sol particle (⊙). The solid bars symbolize the wall of a microtitration plate well.

TABLE 1

Comparison of molecular dye labeling with dye sol labeling, based on theoretical considerations (cf. Fig. 1).

	r(nm)	dye molecules/ IgG molecule	dye molecules/ immunocomplex	
-AbAgAb*		≤ 200	≤ 200	
-AbAgAb-(dye sol particle, Ab)	100	5500	6.10^6	23.10^3
	50	3000	8.10^5	11.10^3
	25	1500	1.10^5	6.10^3

The number of dye molecules/IgG molecule is already larger for sol labeling than for molecular labeling (Table 1, second column). Only one antibody/dye sol particle is relevant, however, to the formation of the final sandwich immunocomplex (Fig. 1); the number of dye molecules/immunocomplex is therefore considerably higher in case of sol labeling than for molecular labeling (Table 1, first part of last column). A correction should be made for the fact that the dye sol particle label is more space consuming than the molecular label, which gives the values in the second part of the last column (Table 1). But even after this correction the advantage is quite impressive as compared to the 200 dye molecules/IgG molecule in case of molecular labeling! Additionally, it has been demonstrated that the immunoreactivity of the antibodies is drastically reduced by the chemical coupling of 30 dye molecules or more, making the figure 200 already rather unrealistic.

These theoretical considerations were tested in practice with positive results, justifying further experimental evaluation of DIA. Quite a number of parameters play an important role in this type of immunoassay. Dye particle size and shape, adsorptive binding capacity of the dye particles for IgG, and the antibody quality will be discussed particularly.

Another interesting possibility of DIA is the simultaneous determination of two antigens, using the corresponding antibodies labeled with differently coloured dye particles. This will be illustrated by a simultaneous DIA/sandwich for Human Chorionic Gonadotrophin (HCG) and Human Placental Lactogen (HPL).

MATERIALS AND METHODS

A water soluble, reactive dye (Levafix R Brilliant Red E-4BA; Bayer) was used for molecular labeling. Dye sols were prepared from the following disperse dyes: Palanil Red BF, - Luminous Red G, - Yellow 3G, - Luminous Yellow G (BASF), Samaron R Brilliant Red H6GF, - Brilliant Yellow H10GF (Hoechst), Resolin R Brilliant Blue RF (Bayer) and Terasil R Brilliant Flavin 8GFF (Ciba-Geigy); samples of these dyes were gifts from the respective companies.

Human Placental Lactogen (HPL) was purchased from ICN Nutritional Biochemicals (Cleveland, Ohio, U.S.A.), and calibrated against the 1st International Reference Preparation of HPL for immunoassay (National Institute for Biological Standards and Control, Holly Hill, London, U.K.): 1,096 IU/g. Human Chorionic Gonadotrophin (HCG) was a standard preparation from Organon International B.V. (Oss, The Netherlands):

1 000 IU/vial, calibrated against the 2nd International Standard for Chorionic Gonadotrophin (13).

Rabbit anti-HCG IgG was prepared by caprylic acid fractionation of a pool of antisera; further purification by affinity chromatography was performed as described elsewhere (14). Rabbit anti-HPL IgG was prepared by sodium sulphate precipitation of a pool of antisera. IgG concentrations were determined using the value: $A^{1\%,\ 1\ cm}_{280} = 14,5$.

All reagents were of analytical grade quality, and were dissolved in distilled water. Bovine Serum Albumin (BSA), Cohn fraction V, was obtained from Armour (Eastbourne, U.K.).

Polystyrene microtitration (strip) plates were ordered from Greiner (Nürtingen, W.-Germany) and prepacked disposable Sephadex ^{R} G-25 columns (PD-10) were from Pharmacia Fine Chemicals (Uppsala, Sweden). Disposable polystyrene cuvettes (1 cm light path) were obtained from Sarstedt (W.-Germany).

Electron photomicrographs were made using a Philips EM-200 electron microscope.

Spectra of the disperse dye sols and of the dye solutions in ethanol, yielding λ_{max}, were made using a Gilford 250 Spectrophotometer; the same equipment was used for absorbance measurements during the agglutination assays. A Vitatron DCP small volume colorimeter was used during the sandwich assays.

Solutions:

I	: (0,075 mol phosphate + 0,15 mol NaCl)/l, pH 7,4
II	: as I, but with 10 g BSA/l, pH 7,4
III	: (250 g BSA + 5 mmol NaCl + 1 g merthiolate)/l, pH 7,4
IV	: (40 g BSA + 5 mmol NaCl + 1 g merthiolate)/l, pH 7,4
V	: (0,75 mol phosphate + 1,5 mol NaCl + 10 g BSA + 1 g merthiolate)/l, pH 7,4
VI	: (5 mmol NaCl + 1 g merthiolate)/l, pH 7,0
VII	: (0,04 mol phosphate + 0,15 mol NaCl + 1 g BSA)/l, pH 7,4
VIII	: carbonate/bicarbonate buffer 0,2 mol/l, pH 9,0
IX	: (0,04 mol phosphate + 0,15 mol NaCl)/l, pH 7,4
X	: as I, but with 1 g BSA/l, pH 7,4
XI	: (0,2 mol TRIS + 0,15 mol NaCl + 0,5 g Tween ^{R}-20 + 0,1 g merthiolate)/l, pH 7,4

Coating of microtitration plates

The wells of microtitration plates were filled with 0,12 ml of an IgG solution (1,5 - 30 µg/ml l; concentration depending upon the IgG quality used). After incubation

(16 - 20 h, 4°C), the wells were aspirated and filled with 0,15 ml of II, and incubated again (0,5 h, 20 - 22°C). The wells were finally washed with distilled water (3x); the plates were used as such or were dried over silica gel, and stored at 4°C in the presenc of silica gel.

Preparation of disperse dye sols

A dispersion (a ml) of the commercial dyestuff in distilled water (5 g or 5 ml/100 ml) was centrifuged (30 min, 1 000 N/kg). The pellet was discarded, and the supernatant was centrifuged (30 min, 100 000 N/kg); now the supernatant was discarded and the pellet was resuspended in distilled water (a ml). The latter procedure was repeated twice, resuspending the pellet finally to a volume of 1/2 a ml (A).

Disperse dye sol, used in DIA/sandwich: suspension A was centrifuged (30 min, 20 000 N/kg), yielding the dye sol as supernatant; the pellet was discarded.

Disperse dye sol, used in DIA/agglutination: suspension A was centrifuged (30 min, 5 000 N/kg), the pellet was discarded and the supernatant was centrifuged again (30 min, 150 000 N/kg). The supernatant was discarded, and the pellet was resuspende to a volume of 1/2 a ml.

Preparation of dye sol particle - IgG conjugates

For DIA/sandwich: disperse dye sol, IgG solution (1 mg/ml, in aqueous 0,15 mol NaCl/l) and an aqueous NaCl solution were mixed to a final volume of b ml with: 10 mmol NaCl/l, 15 - 30 μg IgG/ml (depending upon the IgG quality used), pH 7,4 and $A_{\lambda max}^{1\,cm}$ = 5,0. The mixture was incubated during 1 h (20 - 22°C); b/5 ml of soluti III was added, followed by a second incubation period of 1 h (20 - 22°C), and subsequently by centrifugation (30 min, 38 000 N/kg). The supernatant was discarded, the pellet was resuspended in solution IV to a total volume of c ml with an $A_{\lambda max}^{1\,cm}$ = 5,0 and c/9 ml of solution V was added yielding the final conjugate.

For DIA/agglutination: The pH of an IgG solution (4-5 mg/ml, in aqueous 5 mmol NaCl/l) was adjusted to 2,0 with 0,05 mol HCl/l; after incubation (1 h, 4°C), the pH was readjusted to 7,4 with 0,05 mol NaOH/l. Disperse dye sol, the latter IgG solution and distilled water were mixed yielding finally: 5 mmol NaCl/l, 33 μg IgG/m pH 7,4 and $A_{\lambda max}^{1\,cm}$ = 5,0. After incubation (2 h, 20 - 22°C), the mixture was centrifuged (30 min, 100 000 N/kg), the supernatant was discarded and the pellet was resuspended in solution VI and diluted to an $A_{\lambda max}^{1\,cm}$ = 2,2 - 2,5, yielding the final conjugat

Molecular dye labeling of IgG

IgG (4 mg) and Levafix Brilliant Red E-4BA (5 mg) were dissolved each in 1,3 ml of solution VIII; both solutions were mixed and the mixture was incubated (17 h, 20 - 22°C). The pH of the reaction mixture was adjusted to 7,4, and the dye labeled IgG was isolated by gel chromatography (2,5 ml reaction mixture/PD-10 column, equilibrated in solution IX).

Sandwich assay procedure

Samples (0,1 ml of antigen dissolved in solution X) were pipetted into the antibody coated wells of a microtitration plate, and the plate was incubated (3 h, 37°C). The wells were aspirated and washed with solution X (3x 0,2 ml). Conjugate (0,1 ml of dye sol or molecularly labeled IgG) was added and the plate was incubated (16 h, 37°C). The wells were aspirated and washed again (3x 0,2 ml of solution XI). Finally, ethanol (0,125 ml) was added to the wells and the absorbance was measured at the appropriate wavelength; 0,1 mol NaOH/l (0,125 ml) was substituted for the ethanol in case of the molecularly labeled IgG.

Agglutination assay procedure

Conjugate (2,0 ml), an aqueous solution of $MgSO_4$ (0,2 ml; 21 mmol/l) and sample (0,2 ml of antigen dissolved in solution VII) were pipetted into a cuvette, and were thoroughly mixed. $A_{\lambda max}^{1\ cm}$ was determined after incubation (2 1/4 h, 20 - 22°C) avoiding agitation of the cuvette.

The same procedure was also performed using small centrifuge tubes, instead of cuvettes, and applying 2 h incubation (20 - 22°C) followed by centrifugation (15 min, 10 000 N/kg, 20 - 22°C). Samples (2,0 ml) of the supernatants were pipetted into cuvettes and the absorbances were measured.

RESULTS

The dye sol particles were used as physical label in an agglutination assay of Human Chorionic Gonadotrophin (HCG; Fig. 2). A steeper standard curve was obtained by applying centrifugation during the assay; this also enabled a reduction of the total test time; half maximum effect: $(A_{[HCG]=0} - A_{[HCG]=5})/2$, was obtained at 0,9 IU HCG/ml after 2,25 h, as compared to a value of 0,6 IU HCG/ml after 20 h at normal gravity conditions!

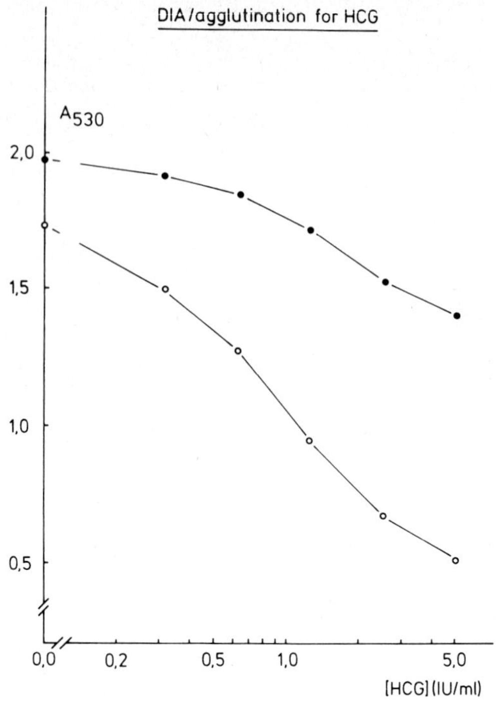

Fig. 2. Decreasing absorbance of a Palanil ^{R} Red BF/anti-HCG conjugate due to agglutination by HCG, after 2 1/4 h at normal gravity (●) and after 2 h followed by 15 min centrifugation at 10^4 N/kg (o).

An example of the application of dye sol particles as chemical label is given in Fig. 3, which also illustrates the advantage of dye sol labeling over molecular dye labeling. A high number of dye molecules covalently coupled per antibody molecule (yielding maximum absorbance) was incompatible with the retention of original immunoreactivity; an additional disadvantage of molecular labeling! A compromise of 19 dye molecules/IgG molecule was chosen; the immunoreactivity of this conjugate was by about a factor of 8 lower than of native anti-HCG, as determined by enzyme-immunoassay.

The best results were obtained using dye sol particles with a diameter of about 200 nm or less. The shape of the particles appeared to be very important: some currently occurring structures are given in Fig. 4. Needle-like structures were less suitable, whereas a gravel-like structure yielded good conjugates. Spherically shaped particles would probably be optimal.

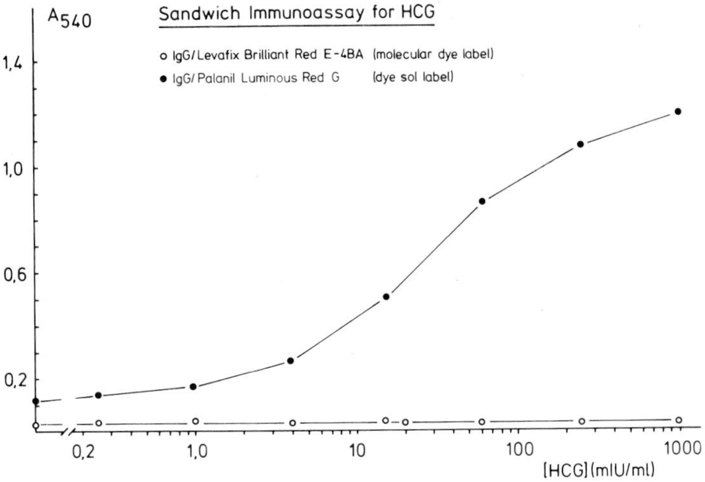

Fig. 3. Comparison of dye sol particle and molecular dye labeling.

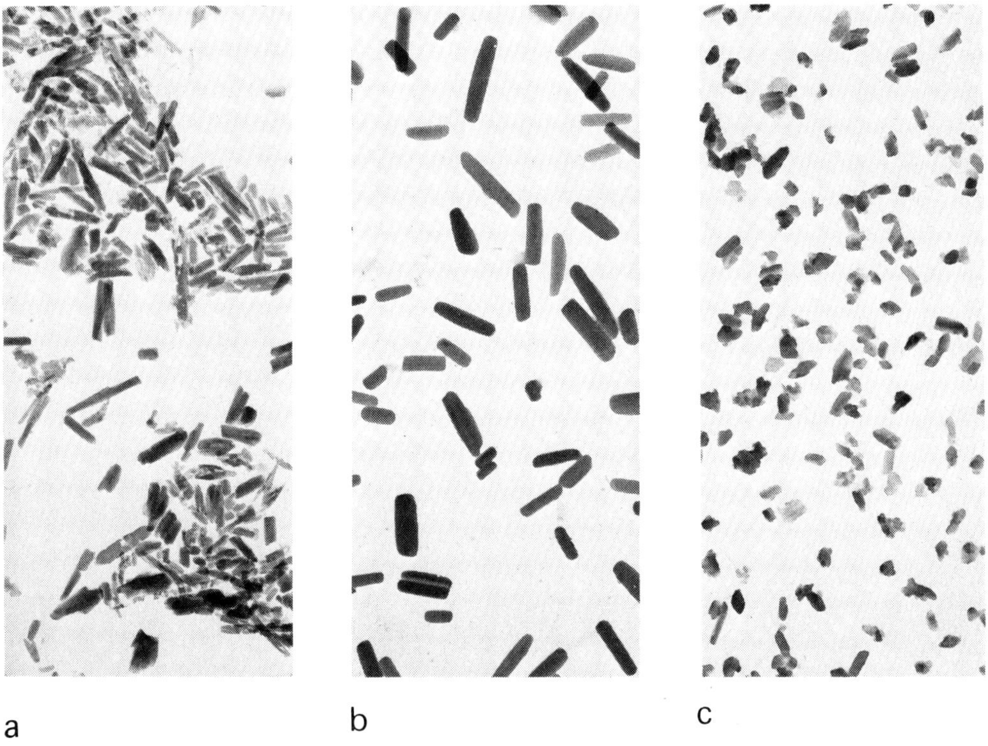

Fig. 4. Electron photomicrographs of disperse dye sols; (a) Cibacet R Violet 2R, (b) Resolin R Brilliant Blue RRL, (c) Palanil R Luminous Red G; magnification factor 16 800.

The immunoreactivity of the conjugates is also determined by the amount of IgG which can be adsorbed onto the dye particle surface. Dyes with suitable properties in this respect can be selected only by screening. An example is given in Fig. 5 for five different disperse dyes, tested in the HCG/anti-HCG system, illustrating the large differences that can be encountered.

The quality of the applied antibodies affects the final assay results with respect to detection limit and detection range. This is demonstrated in Fig. 6, which gives the results of six independent duplicate experiments with a DIA/sandwich for HCG, using conjugates based on total rabbit anti-HCG IgG and on the corresponding antibodies purified by affinity chromatography (14). The standard curves were shifted by about a factor of 5 to the lower concentration range and the detection limit was improved by about a factor of 8.

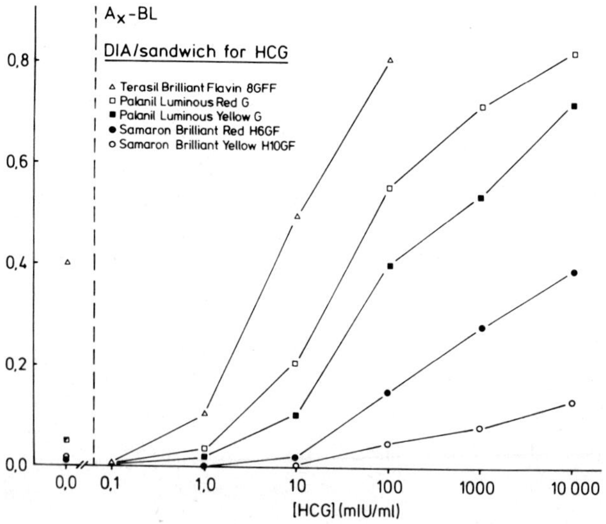

Fig. 5. Screening of disperse dye sols on their applicability as label (BL: response a 0 mIU HCG/ml); x = 443 (△ ■ ○), 510 (●), 540 nm(□).

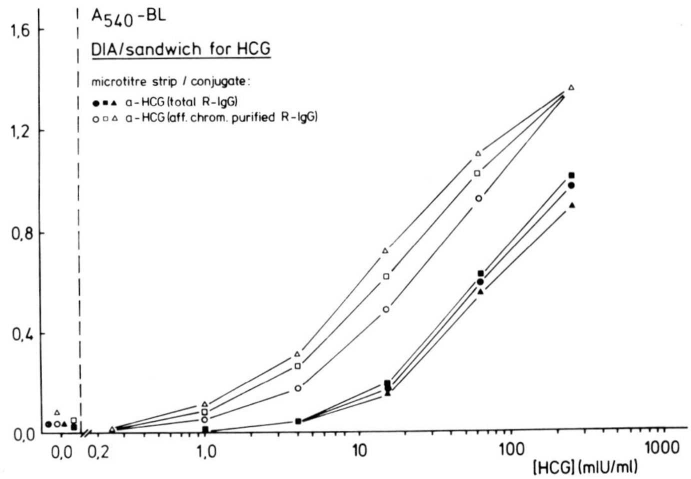

Fig. 6. Standard curves for HCG obtained using Palanil [R] Luminous Red G/anti-HCG conjugates.

The DIA/sandwich was initially evaluated using the HCG/anti-HCG system, but other applications were investigated subsequently. Examples are summarized in Table 2.

TABLE 2

Applications of the Disperse(d) Dye Immunoassay (sandwich system).

ANTIGEN/ANTIBODY	DETECTION LIMIT *	DETECTION RANGE **
Human Placental Lactogen (HPL)	1 - 2 ng/ml (50 - 80)	2 - 100 ng/ml (90 - 4500)
Human Chorionic Gonadotrophin (HCG)	0,25 - 1,0 mIU/ml (1 - 5)	4 - 1000 mIU/ml (20 - 4700)
Prolactin (PRL)	2 - 3 ng/ml (80 - 100)	6 - 100 ng/ml (300 - 4500)
Testosterone***	0,2 - 0,4 pmol/ml	0,5 - 5 pmol/ml
Anti-Rubella	2,5 IU/ml	3 - 50 IU/ml

* defined as the amount of antigen/antibody yielding a response equal to (blank + 2xSD); values in parentheses give the approximate concentration in fmol/ml

** linear part of the standard curve (log dose/response)

*** determined by sandwich inhibition

The possibility of determining two different antigens simultaneously was investigated for samples containing both HCG and HPL. Microtitration plates coated with a 1:1 mixture of anti-HCG and anti-HPL (8) were used in combination with blue labeled anti-HCG and yellow labeled anti-HPL, yielding of course a greenly coloured conjugate mixture. Quantitative results are given in Fig. 7. Samples, containing both antigens, were incubated followed by the addition of the single conjugates or of the conjugate mixture, and the absorbance was measured finally at the two different wavelengths. The response of the single conjugates was always somewhat higher than of the combined conjugate, probably due to steric factors.

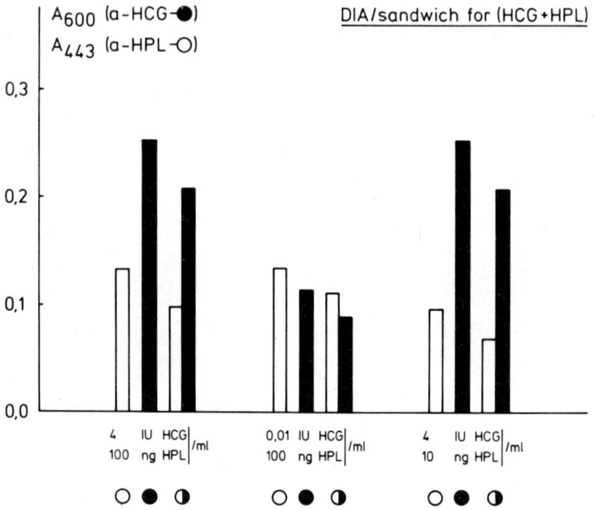

Fig. 7. Simultaneous determination of HCG and HPL using a 1:1 mixture (◑) of Resolin R Brilliant Blue RRL/anti-HCG (●) and Palanil R Yellow 3G/anti-HPL (○); the response of the single conjugates is also given.

DISCUSSION

The application of disperse dye sol particles as label in immunoassays, particularly as chemical label, presents an attractive alternative to other types of labeled immunoassays. For the application as physical label gold sol particles appeared to be preferred (15).

The organic-synthetic, non-radioactive, dye label can be easily prepared from a wide range of commercially available disperse dyes. The labeling of antibodies or antigens is performed by simple physical adsorption, and several detection methods can be applied to the final determination of the label: visual observation, colorimetry, fluorimetry, (carbon rod) atomic absorption spectrophotometry (in case of metal-complex dyes). The assay procedure is simple, with a minimum number of reagents and actions per assay. As compared to enzyme-immunoassay, for example, DIA has the advantage that the enzyme/(substrate-chromogen) incubation can be omitted. The HCG detection limit for DIA is at least the same as for radio- and enzyme-immunoassay. Furthermore, it is possible to determine two antigens simultaneously.

In addition, it was shown by recent experiments that the incubation periods for DIA/sandwich (sample: 3 h/37°C; conjugate: 16 h/37°C) can be decreased (sample: 0,5 h/37°C; conjugate: 2 h/37°C) by using a higher conjugate concentration ($\leftrightarrow A_{\lambda max}^{1\,cm}$ = 15 - 20). The detection limit, in case of HCG, changed from 0,25 - 1,0 mIU/ml to 1 - 2 mIU/ml.

In colorimetric sandwich type immunoassays, sols of disperse dyes could have advantages to metal sols (8), due to the considerably higher molar absorbances of the former; for example: gold sol, particle size ~50 nm: ε = 3 000 - 4 000 $l.mol^{-1}.cm^{-1}$ (16); disperse dyes: ε = 5 000 - 80 000 $l.mol^{-1}.cm^{-1}$ (17). Additionally, the absorbance is increased by dissolving the dye sol particles into an organic solvent for final determination.

ACKNOWLEDGEMENTS

The authors gratefully acknowledge Dr. J. Wester (Pharmacological R & D Laboratories) for preparing the electron photomicrographs, Mr. J. Koopman for performing a substantial part of the experimental work, Mr. A. van Sommeren for fruitful discussions, and Mrs. A. Romme and the SDG Photography Group for the technical preparation of the manuscript.

REFERENCES

1 R.S. Yalow and S.A. Berson, J. Clin. Invest., 39 (1960) 1157 - 1175.
2 B.M. Jaffe and H.R. Behrman (Eds.), Methods of Hormone Radioimmunoassay, Academic Press, New York, 1979.
3 K.E. Rubenstein, R.S. Schneider and E.F. Ullman, Biochem. Biophys. Res. Commun., 47 (1972) 846 - 851.
4 B.K. van Weemen and A.H.W.M. Schuurs, FEBS Lett., 15 (1971) 232 - 235.
5 E. Engvall and P. Perlmann, Immunochemistry, 8 (1971) 871 - 874.
6 A.H.W.M. Schuurs and B.K. van Weemen, Clin. Chim. Acta, 81 (1977) 1 - 40.
7 M. Oellerich, J. Clin. Chem. Clin. Biochem., 18 (1980) 197 - 208.
8 J.H.W. Leuvering, P.J.H.M. Thal, M. van der Waart and A.H.W.M. Schuurs,

J. Immunoassay, 1 (1980) 77 - 91.
9 J.M. Straley, in K. Venkataraman (Ed.), The Chemistry of Synthetic Dyes, Vol. III, Academic Press, New York, 1970, Ch. VIII, pp. 385 - 462.
10 H. Blackshaw and R. Brightman, Dictionary of Deying and Textile Printing, George Newnes Ltd., London, 1961, p. 62.
11 A.H.W.M. Schuurs, T.C.J. Gribnau and J.H.W. Leuvering; these Proceedings
12 D.M. Crothers and H. Metzger, Immunochemistry, 9 (1972) 341 - 357.
13 D.R. Bangham and B. Grab, Bull. Wld Hlth Org. 31 (1964) 111 - 125.
14 T. Gribnau and A. van Sommeren, Abstract Volume 4th International Symposiur on Affinity Chromatography and Related Techniques, Veldhoven, June 22 - 26, 1981, p. A-29; manuscript submitted for publication (J. Chromatogr.).
15 J.H.W. Leuvering, P.J.H.M. Thal, M. van der Waart and A.H.W.M. Schuurs, J. Immunol. Meth., accepted for publication.
16 J.H. Leuvering, personal communication.
17 K. Venkataraman (Ed.), The Analytical Chemistry of Synthetic Dyes, Wiley & Sons, New York, 1977, p. 379.

T.C.J. Gribnau, J. Visser and R.J.F. Nivard (Editors),
Affinity Chromatography and Related Techniques

AFFINITY CHROMATOGRAPHY AND AFFINITY ELECTROPHORESIS:TOOLS TO INVESTIGATE PROTEIN INTERACTIONS AND ENZYME MUTANTS

J.VISSER, H.C.M. KESTER, A.C.G. DERKSEN and J.H.A.A. UITZETTER
Dept. of Genetics, Agricultural University, Wageningen, The Netherlands

ABSTRACT

Multienzyme complexes isolated from pro-and eukaryotes which are involved in α-ketoacid decarboxylation,bind to ethanol-Sepharose 2B.The pyruvate and the α-ketoglutarate dehydrogenase complex can be separated on this adsorbent. Since in the pyruvate dehydrogenase complex the core protein E2 is involved in the binding to the affinity matrix,this can be used to analyze mutants of the complex with respect to the state of assembling.

Reactive dyes have been applied as ligands in crossed affinity immunoelectrophoresis to analyze the selective binding of individual components of the complex. It is a rapid method to select conditions to separate individual components of the complex and it may become interesting to refine mutant analysis.

For the purification of a fungal pyruvate kinase Mikacion Brilliant Yellow 6GS and a dextran conjugate of Cibacron blue F3G-A have been used.These dyes seem to bind to different sites;the interaction between enzyme and Cibacron Blue is disrupted by FDP and PEP. With the yellow dye ATP and PEP elute the enzyme.The selection of dyes which bind to different sites forms a useful basis for purification and mutant analysis.

INTRODUCTION

Affinity chromatography is a useful technique to distinguish between enzymes of identical function but different in composition and/or activity.Relevant examples can be found in comparative biochemical studies and in isoenzyme studies.Differential behaviour of isoenzymes towards affinity matrices is well established(ref.1,2).The application of affinity techniques in isoenzyme studies has recently been reviewed (ref.3).An area which has hardly been investigated is the study of mutant proteins by affinity techniques.Such a methodology is useful to isolate mutant proteins for further investigation but may also reveal nature and effects of certain mutations.The availability of mutant material is often limited whereas the stability of the mutated gene product is often less than that of the wild type.Thus methods are required which are sufficiently rapid and specific. Our studies with bacterial and fungal pyruvate dehydrogenase multienzyme complexes and more recently with a fungal pyruvate kinase illustrate well how affinity techniques contribute in biochemical genetic studies. Although applied to microbial systems in this case,the approach is also relevant for certain areas in biomedical research.

AFFINITY CHROMATOGRAPHY OF THE PYRUVATE DEHYDROGENASE COMPLEX

Choice of an appropriate affinity matrix

The pyruvate dehydrogenase complex consists of multiple copies of three enzyme components each catalyzing part of the following overall reaction.

$$\text{pyruvate} + \text{CoA} + \text{NAD}^+ \longrightarrow CO_2\uparrow + \text{acetylCoA} + \text{NADH} + H^+ \qquad (1)$$

The complex of Escherichia coli has been thoroughly investigated(ref.4).The three different enzyme components are: pyruvate decarboxylase(E1),a TPP-dependent enzyme dihydrolipoamide acetyltransferase(E2) and lipoamide dehydrogenase(E3) which is a NAD^+ dependent flavoprotein . The strucural core of the complex is formed by 24 poly peptide chains of E2 each of which contains two covalently bound lipoyl moieties. T α-ketoglutarate dehydrogenase complex catalyzes a similar reaction leading to succinylCoA.

For a system in which for each of the different enzymes another specific cofacto is required,the design of biospecific affinity matrices seems the method of choice particularly to analyze various mutants with a defect in one of the components. Thiamine pyrophosphate,immobilized through a short spacer arm,has been used to pur the complex from E. coli although the affinity matrix is rather labile(ref.5).Compl xes of other prokaryotes as well as those of mammals bind too.For elution an incr in ionic strength is required whereas cofactor elution fails.The isolated E1 compo of the E. coli complex does not bind. With other systems there are nevertheless so arguments in favour of a biospecific contribution to the binding.The pig heart com fails to bind in TPP(1 mM) whereas in the absence of cofactor the inactive,phosphorylated form of the enzyme,known to have less affinity for TPP(ref.6),is almost quantitatively recovered in the eluate.

A number of NAD^+ derivatives have been tested,substituted at position N-6 or C-8 with spacer arms of variable length and hydrophobicity.Some matrices bind but E. coli complex cannot be recovered by cofactor elution.Lowe(ref.7) also failed to find a nucleotide specific affinity system for the flavoprotein E3 itself. We also failed to demonstrate a biospecific interaction for the pig heart flavoprotein wit N-lipoyllysyl Sepharose(ref.8).Obviously,the initial approach to study mutants of complex had to be left. CoA and AMP derivatives were therefore not further investiated.

Hydrophobic interactions contribute to the formation of the highly ordered structure of the pyruvate dehydrogenase complex. Screening various homologous series of alkyl and substituted alkyl compounds indicated that with the alkyl series the overall enzyme activity is hardly recovered but that with appropriate ligands the complex remains intact. From the series $CH_2OH\text{-}(CH_2)_n\text{-}NH$ -Sepharose 2B, ethanol-Separose prepared through a bisoxirane activation,was chosen(ref.9). This matrix is n used as an essential step in our purification scheme of the pyruvate dehydrogenase complexes from various sources viz E.coli(ref.9), Bacillus subtilis, B. stearo-

thermophilus and other *Bacillus spp.*(ref.10), *Pseudomonas aeruginosa* PAO(ref.11) and *Aspergillus nidulans*, a hyphal fungus.

Differential behaviour of α- ketoacid dehydrogenase complexes on ethanol-Sepharose 2B

Pyruvate and α-ketoglutarate dehydrogenase complexes are usually separated by isoelectric precipitation.The behaviour of both complexes on the affinity matrix is relevant for mutant analysis since in *E. coli* a single *lpd* gene codes for the flavoprotein E3 which is present in both complexes(ref.12). It turns out that the α-ketoglutarate dehydrogenase complex is more weakly bound and that the E3 activity becomes distributed between the two complexes as shown in Fig.1 . This is a general phenomenon also observed in the case of *B. subtilis* and *P. fluorescens*(data not shown) and of P. aeruginosa(ref.11).

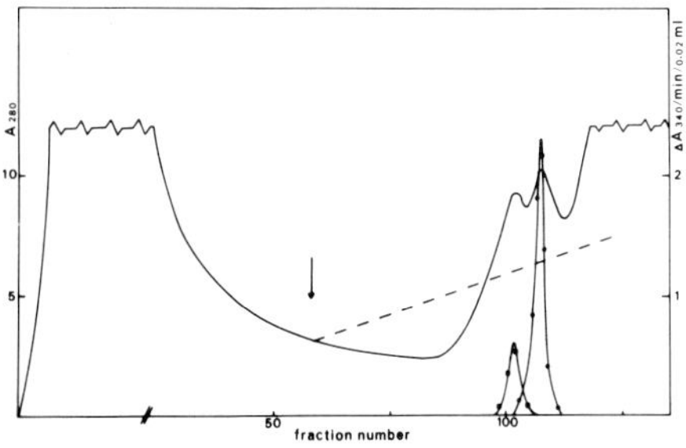

Fig.1 Separation of α-ketoacid dehydrogenase multienzyme complexes of *E. coli* on ethanol-Sepharose 2B. Solid line: absorbance at 280 nm; ■ - ■ α-ketoglutarate dehydrogenase activity; ● - ● pyruvate dehydrogenase activity. The arrow indicates the start of a salt gradient (0-0.2M sodium chloride).

Interaction of component enzymes with ethanol-Sepharose 2B

The pyruvate dehydrogenase complex has been dissociated as described(ref.13). The isolated E3 has no affinity for the adsorbent whereas E1 is only weakly bound. In the presence of some residual E2 protein free components become separated from subcomplexes or from some residual complex which can have a variable chain stoichiometry. All this implies that the presence of E2 is essential for the binding of the total complex.

Pyruvate dehydrogenase complex mutants

In collaboration with Dr. J.R. Guest(Dept.of Microbiology,University of Sheffield, U.K.) we have investigated how different mutants of the multienzyme complex behave on the affinity system described. Structural genes for E1 and E2, *aceE* and *aceF*, constitute an operon at approx. 2.6 min on the *E.coli* linkage map and they are closely linked with *lpd*,the structural gene for the flavoprotein E3(ref.14).Extracts of various point and deletion mutations have been analyzed on ethanol-Sepharose 2B to investigate whether the complex or part of it still assembles.The distribution of component enzymes over the fractions can be measured following the partial enzyme activities or a precipitation reaction with specific antibodies(anti-complex, anti-E1 and anti-E3).

The chromatography pattern of crude extracts of deletion strains lacking both E1 ar E2,like for instance in strain KΔ15, is shown in Fig.2 Frontal elution of uncomple E3 is observed whereas a second E3 enzyme peak is detected in the salt gradient coinciding with overall α -ketoglutarate dehydrogenase complex activity(ref.15).

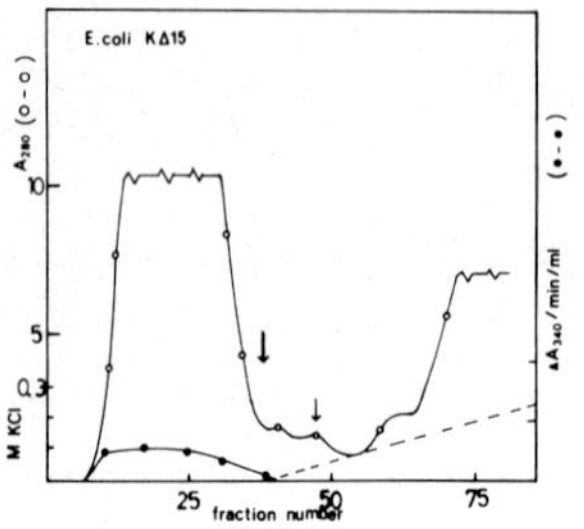

Fig.2 Elution profile of a crude extract of E. coli mutant KΔ15 on ethanol-Sepharose 2B. Absorbance at 280nm o - o ; uncomplexed lipoamid dehydrogenase activity • - • ; The solid arrow indicates the start of salt gradient whereas the thin arro indicates the peak fraction of α-ke glutarate dehydrogenase complex act vity.

Strain A2T3 lacking E1 activity and strain A10 defective in E2 have also been a lyzed since these strains are frequently used in reconstitution experiments.The elution pattern of A2T3 is shown in Fig. 3 and reminds of that of wild type comple Surprisingly, behaviour of A10 on the affinity matrix is also similar although thi mutation affects the core protein.Gel filtration on Biogel A-50-M indicates that t assembled complex of strain A2T3 looses E1 more readily than wild type complex. Th complex isolated from A10 falls apart during gel filtration. It explains why this strain is effective in reconstitution experiments with other mutants(ref.16).

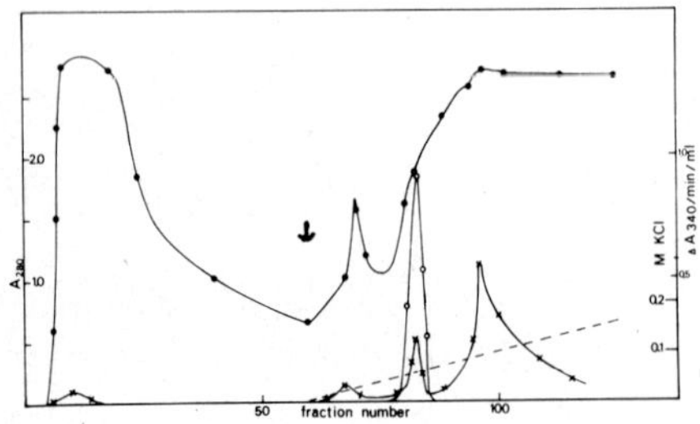

Fig.3 Elution profile of a crude ext of E. coli mutant A2T3 on ethanol-Se arose 2B. Absorbance at 280 nm • - • lipoamide dehydrogenase activity x - α -ketoglutarate dehydrogenase acti o - o . The start of a linear salt gradient is indicated by an arrow.

All lpd mutants tested demonstrate wild type behaviour on ethanol-Sepharose 2B. Further purification and analysis by SDS-PAGE reveal interesting differences between the mutants. As expected,elution behaviour of the system does not depend on E3 which is confirmed by mutant Hlpd8. This mutant,previously reported as immunologically inactive(ref.17), lacks in purified form a flavin spectrum but still contains a residual amount of E3 protein(15%). Binding of this complex is not different from that of the wild type complex. Other mutants analyzed synthesize amounts of E3 comparable with those found in wild type and are clearly missense mutants. In lpd7 which maps rather closely to lpd8 also no flavin spectrum is detected in the final preparation whereas in others(lpd5, lpd9) a normal flavin spectrum is observed.

One of the conclusions which can be drawn from affinity chromatography data is thus whether complex mutants assemble. At the same time partial activities of the components can be assayed more easily and spectral properties can be studied whereas the presence and relative amounts of the subunits can be verified by immuno-detection and SDS-PAGE. Similarly, lesions of ace mutants in *P. aeruginosa* have been defined using this approach.Assembled,inactive complexes were found which lacked E1 or E2 enzyme activity(ref.11).

Gene-protein relationship in pyruvate dehydrogenase complex mutants of Aspergillus

The fungus *A. nidulans* contains a pyruvate dehydrogenase complex typical for a eukaryote in which α- and β- E1 subunits are present whereas the enzyme activity is regulated through a proces of (de-)phosphorylation. Three classes of complex mutants have been found *viz* pdhA, pdhB and pdhC, each located on a different chromosome (ref. 18). The eukaryote complex also binds to the matrix and representative mutants of each class have been analyzed. In Fig.4 the behaviour of a pdhA mutant is shown. The components present hardly demonstrate some affinity for the matrix. A mutation in E2 is therefore expected which could be confirmed by enzyme assays of these fractions. Similarly, the pdhC and pdhB genes were correlated with the α- and β - E1 component, respectively(ref.19). In these mutants elution of the (partially) assembled complexes occurs as with the wild type complex.

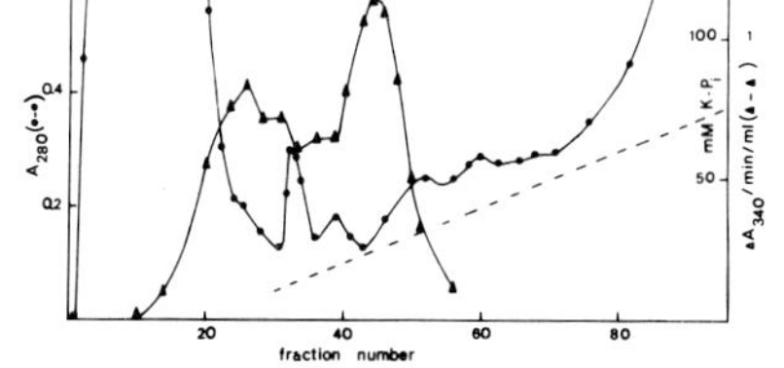

Fig.4 Affinity chromatography of pdhA1 pyruvate dehydrogenase complex. A phosphate buffer gradient is applied. The fractions were analyzed for their ability to restore overall enzyme complex activity upon incubating 0.15 ml samples with an aliquot of an extract obtained from pdhC1 for 10 min at room temperature. Absorbance at 280 nm ● - ●; ▲ - ▲ restored overall activity.

CROSSED AFFINITY IMMUNOELECTROPHORESIS

Affinity chromatography based upon reactive dyes as ligands continues to receive attention. We have considered the possibility to use these dyes for the dissociatio of the pyruvate dehydrogenase complex. There are various ways to find the ligand o choice like for instance small scale column chromatography(ref.20), determination of enzyme inhibition by dyes in solution or protection by other dyes against enzym inactivation by dichlorotriazine dyes(ref.21). We have used crossed affinity immuno electrophoresis,a technique described by Bøg-Hansen (ref.22), to establish interact of dyes with the various components of the pyruvate dehydrogenase complex in order to design a preparative method or an affinity chromatography method to dissociate complex. Amongst the various dyes tested Procion Blue MX-R, Procion Blue MX-3G, Procion Yellow MX-4G and Mikacion Brilliant Yellow 6GS were analyzed in more detai

The crossed immunoelectrophoretic pattern which the E. coli multienzyme complex forms with antibodies prepared against intact complex or separate components depen amongst others on the state of assembly of the complex. Antiserum prepared against the complex contains both anti-E1 and anti-E3 but very little,if any, anti-E2. The control pattern(Fig.5a) consists at least out of two precipitin lines with a diff zone in between as can be seen most clearly at the skewed, tailing end of the patt In Fig.5b,c the effect is shown of agarose-linked Mikacion Brilliant Yellow 6GS an Procion Blue MX-R in the first dimension.In the presence of these dyes two precipi lines separate. The outer one is also formed when anti-E1 antibodies are used. In profile of the inner one the presence of E3 is established by enzyme activity stai and by using anti-E3. Interaction with the dye is furthermore indicated by the tai

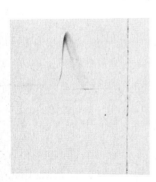

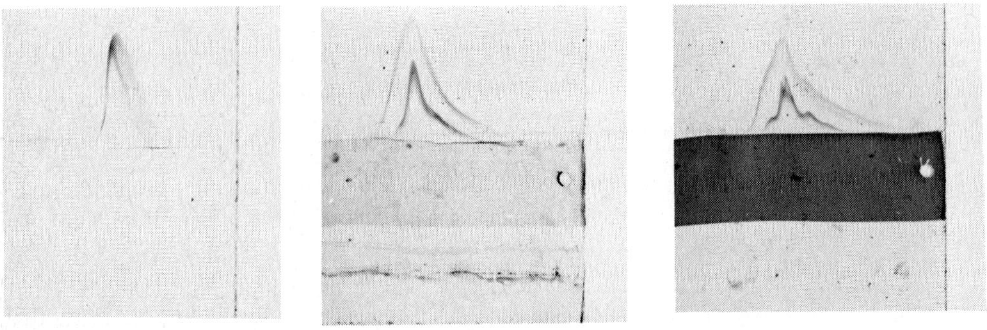

Fig.5 Crossed (affinity) immunoelectrophoretic patterns of E. coli pyruvate de-hydrogenase complex(0.15 mg/ml). In the second dimension the agarose contains 0.5 (v/v) antiserum prepared against the total complex.The first dimension contains: a) agarose ; b) Mikacion Brilliant Yellow 6GS - agarose; c) Procion Blue MX-R -ag

of the precipitin lines and by the various peaks which arise particularly in the inner precipitate. In the case of Procion Blue MX-R this is more outspoken.

The example shown illustrates the principle well although in the cases shown the strength with which individual components interact with the dye is not sufficiently different to base a preparative method on.

Another illustration is given in Fig.6 for the pyruvate dehydrogenase complex of *Bacillus stearothermophilus*.This complex differs in composition compared to that of *E. coli* and the crossed immunoelectrophoretic pattern is also quite distinct.The main precipitation pattern shows various peaks and shoulders indicative of a partial desintegration of the complex.The second faint precipitin line represents a reaction of the flavoprotein(Fig.6a). In Fig.6b and 6c the difference in resolving power is illustrated between dye-agaroses either used in the first dimension or in the second dimension as intermediary gel. Particularly in the case of Procion Blue MX-3G one of the components(E3) remains practically unretarded.In Fig.6d and 6e a similar comparison is made for Procion Yellow MX-4G.

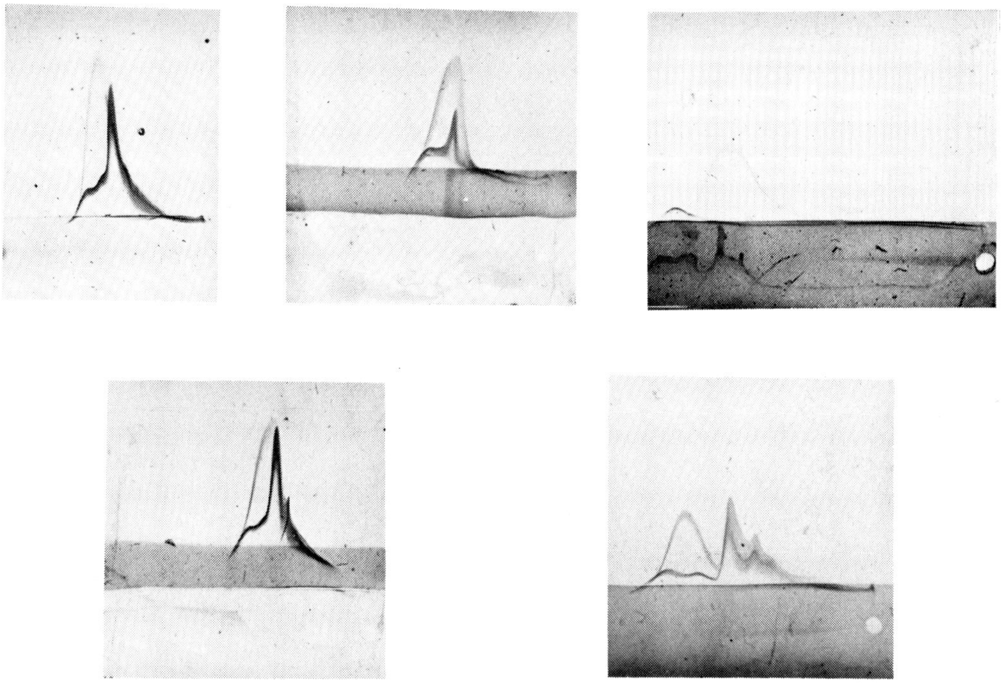

Fig.6 Crossed affinity immunoelectrophoresis of *B. stearothermophilus* pyruvate dehydrogenase complex(0.25 mg/ml). The antibody gel contains 1%(v/v) antiserum prepared against the total complex. The different pictures correspond to: a) agarose in the first dimension; b) Procion Blue MX-3G - agarose as intermediary gel in the second dimension; c) same dye in the first dimension; d) Procion Yellow MX-4G - agarose

as intermediary gel in the second dimension; e) same dye in the first dimension.

An application which we found useful for further analysis of the interaction of various dyes with individual components of the complex is to modify the multienzym complex with dichlorotriazine dyes like for instance Procion Blue MX-R.This compou destroys the overall enzyme activity of the *E. coli* complex rapidly at neutral pH although E3 and E2 remain partially active(80% and 20%).Modification of the compl can be analyzed by crossed (affinity) immunoelectrophoresis. In our example we fin that E1 becomes extensively modified and that dissociation from the complex increa upon modification. The separation is complete when analysis by crossed affinity im electrophoresis using the same dye as ligand is combined with the chemical modific ion of the protein.

AFFINITY CHROMATOGRAPHY OF PYRUVATE KINASE

In various purification procedures for pyruvate kinase,using different sources like for instance human erythrocytes,liver or kidney,rabbit muscle or brewer's ye Cibacron Blue F3G-A has been used as a ligand.This dye can be bound to the matrix direct coupling or by coupling of a dextran conjugate of the dye(refs. 20,23,24). Fructose 1,6 diphosphate is generally found to be the most effective eluant.In hyp fungi only the pyruvate kinase of *Neurospora crassa* has been studied(ref.25).It ha been shown that FDP behaves as an allosteric effector for this enzyme which compet with Cibacron Blue F3G-A for binding(ref.26). In *Aspergillus nidulans* a number of pyruvate kinase mutants have been isolated which are phenotypically different with respect to their sucrose tolerance on a mixed carbon source such as acetate/sucros It is unknown whether the protein structure itself is related to this phenomenon a therefor interest exists in the development of a purification method also useful mutant proteins.

The purification method applied for this enzyme is based on a combination of di ferent reactive triazine dyes coupled to Sepharose.Since the various polysulphona aromatic chromophores used seem to bind with different specificity,a highly select strategy can be developed for a single enzyme.Amongst the dyes screened by small s chromatography were Procion Red H-3B,Procion Orange MX-G, Procion Yellow MX-4G, Pr cion Olive MX-3G, Procion Navy MX-RB, Cibacron Blue-dextran and Mikacion Brilliant Yellow 6GS. The fungal enzyme is bound to most dyes. However,unlike the rabbit mu enzyme, addition of ethylene glycol (5% v/v) to all buffers applied, is necessary optimize enzyme recovery.

Mikacion Brilliant Yellow 6GS (C.I. 18971,reactive yellow 1) has been selected a ligand for further analysis. This choice is based on elution pattern, total rec , specific activity and SDS-PAGE pattern. The ligand concentration of the matrix been shown to be an important factor and may have to be chosen differently for di ferent organisms. Besides influencing the capacity,the ligand concentration also influences the total recovery and the specific activity of the enzyme finally obtained in this step. A concentration of 30 μmoles/g dry weight agarose has been chosen for the fungal enzyme. The rabbit muscle enzyme requires a 4-5 fold concentration of the ligand, even in the absence of ethylene glycol. In Table 1 it is shown from the bio-specific elution data that the enzyme is recognized by various dyes in a different, rather specific fashion.

TABLE 1

Elution of pyruvate kinase of A. nidulans from Mikacion Brilliant Yellow 6GS-Sepharose CL-4B and from Blue dextran 2000 - Sepharose 4B. Crude extract (4-5 units) in 50 mM potassium phosphate buffer pH 7.5 containing EDTA (0.5 mM), magnesium chloride (5 mM), β-mercaptoethanol (5 mM) and ethylene glycol (5% v/v) is loaded on 1 ml columns. Elution occurs with the same buffer containing 0.1 M and 0.15 M NaCl respectively.

ligand added	Brilliant Yellow - Seph (30 μmoles/g dry weight)	Blue dextran- Seph (14.5 μmoles/g dry weight)
none	0%	0%
ATP, 10 mM	50 %	10%
FDP, 10 mM	15 %	53%
PEP, 10 mM	41 %	108%

With Cibacron Blue F3G-A both FDP and PEP interfere with the binding whereas ATP is not effective. With the Brilliant Yellow matrix however, ATP and PEP are effective and FDP now has hardly influence. Increasing the salt concentration leads to higher recoveries. Interestingly, with Procion Red H-3B these effectors and substrates or products do not elute the enzyme; this can only be realized by the addition of salt. It remains to be seen whether mutations, situated in different regions of the protein, lead to specific differences in chromophore-protein interaction in comparison to the wild type.

The final purification protocol is shown in Table 2.

TABLE 2

Purification of pyruvate kinase of Aspergillus nidulans

step	volume (ml)	protein (mg/ml)	specific activity (μmoles/min/mg)	yield (%)
10,000 xg supernatant of crude extract	192	8.75	0.29	100
Brilliant Yellow 6GS - Sepharose CL-4B	59	0.84	8.1	82
Blue dextran-Sepharose 4B	30	0.36	23.1	53

The Brilliant Yellow column has been eluted by a combination of salt(0.5M NaCl) and substrate(1mM) to avoid unnecessary dilution of the enzyme.In the second step some contaminating proteins are succesfully removed by adding L-alanine(50 mM) to elution buffer before the final elution of the enzyme with 0.3 M NaCl. This effect is shown in Fig.7.

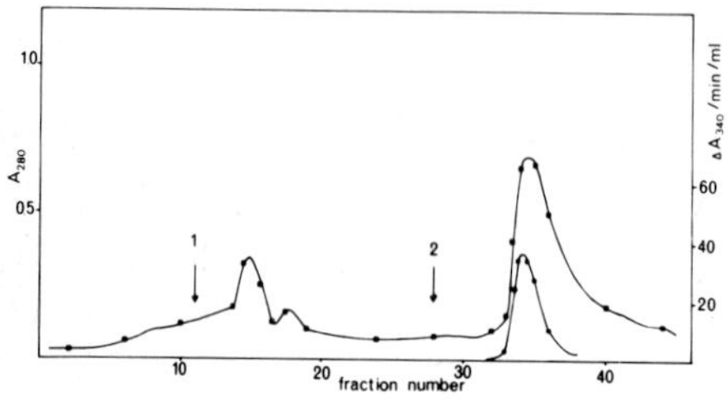

Fig.7 Elution profile of pyruvate kinase of A. nidulans on Blue dextran-Sepharose 400 units of enzyme have been applied to the affinity column(Ø 2cm, heighth 9.5cm in the same buffer mentioned in TABLE 1. Elution conditions are: arrow 1,0.1 M NaC and 50mM L-alanine added; arrow 2, 0.3M NaCl added.

The enzyme is finally purified to homogeneity by a gel permeation chromatograph step on Sephacryl S-300. The specific activity amounts 67 units/mg of protein.The dialysis and concentration step before Sephacryl chromatography result in consider losses (10-15% final yield) and need further improvement.

ACKNOWLEDGEMENTS

We are indebted to Dr. C.V. Stead, ICI Manchester,U.K. for gifts of Procion dye and to Nippon Kayaku Co Ltd., Tokyo,Japan for a sample of Mikacion Brilliant Yello 6GS. The assistence of Mr J. Haas, Centrum voor Kleine Proefdieren,Agricultural University,Wageningen, in preparing various antibodies is highly appreciated.Exper mental work of Mrs M. Strating and A. v.d Ouweland has been included in this paper We thank Dr. P.D.G. Dean for some stimulating discussions.

REFERENCES

1 P. O'Carra and S. Barry, FEBS Letters, 21(1972)281-285.
2 P. Brodelius and K. Mosbach, FEBS Letters, 35(1973)223-226.
3 D.M. Swallow, in M.C. Rattazzi,J.G. Scandalios and G.S. Whitt(Eds.),Isozymes,Cu rent Topics in Biological and Medical Research,Vol.I, Alan R.Liss Inc.,New York 1977,pp.159-187.
4 D.L. Bates, M.J. Danson, G.Hale, E.A. Hooper and R.N. Perham, Nature, 268(1977) 313-316.
5 J. Visser, M. Strating and W. van Dongen, Biochim.Biophys.Acta, 524(1978)37-44.
6 J.R. Butler, F.H. Pettit, P.F. Davis and L.J. Reed, Biochem.Biophys.Res.Commun.

74(1977)1667-1674.
7 C.R. Lowe,in O.Hoffmann-Ostenhof et al(Eds.),Affinity Chromatography,Pergamon Press Ltd., Oxford, 1978,pp 39-53.
8 J. Visser and M. Strating, Biochim.Biophys.Acta, 384(1975)69-80.
9 J. Visser, W. van Dongen and M. Strating, FEBS Letters, 85(1978)81-85.
10 J. Visser, H.C.M. Kester and A. Huigen, FEMS Microbiol.Letters, 9(1980)227-232.
11 K. Jeyaseelan, J.R. Guest and J. Visser, J.General Microbiology, 120(1980)393-402.
12 J.R. Guest and I.T. Creaghan, J.General Microbiology, 75(1973)197-210.
13 L.J. Reed and C.R. Willms, in S.P. Colowick and N.O. Kaplan(Eds.),Methods in Enzymology, Vol.IX, Academic Press,New York,1965, pp247-265.
14 J.R. Guest, J.General Microbiology, 80(1974)523-532.
15 J. Visser, M.Strating, H.C.M.Kester, K.Jeyaseelan and J.R. Guest, manuscript in preparation.
16 J.R. Guest anf I.T. Creaghan, J.General Microbiology,75(1973)197-210.
17 J.R. Guest and I.T. Creaghan, J.General Microbiology,81(1974)237-245.
18 C.J. Bos, M. Slakhorst, J. Visser and C.F. Roberts, submitted for publication.
19 J. Visser, M. Strating, C.J. Bos and C.F. Roberts,unpublished results.
20 P.D.G. Dean and D.H. Watson, J.Chromatography, 165(1979)301-319.
21 Y.D. Clonis and C.R. Lowe, Biochem. J., 191(1980)247-251.
22 T.C. Bøg Hansen, Anal.Biochem., 56(1973)480-488.
23 R.L. Easterday and I.M. Easterday, in R.B. Dunlap(Ed.), Immobilized Biochemicals and Affinity Chromatography, Adv. Exp. Medicine and Biology,Vol. 42, Plenum, New York,1974, pp. 123-133.
24 J. Marie, A. Kahn, and P. Boivin, Biochim.Biophys.Acta, 481(1977)96-104.
25 M. Kapoor, Int. J. Biochem., 7(1976)439-443.
26 M. Kapoor and M.D. O'Brien, Can.J. Microbiology, 26(1980)613-621.

T.C.J. Gribnau, J. Visser and R.J.F. Nivard (Editors),
Affinity Chromatography and Related Techniques

NUCLEIC ACID INTERACTING DYES SUITABLE FOR AFFINITY CHROMATOGRAPHY, PARTITIONING AND AFFINITY ELECTROPHORESIS

Werner Müller, Hans Bünemann, Hans-Jürgen Schuetz and Antonin Eigel
Fakultät für Biologie, Universität Bielefeld, 4800 Bielefeld, G.F.R.

ABSTRACT

The application of DNA-interacting dyes of distinct base and sequence specificity for effective base composition dependent DNA fractionations by affinity techniques is reported. The methods described include adsorption chromatography, two phase partition, liquid-liquid chromatography, and gel electrophoretic procedures.

INTRODUCTION

The fractionation of nucleic acids with respect to size or base composition is performed in general by centrifugation or electrophoretic methods. Both techniques are relatively simple and widely used; they suffer, however, from limited resolution and capacity, especially when species or fragments larger than 5×10^6 d are to be separated. We therefore tried to develop more efficient methods by making use of the principles underlying the affinity techniques common in protein chemistry.

The application of affinity techniques requires the availability of specific ligands which may either be bound in a suitable way to a support, thus forming a matrix for adsorption chromatography, or which are able to change the properties of the target macromolecules strongly enough to allow a separation from the uncomplexed species by simple means. In the field of proteins the ligand in general interacts specifically with the active site of an enzyme. The structure of the ligand is therefore related in most cases to the structure of the natural substrate of the enzyme. For nucleic acids, the base composition, the base sequence and the tertiary structure yield elements by which a fractionation with the aid of ligands is possible.

Several years ago we had selected and newly synthesized numerous basic dyes which show more or less pronounced base pair and sequence specificities on interacting with DNA (1, 2, 3). Chemical modification of some of these dyes allowed us to incorporate them into polymers, yielding suitable resins for DNA affinity chromatography, ("affinity gels") (4, 5, 6). Binding of the same dyes to one end of polyethylene glycol chains produces macroligands for DNA which strongly increase the affinity of complexed DNA for the PEG rich phase of PEG-dextran two-phase systems on one hand and strongly increase the frictional coefficient of DNA in agarose or polyacrylamide gels on the other. The first effect allows one to fractionate large DNA quantities according to base composition by partitioning in PEG-dextran two-phase systems (7) or to separate DNA mixtures by two phase partition chromatography with high resolution (8). The effect on the frictional coefficient forms the basis for an efficient DNA fractionation by base composition in gel electrophoresis (9).

In the present paper these DNA fractionations are described.

RESULTS

Base and sequence specific DNA ligands used for the development of affinity techniques for DNA

Chemical modifications of selected base and sequence specific dyes yielded the following derivatives suitable for incorporation into chromatography resins or transformation into macroligands (4, 9):

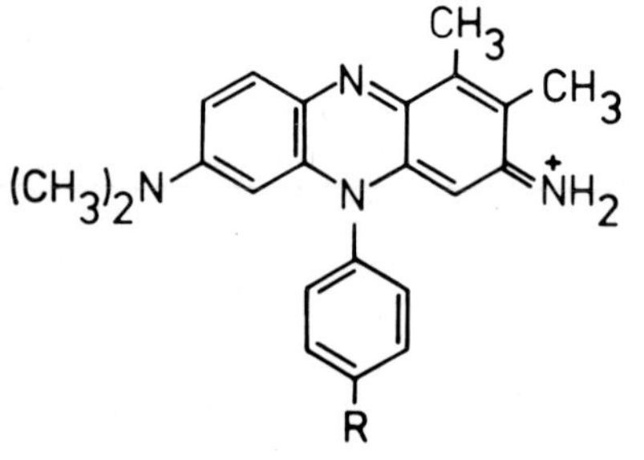

Ia: R = NHCOCH = CH_2

Ib: R = COO^-

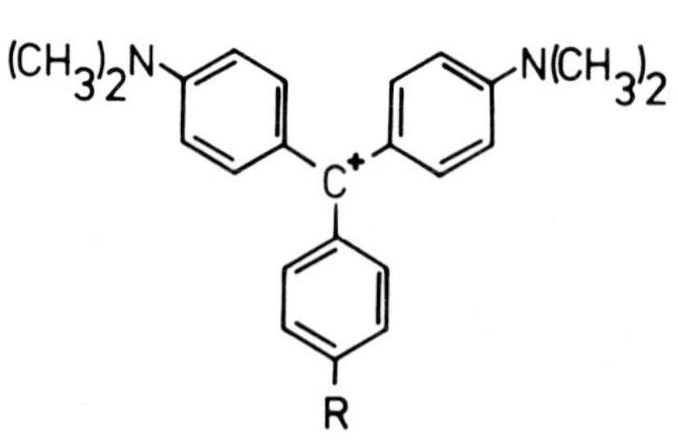

IIa: R = NHCOCH = CH_2

IIb: R = COO^-

Compound I is highly specific for adjacent GC pairs (2), while II strongly prefers two adjacent AT pairs (3) in alterning arrangement (9). For specific complexation of non-alternating sequences of 3 adjacent AT pairs the bis-benzimidazol derivative III known as "Hoechst 33258" could be used without further modification (3, 9):

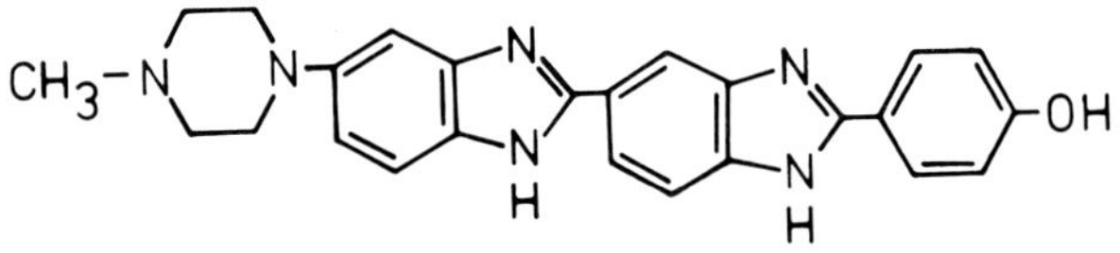

III

For discrimination between closed circular and linear or open circular DNA molecules the following intercalating ligands have been employed:

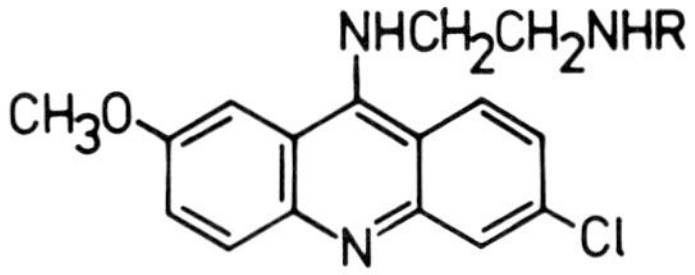

IV: R = COCH=CH$_2$

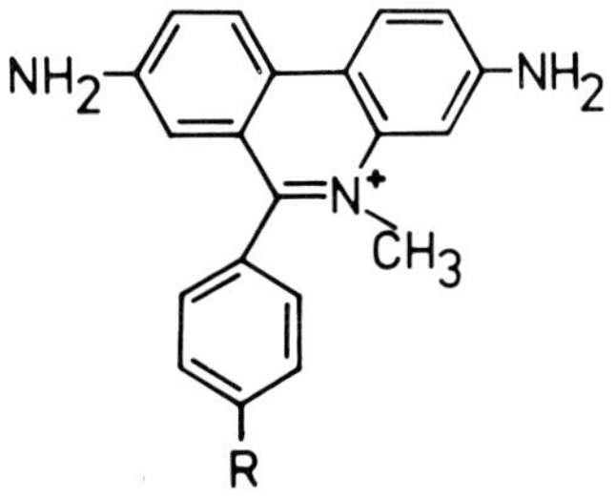

V: R = COO$^-$

IV forms the effective group of the structure-specific "yellow affinity gel"; V has been transformed into a macroligand and could be used successfully in partition techniques.

Incorporation of base and sequence specific DNA ligands into chromatography resins and their application

The acroylamino derivatives Ia, IIa and IVa may be incorporated into resins suitable for absorption chromatography by a two-step polymerization procedure (4, 5). In the first step bisacrylamide is homopolymerized in methanol-water with ammonium peroxydisulfate and N,N,N',N'-tetraethylethylene diamine to form a white solid with about 1 mmol of unreacted double bonds/gram of polymer available on the surface. In a second polymerization step linear polyacrylamide chains containing the dye subunits in statistical positions are grafted on the available double bonds by a mixed polymerization of acrylamide and acroylamino-dyes using mercaptoethanol and sodium peroxide as catalysts under nitrogen.

The resulting material binds DNA at low ionic strength and releases it at higher salt concentrations, the elution position in a salt gradient being dependent on the base composition of the DNA. Thus "red gels" (containing Ia) release AT rich DNA at lower salt than GC rich species while "green gels" (containing IIa) release the DNA in the opposite order. "Yellow gels" (containing IV) bind closed circular DNA more strongly than linear duplexes since the binding ratio of the fixed DNA never reaches the critical binding ratio ν_c. Thus the linear form is released earlier in a gradient than the closed circular form.

Transforming base and sequence specific dyes into macroligands and their application in partition and liquid-liquid chromatography

The carboxyl derivatives Ib, IIb and V may easily be esterified with excess ethylene glycol (PEG) to yield PEG monoesters in excellent yields. The esterification is carried out in a melt of PEG (a $\overline{M}_w$-species of 6000 - 7500 was used in most cases) in the presence of large amounts of imidazol using toluenesulfonyl chloride as condensing agent at 90°C. Compound III ("Hoechst 33258") is coupled to PEG using monobromo-PEG and potassium carbonate in dimethylsulfoxide at 80°C by formation of an ether bond to the OH-group of the dye. Removal of the imidazol and unreacted PEG is performed by ion exchange chromatography on CM- or SP-sephadex gels. More details of the preparations are given in ref. 9.

The macroligands obtained all show a marked preference for the PEG-rich upper phase of PEG-dextran two-phase systems. Partition coefficients were found between 4 and 5 at 37°C at ionic strengths around 0.15 in systems containing PEG of 8000 to 9000 $\overline{M}_w$ units and dextran of $\overline{M}_w \sim 500000$. In the absence of macroligands the partitioning of DNA in such systems depends strongly on the presence of salts. Albertsson (10), who studied the behavior of polynucleotides in such systems in detail, showed that Li^+ ions and polyvalent anions cause DNA to move to the upper phase, while K^+-, Cs^+- and Rb^+-ions excert the opposite effects. In the presence of macroligands the salt effects may be overcome; binding of 1 macroligand per 50 base pairs is sufficient to raise the partition coefficient of DNA by roughly three orders of magnitude in a system containing 30 mM KCl (7).

When several DNA species are present in the two-phase system the base specificity of the macroligands obtained from I - III may be used for effective separations according to base composition. We could show for example, that if two DNAs, within the range 40 % - 60 % GC content, differ by only 1 % in base composition, the PEG ester of Ib produces a 30 % larger shift in the partition coefficient of the more GC rich DNA (7). This implies that two DNAs differing by 20 % in base composition present in equal amounts may be separated to yield 90 % pure components by one partition step. Effective prefractionations of DNAs from eucaryotic systems in large scales are possible this way. Multistep extractions and countercurrent distributions carried out over 10 - 20 steps by hand resulted in substantial enrichments of the different main satellite DNAs of calf thymus DNA (7).

Extension of the principle of two phase partition to liquid-liquid chromatography systems is possible because inert materials were found to bind the dextran rich lower phase strongly enough to act as supports for the stationary phase. These materials comprise cellulose (11), polyacrylamide gels and celite, of which cellulose shows the highest binding capacity and the best accessibility of the bound phase.

Chromatographic columns prepared from dextran phase-coated supports using the PEG rich upper phase as mobile phase separate in the absence of the macroligands DNA mixtures according to size (11). When the macroligands are added to the mobile phase, the sizing effect is reduced and very efficient separations according to base composition may be obtained. Two main procedures for using the macroligand have been employed. In the first, the DNA mixture loaded on the column is eluted in a gradient of increasing macroligand concentration at constant salt in the mobile phase. The alternative procedure uses a salt gradient from K^+ to Li^+ at constant ionic strength and constant macroligand concentration. While the second procedure is always applicable, the first requires that the components which bind the macroligand more strongly are present in smaller amounts in the sample than the bulk DNA. A theoretical treatment of gradient techniques applied to two-phase columns is in progress (12).

A slight disadvantage of the columns is their low permissible flow velocity. 5 ml/h per cm^2 of cross section was found to be the upper limit for retaining maximum resolution, probably due to the limited diffusion rate of the DNA (DNA samples of $\overline{M}_w \sim 30 - 40 \times 10^6$ have been used routinously) in the rather viscous medium of the phase pair. In order to increase the diffusion rates, the temperature at which the columns are run is kept at 37° C. Higher temperatures do not yield any further improvement since there are several counteracting effects, such as changes in the binding of lower phase to the support, a decrease in affinity of the macroligand for the mobile phase as well as in the base specificity of the macroligand.

Application of macroligands for base and sequence specific DNA separation in agarose and polyacrylamide gels.

The binding of macroligands to DNA results in a branching of the linear DNA molecule. This branching reduces strongly the electrophoretic mobility in gels containing macroligands mainly by increasin the frictional coefficient of the DNA (9). Since the amount of bound base and sequence specific macroligands depends on the base composition of the DNA, a separation of DNA species according to base composition is possible.

Test runs performed with DNA fragments of known base composition and various sizes revealed that DNA fragments differing by 0.5 % in base composition may be resolved after travel distances of $\sim$10 cm. The reproducibility of the retardation effects observed yields a simple way to determine DNA base compositions with very small amounts at a precision of $\pm$ 1 %. The formal treatment of the method as presented in ref. (9) leads to the following relation between the electrophoretic DNA retardation (v_o/v - 1) and the molar fraction χ of GC or AT pairs in the DNA studied:

$$(v_o/v - 1) = \frac{K}{f_o}\left[\chi^n_{GC(or\ AT)}\right]^{\beta}$$

v_o and f_o design the mobility and the frictional coefficient of the uncomplexed DNA, v the mobility of the DNA in the presence of the macroligand, K a constant, n the number of adjacent base pairs of the same kind required for binding of the macroligand and ß an experimental factor allowing for a non-linear power dependence of the frictional coefficient of DNA on the number of bound PEG chains. ß and K/f_0 are determined in a reference run using DNA fragments of known base composition. The n-values for the three macroligands have been determined by separate binding studies (9).

The high DNA resolution with respect to base composition obtained by adding macroligands to the gels caused us to develop a two-dimensional gel technique for the analysis of more complex mixtures of fragments. The simple procedure which does not require any special facilities except thermostating the gels during the second run starts with a size separation of the fragments. (First dimension run.) The gel strip containing the separated bands is used as starting line for the second dimension gel to which the macroligand was added. Thus "diagonal" gels are obtained. Alternatively, the bands formed in the sizing run are cut out and the gel pieces are embedded separately side by side into the second dimension gel when this is poured. Since the bands may move on in the same direction as in the first run, smaller mobility differences are detectable this way. For more technical details see ref. (9).

CONCLUSIONS

The base and sequence specific DNA binding dyes used as adsorption elements in chromatography resins or in the form of macroligands in partition techniques, liquid-liquid partition chromatography or gel electrophoresis form powerful elements for affinity separation techniques for DNA. The resolutions obtained in the two-phase partition technique is comparable to classical density gradient centrifugations in the presence of base specific ligands. The advantage of the partition method consists in enormous capacity which allows to handle up to 500 µg of DNA (20 - 40 x 10^6 d) per gram of two-phase system.

The resolving power of the other procedures described which allows one to separate DNA fragments differing by 2 - 3 % in base composition by adsorption chromatography, 0.5 % by gel electrophoresis and 0.2 - 0.3 % by liquid-liquid chromatography is superior to all other techniques known to us. Since neither the preparations of the adsorption chromatography gels (which are available from Boehringer, Mannheim) nor the preparations of the macroligands (which might become soon available on the market) are too complicated (4, 9), the techniques are fully accessible.

The application of the described fractionation procedures to RNA mixtures is in progress.

ACKNOWLEDGEMENTS

We are grateful to D. M. Crothers for many helpful discussions. This work was supported by grants from the German Federal Government, from the Deutsche Forschungsgemeinschaft and from the NIH.

REFERENCES

1 W. Müller and D. M. Crothers, Eur. J. Biochem. 54 (1975) 267-277.
2 W. Müller, H. Bünemann and N. Dattagupta, Eur. J. Biochem. 54 (1975) 279-291.
3 W. Müller and F. Gautier, Eur. J. Biochem. 54 (1975) 385-394.
4 H. Bünemann and W. Müller, Nucleic Acids Res. 5 (1978) 1059-1074.
5 H. Bünemann and W. Müller, in O. Hoffmann-Ostenhof (Ed.), Proceedings of the International Symposium of Affinity Chromatography, Pergamon Press, Oxford, 1978, p.p. 353-356.
6 H. Bünemann and W. Müller, Naturwiss. 64 (1977) 632-633.
7 W. Müller and A. Eigel, submitted for publication.
8 W. Müller, manuscript in preparation.
9 W. Müller, I. Hattesohl, H.-J. Schuetz and G. Meyer, Nucleic Acids Res. 9 (1981) 95-119.
10 P.-Å. Albertsson, in Partition of Cell Particles and Macromolecules, 2nd edn., Almquist and Wiksell (Eds.), Stockholm (1971).
11 W. Müller, H.-J. Schuetz, C. Guerrier-Takada, P. E. Cole and R. Potts, Nucleic Acids Res. 7 (1979) 2483-2499.
12 D. M. Crothers and W. Müller, manuscript in preparation.

T.C.J. Gribnau, J. Visser and R.J.F. Nivard (Editors),
Affinity Chromatography and Related Techniques

IMMOBILIZED ACRIFLAVIN FOR AROMATIC INTERACTION CHROMATOGRAPHY : SEPARATION OF NUCLEOTIDES, OLIGONUCLEOTIDES AND NUCLEIC ACIDS

J.M. EGLY* and E. BOSCHETTI**

* INSERM, U-184, Laboratoire de Génétique Moléculaire des Eucaryotes du C.N.R.S., 11, rue Humann, 67085 Strasbourg Cedex, France.

** Réactifs IBF-Pharmindustrie, 35 avenue Jean-Jaurès, Villeneuve-la-Garenne, France.

SUMMARY

Using aromatic ligands anchored on insoluble polysaccharide matrices it is possible to separate many classes of synthetic and naturally occurring substances according to their structure.

With highly substituted acriflavine-agarose, we have developed a simple and efficient technique for the separation of nucleotides and oligonucleotides (chemically or enzymaticly synthesized, hydrolysis products...). In the addition of aromatic interactions, the separation power is improved by some electrostatic effects (positive charge of acriflavine - negative charge of phosphate groups). On the other hand, with lowly substituted acriflavine-agarose, we have perfected a new method for the separation of single strand from double stranded nucleic acid.

INTRODUCTION

Charge-transfer interactions are a field of growing interest in molecular biology and genetic. Among these interaction studies, one basic separation method has been developed : the aromatic interaction chromatography (AIC) between aromatic or pseudo aromatic compounds [1,3]. The principle of this chromatography is based on a charge transfer interaction between aromatic solutes and an aromatic ligand immobilized on an insoluble and non ionic chromatographic matrix. During the chromatographic process the aromatic components of the mixture interact with the aromatic ligand which promotes a certain delay in the elution and thus proportionally to the intensity of the interaction. The most important point of this type of chromatography is the proper selection of the ligand which has been recently discussed [4].

In the field of nucleotides and nucleic acids in general, the choice of the acriflavine is justified by the formation of complexes between purines and flavins

[5,6] already exploited for the separation of DNA fragments [7] of oligonucleotides [8] and single stranded from double stranded nucleic acids [10].

In this paper we describe the use of immobilized acriflavine for the separation of nucleotides and their related polymers on higly or weakly acriflavine substituted gels.

MATERIALS AND METHODS

Chemicals

Nucleotides and Nuclease S1 were purchased from Boehringer (Mannheim, West-Germany) ; Oligonucleotides from Collaborative Research Inc. (Waltham, Mass., USA) Poly-A from New England Nuclear Co. (Boston, Mass., USA) ; Poly-C from Miles Laboratories (Slough, England). Acriflavine, Acriflavine-Ultrogel A4R type I and type II respectively highly and weakly substituted were prepared by a modification of the original method described by Egly et al [2] or obtained from Reactifs-IBF (Villeneuve-la-Garenne).

All other chemicals reagents were of analytical grade.

Synthesis of acriflavine-agarose adsorbents

Acriflavine-agarose gels were synthesized as described elsewhere [2] where the dextran gel was replaced by a 4 % agarose gel. The acriflavine-agarose gels thus obtained were extensively washed with 10-20 volumes of ethanol-water (v/v), with water, with 10 % acetic acid, with 1 M sodium chloride and finally with the adequate chromatographic buffer. The details of the chromatography experiments are given on the figure legend.

RESULTS AND DISCUSSION

Separation of nucleotides

It is known that acriflavine interacts strongly with purines and particularly with guanine and adenine. Furthermore this acridine compound shows a cationic character due to the positive charge of N_9 and is able to provide some electrostatic effect. Thus it is possible to distinguish first purines from pyrimidines as represented Fig. 1a and secondly to separate a mixture of adenosine, AMP, ADP and ATP (Fig. 1b). In this latter case the separation is effected at low ionic strength to favorize the electrostatic effect between acriflavine and phospho-group [4].

These interesting properties were exploited for the separation of cyclic nucleotides from their corresponding 5' nucleotides on the basis either to their different charge transfer interaction or to their different sterical approach during the complexe formation. The good separation efficiency for cyclic AMP from 5'-AMP has been applied to the determination of 3'-5' cyclic nucleotide phosphodiesterase activity [12].

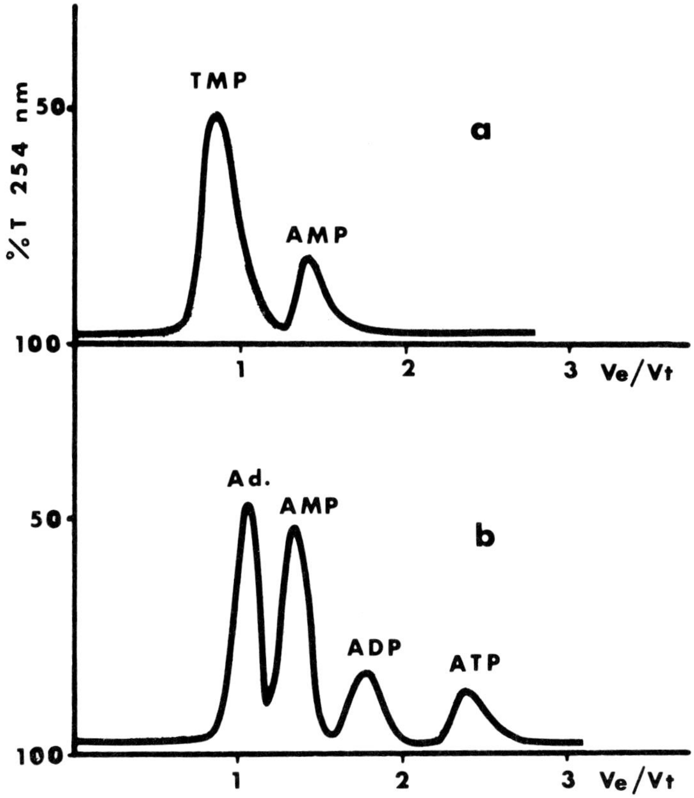

Fig. 1. Separation of mononucleotides on Acriflavine-Ultrogel A4R type I. Column 1.6 x 15 cm ; buffer 0.1 M ethylmorpholine, pH 7 ; flow-rate 26 cm . h^{-1} ; temperature 20° C. The separations were performed on isocratic conditions.
Ad. = adenosine

Separation of oligonucleotides

Usually chemically or enzymaticly synthesized oligonucleotides are separated by ion exchange chromatography on cationic resins, but due to the low difference of charge between two consecutive oligomers the separation are often incomplete.

The chromatographic behaviour of oligonucleotides on immobilized acriflavine is very particular because these molecules associate in addition to the electrostatic and the charge transfer interactions, the multipoint attachment phenomenon according to their molecular size. In this optic, increasing the degree of substitution of the acriflavine, one increases the interaction probabilities between the solute and the substituted gel. As a consequence, one observes an increase of the adsorption strength and this, proportionally to the size of the oligonucleotide. With

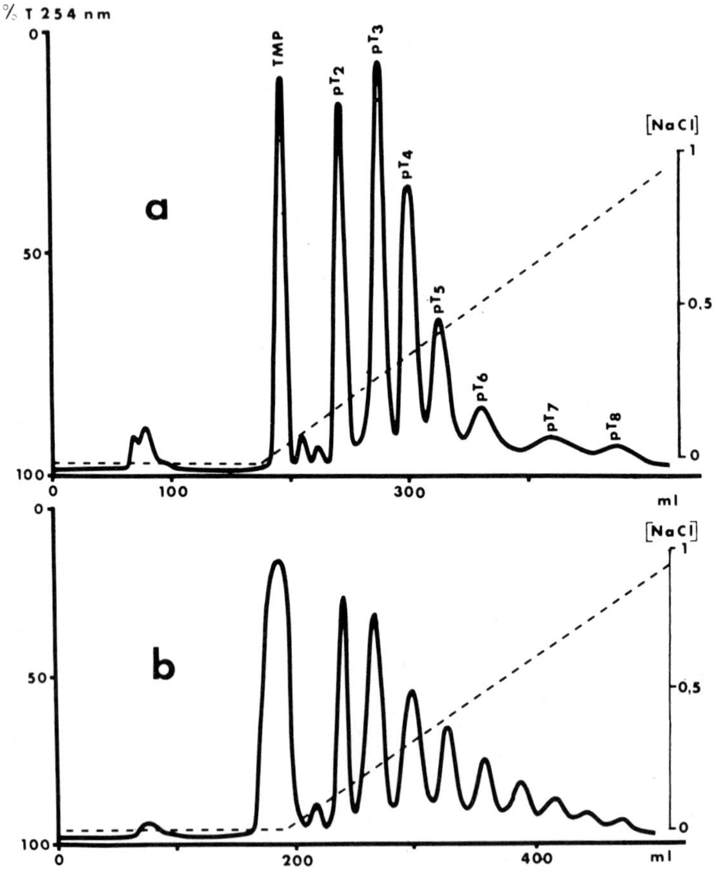

Fig. 2. Separation of oligonucleotides mixtures on acriflavine-Ultrogel A4R type I. a : synthetic oligonucleotides obtained from 5'-TMP. b : the sample was thymus DNA hydrolyzate obtained after treatment with 60 % formic acid at 37° C for 20 h. Experimental conditions : column dimension : 1.6 x 60 cm ; buffers : 0.01 M sodium acetate, pH 4.5 (a) and 0.01 M ethylmorpholine, pH 5 (b) ; elution gradient sodium chloride in the starting buffer from zero M to 1 M ; gradient slope : 4.7 mM . cm^{-1} ; flow-rate 17 cm . h^{-1} ; temperature 20° C.

highly substituted acriflavine-agarose gel the separation efficiency is generally maximum at a pH where both charge transfer and electrostatic interaction add up : the optimal pH zone is around 4-5 [8]. Fig. 2a and b shows elution profiles of oligo-dT serie and a mixture of oligodeoxynucleotides obtained by acidic hydrolysis of thymus DNA.

This technique seems to represent a potential tool for analytical or preparative scale of sequenced oligonucleotides and is a serious competitor for high pressure liquid chromatography.

Separation of single stranded from double stranded nucleic acids

It was suggested that the interaction between acridine and DNA follows two binding mechanisms [9] : a strong one when the molecule intercalates between two adjacent base layers in the DNA helix and a weaker one which involves mainly electrostatic interaction between the positive charge of acriflavine and the negative charges of DNA phosphates.

In chromatography, the interaction between double stranded DNA and acridine drugs are prevented first for lowly substituted gels [10] and secondly when increasing salt concentration as observed for ethidium bromide [11]. In such conditions, it is possible to discriminate between single stranded and double stranded nucleic acid.

TABLE 1

Interaction of polynucleotides and nucleic acids on immobilized acriflavine*.

	Solute	Non adsorbed on the column %	Adsorbed and eluted %
single stranded	Poly A	20	50
	cDNA**	20	66
double stranded	DNA***	49	46
	PolyA-PolyU	63	30
	cDNA-RNAov****	58	38
Double stranded treated by Nuclease S1	DNA	92	7
	cDNA-RNAov	88	10

* The substitution degree was 5 μmoles of acriflavine per mol of gel.
** Complementary DNA from total mRNA from chicken oviduct.
*** Crude DNA from plasmacytoma cells.
**** For hybridization conditions and S1 treatment see Ref. 10.

From Table 1, three main results are evidenced. Firstly single stranded nucleic acid e.g. poly A or cDNA are adsorbed on the acriflavine gel. Secondly once these nucleic acid are hybridized e.g. polyA-polyU or cDNA-RNAov, they partially lost their adsorption properties towards the subsituted gel. In this case as observed for DNA between 49 and 63 % of the nucleic acid pass through the column. Thirdly when one chromatographied DNA or cDNA-RNAov hybrid pretreated by nuclease S1 to hydrolyze the single stranded region they directly pass through the column. As a control the hybrid cDNA-RNAov was chromatographied on an acriflavine-agarose gel column and the two obtained fractions were treated by nuclease S1. Less than 8 % of labelled material was hydrolyzed in the first peak whereas up to 65 % of material was hydrolyzed in the second peak.

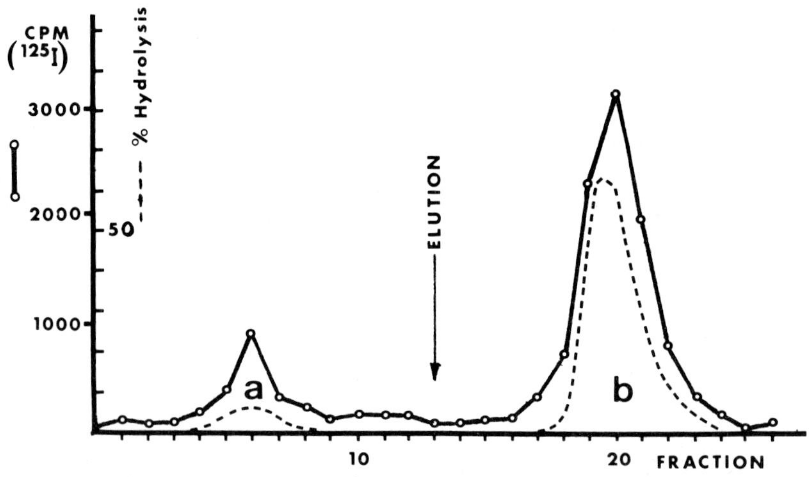

Fig. 3. Separation of cDNA-RNA hybrid from single strand on acriflavine-Ultrogel type II. The sample (20 μg) was constituted of a mixture of hybrid, cDNA and mRNA. Column : 0.3 x 2.8 cm ; initial buffer 10 mM Tris-HCl, pH 7.6 + 0.5 M KCl ; elution buffer 10 mM Tris-HCl, pH 7.6 + 0.5 % SDS ; flow-rate 15 ml . h^{-1} ; temperature 20° C.
a : cDNA-^{125}I RNAov hybrid (nuclease S1 hydrolysis : 8 %)
b : single stranded nucleic acids (nuclease S1 hydrolysis : 68 %).

CONCLUSION

As a consequence of the natural complexe interaction between acriflavine and nucleotides it is possible to perfect methods for the purification of different nucleotides or nucleic acids. However the success of the separations is dependent on the knowledge of the different parameters governing the interaction mechanisms. The development of aromatic interaction chromatography using acriflavine as a ligand offers several advantages and the methodologies described here seem to be a promizing approach either as a preparative scale for oligonucleotides obtention or as an analytical scale for the purification of small quantities of hybrids or sequenced oligonucleotides.

REFERENCES

1 J.M. Egly and J. Porath, in Hoffmann-Ostenhof, O., Breitenbach, M., Koller, F., Draft, D., Scheiner, O., Eds., Proceedings of an International Symposium at Vienna, 1977 , Affinity Chromatography, Pergamon Press, 1978, pp 5-22.
2 J.M. Egly, FEBS Lett., 93 (1978) 369-372.
3 J.M. Egly and J. Porath, J. Chromatogr. 168 (1979) 35-47.
4 J.M. Egly, J. Chromatogr. Library, Elsevier Publ. Co. (1981) in press.
5 G. Weber, Biochem. J. 47 (1950) 114-121.
6 M.A. Slifkin, Biochim. Biophys. Acta 103 (1965) 365-373.
7 H. Bunemann and W. Muller, Nucl. Acids Res. 5 (1978) 1059-1074.
8 E. Boschetti, P. Girot and J.M. Egly, FEBS Lett. submitted
9 L.S. Lerman, J. Mol. Biol. 10 51964) 367-380.
10 J.M. Egly, J.L. Plassat and E. Boschetti, Anal. Biochem. submitted.
11 J.B. Le Pecq and C. Paoletti, J. Mol. Biol. 27 51967) 87-106.
12 C. Rochette-Egly and J.M. Egly, J. Cyclic Nucl. Res. 6 (1980) 335-345.

CHAPTER VI

APPLICATIONS

- HIGH PERFORMANCE LIQUID (AFFINITY/HYDROPHOBIC) CHROMATOGRAPHY
- AFFINITY PARTITION
- PEPTIDE SYNTHESIS

4th INTERNATIONAL SYMPOSIUM ON AFFINITY CHROMATOGRAPHY AND RELATED TECHNIQUES

T.C.J. Gribnau, J. Visser and R.J.F. Nivard (Editors),
Affinity Chromatography and Related Techniques

APPLICATION OF HIGH PERFORMANCE LIQUID CHROMATOGRAPHY TO THE SEPARATION AND ISOLATION OF BIOPOLYMERS

K.K. UNGER and P. ROUMELIOTIS
Institut für Anorganische Chemie und Analytische Chemie
Johannes Gutenberg-Universität
65oo Mainz, G.F.R.

SUMMARY

This paper reports on the application of High Performance Liquid Chromatography (HPLC) to the separation of proteins, peptides and nucleotides. Several criteria need to be considered in order to adapt HPLC to the analysis and isolation of biopolymers. A survey is given on the composition, selectivity and performance of size exclusion, reverse phase and ion-exchange systems. The various aspects of developing a separation, i.e. sample pre-treatment, choice of phase system, column packing, size and column dimension, column operation and detection are broadly discussed. The potential of various modes of HPLC is illustrated by numerous examples.

---

1. INTRODUCTION

Biologically active substances from natural sources may be isolated and characterized by a variety of well-established methods, i.e. centrifugation, electrophoresis, dialysis, ultrafiltration, classical column chromatography and affinity chromatography, each providing a characteristic selectivity, fraction capacity, sample load, speed and convenience (ref. 1,2). Most of the above methods are laborious and time-consuming.

During the past decade efforts have been directed at adapting High Performance Liquid Chromatography (HPLC) to the separation of biomolecules in clinical and pharmaceutical analysis (ref. 3). HPLC offers two particular advantages: rapid <u>and</u> high resolution separations both in the analytical and semi-preparative mode. This paper aims to survey the most important contributions and aspects in this field from a chromatographic point of view. Compounds containing biomolecules are numerous and discussion here is confined to biopolymers i.e. enzymes, proteins, peptides and nucleotides.

2. CHARACTERISTICS OF HPLC PHASE SYSTEMS AND THEIR SUITABILITY FOR THE SEPARATION OF BIOPOLYMERS

Biomolecules of living organisms require a much more careful and sometimes distinctly different treatment in fractionation, separation and identification than do mixtures of synthetic products. Work usually begins with controlled fragmentation of cells and tissues into smaller polymeric units such as peptides, proteins and nucleotides by means of specific cleavage reactions. The digests obtained are buffered solutions and hence buffers are also the potential eluents for further HPLC separations. Organic solvents or modifiers may be added to some extent, but care must be taken not to affect the biological activity of the solutes. Maintenance of the chemical structure and biological activity is one of the most important aspects in the separation of biopolymers by means of HPLC. In addition, the great complexity and diversity in molecular structure of biopolymers requires an extremely high resolving power of the HPLC phase system.

Given these considerations it is obvious that in adapting traditional HPLC to the separation of biopolymers the specific molecular properties of the solutes and the conditions under which the solutes can be "handled" must be considered, i.e. (i) the existing HPLC phase systems must be critically examined with respect to their suitability and limitations, (ii) appropriate conditions for high selectivity and performance must be established and (iii) the process variables and techniques with regard to speed and convenience must be optimized.

2.1 Separation criteria

Biopolymer analysis may be carried out with widely differing objectives. Nevertheless it would appear useful to establish some generally valid criteria for the setting-up of an HPLC separation. Degradation and fractionation of the crude should be undertaken by means that are highly reproducible and with a minimum number of steps yielding sufficiently chemically stable intermediates of appropriate concentration. The digest sample on a HPLC column should contain less than 100 constitutions (except for size exclusion columns), otherwise an additional subfractionation is required. Secondly, the HPLC phase system selected should offer a maximum resolution. As chromatographic resolution (ref. 4) is governed by the selectivity of the phase system, attention must be paid to the type of packing and to the eluent composition. A high efficiency of the phase system may be achieved by employing microparticulate packings, appropriate column lengths and optimum flow rates of the eluent. Furthermore, the column should have a long life-time. This last point is critical in HPLC of biopolymers; silica-based packings are not sufficiently stable towards buffered eluents. The batch-to-batch reproducibility of the employed packing should be carefully controlled by chromatographic means. The selectivity of the column, at constant eluent composition, may be altered by substituting the packing with a similar packing from another supplier, differing in surface structure and pore size. Thirdly, high recovery and maintenance of the biological activity of solutes during elution is a "conditio sine qua non" in the resolution of native biopolymers. To this end both quantities should be assessed by appropriate measurements. Recovery and stability data obtained under one set of conditions cannot be assumed to apply to other eluent compositions and therefore need to be checked again.

2.2 Composition, selectivity and performance of phase systems (ref. 3)

There are three major HPLC phase systems which can be applied successfully to biopolymer separation, each providing a characteristic selectivity.

(i) size exclusion phase systems (ref. 5,6)

As the name implies, retention is governed by a size exclusion mechanism, i.e. solutes are eluted in the sequence of decreasing molecular weight. Secondary effects based on ionic and hydrophobic interactions may additionally affect retention under some conditions, causing deviations from the predictable elution sequence. The packings are either polar cross-linked organic gels or polar chemically modified silicas (ref. 6). The latter carry bonded polymeric or monomeric layers of polyether or the glycerolpropyl type and differ in carbon load, surface concentration, content of residual silanol groups, pore size of the packing and column porosity. The packing properties in combination with a given eluent composition control the selectivity in size exclusion. Sufficient chemical stability was observed for some, but not all, types of packing. Compared to other modes of HPLC, selectivity and peak capacity in size exclusion is rather poor. It can be improved to some extent by employing highly efficient packings and/or by utilizing the secondary effects influencing retention (ref. 7). As a rule of thumb biopolymer differing in MW by a factor of two and sometimes less can be separate

(ii) reverse phase systems

Reverse phase systems composed of n-alkyl bonded silicas and polar eluents are capable of discriminating between biopolymers, e.g. proteins, peptides and nucleotides, in accordance with their hydrophobic character. The retention is based on solvophobic interactions arising between the hydrophobic part of the solutes and the surface of the hydrophobic ligand (ref. 8). Hence, retention and selectivity are controlled by both the properties of the reverse phase packing, i.e. chain length of bonded n-alkyl group, the carbon load, the surface concentration and the content of residual silanol groups (ref. 9) and the solvent strength of the eluent, its pH and ionic strength (ref. 8). It was found that the addition of organic modifier e.g. n-propanol, to the buffered hydro-organic eluent improves selectivity and peak symmetry (ref. 10). Reverse phase columns are preferably run under non-isocratic, i.e. gradient elution condition When the pores of the packing are sufficiently wide, peptides can be separated up to 50 000 daltons. An improvement of the selectivity, apart from changing the solvent strength, pH and the ionic strength, may be accomplished by adding an ion pair reagent in a low concentration. An uncharged ion pair is thus formed between the charged solute and the oppositely charged ion pair. Anions, e.g. phosphate, acetate, formate, trifluoracetate, are commonly employed in the separation of peptides (ref. 10). This mode is particularly suited to hydrophilic peptides which attain a sufficiently hydrophobic character by ion pair formation. Strongly hydrophobic proteins are highly retained. In this case reverse phase packing with short n-alkyl chains were employed.

(iii) ion exchange phase systems (ref. 11,12)

In principle, peptides, proteins and nucleotides can be separated. The retention is based on an ion exchange distribution process between the charged solutes and the ionic surface functional groups as far as they are accessible to the solutes. In some cases hydrophobic interactions and complexation are also involved. The ion exchangers are either of the pellicular or totally porous type composed of organic gels, or chemically modified silicas (ref. 9). At a given composition of the phase system, retention is a function of the charge of the solute (ref. 13). The most dominating parameter of the eluent composition controlling retention and selectivity are the pH, the type, valence and concentration of electrolyte ions and

and the type of buffer. A certain draw-back in operating silica-based ion exchange columns is the limited chemical stability, which is associated with a continuous decrease of the ion exchange capacity. In order to elute solutes with a wide range of pk or pI values, gradients in pH or in electrolyte concentration are applied.

Highly efficient columns can be packed with each of the phase systems mentioned above, or pre-packed columns are commercially available complete with a test protocol. With 5 μm particles up to 20 000 theoretical plates can be achieved on a 250 mm column under isocratic conditions. Column coupling provides an additional technique to enhance efficiency. Peak symmetry often creates difficulties but can be improved by adding suitable modifiers to the eluent.

2.3 Designing a separation

Depending on the nature of the biopolymers and on the objective of the separation, a series of practical aspects need to be considered:

(i) sample pre-treatment (ref. 14)

The sample to be analyzed should be free of any fines, colloidal constituents or precipitates. Guard columns may be employed to protect the column itself, but they must not impair total column efficiency. Highly reproducible and well-defined clean-up procedures are required for quantitative analysis. This is preferably done using automatic devices.

(ii) choice of the type and size of packing and of column dimensions

Having decided on the specific mode of HPLC for the separation, i.e. size exclusion, reverse phase, ion exchange, the packing and eluent composition should be chosen. Additional experiments may be required in order to transfer separation conditions from one packing to another of similar composition due to differences in surface properties, pore size and porosity between packings. Columns packed with 10 μm are adequate to start a separation at this stage. Columns with 5 μm particles are difficult to pack without experience and in this case pre-packed columns should be employed. The length and inner diameter of the column are determined by the required column efficiency and column loadability.

At a given particle size and at optimum linear velocity of the eluent the column efficiency, expressed in terms of the number of theoretical plates, can be increased by using longer columns. This, on the other hand, lengthens the analysis time. A reasonable compromise might be a column length of 250 to 300 mm. In size exclusion two columns of this length are often assembled in order to match a wide molecular weight range and to increase resolution (ref. 15). Well-packed 5 µm particle columns will usually generate 10 000 to 20 000 theoretical plates, depending on the capacity factor of the solute.

Linear sample capacity as a measure of column loadability varies slightly from phase system to phase system, but generally ranges from 0.1 to 1 mg solute per gram packing. Analytical columns of 250 x 4 mm i.d. contain about 2 g of the packing material and thus permit a sample load of a maximum of 2 mg per solute and injection. For columns of 250 mm length and 10 mm i.d. the mass of the packing is higher by one order and hence analytical as well as semi-preparati high performance separations can be carried out. Large bore microparticulate columns of 22 to 30 mm i.d. are also commercially available. Although their cost is extremely high due to the large amount of packing material, they are useful for semi-preparative high performance separations of expensive compounds.

(iii) adjusting retention and controlling the selectivity or phase systems

The most convenient method of adjusting retention and achieving high selectivity is to vary the eluent composition at a given type of packing, i.e. column. In size exclusion, biopolymers are eluted within a limited volume range, namely between the interstitial column volume V_o and $V_o + V_i$, V_i being the total pore volume of the column packing. Selectivity in size exclusion is mainly determined by the pore size distribution and the pore volume of the packing. Changes in the mobile phase composition may have small effects on the selectivity through ionic and/or hydrophobic interactions (ref. 7), and in some cases column efficiency and peak shape can be improved in this way also.

For developing a separation in reverse phase chromatography it is recommended to start under isocratic conditions, whereby the selectivity and column efficiency can be measured exactly, which is not the case with a gradient elution. In the reverse phase mode ternary solvent mixtures composed of an electrolyte, an organic solvent and an organic modifier were found to provide the highest selectivity. In order to increase retention an ion pairing reagent may be added to the eluent. The effect of the solvent strength and other mobile phase parameters on retention of biological solutes has been examined extensively (ref. 8,10). The optimization of k' requires the application of a gradient in the eluent composition, as in most cases the sample is of a complex composition and substances are eluted over a wide capacity factor range. A gradient elution increases the peak capacity, the resolving power of the column and it reduces analysis time (ref. 16). However, the conditions chosen should maintain a selectivity comparable to that of the isocratic system.

(iv) column operation

The most important process variable in column operation is the flow rate or linear velocity of the eluent. Optimum linear velocities for achieving maximum efficiency are approximately 1 mm/s for 5 μm and 0.5 mm/s for 10 μm reverse phase particles (cf. ref. 17). This corresponds to a volume flow rate of about 0.1 to 0.5 ml/min for a 4 mm i.d. column. As the solutes have a high molecular weight, MW diffusivity will increase with MW and hence column efficiency is very much a function of MW. For instance, the efficiency for proteins was estimated at 5 000 to 9 000 theoretical plates on a 250 mm LiChrosorb^{R} Diol 5 μm particle column, while monomeric compounds may generate 20 000 to 25 000 theoretical plates (ref. 18). It was mentioned that gradient elution is a useful tool in resolving complex mixtures. The following types of gradients can be distinguished:

change in the concentration of the organic solvent or modifier,
change in the lipophilic character of the counter ion in ion pair reverse phase chromatography,
change in concentration of the counter ion in ion pair reverse phase chromatography.

Band migration, resolution and peak width in a gradient elution are largely controlled by the shape and the steepness of the gradient (ref. 16). In ion exchange chromatography a gradient in pH and electrolyte concentration can be applied in addition to a solvent gradient.

Another technique to improve selectivity and to reduce the separation time in analyzing complex mixtures is column switching. This technique calls for columns containing packings of different selectivity which are connected by a switching valve and run under isocratic conditions with the same eluent composition. Depending on the configuration of the column set, certain parts of the chromatogram on the main column may be cut and are eluted with a higher resolution on a second or third column. However, the application of this stationary phase programming as an alternative to mobile phase programming in gradient elution calls for a great deal of experience in handling HPLC columns.

2.4 Detection

The available detectors (UV photometric, fluorescence refractive index and electrochemical) are not specific to the detection of peptides. Noticeable progress in the detection of biopolymers in HPLC was made through the development of appropriate reaction detectors. Basically, a highly specific reagent such as a fluorophor or an enzyme is added to the column effluent, which reacts with the compound(s) to be monitored in tubular reactors, packed bed reactors or air-segmented liquid flow systems. The reaction products are detected appropriately by means of fluorescence, UV etc. The reactor should be designed in such a way that additional peak broadening is minimized and the performance of the analytical column maintained. These aspects are broadly discussed by Deelder et al (ref. 19), Frei and Scholten (ref. 20) and Huber et al (ref. 21). Fluophores such as fluorescamine and o-phthalaldehyde were found to be sensitive reactants in the detection of proteins and peptides (ref. 3). Research has also focussed on specific enzyme detectors, where the coupling enzyme is either pumped continuously through the reactor (ref. 22-24) or immobilized (ref. 25).

3. APPLICATION OF HPLC TO THE SEPARATION OF PROTEINS, PEPTIDES AND NUCLEOTIDES

3.1 Proteins and peptides

Two extensive review articles on this subject were published last year (ref. 3,10) and recent contributions will be taken into account. Coming out shortly are the Proceedings of the International Conferenc on High Pressure Liquid Chromatography in Protein and Peptide Chemistry, held in January 1981 in Munich and edited by Walter de Gruyter & Co., Berlin, with F. Lottspeich, A. Henschen and K.H. Hupe as editors.

A number of papers deal with the size exclusion of proteins in the high performance mode, reporting on column characterisation, the retention mechanism and separations. Hearn et al (ref. 26) studied the size exclusion separation of human thyroglobulin and aldehyde dehydrogenase on μ Bondagel E and Sephacryl S-200 as a function of the pH and the salt concentration of the eluent. A comparative size exclusion separation of insulin under denaturating conditions was performed on Waters Bondagel E-125, Waters I-125 and Toyosoda G 2000 SW and G 3000 SW (ref. 27). The results are compared to those obtained on soft gels, e.g. Sephadex and Biogel.

A number of papers report on the application of TSK-Gel SW type columns to proteins (ref. 28-31). Kato et al (ref. 30) recommend the use of sodium dodecyl sulphate in sodium phosphate solutions as eluent for protein separation on TSK-Gel SW columns. Proteins are also observed to separate on TSK-Gel SW columns in 6 M guanidine hydrochloride as eluent (ref. 31). Enzymes were reported to be purified on these columns in 100 mg quantities (ref. 32). The size exclusion separation of proteins on LiChrosorb^R Diol and diol-modified silicas with a wide range of eluent compositions demonstrated that size exclusion is the dominant mechanism, but is superimposed by ionic and diol-ligand/solute interactions (ref. 7).

Reverse Phase Chromatography has become a potential technique for analyzing and isolating peptides. Bovine proinsulin C-peptide fragments were isolated on 5 to 10 μm microparticulate Reverse Phase columns in a relatively short time compared to traditional packings (ref. 33). Somastatin, a tetradecapeptide, was determined quantitatively down to 10 - 20 ng on an Ion-Pair Reverse Phase column (ref. 34). Bishop et al (ref. 35) performed preparative reversed phase high performance separations of a synthetic underivatised peptide mixture on the Waters Prep LC System 500 instrument. Separated on a n-octadecyl bonded silica with a solution of triethylammoniumphosphate of pH 3.2, acetonitrile and iso-propanol as modifier (ref. 36) were the following: C-apolipo proteins isolated from human very low density lipoproteins, the polymorphs of apolipoprotein A-I, the tryptic fragments of apolipoprotein C-II, the complex mixture generated by partial protolosis of apolipoprotein B, the tryptic fragments of an ^{3}H- and ^{14}C-labelled β-chain of murine IA alloantigen and tryptic fragments of a carboxymethylated lambda chain isolated from human immunoglobulin G. Peaks were detected by amino acid analysis and radio-activity counting. In a subsequent paper Hancock and Sparrow (ref. 37) stated that peptides are retained by a mixed mechanism controlled by the relative ratio of hydrocarbon to silanol groups on the surface of the n-octadecyl-bonded silica. The influence of silanol groups on retention can be suppressed by the addition of n-alkylammoniumphosphate to the eluent. The effectiveness of anionic hydrophobic counterions such as trifluoroacetic acid, pentafluoropropanoic acid, heptafluorobutyric acid and undecafluorocaproic acid in Ion-Pair Reverse Phase Chromatography was studied for its application to the purification of peptides (ref. 38). As the ion pair reagents are completely volatile, bioassay, radioimmunoassay and amino acid analysis of column fractions is easi performed. Lin et al (ref. 39) studied the effect of chain length an concentration of aliphatic carboxylic acids as counterions in Ion-Pa Reverse Phase Chromatography on the retention of basic, acidic and neutral dipeptides and found specific dependence of retention on the nature of solutes. Acidic and neutral dipeptides were identified by means of gas-chromatographic-mass-spectrometric techniques. A select phase system for resolving tyrosinyl peptides can be built up by conditioning a microparticulate silica column with small amounts of anionic surfactants, e.g. sodium dodecyl sulphate (SDS) in methanol/ water as eluent (ref. 40). The retention of peptides on this dynamic coated cation exchange column was shown to be controlled by the concentration of surfactant and the pH of the eluent. Un-derivatised peptides were separated on a silica-based weak anion, exchanger, MicroPak AX-10, in triethylammonium acetate buffer and acetonitrile as eluent (ref. 41).

3.2 Nucleotides

Oligomeric and polymeric ribonucleotides and deoxyribonucleotides can be separated on all three HPLC phase systems (size exclusion, reverse phase, anion-exchange) by different mechanism. This is illustrated in Fig. 1 for uridine and oligoribo uridylic acids. On the LiChrosorb^{R} Diol column (Fig. 1a) the elution order follows decreasing molecular weight. Uridine is eluted last. The elution patte is completely changed when an ethylammonium dimethyl group is substituted for a proton of the diol group (see Fig. 1b). Retention increases with the charge of the solutes and uridine is eluted first (ref. 13). Oligo uridylic acids are shown to be separated by Ion-Pai Reverse Phase Chromatography employing an n-octadecyl bonded silica and tetraalkylammonium as counter ion (Fig. 1c,d). The retention is controlled by the type, i.e. n-alkyl chain length, of the ion reagen its concentration and the eluent composition (ref. 42). Oligo deoxyribonucleotides are resolved on a Waters I-125 Protein Analysis Column in a triethylammoniumacetate buffer by a size exclusion mechanism (ref. 43). Maybaum et al (ref. 44) report on HPLC assay fo determining tissue pools, uridine, deoxyuridine, cytidine, deoxycyti mono- and triphosphates by anion exchange chromatography followed by preparative reverse phase chromatography after conversion of nucleot to nucleosides by acid phosphatase.

A HPLC phase system based on a strong anion exchanger and run wit a buffer gradient was developed for the separation of 5'-nucleotides and nucleotide sugars present in liver tissue (ref. 45). The paper also includes a comparative study on extraction procedures prior to HPLC separation. As a continuation of previous studies Grushka and Chow (ref. 46) performed selective separations of di- and triphosphate nucleotides as magnesium complexes on a dithiocarbamate bonded silica column in a phosphate buffer as eluent. Retention is controlled by the concentration of Mg^{2+} and the pH and corresponds to predictions of the assumed retention mechanism. Ion-Pair Reverse Phase Chromatography employing tetrabutylammonium as counterion was applied to the separation of mixtures of 5-fluorouracil, its deoxyribo- and ribonucleosides and nucleotides (ref. 47) as well adenine nucleotides (ref. 48).

4. CONCLUSIONS

It was demonstrated that a present HPLC offers three different kinds of phase system, each providing a specific selectivity and a high performance. Variation of the composition of the stationary and mobile phases supplies manifold options to control the retention and selectivity of biopolymers of a wide range of molecular structures. The optimization of a given separation, however, is still based on empirical procedures and knowledge. Attemps should therefore be made to cast more light on the retention mechanism of biopolymers on HPLC phase systems in order to predict and to evaluate a separation scheme. This work should be carried out by chromatographers in cooperation with biochemists, as comprehensive knowledge is required to understand the complex structure of the solutes and the resulting interactions. At the same time, more thought than at present must be given to the role of the surface structure of the packing and the way it affects retention, selectivity and performance. This would also help to synthesize better and more stable supports. Activities must be concerned also with the still wide-open field of detection, as we believe that the potential offered by post column reactors in the specific analysis of biopolymers has up to now been explored to a small extent only. In conclusion, HPLC is bound to become increasingly important as a powerful technique for the separation, purification and isolation of biopolymers.

Fig. 1 : Separation of uridine and oligoribo-uridylic acids on various phase systems

a) packing: LiChrosorb[R] Diol, dp = 5 µm

surface composition: $\equiv Si-(CH_2)_3-O-CH_2-\underset{OH}{CH}-\underset{OH}{CH_2}$

column: 300 x 4 mm

eluent: phosphate buffer,
0.01 m KH_2PO_4/H_3PO_4, pH 7.0

flow rate: 1.0 ml/min

b) packing: Silica DMA (prepared by ourselves), dp = 5 µm

surface composition: $\equiv Si-(CH_2)_3-O-CH_2-\underset{O}{CH}-\underset{OH}{CH_2}$
$\quad$ O–$CH_2-CH_2-N(CH_3)_2{}^{\oplus}X^{\ominus}$

eluent: phosphate buffer,
0.1 M K_2HPO_4/H_3PO_4, pH 7.0

flow rate: 1.0 ml/min

c) packing: LiChrosorb[R] RP-18, dp = 5 µm

surface composition: $\equiv Si-(CH_2)_{17}-CH_3$

column: 100 x 4 mm

eluent: methanol/water 10/90 (v/v) + 0.1 mol
tetramethylammoniumchloride, pH 7.0

flow rate: 1.0 ml/min

d) same as c) except for eluent composition

eluent: methanol/water 10/90 (v/v) + 0.1 mol
tetraethylammoniumchloride, pH 7.0

solutes: 1 = uridine, 2 = UpU, 3 = $(Up)_2U$, 4 = $(Up)_3U$
5 = $(Up)_4U$, 6 = $(Up)_5U$, 7 = $(Up)_6U$, 8 = $(Up)_7U$

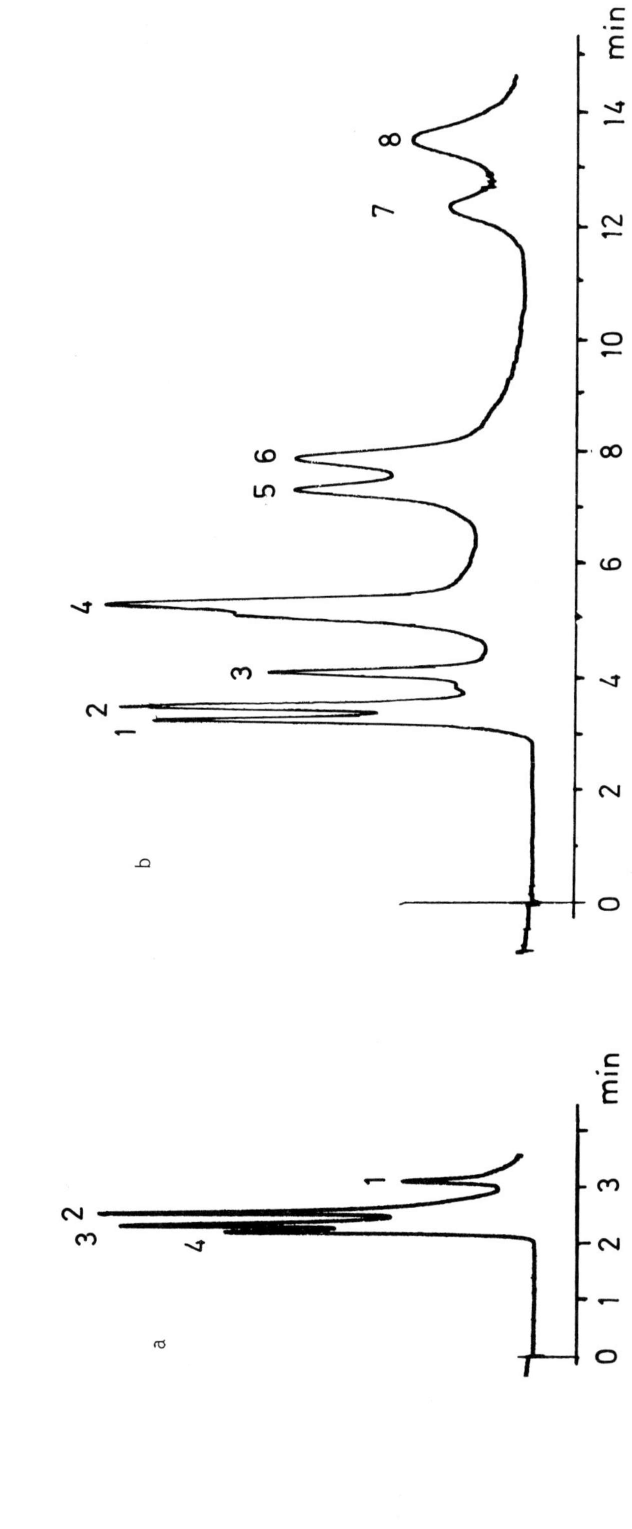
b
1
2
3
4
5 6
7
8
0
2
4
6
8
10
12
14
min
a
3 2
4
1
0
1
2
3
min

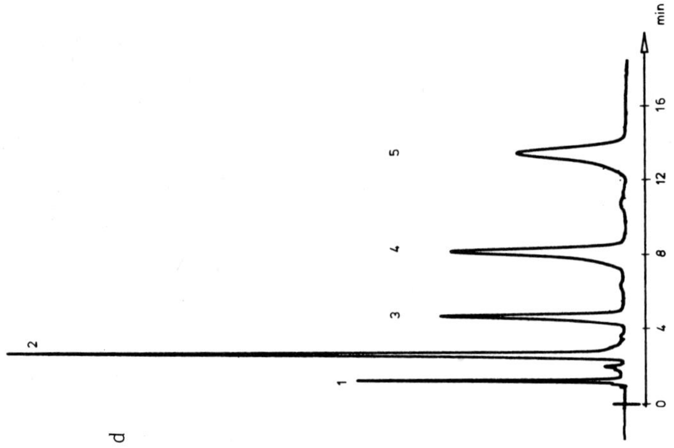
d
1
2
3
4
5
0
4
8
12
16
min

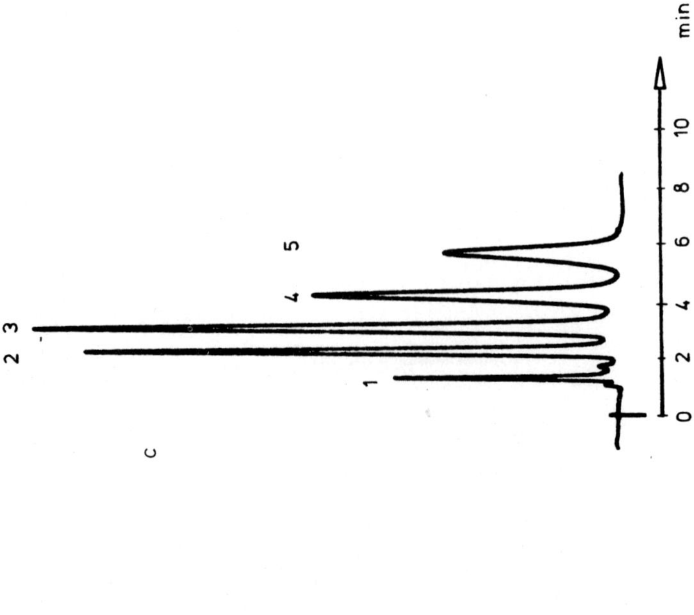
c
1
2
3
4
5
0
2
4
6
8
10
min

5. REFERENCES

1 A.L. Lehninger, Biochemistry, Worth Publisher Inc., New York 1970
2 D.E. Metzler, Biochemistry, Academic Press, New York 1977
3 F.E. Regnier and K.M. Gooding, Anal.Biochem., 103(1980)1-25
4 L.R. Snyder and J.J. Kirkland, Introduction to Modern Liquid Chromatography, Wiley Interscience, New York 1979, p.36
5 H.G. Barth, J.Chromatogr.Sci., 18(1980)409-429
6 E. Pfankoch, K.C. Lu, F.E. Regnier, J.Chromatogr.Sci., 18(1980) 430-441
7 P. Roumeliotis and K.K. Unger, J.Chromatogr., to be published
8 C. Horvath and W. Melander, Internat.Laboratory, Nov/Dec.(1978)11-35
9 D.E. Majors, J.Chromatogr.Sci., 18(1980)500
10 M.T.W. Hearn, J.Liquid Chrom., 3(1980)1255-1276
11 F.M. Rabel in Advances in Chromatography, J.Giddings, E. Grushka, J. Cazes and Ph. Brown (Editors), Vol.17, Marcel Dekker, New York 1979, pp.53-100
12 R. Wood, L. Cummings and T. Jupille, J.Chromatogr.Sci., 18(1980) 551-558
13 W. Jost, K.K. Unger, R. Lipecky and H.G. Gassen, J.Chromatogr., 185(1979)403-412
14 L.R. Snyder and J.J. Kirkland, Introduction to Modern Liquid Chromatography, Wiley Interscience, New York 1979, pp.720-751
15 W.W. Yau, J.J. Kirkland and D.D. Bly, Modern Size Exclusion Liquid Chromatography, John Wiley, New York 1979, p.114
16 L.R. Snyder and J.J. Kirkland, Introduction to Modern Liquid Chromatography, Wiley Interscience, New York 1979, pp.662-719
17 P.A. Bristow and J.H. Knox, Chromatographia 10(1977)279-289
18 P. Roumeliots and K.K. Unger, J.Chromatogr., 185(1979)445-452
19 R.S. Deelder, M.G.F. Kroll, A.J.B. Beeren and J.H.M. van den Berg, J.Chromatogr.,149(1978)669-682
20 R.W. Frei and A.H.M.T. Scholten, J.Chromatogr.Sci., 17(1979)152-159
21 J.F.K. Huber, K.M. Jonker and H. Poppe, Anal.Chem., 52(1980)2-9
22 S.H. Chang, K.M. Gooding and F.E. Regnier, J.Chromatogr., 125(1976) 103-114
23 T.D. Schlabach, S.H. Chang, K.M. Gooding and F.E. Regnier, J. Chromatogr., 134(1978)91-106
24 T.D. Schlabach, A.J. Alpert and F.E. Regnier, Clin.Chem., 24(1978) 1351-1360
25 T.D. Schlabach and F.E. Regnier, J.Chromatogr., 158(1978)349-364
26 M.T.W. Hearn, B. Grego, C.A. Bishop and W.S. Hancock, J.Liquid Chromatogr., 3(10) 1980, 1549-1560
27 B.S. Welinder, J.Liquid Chromatogr., 3(a)(1980)1399-1416
28 Y. Kato, K. Komiya, H. Sasaki and T. Hashimoto, J.Chromatogr., 190(1980)297-303
29 M.E. Himmel and Ph.G. Squire, Int.J.Peptide Protein Res., 14(1981) 365-373
30 Y. Kato, K. Komiya, H. Sasaki and T. Hashimoto, J. Chromatogr., 193(1980)29-36
31 Y. Kato, K. Komiya, H. Sasaki and T. Hashimoto, J. Chromatogr., 193(1980)458-463
32 Y. Kato, K. Komiya, Y. Sawada, H. Sasaki and T. Hashimoto, J. Chromatogr., 190(1980)305-310
33 J.G.R. Hurrell, R.J. Fleming and M.T.W. Hearn, J.Liquid Chromatogr., 3(4)(198o)473-494
34 M. Abrahamson and K. Gröningson, J.Liquid Chromatogr., 3(4)(1980) 495-511
35 C.A. Bishop, D.R.K. Harding, L.J. Meyer, W.S. Hancock and M.T.W. Hearn, J. Chromatogr., 192(1980)222-227
36 W.S. Hancock, J.D. Carpa, W.A. Bradley and J.T. Sparrow, J. Chromatogr., 206(1981)59-70

37 W.S. Hancock and J.T. Sparrow, J.Chromatogr., 206(1981)71-82
38 H.P.J. Bennett, C.A. Browne and S. Solomon, J.Liquid Chromatogr., 3(9)(1980)1353-1365
39 S. Lin, L.A. Smith and R.M. Caprioli, J.Chromatogr., 197(1980)31-4
40 R.W. Wall, J.Chromatogr., 194(1980)353-363
41 M. Dizdaroglu and M.G. Simic, J.Chromatogr., 195(1980)119-126
42 W. Jost, K.K. Unger and G. Schill, Anal.Biochem., to be published
43 D. Molko, R. Derbyshire, A. Guy, A. Roget, R. Teoule and A. Boucherle, J.Chromatogr., 206(1981)493-500
44 J. Maybaum, F.K. Klein and W. Sadee, J.Chromatogr., 188(1980)149-1
45 T.L. Riss, N.L. Zorich, M.D. Williams and A. Richardson, J.Liquid Chromatogr., 3(1)(1980)133-158
46 E. Grushka and F.K. Chow, J.Chromatogr., 199(1980)283-293
47 C.F. Gelijkens and A.P. de Leenheer, J.Chromatogr., 194(1980)305-3
48 J.D. Schwenn and H.G. Jender, J.Chromatogr., 193(1980)285-290

T.C.J. Gribnau, J. Visser and R.J.F. Nivard (Editors),
Affinity Chromatography and Related Techniques

HIGH-PRESSURE HYDROPHOBIC CHROMATOGRAPHY OF PROTEINS

A.H. NISHIKAWA, S.K. ROY and R. PUCHALSKI
Biopolymer Research Dept., Hoffmann-La Roche Inc., Nutley, NJ 07110 (U.S.A.)

ABSTRACT

Commercially available RP-8, 'reversed phase' C-8 coated, silica gels have a tendency to bind a variety of proteins very avidly at neutral pH. Gradient elution systems at pH 2.1 allow better recovery of proteins. However, many proteins are denatured or otherwise altered at low pH. To avoid these problems attempts have been made to develop hybrid bonded-phase silica gels, which contain both hydrophilic and hydrophobic ligands. Hybrid gels with a low ratio of hydrophobic groups to hydrophilic groups for the first time have enabled the chromatography of several test proteins at pH 6.2. These new hybrid gels still show a sensitivity to pH on the elution of protein bands. However, the results indicate that hybrid bonded phase silicas have great promise for hydrophobic HPLC of proteins.

INTRODUCTION

High-performance liquid chromatography (HPLC) has made a tremendous impact in the past decade on several areas of bioanalytical chemistry due to its high speed and resolving power. By contrast, it has only made a modest start in the analysis and resolution of water-soluble macromolecules -- in particular proteins. To be sure some pioneering efforts in this area have been recorded -- such as that on ion-exchange and hydrophobic chromatography by Regnier and colleagues (ref. 1), size-exclusion chromatography by Engelhardt and Mathes (ref. 2) and on affinity chromatography by Mosbach and collaborators (ref. 3).

Fine-particle technology with rigid silica gels has made possible chromatography columns with unprecedented high separation efficiencies -- impractical to achieve with organic soft-gel supports. The mechanical stability of silica permits the tolerance of high pressures (reaching to 5,000 or 10,000 psi in some cases) attained with very fine particles (now 3 μ particles are commercially available). Many of the properties of silica have contributed to the development of HPLC of small molecules as well as a number of macromolecular solutes. However, one has proven to be a drawback: the strong surface

+ Abbreviations used: BLG = beta-lactoglobulin, INS A = insulin A chain oxidized, LZM = egg white lysozyme, MYO = equine myoglobin, PrOH = n-propanol activity of silanol groups. While many low molecular weight solutes (including oligopeptides) can be chromatographed in their presence, macromolecular solutes including many proteins could not. The large and complex surface area of proteins results in non-specific, multiple-contact binding to bare silica. The resulting interactions are frequently so strong that not only denaturation but also irreversible adsorption has been observed.

The development of HPLC of proteins in general has been tied to the development of a variety of surface coatings for silica gel. The great success of covalently 'bonded phase' silica gel coatings for the HPLC of small organic molecules, however, has not yet been matched with protein solutes. To be sure, the results with hydrophilic bonded-phases for size-exclusion chromatography and the ionic-hydrophilic coatings for ion-exchange chromatography have been encouraging in protein application. The hydrocarbon coated bonded phase silica gels, which gave rise to 'reversed-phase' HPLC of small organic molecules, are being examined by several laboratories as column packings for hydrophobic chromatography of proteins. While some successes have been noted, many problems remain. This paper will try to address some of these problems and point the way to some possible solutions.

2. Some problems with hydrophobic HPLC of proteins

2.1 High-pressure. While uniformly sized, fine particles afford columns with high efficiencies, this is attended by high hydrostatic pressures at practical flow rates. Of course this is of no consequence to low molecular weight solutes. However, as has been observed in ultracentrifuge studies, high pressures ($\geq$ 4,500 psi) can affect monomer-multimer equilibria of multi-subunit proteins (refs. 4 and 5) and in some instances result in denaturation or other alteration of conformational structure of proteins. Thus it is of some concern in HPLC as to how high an ambient pressure should be allowed with proteins. Furthermore due to the presence of large surface areas in bonded phase silica gels, the risk of pressure-induced denaturation of proteins may be greater in HPLC than when solutions are simply compressed to high pressures.

2.2 Imperfectly coated silica surfaces. There is a long-standing observation that many protein solutions, especially when dilute, upon passing through sintered glass filters result in loss of proteins due to adsorption to the sintered surface. This phenomenon is exaggerated many fold in packed beds of fine silica gel particles, where the surface area is frequently in excess of 100 m^2/g. It is probably due to this high surface area that imperfections in the bonded phases of silica gels become so readily apparent. In addition, with hydrocarbon coatings, proteins seem to bind more avidly to areas where silanol groups are situated in abundance next to hydrophobic ligands. That is, proteins may bind more tightly to mosaic patches of silanol and hydrophobic surfaces.

The problem of imperfect coverage in presently available commercial packings can be illustrated by two relevant examples.

First, the hydrophilic bonded phases of the 'glycerol' or 'glycol' variety:

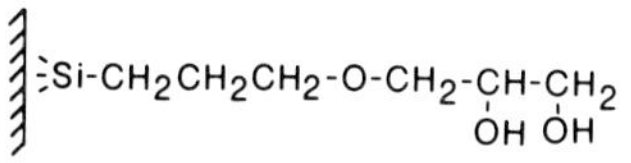

have performed reasonably well in size-exclusion chromatography (SEC) of proteins. Yet chromatographic anomalies have been observed which could be corrected somewhat only by increasing ionic strength and/or adding high concentrations of ethylene glycol to the eluant (ref. 6).

Second, the hydrocarbon coated silica gels, which have had great success in the 'reversed-phase' chromatography of small organic molecules, show a persistent recovery problem with proteins at pH 7. As noted by O'Hara and Nice, the recovery of proteins or peptides larger than a pentadecapeptide is somewhat unpredictable (ref. 7). With RP-8 ('reversed-phase', octyl silane coated) silica gels, the recovery of more types of proteins improves if the packing is used at pH 2.1 (refs. 7 and 8). In addition, band sharpening and improved separation are observed. These have been attributed to the protonation at low pH of the silanol groups ($pk_a \approx 7$) remaining uncoated in the bonded phase gel preparation. This explanation is supported by the observation that at higher pH, proteins with high isoelectric points (e.g. egg white lysozyme) bind more avidly than do acidic proteins.

Operating RP-8 columns at pH 2.1 is not the end to problems, however, since there is increased risk of hydrolytic destruction of the silica support itself at low pH. This of course leads to shortened use life and its attendent high costs. Furthermore, many proteins are dissociated or denatured at such low pH conditions making the recovery of fractions futile. Finally, any separation observed under these conditions may be due to artificial properties exhibited by the proteins and may not readily translate to preparatory scale operations.

Thus a major task of preparing materials for hydrophobic HPLC of proteins is to first obtain bonded phases of high coverage of silanol groups with protein-compatible ligands. A related task is to prevent attrition of the coated surfaces either by appropriate selection of eluents (not always possible) or by using bonding chemistries which yield durable surfaces (this is yet to be developed).

2.3 Proper ligand density. When one observes the wetability of various bonded phase silicas, those bearing a hydrophobic coating are distinctly less easily wetted by water than those with hydrophilic coatings. That RP-8 silica gels can be wetted at all, points to the imperfectness in the bonded phase as suggested earlier.

Hydrophobic chromatography in soft gel supports has already shown the importance of ligand density (as well as bulk size of ligand) in the binding and separation of proteins (refs. 9 and 10). Thus it is not surprising that successful hydrophobic HPLC of proteins would also be affected by ligand density in the bonded phase silica gel.

Commercially prepared RP-8 (as well as RP-18) has a very high density of hydrophobic ligands. Indeed to obtain good performance uniformity and chemical stability of the support matrix, efforts have been expended to maximize coverage of the silica surface. For protein analytes that is not necessarily a desirable end. In the extreme case, a completely coated surface would probably not wet and thus protein adsorption would not take place at all. Short of that, the RP-8 preparations which can be wetted by buffer tend to bind a variety of proteins very tightly - some irreversibly. Clearly a new approach is needed to prepare sorbents which are more compatible for use with protein analytes.

3. Hybrid sorbents

The experience gained with hydrophobic chromatography on soft gels, suggests that for HPLC of proteins, well coated hybrid silica surfaces comprised of hydrophilic and hydrophobic ligands should be of interest. Several approaches suggest themselves for preparing such sorbents.

3.1 Glycidoxysilane coated silica gel

One approach to creating a hybrid surface coating is to prepare first a reactive coating which can then be modified to the desired end. The scheme shown in Fig. 1 was an approach attempted early on. The epoxy coated intermediate was treated with octanol and boron trifluoride etherate followed by treatment with dilute HCl solution.

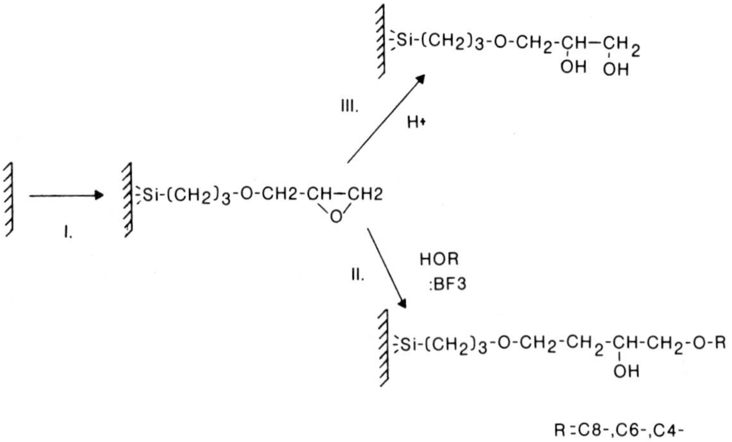

Fig. 1. Scheme for glycidoxysilanization and alkyl ether formation.

One such preparation (diol-C8) was evaluated with a set of test proteins which included: myoglobin, β-lactoglobulin, lysozyme, chymotrypsinogen A, and oxidized insulin A chain. The starting buffer was 0.1 M MOPS/0.2 M KSCN at pH 6.2. The gradient eluent was the buffer combined with acetonitrile to 60% (v/v). Fig. 2A shows the separation which was attained with a linear gradient at 1 ml/min.

By comparison, an RP-8 Hibar-II (E. Merck) column when tested with the same set of proteins and gradient elution system permitted recovery of only chymotrypsinogen A and insulin A chain. The other three proteins were not recovered even after extensive washing with 60% acetonitrile/buffer (see Fig. 2B).

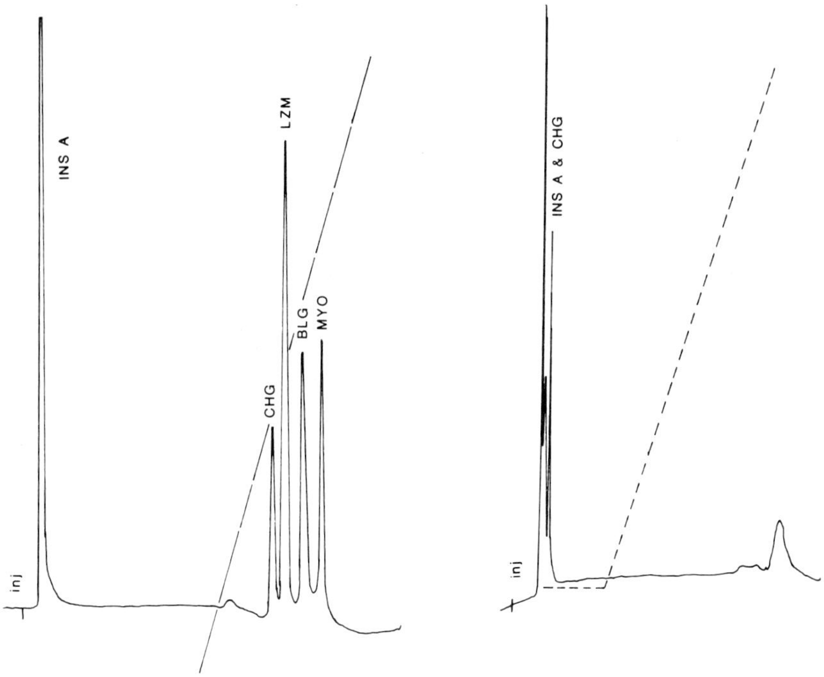

Fig. 2A (left). Chromatography of test proteins on diol-C8 silica gel. Column dimensions: 4.6 mm diam X 250 mm and 1 ml/min flow rate used in all experiments. Buffer: 0.1 MOPS/ 0.2 M/0.2 M KSCN, pH 6.2. Solvent soln: buffer + 60% acetonitrile. Starting conditions: 100% buffer-0% solvent soln. Linear gradient of 30 min to 0% buffer-100% solvent soln. Fig. 2B (right). Chromatography of test proteins on Hi-Bar II RP-8. Buffer: 0.1 M MOPS/ 0.2 M KSCN pH 6.2. Solvent soln: buffer + 60% acetonitrile. Starting conditions: 83% buffer-17% solvent soln. Linear gradient of 30 min to 0% buffer-100% solvent soln. Proteins injected: INS A, CHG, LZM, BLG, MYO. Unlabelled peak is artifact. Dashed line corresponds to gradient through the column.

A further contrast was seen with Whatman Partisil C-8. At pH 6.2 only egg white lysozyme and oxidized insulin A chain readily eluted from the column. As seen in Fig. 3 when the eluent was at pH 2.1, recovery improvements were seen with the elution of myoglobin and β-lactoglobulin. However, even at low pH chymotrypsinogen A was not recovered.

Thus the hybrid coating of silica obtained with the glycidoxysilane approach does seem promising. But reproducibility of the coating and its corrosion resistance to solutions of salt and buffer ions is not yet reliable.

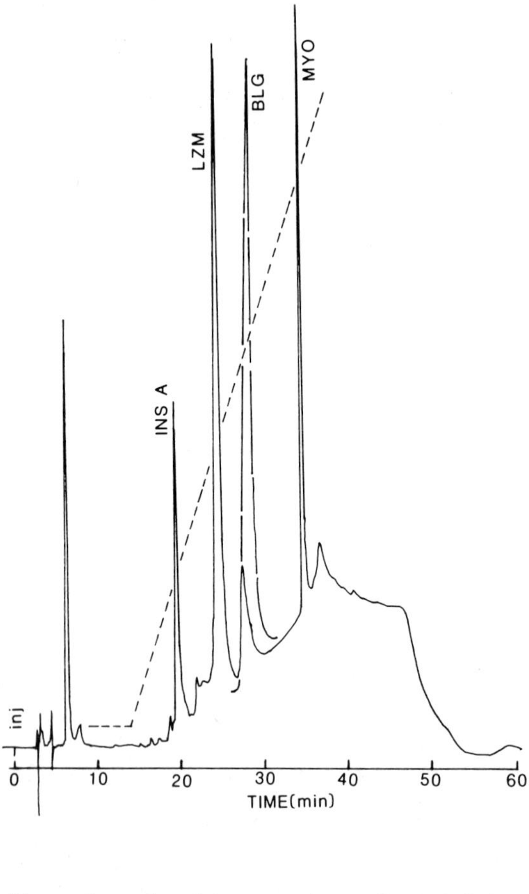

Fig. 3. Chromatography of test proteins on Partisil C-8. Non-marked peaks are artifacts. BLG peak obtained in separate run and superimposed. CHG did not elute. Buffer: 0.2 M phosphate, pH 2.1. Solvent soln: buffer + 60% acetonitrile. Start conditions: 17% solvent soln-83% buffer. Linear gradient of 30 min to 95% solvent soln-5% buffer.

3.2 Gamma-aminopropyl silane coated silica gel

This second approach is conceptually analogous to the first in that a hydrophilic coating was first prepared then subsequently modified. As shown in Fig. 4 the scheme differs only in details.

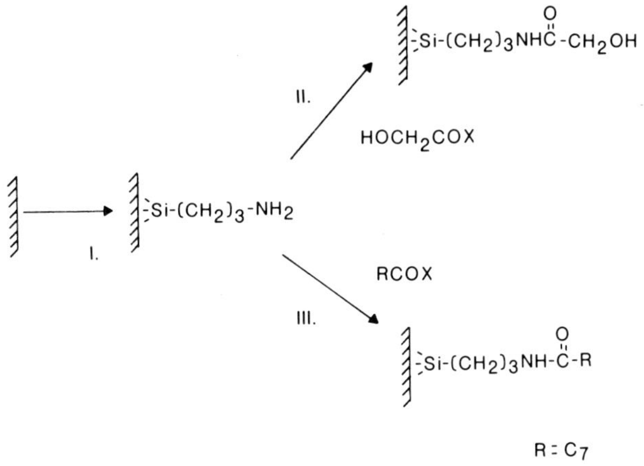

Fig. 4. Scheme for aminopropylsilanization and acylation.

3.3 Mixed silanization of silica gels

A third approach to creating a hybrid coated silica gel is to react the silica surface with two different silanizing reagents either simultaneously or in sequence. The scheme is outlined in Fig. 5.

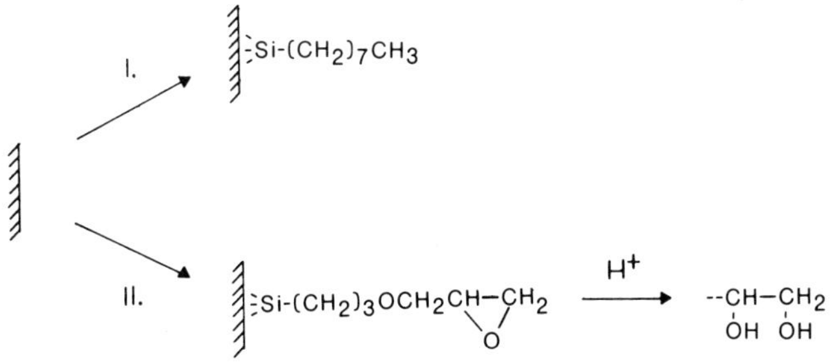

Fig. 5. Scheme for mixed silanization.

In the simultaneous treatment, the ratio of hydrophobic to hydrophilic ligands might be controlled by adjusting the relative amounts of the two reagents in the coating mixture. However in practice the results were not satisfactory. A sequential treatment procedure appeared more promising. Here the silica gel was first treated with the hydrophobic reagent followed by separate treatment with the reagent introducing hydrophilic ligands.

3.3.1 A diol-C8 gel (GEL-38) exemplifying the sequential approach was made by treating LiChrosorb SI-100, 10 μ particles, with a 10% octyltriethoxysilane solution in toluene at 80°C for 8 hrs. This was followed by treatment with a 10% gamma-glycidoxypropyl-dimethyl methoxysilane solution in toluene at 80°C for 10 hrs. The epoxides were hydrolyzed at pH 2 (dil. HCl)for 3 hrs at room temperature. Analysis of GEL-38 revealed 1.02 μeq of octyl group/m^2 and 0.83 μeq of diol group/m^2. It was slurry packed in chloroform at 10,000 psi into a 4.6 mm X 250 mm SSI column.

Protein profile. The first evaluation of GEL-38 was at pH 6.2 with our standard test battery of five proteins. Fig. 6 shows a chromatogram yielding oxidized insulin A chain, lysozyme beta-lactoglobulin, and myoglobin. In other tests bovine serum albumin was seen to elute at 39.6% acetonitrile. In contrast to other types of diol-C8 gels, GEL-38 did not elute chymotrypsinogen A nor bovine IgG.

pH effects on elution. On Fig. 6 are also shown gradient elutions of test proteins on GEL-38 at lower values of pH. The most conspicuous change involves lysozyme, which exhibits a wide band at pH 6.2 but a sharp one at pH 3.1. At pH 2.1 (not shown here to avoid further complication of Fig. 6) the lysozyme band was further sharpened. At lower pH values shifts in the separation of the different proteins are noticeable. This is shown more clearly on Table I. For example the separation of lysozyme from insulin A chain is poorest at pH 4.1, acceptable at pH 6.2 and very good at pH 2.1 or 3.1. The emergence of myoglobin as two bands at lower pH values is an unexpected phenomenon, which is unexplainable at the moment. It serves to raise caution when examining crude extracts of proteins containing unknown components.

In contrast to the other proteins in the standard mixture, beta-lactoglobulin manages to elute at about the same volume regardless of the pH. This is somewhat unexpected in that as pH decreases we anticipate changes in conformation which in turn should be reflected in changes in the hydrophobic chromatography of a protein.

Other organic modifiers. With the observation that chymotrypsinogen A and bovine IgG would not elute from GEL-38 with acetonitrile gradients, it was decided to investigate other organic modifiers. At pH 3.1 the gradient elution of three solvent systems are compared in Fig. 7. Acetonitrile by itself gives sharp bands. The MYO-I band seems to have disappeared - possibly into the trailing shoulder of the BCG peak. The inclusion of distilled cyclohexanol in the acetonitrile phase had the effect of increasing band

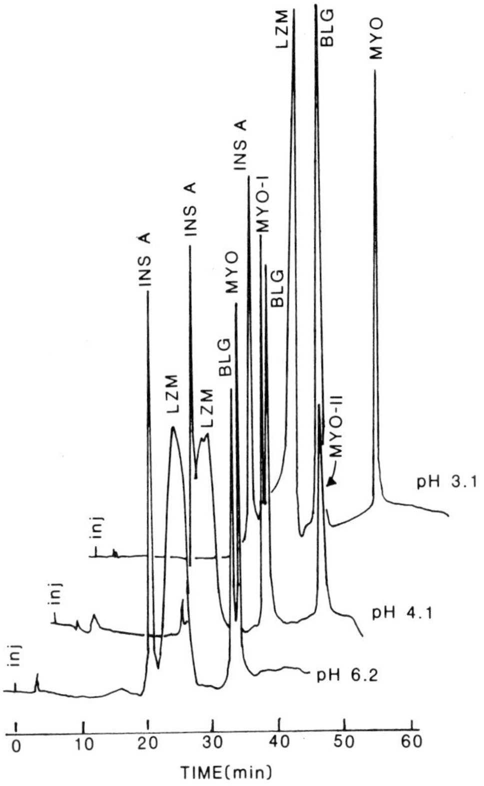

Fig. 6. Effect of pH on protein elution from GEL-38. Buffer: 0.05 M phosphate/0.2 M guanidinium hydrochloride adjusted to pH indicated. Solvent soln: buffer + 60% acetonitrile. Starting conditions: 95% buffer-5% solvent soln. Linear gradient of 35 min to 5% buffer-95% solvent soln.

TABLE I

Elution of test proteins as function of pH

pH	INS	LZM	MYO-I	BLG	MYO-II
			Elution vol (mL)		
2.1	23.5	29	32	33	42.5
3.1	23.5	30	?	33.5	42.5
4.1	22	23.5	31.5	32.5	40.5
6.2	20	24	-	33	33.5

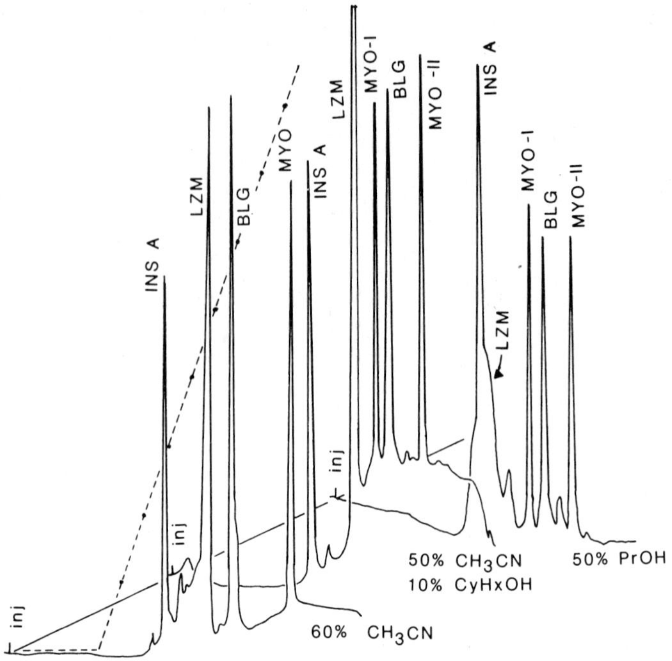

Fig. 7. Solvent effects on protein elution from GEL-38. Buffer: 0.05 phosphate/0.2 M guanidium hydrochloride, pH 3.1. Solvent solution: buffer + solvent indicated. Starting conditions: 95% buffer-5% solvent soln. Linear gradient of 35 min to 5% buffer-95% solvent soln. Actual gradient through column indicated by dotted line. Injected was 20 ul soln containing proteins each at 1 mg/ml.

sharpening and earlier elution of each of the bands but no effect on recovery of chymotrypsinogen A. At pH 6.2 (data not shown) the addition of cyclohexanol to acetonitrile resulted in sharpening the lysozyme band by 1/2 but the separation of beta-lactoglobulin from myoglobin became poorer. Propanol has been studied as a solvent for proteins for some time (refs. 11 and 12). In the present work it yielded sharp protein bands, but unfortunately the accelerated elution of lysozyme resulted in its merging with the insulin A band. Propanol did not permit recovery of chymotrypsinogen A or bovine IgG.

4. Problems with silanized silica surfaces.

The advent of bonded-phase silica gels has resulted in a virtual explosion in the application of HPLC to a wide range of solutes including proteins. However, with these macrosolutes significant problems still remain. In general, these problems center around the imperfectness of bonded phase coatings of silica.

4.1 Monolayers vs. polymers.

One phenomenon associated with the imperfect coating problem is the formation of polymers during the silanization procedure. An illustrative scheme is shown in Fig. 8.

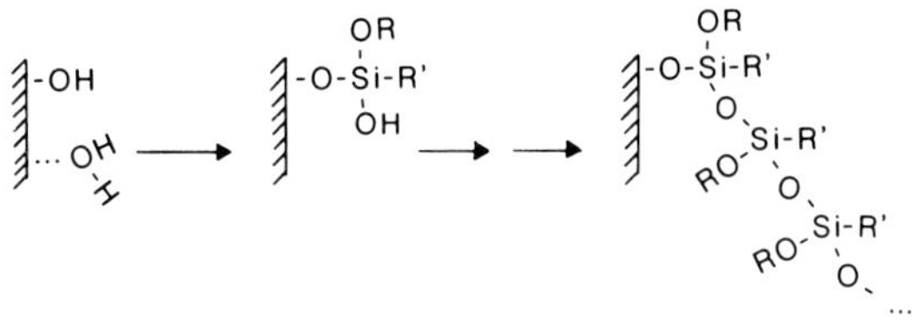

Fig. 8. Scheme for polymer formation during silization.

As pointed out by Unger and coworkers (ref. 13) polymer formation can occur when physisorbed water is present during the silanization procedure. Polymer formation results in a high carbon loading and diminution of porosity as well as decline in the available surface area. Extensive polymer formation may yield a bonded phase gel which appears well coated (e.g. negative to methyl red binding). However the use life of such preparations is unpredictable and often quite short especially in the presence of buffers containing high concentrations of salt ($>$0.2 M). The performance of such gels usually falls precipitously since the detachment of polymers from the silica surface leaves relatively large gaping patches of naked silanols which avidly bind protein solutes. Frequently a given column may yield a score or more successful chromatographic runs, then suddenly in the next one or two runs show a dramatic change where the applied sample no longer elutes. By contrast a silica gel surface uniformly coated with a covalent monolayer is expected to deteriorate more gradually.

4.2 Surface coverage. The literature on the chemistry of silica surfaces points to a theoretical maximum density of 8 μeq of silanol groups per sq. meter (ref. 13). However, studies on chemical modification of silica surfaces has usually resulted in introducing only 4-5 μeq of ligands/m^2 (ref. 15). While 'steric hindrance' during covalent surface reaction has been suggested, the exact reasons for this difference are not known. GEL-38, which we have evaluated in some detail, possesses relatively low surface coverage of about 23.1%. Hydrophobic ligands comprise about 12.8% of the surface. This low density of a non-polar ligand has afforded a bonded-phase silica gel which can permit recovery of a number of proteins at neutral pH. Yet the unexplainable binding of chymotrypsinogen A and bovine IgG points to the need for further improvements in coating methodology for obtaining better silica gels for hydrophobic HPLC of proteins.

REFERENCES

1 S.H. Chang, K.M. Gooding and F.E. Regnier, J. Chromatogr. 120(1976)321.
2 H. Engelhardt and D. Mathes, J. Chromatogr. 142(1977)311.
3 S. Ohlson, L. Hansson, P.-O. Larsson and K. Mosbach, FEBS Lett. 93(1978)5.
4 R. Josephs and W.F. Harrington, Biochemistry 7(1968)2834.
5 R.M. Arnold and L.J. Albright, Biochim. Biophys. Acta 238(1971)347.
6 D.E. Schmidt, Jr., R.W. Giese, D. Conron and B.L. Karger, Anal. Chem. 52(1980)177.
7 M.J. O'Hara and E.C. Nice, J. Chromatogr.171(1979)209.
8 I. Molnar and C. Horvath, J. Chromatogr.142(1977)623.
9 B.H.J. Hofstee, Prep. Biochem. 5(1975)5.
10 S. Shaltiel, Meth. Enzymol. 34(1974)126.
11 H.B. Bull and K. Breese, Biopolymers 17(1978)2121.
12 M. Rubinstein, Anal. Biochem. 98(1979)1.
13 K.K. Unger, N. Becker and P. Roumeliotis, J. Chromatogr. 125(1976)115.
14 L. Boksanyi, O. Liardon and E. Sz. Kovats, Adv. Colloid Interface Sci. 6(1976)95.
15 H.P. Boehm, Adv. Catalysis 16(1966)179.

T.C.J. Gribnau, J. Visser and R.J.F. Nivard (Editors),
Affinity Chromatography and Related Techniques

HIGH (INTERMEDIATE) PERFORMANCE HYDROPHOBIC INTERACTION CHROMATOGRAPHY OF BIOPOLYMERS

S. HJERTÉN, K. YAO and V. PATEL
Institute of Biochemistry, Biomedical Center, University of Uppsala,
Box 576, S-751 23 Uppsala, Sweden

ABSTRACT

Columns packed with agarose gel spheres of high concentration (9-15%) afford high flow rates even at relatively small particle diameters. Such rigid (cross-linked) gels can therefore be used in high or intermediate performance liquid chromatography, as illustrated by a hydrophobic interaction chromatography experiment with plasma proteins on octyl-agarose (other examples are found in ref. 1). Silica spheres derivatized with short alkyl chains (for instance hexyl groups) can be useful for high performance hydrophobic interaction chromatography of transfer RNA. In all of the experiments presented desorption is achieved by decreasing the ionic strength of the buffer and in the absence of organic solvents to avoid denaturation (compare high performance reversed phase chromatography, where elution of the solutes, including peptides and proteins, is performed in the presence of organic solvents).

INTRODUCTION

In reversed phase chromatography the stationary phase is less polar than the mobile phase. The two phases often contain a mixture of water and an organic solvent, the concentration of the latter being higher in the stationary phase. This chromatographic technique has been applied to the fractionation of, for instance, proteins on columns of silica particles to which nonpolar groups (commonly octyl or octadecyl chains) have been attached. The elution order is governed mainly by the hydrophobicities of the proteins, although adsorption to the matrix (silica) also can play an important role (ref 2). Organic solvents such as methanol, propanol and acetonitrile are used for desorption. Furthermore, most experiments with proteins have been conducted at low pH (in the range 2-4). In such experiments there is an obvious risk of denaturation, as stressed by Nice, Capp and O'Hare (ref. 3). However, it is well established that agarose gels show little or no adsorption of proteins, and when the gel is derivatized with nonpolar groups proteins can easily be desorbed by decreasing the ionic strength at neutral pH and without employing denaturing organic solvents. The separation mechanism of this technique, called hydrophobic interaction chromatography, is based on differences in surface hydrophobicities of proteins (ref. 4) - just as in reversed phase chromatography (the main difference between these two methods is probably that in the former method the hydrophobic groups act directly as adsorption sites for the proteins, whereas in the latter method the more closely situated hydrophobic groups probably serve to bind the organic solvent, which thus forms a continuous hydrophob stationary phase). In this paper we describe high or intermediate performance hydrophobic interaction chromatography of proteins on columns made up of amphiphili agarose gel spheres. The agarose concentration in the gel spheres is high (9-15%), so that the diameters of the (cross-linked) spheres can be made small to permit high resolution at high flow rates.

We have also investigated whether transfer RNA, which is less hydrophobic than proteins (ref. 5), can be fractionated in the absence of organic solvents by high performance hydrophobic interaction chromatography on silica spheres to which nonpolar groups were attached. This seems to be possible, provided that short alkyl chains (hexyl groups) are used as ligands.

MATERIALS AND METHODS

Packing. Agarose gel spheres were prepared according to a method previously described (ref. 6) and were introduced in the form of a slurry into a Plexiglass tube the outlet of which was covered with a stainless steel mesh sheet or a porous polyethylene disc covered with a filter paper of low porosity. A peristaltic pump was used to pump water or buffer through the bed during the packing procedure. Hexylsilica (Spherisorb; diameter: 5 µm) was a product, manufactured by Deeside Industrial Estate Queensferry, Clwyd, U.K. The silica particles were suspended in acetone and packed into a stainless steel tube at a pressure of 300 atm. in chloroform.

Coupling of nonpolar groups to agarose was performed by a method which will be described elsewhere (the method dealt with in ref. 7 can also be used).

Equipment. All experiments were run using a Varian 5000 Liquid Chromatograph. The distributions of protein and nucleic acid in the chromatograms were determined by fluorescence (excitation: 280 nm) and absorption (260 nm) measurements respectively, with an FS 970 LC Fluorimeter and a Spectroflow monitor SF 770 from Schoeffel Instrument Corp. (24 Booker St., Westwood, N.J., 07675 U.S.A.). The conductivity of the effluent was monitored by an LDC Conducto Monitor from Laboratory Data Control, P.O. Box 10234, Riviera Beach, Flo. 33404, U.S.A.

EXPERIMENTS AND RESULTS

Hydrophobic interaction chromatography of human plasma

A Plexiglass column (6 mm i.d. x 400 mm) was packed with 12% agarose beads (diameter: 100 μm) to which octyl groups had been coupled. The bed was equilibrated with 0.02 M sodium phosphate buffer, pH 6.8, containing 1.3 M ammonium sulfate (buffer A). About 10 μl of undialyzed normal human plasma was applied. Buffer A was pumped through the column for 10 min at a flow rate of 0.5 ml/min. The elution was then continued for 60 min by decreasing continuously the concentration of the ammonium sulfate to zero with the aid of a linear, reverse salt gradient formed from buffer A and 0.02 M sodium phosphate, pH 6.8. The protein distribution in the effluent was determined by fluorescence measurements (Fig. 1a). The decrease in

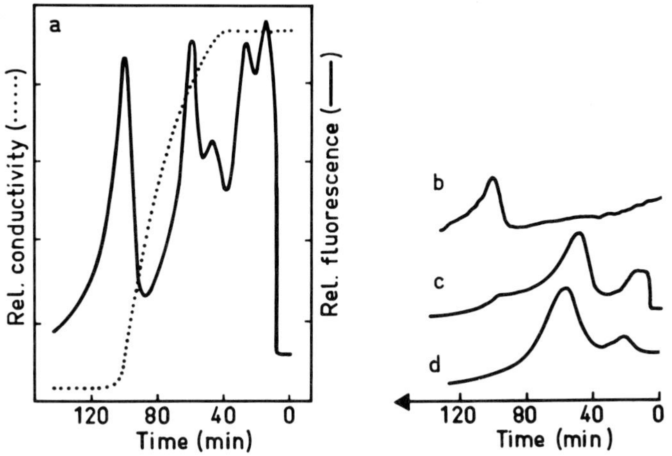

Fig. 1. Hydrophobic interaction chromatography of plasma proteins on a column of 12% agarose. a) Sample: human plasma. b) Sample: chiefly albumin (see Fig. 2b). c) Sample: chiefly α_2-globulin (see Fig. 2c). d) Sample: chiefly fibrinogen (see Fig. 2d).

the ammonium sulfate concentration was followed by conductivity measurements (Fig. 1a). The small amount of sample applied prevented a simple analysis of the chromatographic fractions. Therefore, to get some information about the content of the fractions, plasma was fractionated by agarose suspension electrophoresis (ref. 8) and 100 μl of some of these fractions (b, c and d in Fig. 2) were used as samples for hydrophobic interaction chromatography under conditions identical to those used in the chromatographic run described. The result, given in Fig. 1b, c and d, agrees with that reported for stepwise elution of plasma proteins adsorbed to Octyl-Sepharose® CL-6B (from Pharmacia Fine Chemicals, Uppsala, Sweden) in the sense

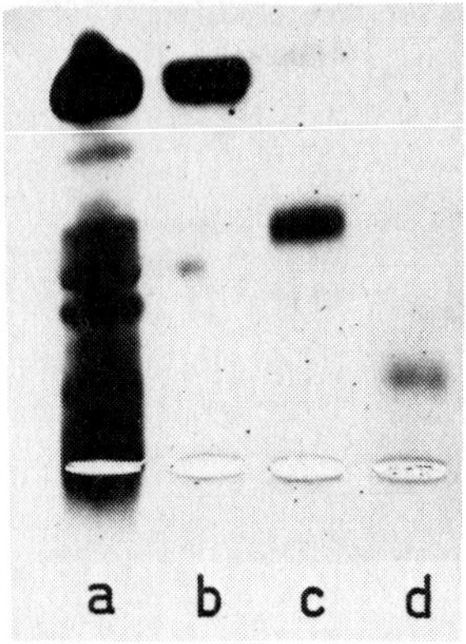

Fig. 2. Analysis by agarose electrophoresis of the samples a, b, c and d in Fig. 1. Buffer: 0.075 M veronal, pH 8.6, containing 1% non-cross-linked polyacrylamide to suppress electroendosmosis and increase the resolution (ref. 14).

that the elution order of the proteins is roughly the same (ref. 7). It is not surprising that albumin was the most strongly adsorbed of the plasma proteins, sinc albumin is known to transport fatty acids in the blood by adsorbing them to hydrophobic sites. It should be emphasized that a similar salt gradient elution on the commercial Octyl-Sepharose CL-6B gave a lower resolution than the experiment shown in Fig. 1a.

Hydrophobic interaction chromatography of tRNA from *E. coli*

A hexyl silica column (4 mm i.d. x 300 mm) was equilibrated with 0.02 M sodium phosphate (pH 6.8) containing 0.2 M Na_2SO_4 (buffer R_1). About 10 µl of a tRNA solution (A_{260}^{1cm} = 509) was applied. Elution was performed with two linear, reverse salt gradients formed from buffer R_1 and 0.02 M sodium phosphate, pH 6.8 (buffer R_2 in the following way. The initial concentration of buffer R_2 in the first gradient used was 0% (= 100% R_1) and the final concentration 60% (= 40% R_1) which was reache after 40 min; for the second gradient the corresponding figures were: 60% $R_2 \rightarrow$100% F 80 min. The form of the salt gradient was recorded by continuous conductivity determinations. The flow rate was 0.5 ml/min. The distribution of tRNA was determir by absorption measurements at 260 nm. The result is given in Fig. 3a. The experimer was very reproducible; the same chromatogram was obtained when smaller amounts of the sample were injected.

When the experiment was repeated with another reverse salt gradient the chromatogram shown in Fig. 3b was obtained. The gradient was formed from 0.02 M sodium phosphate buffer, pH 6.8, containing 0.8 M ammonium sulfate (Buffer T_1) and buffer R_2: during 60 min the concentration of R_2 changed from 0% (= 100% T_1) to 100% (= 0% T_1). A comparison between Fig. 3a and Fig. 3b shows that an increase

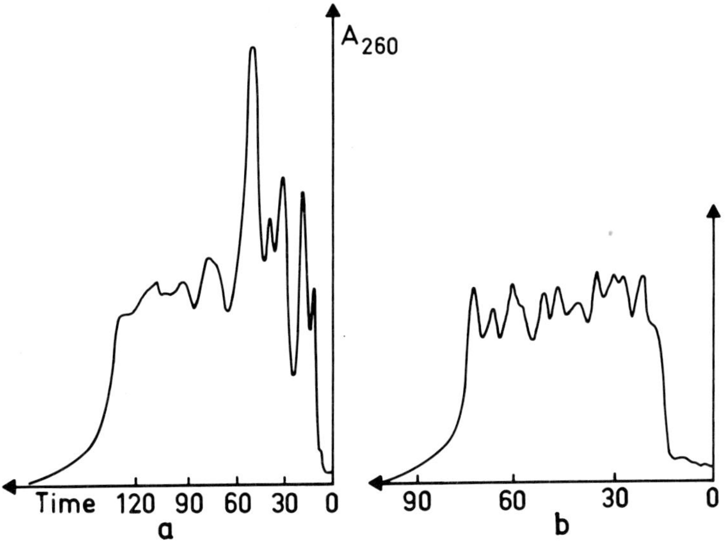

Fig. 3. Hydrophobic interaction chromatography of tRNA from *E. coli* on hexyl silica. The differences between chromatograms a and b are caused by differences in the conditions under which the corresponding experiments were conducted (see the text).

in the ionic strength of the starting buffer, *i.e.* an increase in the hydrophobic interaction between the bed and tRNA, had a profound influence on the appearance of the chromatogram. Another preparation of tRNA (with A_{280}^{1cm} = 30) was used in the experiment shown in Fig. 3b which also may affect the result.

DISCUSSION

High concentration agarose gels (9-15%) are sufficiently porous to permit penetration also of high molecular weight biopolymers. This means that these gels 1) can be used for the fractionation of proteins, nucleic acids, etc. by molecular-sieve chromatography and 2) exhibit a high capacity when used as matrixes in hydrophobic interaction chromatography, ion exchange chromatography, affinity chromatography, etc. These gels are also very rigid and therefore afford surprisingly high flow rates even when the diameter of the gel spheres is relatively small (see ref. 1). Cross-linking increases the flow rate (ref. 9) which, however, probably is below that obtainable with silica beds (if optimal resolving power is desired the maximum obtainable flow rate of silica beds cannot be utilized for the fractionation of macromolecules owing to their slow diffusion rates; see Fig. 1 in ref. 10).

Spheres of agarose are not so expensive to prepare as those of silica and may therefore have a future not only in analytical but also in preparative high (intermediate) performance liquid chromatography (ref. 11).

It is questionable whether the separation of plasma proteins on octyl agarose should be rated as high or intermediate performance. However, in comparison with the commercial Octyl-Sepharose, the octyl agarose used in the experiment described in this paper showed a better resolution, as mentioned above. In a forthcoming paper we will discuss the terminology more thoroughly and describe experiments with gel spheres much smaller than those used in the experiment shown in Fig. 1; preliminary results are given in ref. 1.

Proteins have a much higher affinity for amphiphilic agarose gels than does tRNA. For instance, considerably higher ionic strengths are required for adsorption of tRNA than for serum proteins to Octyl-Sepharose (ref. 4). However, octyl silica could not be used for the fractionation by hydrophobic interaction chromatography of serum proteins or tRNA, since the solutes were irreversibly or too strongly adsorbed. Reduction of the chain length of the ligand to six carbon atoms gave an appropriate adsorption of tRNA, but proteins were too strongly adsorbed. The stronger affinity for silica beds than for agarose beds derivatized with the same ligand is probably due to the much higher ligand concentration of the silica gels, although nonspecific adsorption also may be of importance. High ligand concentrations should in general be avoided when chromatographing macromolecules, particularly when the interaction between the adsorption sites of the macromolecules and the adsorbent is strong. Otherwise there is risk of irreversible adsorption, since a prerequisite for desorption is that all links between the macromolecules of interest and the adsorbent are broken at the same time. This basic requirement for elution may be difficult to realize if the degree of substitution is excessive. The commercial silica beads with attached nonpolar groups usually have too high a degree of substitution with too hydrophobic ligands to be useful for hydrophobic interaction chromatography because they are synthesized for reversed phase chromatography, which requires high ligand density to create a continuous nonpolar stationary phase of an organic solvent. There is a risk that the silanol groups in the silica also interact with the macromolecules in the sample. This may impair the resolution, since in most cases it is complicated to use a separation method based on two fractionation parameters, as has previously been pointed out (ref. 12). Ethylene glycol in the buffer decreases the hydrophobic interaction and can then partly compensate for a high ligand concentration and a high hydrophobicity of the ligands. Ethylene glycol also suppresses the nonspecific interaction to silica (with the silanol groups?). An obvious disadvantage of the use of ethylene glycol is its high viscosity.

For the preparation of individual tRNA species one has to combine different separation methods. In a recent paper describing a new support, naphthoyl Sepharose for tRNA fractionation (ref. 13), we published a diagram which can be used as a guide for a proper selection of the combination of suitable separation methods.

Future experiments will show whether hydrophobic interaction chromatography on hexyl silica should be added to the list of the methods mentioned there.

ACKNOWLEDGEMENTS

We are much indebted to Dr. D. Eaker and Dr. F.E. Regnier for stimulating discussions on HPLC, and to Dr. I. Svensson for valuable comments on tRNA fractionation. We are also grateful to Mrs. Karin Elenbring and Mrs. Irja Johansson for important technical assistance. The work has been supported by the Swedish Natural Science Research Council.

REFERENCES

1 S. Hjertén and K. Yao, J. Chromatogr., October 1981, in press.
2 E. Pfannkock, K.C. Lu and F.E. Regnier, J. Chrom. Sci. 18 (1980) 430-441.
3 E.C. Nice, M. Capp and M.J. O'Hare, J. Chromatogr. 185 (1979) 413-427.
4 S. Hjertén, in J.C. Giddings *et al.* (Eds.), Advances in Chromatography, Vol. 19, Marcel Dekker, New York, 1981, pp. 111-125.
5 S. Hjertén, U. Hellman, I. Svensson and J. Rosengren, in J.-M. Egly (Ed.), Les Colloques de L'INSERM, Affinity Chromatography, INSERM, June 1979, Strasbourg, Vol. 86, pp. 315-320.
6 S. Hjertén, Biochim. Biophys. Acta 79 (1964) 393-398.
7 S. Hjertén, J. Rosengren and S. Påhlman, J. Chromatogr. 101 (1974) 281-288.
8 S. Hjertén, J. Chromatogr. 12 (1963) 510-526.
9 J. Porath, T. Låås and J.-C. Janson, J. Chromatogr. 103 (1975) 49-62.
10 S. Rokushika, T. Ohkawa and H. Hatano, J. Chromatogr. 176 (1979) 456-461.
11 E.F. Regnier, Anal. Biochem. 103 (1980) 1-25.
12 S. Hjertén, in N. Catsimpoolas (Ed.) Hydrophobic Interaction Chromatography of Proteins on Neutral Adsorbents, Vol. 2, Methods of Protein Separation, Plenum Publ. Corp., New York, 1976, pp. 233-243.
13 S. Hjertén, U. Hellman, I. Svensson and J. Rosengren, J. Biochem. Biophys. Methods 1 (1979) 263-273.
14 B.G. Johansson and S. Hjertén, Anal. Biochem. 59 (1974) 200-213.

T.C.J. Gribnau, J. Visser and R.J.F. Nivard (Editors),
Affinity Chromatography and Related Techniques

AFFINITY PARTITION STUDIED WITH GLUCOSE-6-PHOSPHATE DEHYDROGENASE IN AQUEOUS TWO-PHASE SYSTEMS IN RESPONSE TO TRIAZINE DYES

K.H. KRONER, A. CORDES, A. SCHELPER, M. MORR, A.F. BÜCKMANN, and M.-R. KULA
Gesellschaft für Biotechnologische Forschung mbH., Mascheroder Weg 1,
D-3300 Braunschweig-Stöckheim (G.F.R.)

SUMMARY

The partition coefficient of glucose-6-phosphate dehydrogenase in aqueous two-phase systems can be raised from values <1 to >20 by the addition of Cibacron-Blue or Procion-Red covalently bound to polyethylene glycol in a suitable system. Extraction of the enzyme by affinity-partition is possible only at low ionic strength. The capacity depends on the desired yield, the length of the tie-line, the volume ratio, and the degree of substitution of polyethylene glycol in the carrier system with the polyethylene glycol-dye derivative. Extraction of 100-120 U/ml at 95 % yield were realized, but higher values appear possible by further development of these technique.

INTRODUCTION

Aqueous-phase systems composed of polyethylene glycol (PEG) and dextran or of polyethylene glycol and salts have been successfully used for the extraction and purification of enzymes and cell organelles (refs. 1, 2). The partition coefficient K is defined as

$$K = C_{top}/C_{bottom} \tag{1}$$

where C_{top} and C_{bottom} denote the concentration of the compound of interest in the top and bottom phase respectively. Due to the rather high molecular weight of proteins, the molar concentration of enzymes usually encountered is fairly low, so that the partition coefficient of an enzyme is a constant of a given system. However, the aqueous-phase system itself can be easily modified by the variation of one or several of the following parameters:

- the average molecular weight of polymers
- the concentration of polymers and the length of the tie-line
- the kind and the concentration of ions included in the system
- the pH-value
- the temperature

The observed partition coefficient of an enzyme in aqueous two-phase systems is the result of a large number of interactions:

$$\ln K = \ln K_{el} + \ln K_{hphob} + \ln K_{hphil} + \ln K_{conf} + \ln K_{lig} \qquad (2)$$

where the subscripts 'el', 'hphob', 'hphil', 'conf', and 'lig' describe increments due to charge, hydrophobic and hydrophilic forces, conformation and ligand interactions respectively (ref. 3).

In principle each of these interactions can be exploited in an attempt to alter the partition coefficient of an enzyme in the desired direction and improve the selectivity of extraction. The most rational approach seems to be to utilize biospecific interactions of an enzyme with a substrate, product, inhibitor or antibody. The method is called affinity partition in analogy to affinity chromatography. If the ligand by itself shows a one-sided partition in the selected phase system no chemical modification is necessary. Such circumstances can only be expected if the ligand is a macromolecule. For example, from the experiments described by Hustedt and Kula (ref. 4), conditions for the extraction of $tRNA^{ile}$ by isoleucyl-tRNA synthetase can be derived. In the majority of cases, however, a one-sided partition of ligands is not found, so that covalent binding of the ligand to one of the phases forming polymers becomes necessary. It is assumed that the polymer will direct the partition of the ligand in the phase system and the ligand will accumulate in the same phase. This can lead to alterations in the properties of the carrier system especially if charged ligands generate an electric potential across the interface according to equations 3 and 4 or if the hydrophobicity in the system is much altered.

$$\ln K_p = \ln K_p^o + (Z_p F/RT) \cdot \psi \qquad (3)$$

$$\psi = [RT/(Z^+ + Z^-)F] \ln(K^-/K^+) \qquad (4)$$

ψ denotes the interfacial potential, Z_p the net charge of the protein, R the gas constant, F the Faraday-constant, T the absolute temperature, and K_p^o the partition coefficient of the protein in a system, where the interfacial potential is zero or when Z_p is zero. Z^+ and Z^- are the charges of excess ions included in the system and K^- and K^+ the partition coefficients of such ions.

These affects have to be considered when one interprets the results of affinity-partition experiments. Similar complications may also occur with solid affinity resins, but they are much more apparent in an aqueous two-phase system. Affinity partition shares with affinity chromatography the need to link each ligand separately to the support by chemical synthesis. This makes affinity techniques quite expensive; therefore general ligands for groups of proteins like coenzymes, lectins or nucleic acids, etc. were introduced to facilitate the synthetic work otherwise required. In recent years also a number of triazine dyes were tried as affinity reagents for the purification of dehydrogenases, kinases, and synthetases (ref. 5). Since the dyes are cheap enough and the synthesis of Dye-PEG compounds can be carried out without difficulty, we started a detailed investigation of affinity-partition in PEG-dextran systems with glucose-6-phosphate dehydrogenase as a model enzyme. Recovery of the soluble modified polymer is more difficult as compared to the recovery of solid affinity resin. Yet several advantages of the immiscible liquid systems warrant a careful examination of the potential of affinity-partition for enzyme purification: i) approach to equilibrium should be much faster and much less diffusion-controlled, ii) the binding capacity per unit volume is expected to be higher, by which the scale of operation is reduced, iii) continuous processes are possible with aqueous-phase systems (ref. 6), in contrast to chromatography which is a batch process.

The first part of our investigation was directed towards, the assessment of the capacity of aqueous-phase systems for affinity-partition. In addition, the contribution and possible interference of some general parameters of phase systems on the affinity-partition were analyzed.

RESULTS

The reactive chlorine group in the triazine-ring of Cibacron and Procion dyes (ref. 7) is a good leaving group. A new covalent linkage is formed by attack with a suitable nucleophile. Reaction with the terminal hydroxyl group of PEG under alkaline conditions is slow, and the degree of substitution is fairly low ($\sim$5 %). High degrees of substitution at both ends of polyethylene glycol (>80 %) can be obtained by the introduction of an amino group at the terminal. This was achieved by the replacement of the hydroxyl group with bromide (reaction with thionylbromide) and subsequent reaction with hexamethylenediamine or ammonia respectively, as described elsewhere (ref. 8). The coupling of the dye to the polymer was performed in methanol in the presence of triethylamine to neutralize the HCl generated. In the partition experiments described, Cibacron-Blue-3GA was bound to PEG 6000 via a hexamethylenediamine spacer (PEG-Blue), while Procion-Red HE-3B was bound directly to polyethylene glycol 6000-diamine (PEG-Red). Any excess of free dye was removed from the modified polymer by gel filtration. Fig. 1 presents the strutural formulas of the two dyes used. It is known that both dyes acts as competitive inhibitors for the enzyme. This inhibition has to be taken into account if the enzyme concentration is measured in the presence of modified dye.

Partition experiments were carried out at room temperature, mainly in 5 g of a two-phase system containing 7 % PEG 4000 and 5 % Dextran T-500 and various additions, as indicated in the figures. The volume ratio of this system is close to 2.2. PEG was in part replaced by PEG-Dye, and the concentration of modified dye is always expressed as % of total PEG in the system. 4 or 8 units of glucose-6-phosphate dehydrogenase (grade II, from Boehringer; Mannheim, Germany) were added, and the mixture was vigorously shaken on a Vortex for 5-10 seconds. Finally the phases were separated by centrifugation at 1800 g in a swinging-bucket rotor. Samples from upper and lower phase were carefully removed and analyzed seperately after suitable dilution. The concentration of modified dye was determined from absorptior measurements at 625 nm and 540 nm respectively. Glucose-6-phosphate dehydrogenase activity was assayed as described (ref. 9). The partition coefficients were calculated according to equation (1).

In the absence of dye, the partition coefficient of the enzyme (K_E^o) in the carri system varies in the range between 0.18-0.73 depending on the salt and pH selected. The PEG-dye prefers - as expected - the PEG-rich upper phase. The partition coefficient of PEG-Blue (K_B) reaches values between 17 and 140, as shown in Fig. 2 and 3. According to the theory developed by Flanagan and Barondes (ref. 10) the difference of K_E in the presence and absence of a ligand is a function of the partition coefficient of the ligand and the number of binding sites n:

$$\Delta \log K_E = n\ (\log K_L) \qquad (5)$$

High partition coefficients for the affinity ligand are therefore desirable so tha high extraction yields are obtained, but also losses in the other phase during multiple-step operations are minimized. K_B is influenced by the interfacial potent and solvation. In the presence of excess salt and an interfacial potential ψ near zero e.g. in systems containing sodium acetate K_B is nearly constant (Fig. 3).

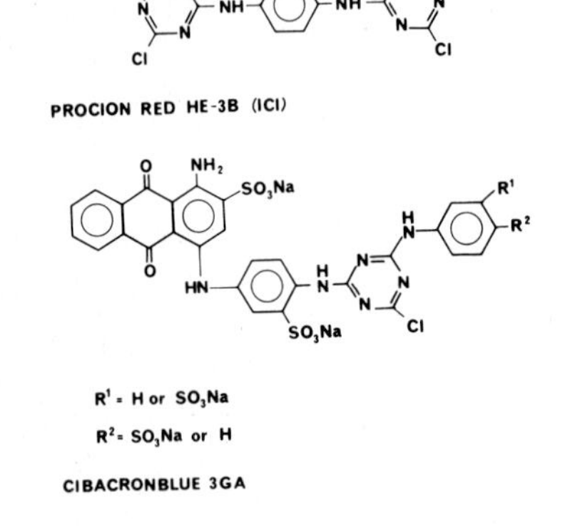

Fig. 1

Structural formulas of Cibacron-Blue 3GA (Ciba-Geigy) and Procion-Red HE-3B (ICI), used as affinity ligands after coupling to polyethylene glycol (PEG 6000).

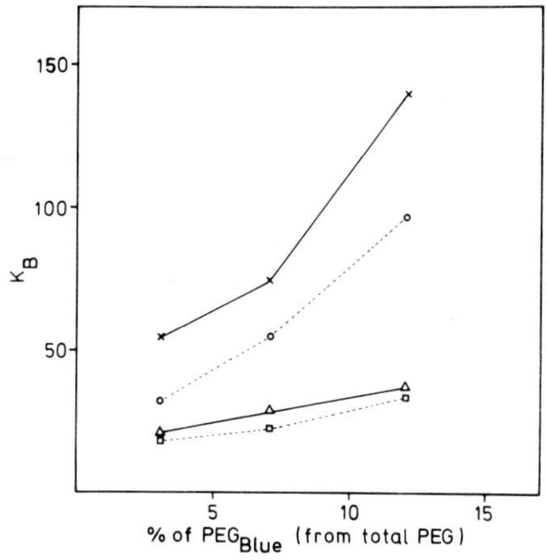

Fig. 2

Dependence of the partition coefficient of PEG-Blue (K_B) on the salt concentration, the pH and the concentration of PEG-Blue.

Phase system: 7 % PEG 4000, 5 % Dextran T-500

Symbols: (x) 0.3 M potassium phosphate pH 7.8
(o) 0.3 M potassium phosphate pH 6.0
(Δ) 0.05 M potassium phosphate pH 7.8
(□) 0.05 M potassium phosphate pH 6.0

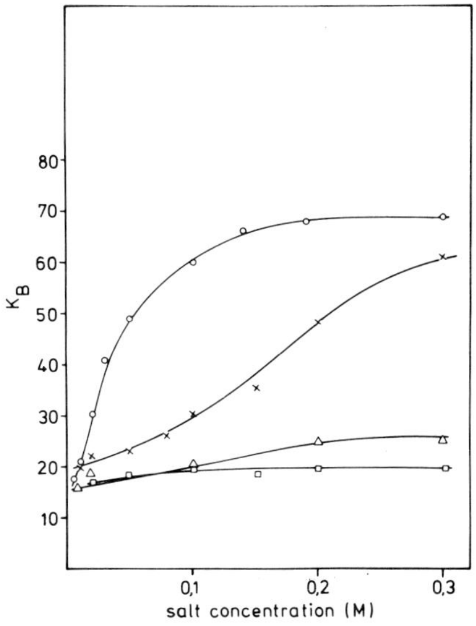

Fig. 3

Effect of different salts on the partition of PEG-Blue

Phase system:

7 % PEG 4000, 5 % Dextran T-500, 25 mg PEG_{6000}-Blue in 5g of total system, pH 7.8

Symbols:

(o) potassium phosphate
(x) ammonium sulphate
(Δ) potassium chloride
(□) sodium acetate

In systems with negatively charged lower phases, e.g. in the presence of phosphate at pH 7.8, K_B rises which can be explained by the negative charges on the dye. The large influence of the salt concentration on K_B, following the Hoffmeister series of chaotropic ions, demonstrates the effects of solvation on partition of PEG-Blue. The magnitude of change in K_B with the dye concentration (Fig. 2) is quite unexpected and must be related to subtle changes in the hydrophobicity of the upper phase with increasing dye concentration. This means that inherent properties are changed, and that two-phase systems containing different amounts of dye are not strictly comparable. It also follows that equation (5) cannot directly apply, because K_B is variable. At high dye concentration equation (5) approximates boundary conditions (see also ref. 11).

Fig. 4 summarizes the effects of different ions and salt concentrations on the affinity-partition of glucose-6-phosphate dehydrogenase in the presence of PEG-Blu (7 % of total PEG). The specific interaction of the enzyme with the inhibitor is suppressed at rather low salt concentrations, especially with ammonium sulfate, potassium phosphate, and potassium chloride; sodium-acetate is much less effective This sensitivity to salt in the enzyme-inhibitor interaction is not observed in affinity chromatography. Elution with salt of specifically bound dehydrogenases is reported to occur in the range from 0.1 to 1 M sodium chloride only (ref. 5). In affinity-partition, salt concentrations should not exceed 10 mM, with the possi exception of sodium-acetate systems. In addition this observation suggests a very simple strategy to re-extract the enzyme after affinity-partition into the salt phase of a secondary PEG/salt system (Table 1). It also shows clearly that PEG/sa systems cannot be employed as carrier systems for affinity partition with group-sp cific dyes.

The dependence of K_E and K_B on the length of the tie-line in a PEG-dextran system is shown in Fig. 5. K_B rises with increasing length of the tie-line and may level off at still higher values. K_E increases also, but reaches a maximum value around 20-22 % and then drops off sharply. The influence of the length of the tie-line is more pronounced in PEG 6000/dextran T-500 systems, where also higher K_E-values are obtained. This is remarkable since in unsubstituted systems partition coefficients for proteins are decreasing with increasing molecular weigh of PEG (refs. 1,2). Affinity-partition in the PEG-6000/Dextran T-500 system are therefore expected to give a higher selectivity of extraction into the upper phase. Similar observations were described for the affinity-partition of albumin in the presence of PEG palmitate (ref. 11).

The value for log K_E reaches a plateau with increasing dye concentration. The plot of Δlog K versus concentration of PEG-Blue (Fig. 6) represents a typical saturation curve which can be linearized in a doubly-reciprocal plot, as shown in Fig. 7. Here the reciprocal of the dye concentration is multiplied with the enzyme concentration. As can be seen, the results of two experiments at different enzyme

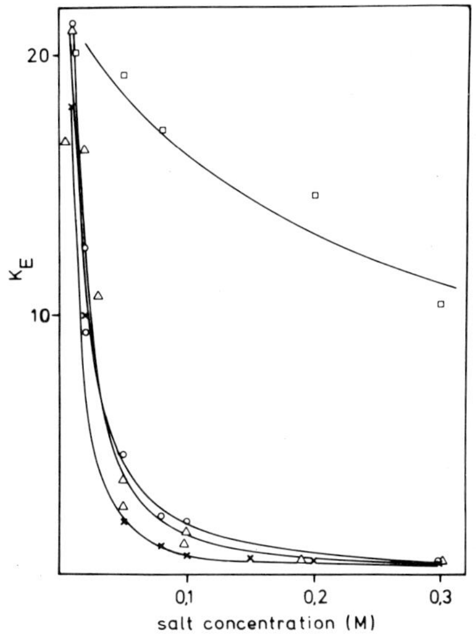

Fig. 4

Effect of different salts on the partition of glucose-6-phosphate dehydrogenase (G6P-DH) in the presence of PEG-Blue.

Phase systems:

7 % PEG 4000, 5 % Dextran T-500, 25 mg PEG_{6000}-Blue in 5 g total system.

Symbols:

(▲) potassium phosphate pH 7.5
(x) ammonium sulphate / TEA pH 7.8
(●) potassium chloride / TEA pH 7.8
(□) sodium acetate pH 7.5
TEA = Triethanolamine buffer 5×10^{-3} M

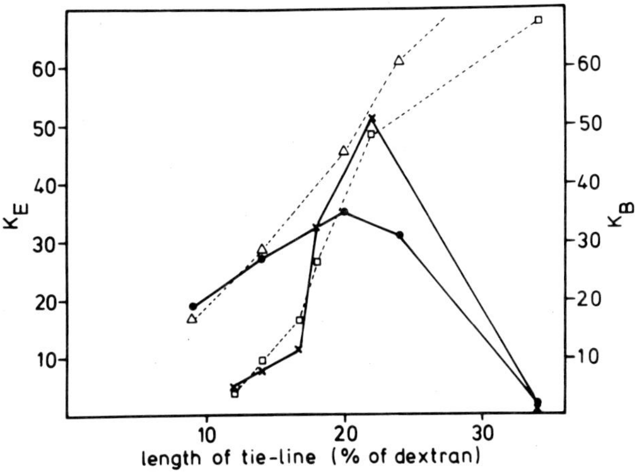

Fig. 5

Dependence of partition of G6P-DH and PEG-Blue on the length of the tie-line and the molecular weight of PEG.

Phase system: PEG and dextran concentrations are variable, 0.05 M sodium acetate, pH 7.5, 25 mg PEG_{6000}-Blue and 4 units of G6P-DH in 5 g total system.

Symbols: (▲) K_B PEG 4000 , (□) K_B PEG 6000 ,
(o) K_E PEG 4000 , (x) K_E PEG 6000 .

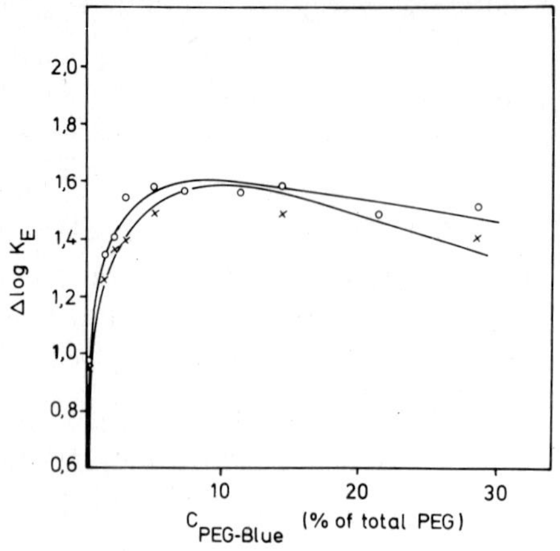

Fig. 6

Δlog K of G6P-DH as function of the concentration of PEG-Blue.

Phase system: 7 % PEG 4000, 5 % Dextran T-500, 0.02 M potassium phosphate, pH 7

Symbols: (o) 4 units G6P-DH, (x) 8 units G6P-DH (in 5 g total system).

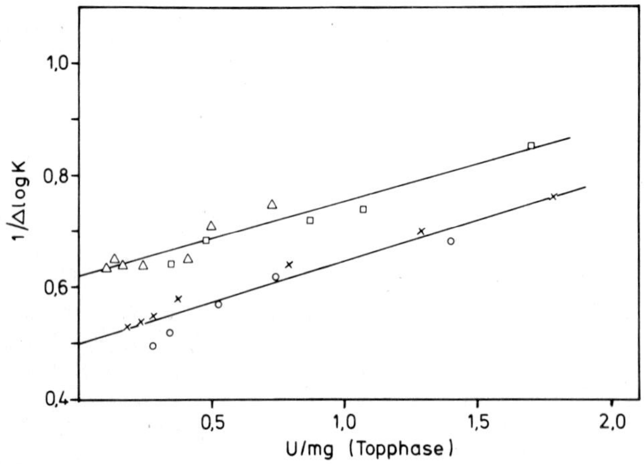

Fig. 7

Flanagan plot (modif for the interaction G6P-DH with PEG-Blue and PEG-Red in PEG/dextran phase sy

Phase system:

Blue - 7 % PEG 4000, 5 % Dextran T-500, 0.02 M potassium phosphate, pH 7.8,
Red - 5.5 % PEG 6000, 7 % Dextran T-500, 0.05 M tris-acetate buffer, pH 7.8, including 0.02 M MgCl , 0.4 mM EDTA.

Symbols: (Δ) and ($\square$) = 4 and 8 units G6P-DH in 5 g total system (Blue)
(x) and (o) = 4 and 8 units G6P-DH in 5 g total system (Red)

levels fit the same line. For comparison, the results of experiments in a different phase system (5.5 % PEG-6000/7 % dextran T-500) with PEG-Red as an affinity ligand are included in Fig. 7.

While the maximal K_E in both systems is different, the slope appears to be identical, which leads to the assumption that the slope could be related to the number of binding sites which should be identical for both ligands. The physical meaning of the slope should not be overinterpreted, since the plot of Fig. 7 is only a formal linearization of the saturation curves; this allows one to asses the capacity of the different systems in a simple way. The maximal K_E value is obtained from the intercept for a high excess of modified dye over enzyme. For practical purposes it is not necessary to work at saturating conditions. Fig. 8 correlates the yield of enzyme with different K_E values for a single extraction into the top phase. If 95 % yield is required a K_E value of ∿8 is sufficient; this raises the capacity from near zero at saturating conditions to 1.16 units/mg PEG-Blue in the standard system. For 90 % yield, the capacity increases to 3.6 units/mg PEG-Blue. It should be noted that the capacity of affinity-partition for a given system is not a constant value, but depends on the K_E values for different saturations, on the volume ratio of the phase system employed and on the desired yield. The volume capacity of PEG-rich top phases for the extraction of glucose-6-phosphate dehydrogenase is plotted in Fig. 9 as a function of the enzyme yield for two different systems. The two-phase system containing PEG-Red yields a higher maximal K_E value. But since this system is made up of 5.5 % PEG-6000 and 7 % Dextran T-500, the maximum amount of modified dye that can be incorporated is smaller as compared to the standard two-phase system used in the experiments with PEG-Blue. Also the volume ratio of the resulting two-phase system is reduced from 2.2 for experiments with PEG-Blue to 1.5 for PEG-Red. The K_E value and the volume ratio both contribute to the final extraction yield. Therefore, despite the 2.5-fold increased maximal K_E values in the PEG-Red system, the volume capacity is smaller. This again demontrates that the capacity of PEG/Dextran systems for affinity-partition is a complex function of a multitude of parameters. The volume capacity around 100-120 units/ml at 95 % yield is already quite high as compared to the capacity of solid supports. But further improvements can be made by selection of the optimal length of the tie-line (see Fig. 5) and by the improvement the volume ratio of the phases. Keeping the length of the tie-line constant corresponds to an increase of the PEG concentration relative to the dextran concentration in the phase system, which again would lead to a higher capacity. It is questionable whether 100 % of PEG can be replaced by a PEG-dye; however, as exploritary experiments show, at least 50 % of PEG can be replaced without too adverse effects on K_E. Table 1 summarizes molecular and technological data for the system studied in comparison with affinity chromatography. Capacity and higher activity yield argue in favour of affinity-partition processes; such processes could also be performed very easy also in large scale. When crude

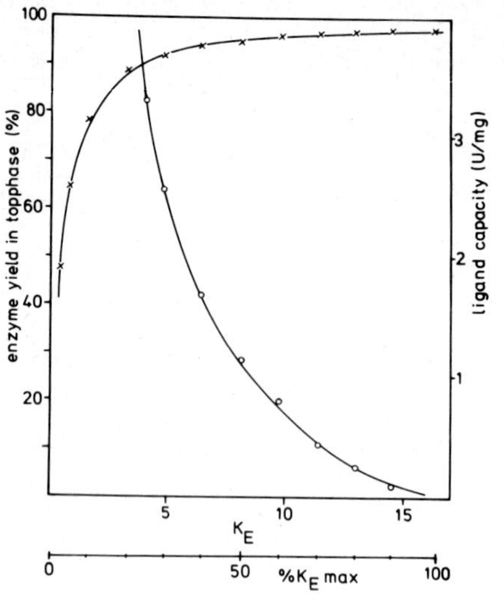

Fig. 8

Dependence of the enzyme recovery and the ligand capacity on the partition coefficient of G6P-DH (K_E) in the presence of PEG-Blue.

Phase system:

7 % PEG 4000, 5 % Dextran T-500, 0.02 M potassium phosphate, pH 7.8

Symbols:

(x) enzyme yield in top phase (according to equ. (2))
(o) ligand capacity (data derived from Flanagan plot, Fig. 7)

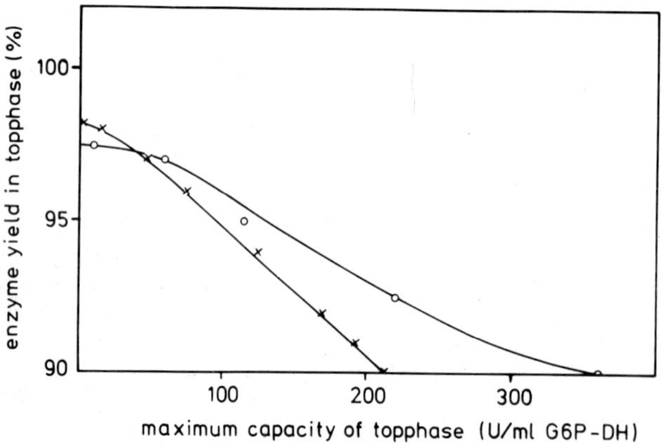

Fig. 9

Volume capacity o
PEG-rich top phas
including PEG-Blu
or PEG-Red.
for G6P-DH.

Phase system: see Fig. 7

Symbols: (o) = PEG-Blue (max. concentration ≈100 mg/ml)
(x) = PEG-Red (max. concentration ≈ 92 mg/ml).

yeast extracts are to be handled, the separation efficiency needs to be investigated.

TABLE 1

Comparison of molecular and technological data between affinity-partition and affinity-chromatography, using Cibacron-Blue 3GA and Procion-Red HE3B as general ligands for the isolation of glucose-6-phosphate dehydrogenase.

	affinity-partition				affinity-chromatography *1		
ligand	K_i (M)	recovery (%)	volume*2 capacity (U/ml)	dye conc. (µMol/ml)	recovery (%)	volume capacity (U/ml)	dye conc. (µMol/ml)
Cibacron-Blue 3GA	$1.8 \cdot 10^{-5}$	90	≈120	20 - 24	60	≈20	2
Procion-Red HE-3B	$2\text{-}5 \cdot 10^{-5}$	95	≈100	12 - 15	85	≈35	4

Recovery corresponds to values after repartition or elution, respectively.

Volume capacity and dye concentration corresponds to 1 ml of top phase or swollen gel, respectively.

*1 values derived from manufacturer data, using G6P-DH enriched yeast extract (Pharmacia, Amicon).

*2 calculated for enzyme yield of 95 % in top phase.

ACKNOWLEDGEMENTS

We thank Dr. Göte Johansson for many interesting and helpful discussions.

REFERENCES

1 P. A. Albertsson, Partition of Cell Particles and Macromolecules, Wiley, New York, 1971
2 M.-R. Kula in L.B. Wingard, E. Katchalski-Katzir and L. Goldstein (Ed's.) Applied Biochemistry and Bioengineering, Vol. 2, Academic Press, New York, 1979, pp 71-95
3 P. A. Albertsson, Endeavour 1 (1977) 69-74
4 H. Hustedt and M.-R. Kula, Eur. J. Biochem. 74 (1977) 191-198
5 Amicon Corporation, Dye-Ligand Chromatography, Lexington, 1980
6 M.-R. Kula, K. H. Kroner, H. Hustedt, and H. Schütte in Biochemical Engineering II, Annals of the New York Academy of Science, 1981 in press
7 K. Venkataraman (Ed.), The Chemistry of Synthetic Dyes, Volume 6, Reactive Dyes, Academic Press, New York, 1972
8 A. F. Bückmann, M. Morr, and G. Johansson, Makromol. Chem. 182 (1981) 1379-1384
9 G. W. Löhr and H. D. Waller in H. U. Bergmeyer (Ed.) Methoden der enzymatischen Analyse, 3rd edition, Verlag Chemie, Weinheim, 1974, pp 673-681
10 S. D. Flanagan and S. H. Barondes, J. Biol. Chem. 250 (1975) 1484-1489
11 G. Johansson, J. Chromatogr. 150 (1978) 63-71

T.C.J. Gribnau, J. Visser and R.J.F. Nivard (Editors),
Affinity Chromatography and Related Techniques

SCALING-UP OF AFFINITY CHROMATOGRAPHY, TECHNOLOGICAL AND ECONOMICAL ASPECTS

JAN-CHRISTER JANSON
Pharmacia Fine Chemicals AB, Uppsala (Sweden) and
Institute of Biochemistry, University of Uppsala, Uppsala (Sweden)

ABSTRACT

In the way affinity chromatography is most often exercised, the technique lends itself ideally both to scaling-up and to automation. The main obstacle to the large scale application of bioaffinity adsorption techniques, from a technological point of view, has been considered to be the limited mechanical stability of current gel matrices. In the author's opinion, there are both technological and chemical ways to cope with this problem. Automation of affinity chromatography experiments is briefly discussed. A comparative economical armchair study of laboratory, pilot and production scale affinity chromatographic purification of fibronectin from human plasma is presented.

INTRODUCTION

The inherent qualities of bioaffinity adsorption techniques, such as high specificity, high binding capacity and high recovery (1), together with the fact that the majority of the applications involve an all or nothing situation for adsorption and desorption, make them ideally suited to scaling-up operations. In some large-scale applications batch procedures are preferred in which the adsorbent is mixed with the process liquor for a couple of hours, collected by filtration or by centrifugation in a basket centrifuge, carefully washed and finally desorbed with displacement buffer either still in the funnel or in the centrifuge, or after having been transfered to a column. Especially when handling viscous extracts or when it is difficult to free the process liquor from particulate matter, batch adsorption is prefered. However, it is highly recommended not to apply the bioaffinity adsorption step too early in the purification procedure. Often the preparation of the adsorbent has meant both a lot of capital investment, time and effort, why to pay off it has to be reused several times. This is why many designers of purification procedures tend to keep traditional methods such as ammonium sulphate precipitation and adsorption to ion exchangers for the early stages of the purification, thus protecting the bioaffinity adsorbent from fouling and from the strain and the risk of losses involved in being mixed with large volumes of often viscous process liquors. In the majority of published purification procedures bioaffinity adsorbents are thus used at a late stage, very often as the last step, when the protein concentration is low and the volume usually small. This fact, together with the convenience in handling the adsorbent, has led to a widespread use of column adsorption and desorption in large-scale work.

When planning to apply bioaffinity adsorption on a large scale, there are certai factors of importance to consider, first of all, the choice of the affinity binding agent, the ligand, and the linkage reagent. Often the procedure is already optimize on a laboratory scale and the scaling-up only means an extrapolation of the laboratory conditions. However, in the small scale one normally can afford to use preactivated matrices and expensive ligands, whereas the cost may prohibit their use on a larger scale. Here there are two alternatives depending on the aim of the large-scale application. If the objective is to prepare more substance for a single research project aiming at, for example, structural studies of a low occurrence enzyme, the best alternative probably is to combine a moderately extrapolated volume scale-up with automation. This will be discussed under the heading automation. If, on the other hand, the main objective is to provide the basis for a subsequent industrial production process, a volume scale-up is necessary and thus more emphasis has to be put on optimization work. This will involve cost-effectiveness studies or ligand synthesis, and purchase of preactivated matrix or custom synthesized adsorbent in bulk versus own capital investment in equipment for large-scale activation and coupling etc..

From a pure technological point of view, the main obstacle to scaling-up of bioaffinity adsorbtion to an industrial scale has for a long time been considered to be the unsufficient mechanical rigidity and stability of the current matrices. The main emphasis of the following chapter will thus be on matters relevant to this fa

TECHNOLOGICAL ASPECTS

The matrix

The basic requirements of a general matrix for affinity chromatography are hydrophilicity, inertness, porosity and rigidity. Unfortunately, there is no material available yet which fulfils all these requirements. Ever since the pioneering work by Cuatrecasas et al, in 1968 (2), 4% agarose has been the most popular general matrix for laboratory scale applications (3) and most of the scaling-up studies published to date (4, 5, 6, 7, 8, 9, 10) have also been based on this material. Th main advantage of agarose is its highly porous, hydrophilic structure, combined with, for most applications, sufficient inertness and chemical stability. The main drawbacks are considered to be its intermediate rigidity and high price. Compared with porous ceramics and silica gels, which are the matrices of choice when mecha-

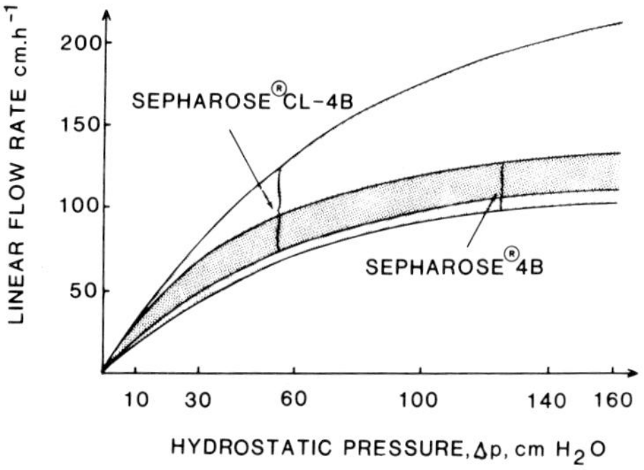

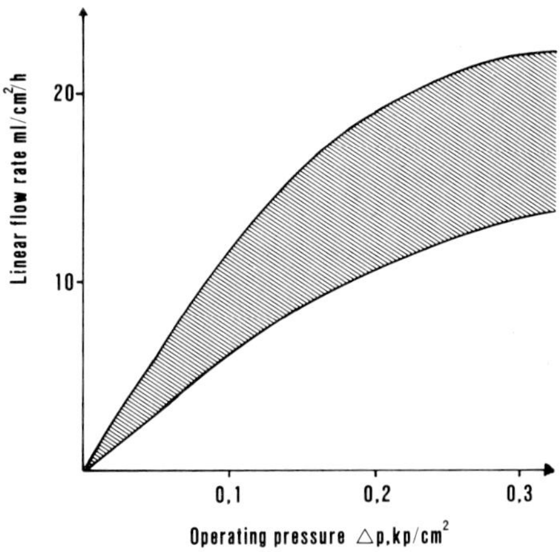

Fig. 1. and Fig. 2. Flow rate as a function of pressure for several production batches of Sepharose 4B packed in laboratory columns (5 x 10 cm) and production columns (37 x 15 cm, KS 370) respectively. In Fig. 1 the result from 14 batches of the cross-linked variety, Sepharose CL-4B is included for comparison.

nical stability is the main requirement, agarose cannot be considered rigid enough. However, to judge from the many applications, the rigidity is enough for the majority of users. In Figures 1 and 2 are shown the results obtained when analyzing flow rate as a function of pressure for several production batches of Sepharose^R 4B packed in laboratory columns and production scale columns respectively. For the laboratory scale columns, where the walls give considerable support to the gel bed, the maximum linear flows obtained varied for six batches between 105 and 135 $cm.h^{-1}$. For the cross-linked variety (11), Sepharose^R CL-4B, the corresponding figures for 14 production batches were 115 and 210 $cm \cdot h^{-1}$. For the production scale column, KS 370 ("the stack"), which is 37 cm in diameter and 15 cm high, the column wall support has no effect (24) and the maximum flow rates are much lower, varying between 14 and 22 $cm \cdot h^{-1}$. This correspond to volumetric flow rates from 15 to 24 $l \cdot h^{-1}$. For most large scale applications this flow rate is quite sufficient, however, one can expect to get still higher values after activation and coupling because cross--linking is one important side effect of most activation procedures. As can be seen in Fig. 1, cross-linking with dibromopropanol results in significantly more rigid agarose gel particles. A much more dramatic improvement of the rigidity of agarose is obtained after cross-linking with divinylsulfone (12). In Fig. 3 is shown the pressure-flow curves obtained after cross-linking of Sepharose 2B with increasing concentrations of divinylsulfone. Still more rigid particles are obtained by cross-linking gels with higher agarose concentration, however, this will lead to lower porosity and thus lower capacity (4). Another way of increasing the maximum flow rates of course is to increase the diameter of the spherical gel particles. A con-

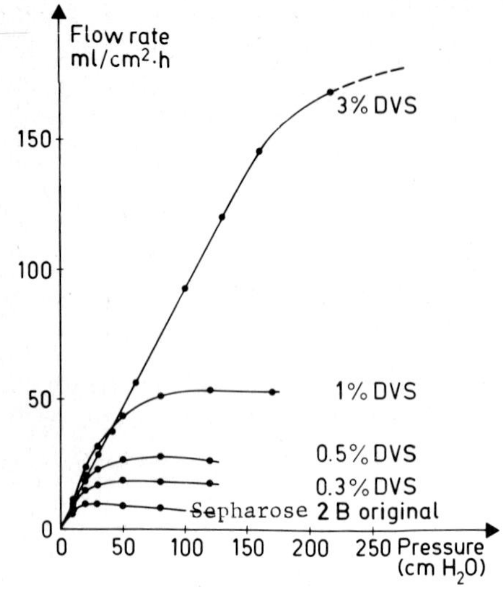

Fig. 3. The effect of cross-linking with divinylsulfone (DVS) on the flow rate as a function of pressure for Sepharose 2B packed in 2.5 x 40 cm columns.

comitant adverse effect of this, however, is a lowering of the efficiency of operation since the time needed for attainment of equilibrium will be increased in proportion to the increased diameter.

Even if agarose is considered to be the most generally applicable matrix, there are other alternatives that, due to their higher rigidity and mechanical stability, have proved useful in special large-scale applications. Thus research workers at Institut Merieux, Marcy L'Etoile, (13) have successfully used immunosorbents based on derivatized porous silica (Spherosil), the same matrix has been used at Diosynth BV, Oss, for the synthesis of large-scale adsorbents based on immobilized Cocanavalin A, antithrombin III and heparin (14). At Boehringer Mannheim, Tutzing, they are currently using adsorbents based on agarose introduced into porous ceramics for the purification of several enzymes on a large scale (15). Turková et al (16) have demonstrated the usefulness of derivatized hydroxyalkyl methacrylate gels (Spheron for large-scale affinity chromatography.

Porosity of the matrix. In a scaling-up study, Robinson et al (4), examined the effect of porosity on binding capacity for β-galactosidase using affinity adsorben based on Sepharose 2B, 4B and 6B. Their study emphasizes the need for a very porou support. Enzyme bound ranged from 380 units on Sepharose 6B to 1890 units on Sepharose 2B. However, the 2% gel was poorer mechanically and this led to a threefold drop in flow rate compared with the 6% gel. As a compromize they chose Sepharose 4 for the large-scale work. Mitra (17), working on a process for the large-scale isolation of antitrombin III based on heparin coupled to cross-linked Sepharose gels, found that the purity of the final product was dependent of the porosity of the matrix used. Thus, the degree of purity obtained using Sepharose CL-2B, CL-4B and CL-6B as matrices, were 72%, 99% and 81.5% of total protein respectively.

AUTOMATION

In the manner affinity chromatography is most often practiced, the technique lends itself ideally to automated routine applications, not only in industrial processes but also on a laboratory scale as was indicated in the introduction. Scaling-up by automation implies reproducibility of all parameters involved in the adsorption and desorption steps, and furthermore, a sufficiently stable binding capacity. Van der Loo and Hamers (18) successfully utilized this technique for the processing of large samples of antisera on a small immunosorbent column. The adsorbent retained more than 50% binding capacity after 200 runs and could still be used after 2 years of storage. Pneumatically activated Chromatronix valves (six 3-way valves and one double4-way valve) were controlled through a home-made cyclic programmer by solenoid valves. Robinson et al (6) designed an automatic, selfrepeating adsorption-desorption process for the large-scale purification of β-galactosidase from E. coli, based on an adsorbent prepared by coupling p-aminophenyl-β-D-galactopyranoside to Sepharose 4B as described by Steers et al (19). The adsorbent was packed in a 15 cm diameter and 10 cm high Perspex column. The column inlet was connected to vessels containing sample, equilibration buffer and displacement buffer via four solenoid valves (Fig. 4).

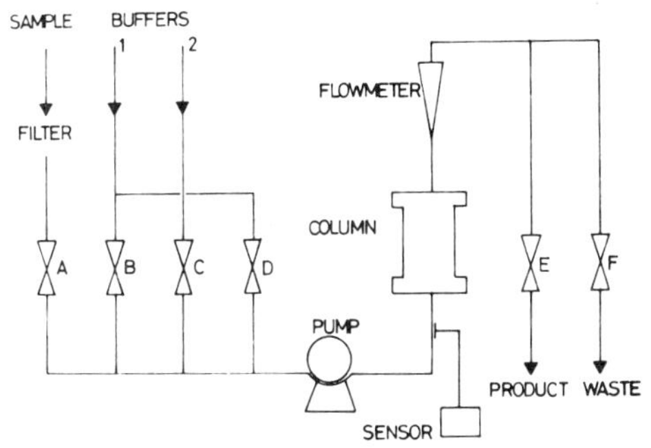

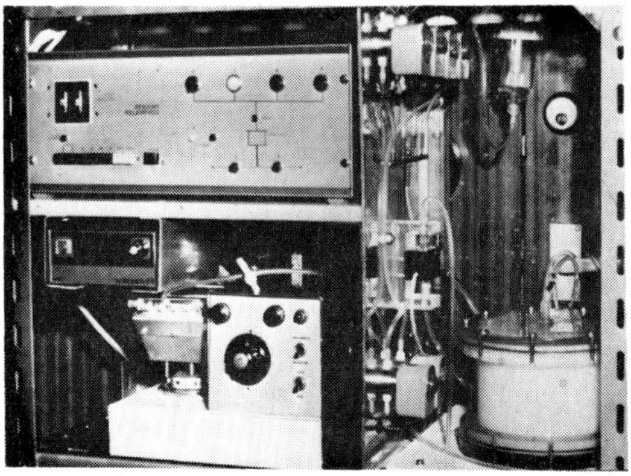

Fig. 4. Flow diagram for the automatic process for purification of β-galactosidase as described in the text. Fig. 5. Photo of the column, pump and control equipment.

The outlet was connected to a product collecting vessel and to waste via two solenoid valves (E,F). All six valves were activated at proper intervals by a multichannel adjustable cam timer (Londex Ltd., London SE 20). The device could be programmed to perform between 1 and 99 complete sequences. The column flow was controlled by a peristaltic pump and indicated on an in-line flow meter. To prevent air from enteri the column, a capacitance-operated monitor device (Fisons Ltd., Loughborough) was connected to the inlet tubing. Fig.5 shows a photograph of the set-up and in Fig.6 is shown a typical elution profile with the consecutive steps in the cycle indicate The system had a useful processing capacity of about 5 g pure enzyme per 2 h cycle.

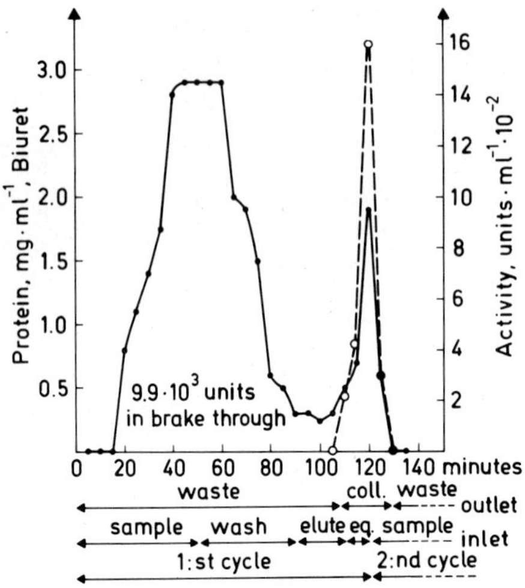

Fig. 6. Elution profile for adsorption and desorption of β-galactosidase. Column: 15x10 cm (1800 ml) packed with p-aminophenyl-β-D-thiogalactopyranoside coupled Sepharose 4B equilibrated in 0.05 M phosphate pH 7.0. Enzyme eluted with 0.1 M borate pH 10.0. Flow rate: 24 $cm \cdot h^{-1}$ (linear flow), 70 $ml \cdot min^{-1}$. Sample: $9.5 \cdot 10^5$ un (87 $units \cdot mg^{-1}$) in 3400 ml. Eluted: $8.9 \cdot 10^5$ units (780 $units \cdot mg^{-1}$) in 1400 ml.

For industrial chromatographic processes, complete, ready mounted system control and display towers are available (20). They contain pump and valve systems for column input and output, in-line filters for the continuous removal of dust particles from the buffer solutions, flow rate and pressure indicators, UV-monitoring and recording equipment, time/volume based control unit and finally a process display wit lamp indicators showing the process path (Fig. 7).

ECONOMICAL ASPECTS

There are two main objectives for the scale-up of affinity chromatography (as fo any chromatographic process). The first is to provide larger amounts of macromolecules for research, and the second is to design a unit operation in an industrial process for the production, for example, of clinical and diagnostic enzymes, cell growth factors and vaccines. In the first case, it is often sufficient to extra-

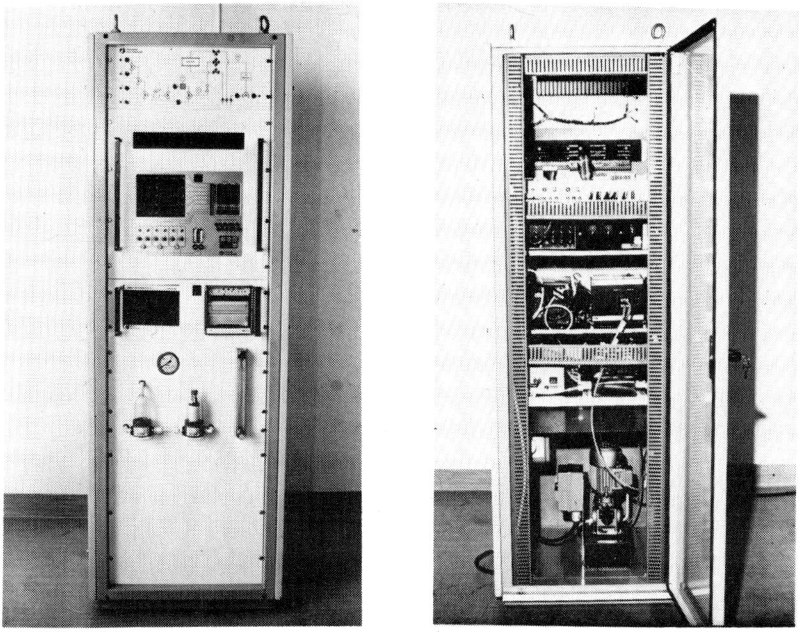

Fig. 7. Complete control and display tower for chromatographic processes. In the left photograph the tower is seen from the front with the system display on top. Below is the Sephamatic[R] Control Unit C-2, the UV-monitor control unit and recorder, and, further below, the pressure and flow rate indicators and the two in-line filters. In the right photograph the same tower is seen from the rear with the UV-monitor in the middle and the pump (Sera) at the bottom. From Pharmacia Fine Chemicals AB.

polate the small scale condition one order of magnitude, for example from 15 ml to 150 ml adsorbent bed volume, and combine this limited scaling-up with an automated, self-repeating chromatographic process design, using a controller of the kind described under the caption automation above. Such a set-up of course means capital investment. However, this equipment need not be confined to one special project, as a large quantity of a specific adsorbent would be. Also it need not be very expensive, the components amount to only a few hundred dollars and their assembly is easy for any electronics engineer. Chromatography controllers are of course also available on the market at prices depending on their versatility and degree of sofistication, from one to several thousand dollars.

If, on the other hand, the objective is to use affinity chromatography in an industrial process, it is natural to start an unbiassed study of the cost-performance characteristics of different alternatives for preparation of the adsorbents involving choice of ligand, linkage reagent and matrix. Here also the effective long-term operation with maximum retention of the immobilized ligand binding capacity is a factor of importance. With few exceptions, ligands for bioaffinity adsorption of enzymes are very complex. Their synthesis involve many steps and they are thus very expensive. Their use on a large scale can only be economically justified for expensive enzymes and other proteins for diagnostic or clinical use. Here the use of general ligands (21) can make the technique economically more feasible. The design of inexpensive general ligand adsorbents and elution methods for more than 60 dehydrogenases and kinases has been described by Lee et al (22).

In order to give a basis for comparison and discussion of various alternatives at the laboratories of the readers of this paper, an armchair scaling-up study has been made for the purification of fibronectin from human plasma based on recent wor by Vuento and Vaheri (23). Fibronectin, a glycoprotein with a molecular weight of 450 kilodalton, has been recognized as an important cell adhesion factor and enhanc the attachment and subsequent growth of many tissue culture cells. The concentratio of fibronectin in human plasma is approximately 300 $mg \cdot l^{-1}$ and as it is fairly expensive (Sigma, BRL: $40 per mg, June 1981) it should be well suited to large scale affinity chromatography. The purification procedure is based on a combination of adsorption to Gelatin-Sepharose 4B and subsequently to Arginine-Sepharose 4B. Befor adsorption to Gelatin-Sepharose 4B, the plasma is allowed to pass through a column of underivatized Sepharose 4B to get rid of non-specifically binding plasma protein The original procedure was based on the processing of 500 ml plasma, using a 1 lite column of Sepharose 4B, a 150 ml column of Gelatin-Sepharose 4B and a 50 ml column of Arginine-Sepharose 4B.

The armchair scaling-up is done in two steps, one order of magnitude in each ste Thus, the so-called pilot scale involves 5 l plasma and the large scale 50 l plasma The result of this study, which only comprises the cost of the chromatography materials according to the dollar pricelist of 1981, is presented in Tables 1,2 and 3.

TABLE 1

Purification of fibronectin from human plasma

Cost of chromatography materials (Lab scale, 500 ml plasma) in US$ (1981)	
Sepharose[R] 4B (1 liter)	99
CNBr-activated Sepharose[R] 4B (4 x 15 g)	324
Gelatin (0.2 g, 100 g package, Sigma)	2
Arginine (500 g)	23
Column K 100/45	1870
Column K 50/30	244
Column K 26/40	113
	2675

Yield per run (average of 10 runs): 77 mg

Price of fibronectin (June 1981) from Sigma and BRL: US$ 40 per mg

TABLE 1 (contin.)

Cost of chromatography material per mg fibronectin (lab scale)

Number of runs	Cost per mg US$	% of price of fibronectin
1	35	88
10	3.60 a)	9
100	0.87 a)	2

TABLE 2

Cost of chromatography materials (Pilot scale, 5 liter plasma) in US$ (1981)

Sepharose 4B (3 liters)	297
CNBr-activated Sepharose 4B (3 x 250 g)	2790
Gelatin (2 g, 100 g package)	2
Arginine (5 kg)	190
Column K 100/45 (3 pcs)	5610
	8889

Yield per run (extrapolated from lab scale): 770 mg

Cost of chromatography material per mg fibronectin (Pilot scale)

Number of runs	Cost per mg US$	% of price of fibronectin
1	12	30
10	1.30 a)	3.3
100	0.60 a)	1.5

TABLE 3

Cost of chromatography materials (Large scale, 50 liter plasma) in US$ (1981)

Sepharose 4B (30 liters)	2600
Gelatin-Sepharose 4B (custom made by Pharmacia) 15 l	Approx. 8500
Arginine-Sepharose 4B (custom made by Pharmacia) 5 l	Approx. 6800
Column KS 370 A (1 pcs)	3310
Column KS 370 B (3 pcs)	7680
	28890

Yield per run (extrapolated from lab scale): 7.7 g

Cost of chromatography material per mg fibronectin (large scale)

Number of runs	Cost per mg US$	% of price of fibronectin
1	3.75	9.4
10	0.46 a)	1.2
100	0.30 a)	0.75

a) Life span of Gelatin-Sepharose 4B : 10 runs (Ref. 23)
Life span of Arginine-Sepharose 4B: 5 runs (Ref. 23)

Perhaps the most evident weakness of this study is the somewhat arbitrary assumption that the price of US$ 40 per mg fibronectin is valid also for gram quantities of this substance. This is of course not the case. However, it was used in the absence of more relevant data. The author has also stuck to the recommendation of Vuento and Vaheri concerning the life span of the two adsorbents without knowing how much optimization work was done by the authors on various procedures for regeneration. In this case the relatively short life span makes an enormous inpact on the economy of the process. Probably the life span for Arginin-coupled Sepharose 4B would increase if it had been attached via an epoxid linkage rather than via a CNBr-activated matrix. Even if the figures in the tables very well speak for themselves, it is worth remarking that, not surprisingly, reuse of the adsorbents is an essential element of a cost-effective process.

REFERENCES

1 C.R. Lowe, in T.S. Work and E. Work (Eds.), Laboratory Techniques in Biochemistry and Molecular Biology, Vol. 7, Part 2, An Introduction to Affinity Chromatography, North Holland, Amsterdam, 1979, pp. 269 - 522.
2 P. Cuatrecasas, M. Wilchek and C.B. Anfinsen, Proc. Nat. Acad. Sci. U.S.A., 61(1968)636
3 M. Wilchek and W.B. Jacoby, in S.P. Colowick and N.O. Kaplan (Eds.), Methods in Enzymology, Vol. 34, Affinity Techniques, Academic Press, New York, 1974, pp. 3 -
4 P.J. Robinson, P. Dunnill and M.D. Lilly, Biochim.Biophys.Acta, 285(1972)28 - 35.
5 M. Matsuda, S. Iwanaga and S. Nakamura, Thrombos.Res., 1(1972)619 - 630.
6 P.J. Robinson, M.A. Wheatly, J.-C. Janson, P. Dunnill and M.D. Lilly, Biotechnol. Bioeng., 16(1974)1103 - 1112.
7 M.J. Holroyde, J.M.E. Chesher, I.P. Trayer and D.G. Walker, Biochem.J., 153(1976) 351 - 361.
8 P.J. Sicard, G. Mialonier, M. Smagghe, F.X. Galen, P. Corvol, C. Devaux and J. Ménard, Prep.Biochem., 8(1978)19 - 36.
9 A.R. Neurath, A.M. Prince and J. Giacalone, Experientia, 34(1978)414 - 415.
10 K.D. Kulbe and R. Schuer, Anal.Biochem., 93(1979)46 - 51.
11 J. Porath, J.-C. Janson and T. Låås, J.Chromatogr., 60(1971)167 - 177.
12 J. Porath, T. Låås and J.-C. Janson, J.Chromatogr., 103(1975)49 - 62.
13 M. Tardy, J.-L. Tayot, M. Roumiantzeff and R. Plan, in R. Epton (Ed.), Chromatography of Synthetic and Biological Polymers, Vol. 2, Ellis Horwood, Chichester, 1978, pp. 298 - 313.
14 F.E.A. van Houdenhoven, Poster presentation at the 1:st Conference on Fermentation Recovery Process Technology, Banff, Canada, June 7 - 12, 1981.
15 D. Jaworek, personal communication.
16 J. Turková, K. Bláha, J Horácek, J. Vajcnar, A. Frydrychová and J. Coupec, J.Chromatogr., (1981) in press.
17 G. Mitra, Poster presentation at the 1:st Conference on Fermentation Recovery Process Technology, Banff, Canada, June 7 - 12, 1981.
18 W. van der Loo and R. Hamers, in H. Peeters (Ed.), Protides of the Biological Fluids, Vol. 23, Pergamon Press Ltd., Oxford, 1976, pp. 603 -608.
19 E. Steers, P. Cuatrecasas and H.B. Pollard, J.Biol.Chem., 246(1971)196 - 200.
20 Pharmacia Fine Chemicals AB, Uppsala, Sweden
21 K. Mosbach, H. Guilford, P.-O. Larsson, R. Olsson and M. Scott, Biochem.J., 125(1971)20.
22 C.-Y. Lee, L.H. Lazarus and N.O. Kaplan, in E.K. Pye and H.H. Weetall (Eds.), Enzyme Engineering, Vol. 3, Plenum Press, New York, 1978, pp. 299 - 337.
23 M. Vuento and A. Vaheri, Biochem.J., 183(1979)331 - 337.
24 J.-C. Janson and P. Dunnill, Proceedings of the Federation of European Biochemical Societies IXth Meeting, Dublin, 1973, North -Holland, Amsterdam, 1974, pp. 81-105.

T.C.J. Gribnau, J. Visser and R.J.F. Nivard (Editors),
Affinity Chromatography and Related Techniques

SPECIFIC SORBENTS FOR HIGH PERFORMANCE LIQUID AFFINITY CHROMATOGRAPHY AND LARGE SCALE ISOLATION OF PROTEINASES

Jaroslava Turková

Institute of Organic Chemistry and Biochemistry, Czechoslovak Academy of Sciences, 166 10 Prague 6 (Czechoslovakia)

ABSTRACT

For an efficient affinity chromatography of proteolytic enzymes suitable affinity sorbents were prepared by attachment of synthetic low molecular weight inhibitors (p-aminobenzamidine for trypsin, Z-Gly-D-Phe for chymotrypsin, ε-aminocaproyl-L-Phe-D-Phe-OMe for carboxylic proteinases) to hydroxyalkylmethacrylate gels derivatized by epichlorhydrin. The coupling of various amino derivatives to the epoxide-containing support has been investigated as a function of pH, coupling time, concentration of the compound bound, degree of epoxidation of the support, and specific structural features of amino compounds. The amounts of the individual substances attached are considerably affected by both the nature of the substance coupled and the characteristics of the solid matrix.

In order to minimalize nonspecific sorption and to utilize the inhibitor attached as much as possible the affinity ligand content of the specific sorbent should be kept at the lowest possible level. The specific sorbent for porcine pepsin with 0.85 µmol of ε-aminocaproyl-L-Phe-D-Phe-OMe per g of dry support sorbed 29.4 mg of pepsin per g of dry sorbent (99% of the attached inhibitor molecules were involved in the specific complex formation). With the increasing content of the affinity ligand attached the portion of inhibitor molecules involved in the specific complex with pepsin sharply decreased (for 4.5 µmol/g already only 26%). In correlation with the specificity of the individual pepsins the same specific sorbent was much less efficient for human and chicken pepsin. The amount of chicken pepsin sorbed increased after its reaction with o-nitrobenzenesulfenyl chloride. For the sorption of pepsin and the carboxylic proteinase of Humicola sp. from diluted solutions the column arrangement only was suitable.

The advantage of sorbents prepared lies in their chemical and mechanical stability, small extent of nonspecific interactions, their ability to undergo almost complete regeneration, and in their suitability for high pressure column or microcolumn operations. By the use of high performance liquid affinity chromatography of pepsin the enzyme concentration in solution was determined within 30 min with a high sensitivity.

INTRODUCTION

Enzymes of perspective application as biocatalysts in industrial practice are mainly microbial enzymes because of the abundance of their resources. Many microorganisms produce proteolytic enzymes in high concentrations in culture media. The simultaneous presence of several types of proteinases and of many other enzymes (amylases, cellulases) results in special requirements in the solid support and affinity ligand for the isolation of the individual enzymes. The support must be not only macroporous and chemically and mechanically stable but also resistant to enzymic and microbial attack. A suitable support is the hydroxyalkyl methacrylate copolymer[x] modified by epichlorohydrine. Synthetic low molecular weight inhibitors are the affinity ligands of choice. Specific sorbents prepared in this manner can be used and reused practically without limits, if the stable bonds between the epoxide groups of the support and the amino groups of peptide inhibitors are used.

RESULTS

Choice of a suitable affinity ligand

To isolate chymotrypsin and trypsin from a crude pancreatic extract (ref. 1) we used original specific adsorbents prepared by coupling two naturally occurring high molecular weight proteinase inhibitors (ovomucoid for trypsin and the polyvalent proteinase inhibitor Antilysine for chymotrypsin) to a hydroxyalkyl methacrylate gel (Spheron) activated with cyanogen bromide (ref. 2-3). Volatile buffers (0.05M formic acid adjusted to pH 8.0 with 5% aqueous ammonia for sorption and 0.1M formic acid adjusted to pH 3.5 with ammonia for desorption) were used. The enzyme fractions after lyophilization contained trypsin and chymotrypsin with less than 5% of salts and with activities higher than those required with commercial preparations standards. After ten runs the capacity of ovomucoid-Spheron dropped to 20% and that of Antilysine-Spheron to 23%. Because of the low stability and the high cost of the naturally occurring protein inhibitors, we focused our attention on the use of specific adsorbents prepared from low molecular weight synthetic inhibitors.

NH_2-Spheron was prepared by coupling hexamethylenediamine to Spheron. Subsequently, N-benzyloxycarbonylglycyl-D-phenylalanine

[x]The hydroxyalkyl methacrylate gels are distributed under the commercial name Spheron (product of Lachema, Brno) or Separon HEMA (product of Laboratory Instruments Works, Prague)

or p-aminobenzamidine were attached to NH_2-Spheron for the isolation of chymotrypsin and trypsin, respectively. Under analogous conditions the results obtained with synthetic peptide inhibitors and with naturally occurring proteinase inhibitors as ligands were identical. Unlike the naturally occurring inhibitors, which undergo denaturation or digestion because of their protein character and thus irreversibly loose their activity, the synthetic low molecular weight inhibitors are completely stable. The capacity of specific sorbents prepared with these inhibitors can be regenerated practically without limits, if stable bonds between the support and the amino groups of the peptide inhibitors are used.

Choice of a suitable support and method of immobilization

Copolymerization of hydroxyalkyl methacrylate with alkylene dimethacrylate yields heavily crosslinked microparticles of a xerogel which by aggregation give rise to macroporous spheroids (ref. 4). The hydroxy groups of the copolymer react in an alkaline medium with epichlorohydrine according to the equation

$$\underset{\overset{|}{CH_2}}{\overset{\overset{|}{CH_2}}{CH_3-C-CO-O-CH_2CH_2-OH}} \xrightarrow{Cl-CH_2-\underset{O}{CH-CH_2}} \underset{\overset{|}{CH_2}}{\overset{\overset{|}{CH_2}}{CH_3-C-CO-O-CH_2CH_2-O-CH_2-\underset{O}{CH-CH_2}}}$$

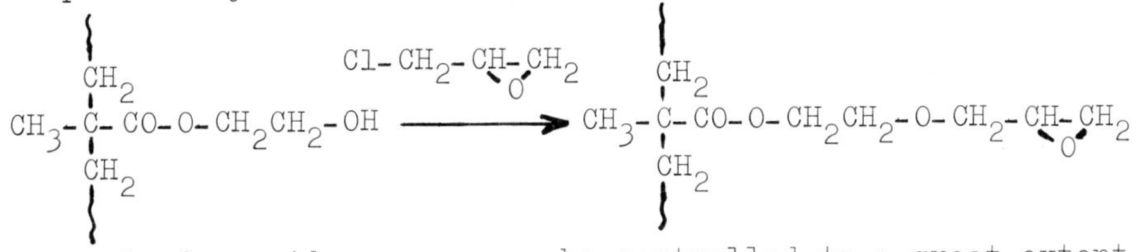

The amount of epoxide groups may be controlled to a great extent by the reaction conditions during the modification (ref. 5).

The epoxide groups of the derivatized support may react with amino, carboxy, hydroxy and sulfhydryl groups and with some aromatic nuclei, such as indole, imidazole, etc, (ref. 6). By coupling of glycyl-L-leucine to a copolymer of glycidyl methacrylate with ethylene dimethacrylate we established that the linkage via the α-amino group was strongly pH-dependent (optimum at pH 9.7) (ref. 7). The coupling of the ε-amino group of acetyl-L-lysyl--glycine methyl ester increases with the increasing pH and does not show a pH extreme. It is by one order of magnitude slower in comparison with the coupling of α-amino groups. The linkage via amino groups is unusually strong. It is not split even after 6-day hydrolysis in 6M HCl at 110°C. The coupling of N-acetyl-L--leucine to the glycidyl methacrylate copolymer is negligible and indicates a pH-independence of the coupling of peptides via carboxyl groups. Fig. 1 shows the pH-profile of the coupling of glycyl-D-phenylalanine to the hydroxyalkyl methacrylate gel (Separon derivatized by epichlorohydrine to a high and a low degree of epoxidation. No decrease of the amount of peptide attached to this support was observed above pH 9.7.

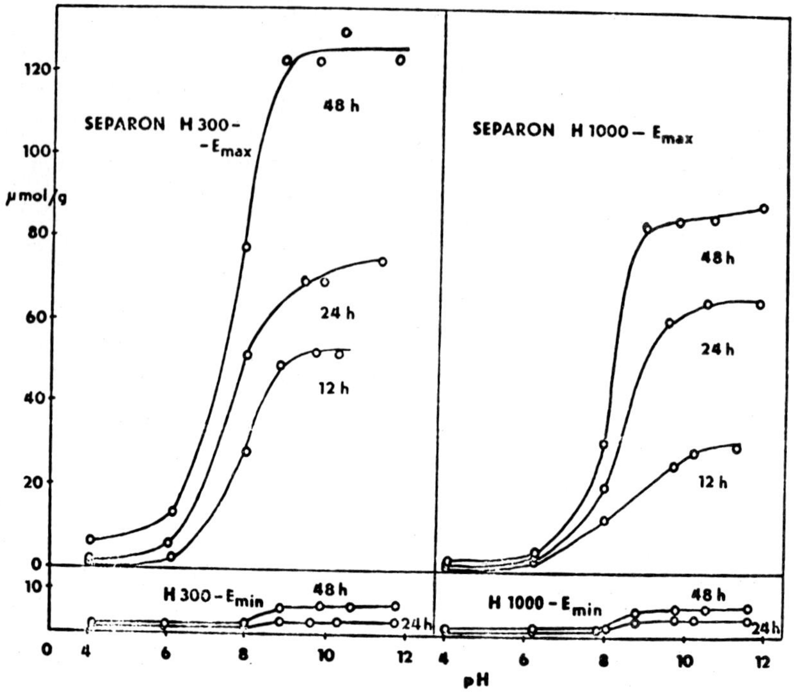

Fig. 1. Coupling of glycyl-D-phenylalanine (in μmol per 1g of dry conjugate) to various types of derivatized Separons as function of pH and time. The hydroxyalkyl methacrylate gel Separon H300 (molecular exclusion limit $3x10^5$ and specific surface $90m^2/g$) was derivatized by epichlorohydrine to a high (E_{max}=1.5mmol of epoxide groups per g of dry gel) and a low (E_{min}=0.24mmol/g) degree; the content of the epoxide groups of derivatized Separon H1000 (molecular weight exclusion limit 10^6 and specific surface $30m^2/g$) was 1.77mmol/g (E_{max}) and 0.14mmol/g (E_{min}).

It is evident from Fig. 2 that the coupling rate is also affected by the structure of the amino component bound. Ammonia, 1,2-diaminoethane, 1,4-diaminobutane and 1,6-diaminohexane were coupled to derivatized Separon during 6 and 40 h. The coupling reaction in the case of 1,6-diaminohexane is complete already after 6 h. The subsequent saturation with 2-aminoethanol did not take place unlike the coupling reactions of other compounds shown in Fig. 2. This may be caused by a local concentration increase of 1,6-diaminohexane on the surface of the support due to hydrophobic interactions which in the case of the hexamethylene chain could play the most important role (ref. 5). The dependence of the coupling on the character of the component coupled and on the character of the support was observed earlier in the case of some proteins attached to various epoxide containing supports (ref. 8).

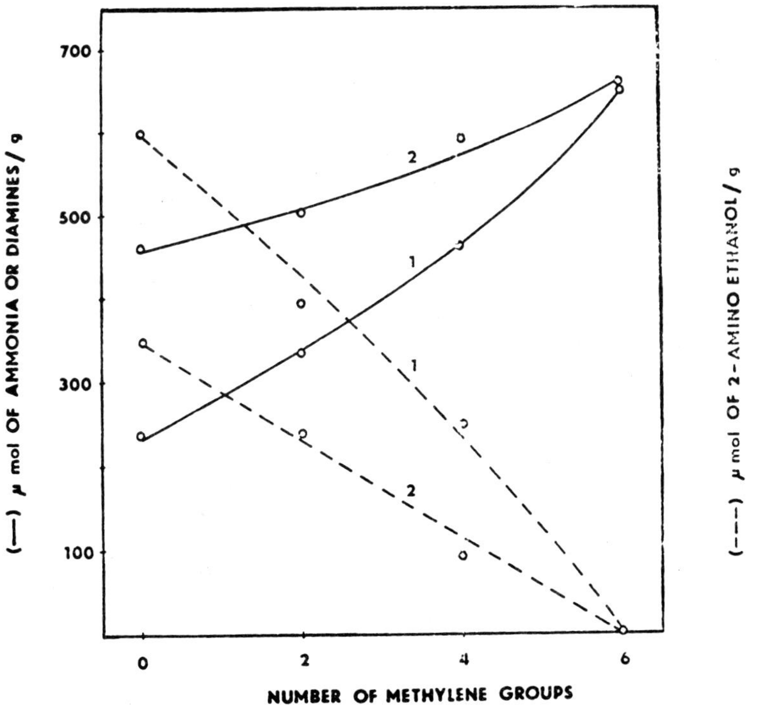

Fig. 2. Coupling of ammonia and diamines $H_2N(CH_2)_nNH_2$ to Separon H1000 E_{max} followed by blocking of the unreacted epoxide groups with 2-aminoethanol. Base solutions (concentration 0.5M) in water, pH adjusted to 11.5. Curves 1 coupling for 6 h, curves 2 coupling for 40 h.

For complete removal of the epoxide groups it is widely recommended (ref. 9) to use additional coupling of 2-aminoethanol. It is evident from Fig. 3 that the coupling reaction of 2-aminoethanol to Separon H1000 E_{max} proceeds slowly and moreover requires a high concentration of 2-aminoethanol and a high pH. We therefore eliminate the unreacted epoxide groups by hydrolysis with 0.1M $HClO_4$ (ref. 5).

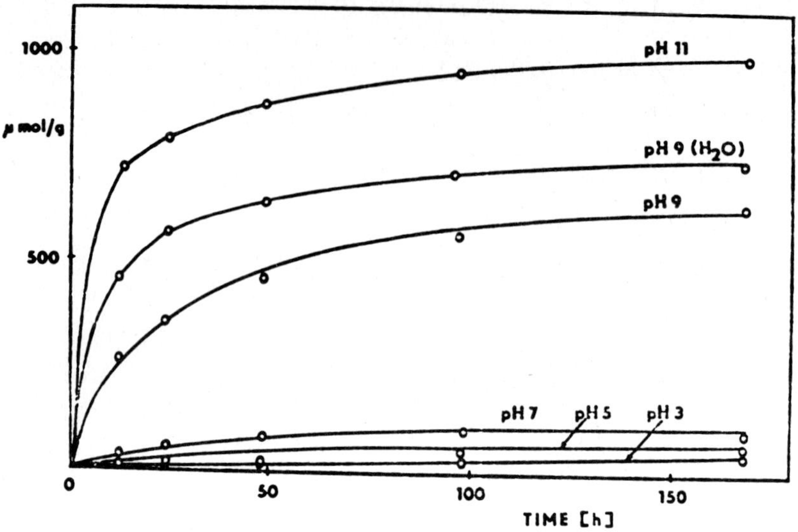

Fig. 3. Coupling of 2-aminoethanol to Separon H1000 E_{max} as a function of coupling time and pH. Britton-Robinson buffers were used for pH 3,5,7,9 and 11, with the exception of pH 9, where an aqueous solution (concentration 1M) was used.

Concentration of the affinity ligand attached to solid matrix

In order to minimalize nonspecific sorption of inert compounds to affinity sorbents and to exploit as much as possible the affinity ligand attached its lowest possible content in the specific sorbent should be used. The importance of the low concentration of the affinity ligand and the effect of the uneven surface of the gel are illustrated in Fig. 4 derived from our own experience and from ideas presented in the papers of Hofstee (ref. 10) and Porath (ref. 11). In the upper part of Fig. 4 the affinity ligand attached to a solid matrix does not form a specific complex with the enzyme isolated (there is no complementarity of the binding site of the enzyme to the immobilized affinant). However, the molecules of this enzyme are sorbed onto the modified support by nonspecific multi-point bonding. In the central part of the figure the specific complementary "one-to-one" bonding of the enzyme isolated to the immobilized ligand is represented. At a low concentration of the affinant attached multiple nonspecific bonds cannot operate. Therefore only biospecific bonding of the molecules via complementary active sites takes place, unless - of course - this bond is not sterically hindered. The figure shows a macroreticular

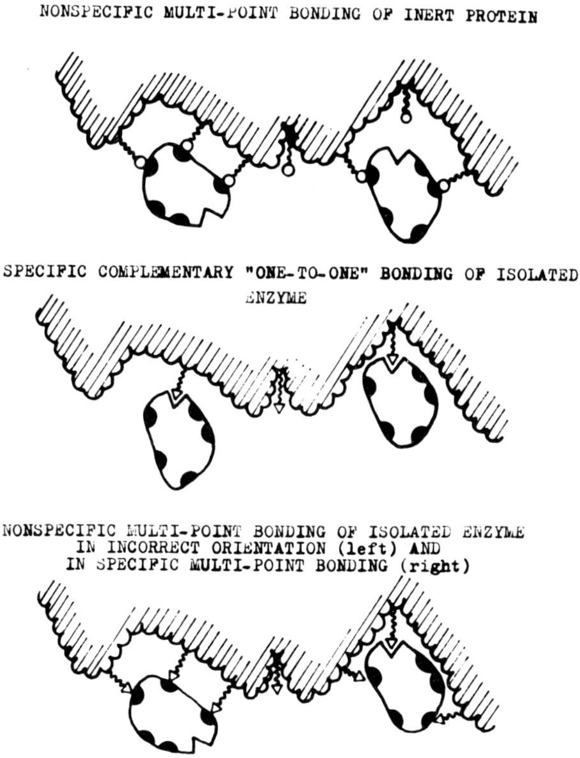

NONSPECIFIC MULTI-POINT BONDING OF INERT PROTEIN
SPECIFIC COMPLEMENTARY "ONE-TO-ONE" BONDING OF ISOLATED ENZYME
NONSPECIFIC MULTI-POINT BONDING OF ISOLATED ENZYME
IN INCORRECT ORIENTATION (left) AND
IN SPECIFIC MULTI-POINT BONDING (right)

Fig. 4. Schematic drawing of effects of concentration of immobilized affinity ligands and of uneven support surface on nonspecific and specific sorption.

hydroxyalkyl methacrylate polymer represented by aggregated globules. Due to the effect of the uneven surface of the gel, well accessible, less accessible and even sterically hindered affinity ligands can be distinguished after the coupling of the affinant through a spacer. These steric hindrances explain not only the low saturation of the immobilized affinant molecules with the isolated enzyme but also their heterogeneity in affinity. In the lower part of the figure multiple bonding of the enzyme is again shown. Although the enzyme contains the complementary site for the immobilized affinity ligand, the orientation of the bonding need not necessarily be correct. The nonspecific bonds can be of the type of electrostatic interactions or of hydrophobic bonds, or a combination of both. The enzyme can be bound nonspecifically to groups of their own affinant or to the spacer or even to the solid matrix. These non-specific multiple bonds may be stronger than one complementary bond between the isolated enzyme and the immobilized affinity ligand. In addition to this specific complementary bond these nonspecific multiple bonds increase the strength of the bonding in a specific complex; this results in the elution of the same enzyme in several fractions (ref. 10) or in the difficulties with the elution of the enzyme from the specific sorbent (ref. 12).

In order to experimentally determine the effect of the concentration of the immobilized inhibitor on the course of affinity chromatography of proteolytic enzymes , we prepared specific sorbents for carboxylic proteinases (ref. 13-14) containing differe concentrations of ε-aminocaproyl-L-Phe-D-Phe-OMe. The inhibitor was coupled to the epoxide-containing Separon H1000 E_{med} (content of epoxide groups 0.8 mmol/g); the resulting concentrations of ε-aminocaproyl-L-Phe-D-Phe-OMe in μmol/g of dry gel were 0.85, 1.2, 2.5, 4.5, and 155. Since, however, nonderivatized Separon was found suitable for hydrophobic chromatography of some proteins (ref. 15), we first examined the behavior of crude porcine, chicken and human pepsins on a column of unmodified Separon H1000 under the conditions of affinity chromatography (ref. 5). In this case the whole proteolytic activity emerged in the front of the equilibration buffer and no material was eluted by the desorption buffer of high ionic strength. We applied then solutions (of the concentration 1g/200ml) of porcine, chicken, or human pepsin to the colum of the affinity sorbents mentioned above continuously until the effluent showed the same activity as the solution applied (Fig. 5). After washing out of unbound pepsin and proteins nonspecifically adsorbed, pepsin was eluted by the sodium acetate buffer containing 1M NaCl. On columns of affinity sorbents containing the inhibitor attached in the concentration range 0.85 - 4.5 μmol/g in all cases we received one fraction of very active pepsin in one sharp peak (Fig. 5A). From absorbance at 278 nm and proteolytic activity measurement we calculated the amount of desorbed pepsin. Contrarily on column of the affinity sorbent containing the inhibitor in concentration of 155 μmol/g, we received several peaks of pepsin showing different specific proteolytic activities (Fig. 5B). This different behavior of the enzyme on affinity sorbents with low and high content of immobilized inhibitor could be due to multiple bonding of the enzyme molecule and inert proteins as demonstrated in Fig. 4.

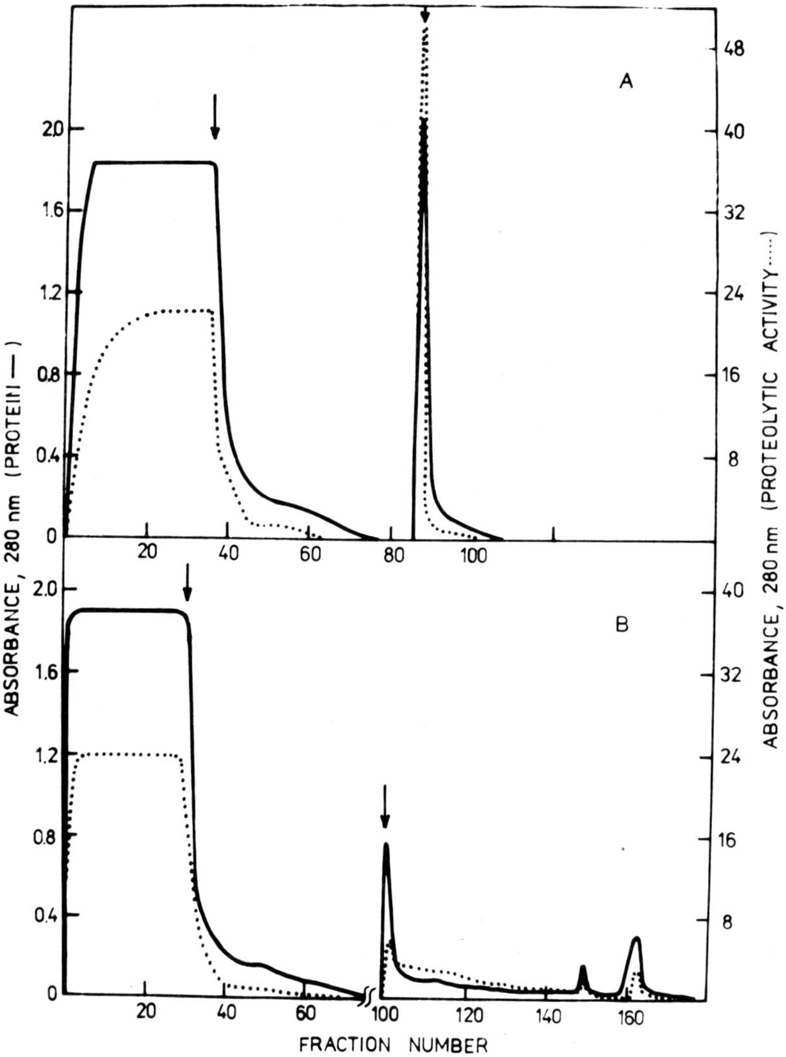

Fig. 5. Affinity chromatography of porcine pepsin on ε-aminocaproyl--L-Phe-D-Phe-OMe-Separon columns with low (A) and high (B) immobilized inhibitor concentrations. The solution of crude porcine pepsin was applied continuously (see text) onto the affinity columns (5ml) equilibrated with 0.1M sodium acetate, pH 4.5. At the site marked by first arrow equilibrated buffer was applied to the columns to remove unbound pepsin and nonspecifically adsorbed proteins. At the second arrow, 0.1M sodium acetate containing 1M NaCl, pH 4.5 was applied. Fractions (5ml) were taken at 4-min intervals. The inhibitor concentration of affinity sorbents in μmol per g of dry support were A: 0.85; B: 155. Protein ——, proteolytic activity ·····.

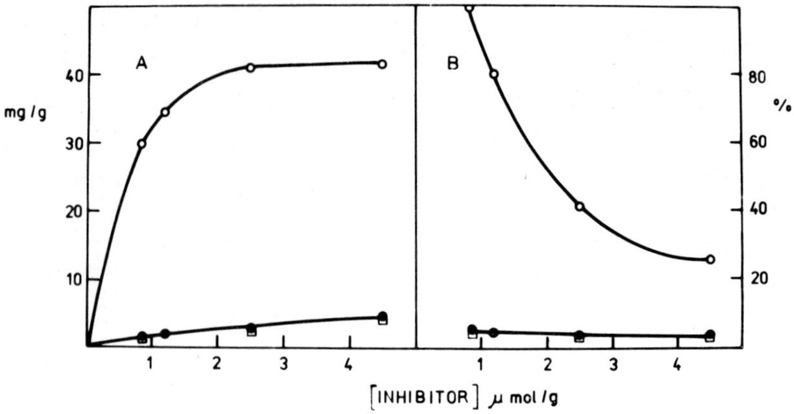

Fig. 6. (A) Capacity of immobilized inhibitor sorbent (ε-aminocaproyl-L-Phe-D-Phe-OMe-Separon) in mg of pepsin per 1 g of dry sorbent and (B) portion of immobilized inhibitor molecules involved in specific complex formation (in %) with respect to immobilized inhibitor concentration (in μmol of inhibitor per 1g of dry sorbent). - o -, porcine pepsin; - ● -, chicken pepsin, and - ◻ -, human pepsin.

Fig. 6A shows the dependence of the amounts of porcine, chicken, and human pepsins eluted on the concentration of ε-aminocaproyl-L-Phe-D-Phe-OMe of the individual affinity sorbents. Fig. 6B illustrates the portion of immobilized inhibitor molecules (in %) involved in the specific complex with the isolated pepsin, again in dependence on the concentration of the inhibitor attached. From the comparison of the curves for the individual pepsins it is evident that ε-aminocaproyl-L-Phe-D-Phe-OMe-Separon is a very good sorbent for porcine pepsin. The specific sorbent containing 0.85 μmol of inhibitor per g of dry support sorbed 29.4mg of porci[ne] pepsin per g of dry sorbent. It follows from the molecular weight of pepsin (35 000) that 99% of the inhibitor molecules attached were involved in the specific complex. With the increasing content of the affinity ligand attached the portion of the inhibitor molecules involved in the specific complex with pepsin sharply decreased. With specific sorbents containing 4.5 μmol inhibitor per g only 26% of the total number of inhibitor molecules attached take part in the sorption of pepsin. In an affinity sorbent with the lowest concentration of the affinity ligand only, all the molecules of the affinity ligand are fully available for the formation of the complex with the enzyme isolated as illustrated in Fig. 4 (central part). This sorbent can be used to advantage the determination of the dissociation constants of the interaction of the immobilized affinity ligand with the enzyme (ref. 16). Only in this case the concentration of the affinity ligand determined from the working capacity of the affinity column was equal to the concentration of the immobilized peptide inhibitor determined by amino acid analysis of acid hydrolysate of the affinity sorbent. The sorption of porcine pepsin to this sorbent proceeds by "one-to-one" bond only; that is by biospecific adsorption as defined by Porath (ref. 17). For chromatography on such sorbents we should use the name "biospecific chromatography".

The molecular weights of chicken and human pepsins do not practically differ from the molecular weight of porcine pepsin, so that similar steric hindrances can be expected. From this viewpoint we can therefore account for the lower sorption of these two pepsins first of all by the lower affinity of the inhibitor immobilized for the binding sites of chicken and human pepsin. This is in good agreement with the described specificity of porcine and chicken pepsins. Z-His-L-Phe-D-Phe-OEt is a good inhibitor of porcine pepsin (K_I=0.27mM) as well as Z-His-L-Phe-L-Phe-OEt (K_m=0.18mM) or Z-His-L-Phe-L-Phe-OMe (K_m=0.33mM) are its good substrates (ref. 18). Becker, Shechter, and Bohak (ref. 19) studied the specificity of chicken pepsin. They could not obtain the value of k_{cat}/K_m with the substrate Z-His-L-Phe-L-Phe-OEt because it was too small ($<$0.1 $M^{-1}sec^{-1}$) for the native enzyme. However, when these authors modified chicken pepsin by o-nitrobenzenesulfenyl chloride the value of k_{cat}/K_m for the same substrate increased to 40 $M^{-1}sec^{-1}$. This modification renders the conformation of the catalytic site in this reaction with the small peptide substrates to a more effective one. During reaction with larger substrates this conformation is probably achieved by the binding of groups in these substrates, which are relatively remote from the site of cleavage.

To prove that the low sorption of the chicken pepsin to be related to the specificity of its binding site we modified chicken pepsin by o-nitrobenzenesulfenyl chloride. Using chromatography on ε-aminocaproyl-L-Phe-D-Phe-OMe-Separon (0.85 μmol/g) we found a four-fold increase of the amount of modified pepsin sorbed. On the basis of these results we explain the variations in the amounts of different pepsins sorbed with the concentration of the immobilized inhibitor by differences in equilibrium constants of the enzyme-immobilized affinity ligand complexes. The course of the curves experimentally determined is in good agreement with the course of analogous curves deduced theoretically by Graves and Wu (ref. 20). The differences in the specificity as a cause of different sorption of the acetylcholinesterase from bovine erythrocytes and the acetylcholinesterase from electric eel to the N-methyl acridinium-Sepharose columns of different ligand concentrations (in μmol/ml: 0.49, 0.99, 1.97 and 2.8) were also considered by Sekar et al. (ref. 21).

Independence of desorbed pepsin on its concentration in the sample applied

A fact of importance for analytical determination of proteinases in animal tissues or fluids by high performance affinity chromatography or for their large scale isolation is that the sorption to columns of a specific sorbent is independent of enzyme concentration of the sample applied (ref. 5). In our laboratory we applied 50mg of crude porcine pepsin dissolved in 10, 100, or 1000ml to the column of ε-aminocaproyl-L-Phe-D-Phe-OMe-Separon. With the increasing dilution of the solution of the enzyme the peak of the first fraction of inert proteins undergoes spreading (and completely disappears at the lowest concentration). On the contrary, the peak of the pepsin fraction (eluted by the desorbing buffer of a high ionic strength) is sharp in all three cases and is in no way affected by the dilution (ref. 5). Analogous results were obtained after the application of ethanol precipitate of the culture medium filtrate of the mold Humicola sp. (ref. 22). In both cases the batchwise arrangement for the sorption of the enzyme from the diluted solution was not suitable since the rate of adsorption was too prolonged with the dilution of the solution sorbed. The extension of the adsorption time for trypsin to the soybean trypsin inhibitor-agarose particles in suspensions of different concentration has already been described by Porath and Kristiansen (ref. 23).

Independence of desorbed fraction of isolated pepsin on the presence of inert protein in the sample applied

Porcine pepsin (50mg) was chromatographed on ε-aminocaproyl--L-Phe-D-Phe-OMe-Separon (4.5 μmol of inhibitor per g of dry support) in the absence and in the presence of 100mg of bovine serum albumin in the sample applied under similar conditions as those described in Fig. 5. All serum albumin was eluted in the first peak with the equilibration buffer. The fractions of desorbed pepsin contained the same amount of the desorbed protein and showed the same proteolytic activity in both cases.

High performance affinity chromatography of pepsin

Affinity chromatography of porcine pepsin was also performed on the analytical level carried out under the conditions of high performance liquid chromatography (Fig. 7). The specific sorbent prepared for this purpose (Separon H1000, particle size 10 μm and with 0.5 μmol/g of ε-aminocaproyl-L-Phe-D-Phe-OMe) appears to be very well suited for a fast and efficient separation. The result allows us to envisage applications of high performance affinity chromatography techniques in the clinical analysis of enzymes.

Concluding remarks

It may be summarized that the hydroxyalkyl methacrylate supports, medium-derivatized with epichlorohydrine, are suitable for the preparation of sorbents with bound low molecular weight peptide inhibitors for efficient affinity chromatography of proteolytic enzymes, both in the classical and in the high performance arrangement. A low concentration only of immobilized inhibitor is necessary to obtain a specific sorbent which guarantees minimum unspecific sorption and maximum exploitation the molecules of the inhibitor attached. The advantage of such sorbents lies in their chemical and mechanical stability, small extent of unspecific interactions, their ability to undergo almost complete regeneration, and possibility of using high pressures in column or microcolumn techniques.

AFFINITY CHROMATOGRAPHY IN COMPARISON WITH SOME OTHER CHROMATOGRAPHIC METHODS AND ITS PROSPECTS

It can be expected, that high performance affinity chromatography will find wide application first of all in clinical biochemistry,

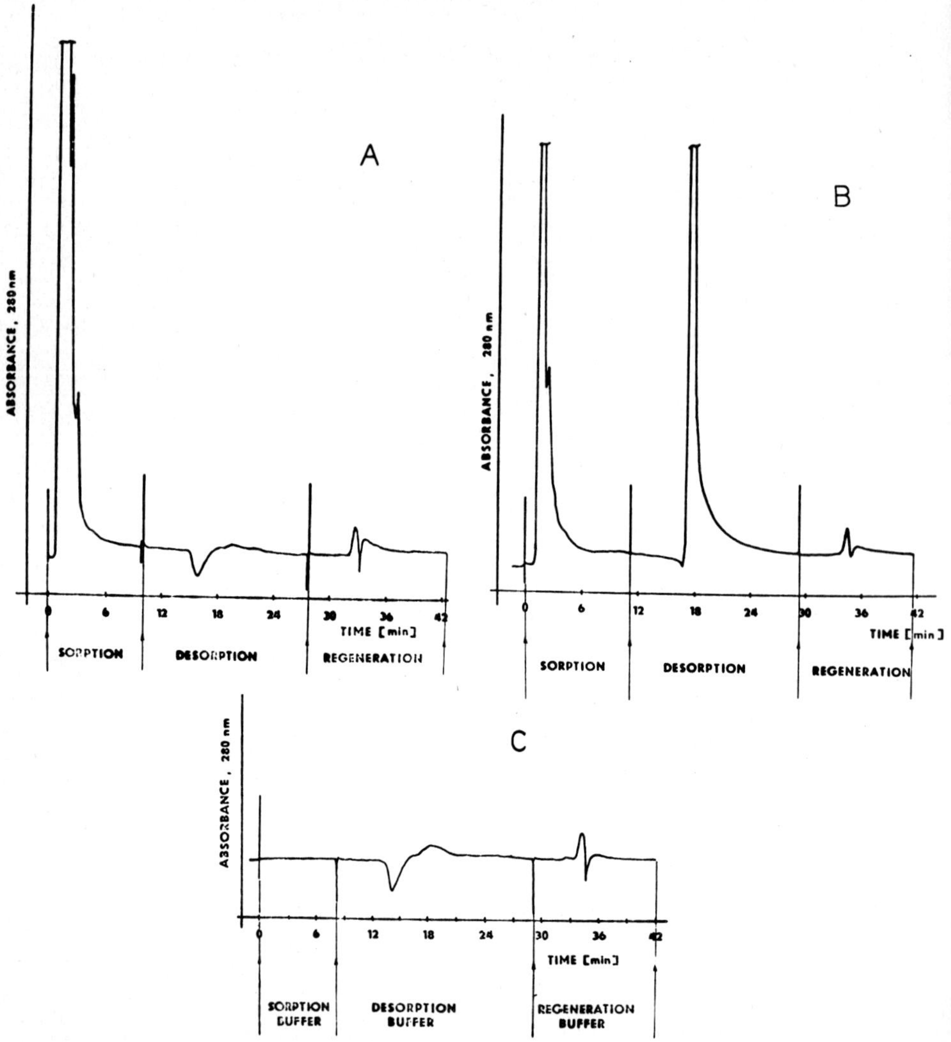

Fig. 7. High performance affinity chromatography of porcine pepsin (A) Chromatogram of pepsin sample on unmodified support (Separon H1000, particle size 10 µm); (B) chromatogram of pepsin on specific sorbent (Separon H1000, particle size 10 µm modified with 0.5 µmol ε-aminocaproyl-L-Phe-D-PheOMe/g carrier) and (C) base line characteristics of the high performance affinity chromatography system. Column: 100 x 4 mm, detector: UV 280 nm. Mobile phase: sorption buffer = 0.1M sodium acetate, pH 4.5; desorption buffer = 0.1M sodium acetate with 1M NaCl, pH 4.5; regeneration buffer = 0.1M sodium acetate with 10% 2-propanol, pH 4.5. Sample: 10 µl of pepsin solution in sorption buffer (0.5mg/µl) in (A) and (B) and 10 µl of sorption buffer without protein in (C). The flow rate was 0.5 ml/min and the pressure was 9 MPa.

where rapid separations of biological macromolecules are needed. Compared to stationary phases used in conventional high performance liquid chromatography, the affinity ligand introduces a far higher specific component into the separation step and thus reduces the need for the prepurification of samples. Compared with classical affinity chromatography this technique is much faster not only for routine separation of macromolecules but also in their analysis of biological samples (ref. 24).

The search for an optimum specific sorbent and an optimum sorption and desorption procedures is often time consuming. Therefore the affinity chromatography is suitable above all in cases in which strong ligand-enzyme interaction is known and where the isolation or the determination of the enzyme is to be repeated many times or where the macromolecule isolated is present in such a low concentration that isolation by any other method would be impossible or would require many steps.

It is expected that the role of affinity chromatography in industrial practice will increase with the increasing demand on the availability of biologically active products, predominantly enzymes. A number of these products will be prepared to advantage by fermentation processes. The latter are characterized by a relatively low concentration of enzymes in the starting material where they moreover exist in mixtures with products of very similar character. The development of efficient biospecific sorbents should provide possibilities of obtaining numerous biologically active, expensive products simultaneously with an increase of their availability and a decrease of their price. The abundance of highly active and inexpensive enzymes should markedly contribute to their extended application in analytical chemistry, medicine, and technology.

REFERENCES

1 J. Turková and A. Seifertová, J. Chromatogr., 148 (1978)293-297.
2 J. Turková, O. Hubálková, M. Křiváková and J. Čoupek, Biochim. Biophys. Acta, 322(1973)1-9.
3 J. Turková, Methods Enzymol., 44(1976)66-83.
4 J. Čoupek, M. Křiváková and S. Pokorný, J. Polym. Sci., Polym. Symp., 42(1973)182-190.
5 J. Turková, K. Bláha, J. Horáček, J. Vajčner, A. Frydrychová and J. Čoupek, J. Chromatogr., in press.
6 J. Turková, in L. Vitale and V. Simeon (Eds.), Industrial and Clinical Enzymology, Pergamon Press, Oxford, 1980, pp.65-76.

7 J. Turková, K. Bláha, M. Malaníková, D. Vančurová, F. Švec and J. Kálal, Biochim. Biophys. Acta, 524(1978)162-169.
8 I. Zemanová, J. Turková, M. Čapka, L.A. Nakhapetyan, F. Švec and J. Kálal, Enzyme Microb. Technol., in press.
9 J. Turková, Affinity Chromatography, Elsevier, Amsterdam 1978, pp. 187-189.
10 B.H.J. Hofstee, in N. Catsimpoolas (Ed.), Methods of Protein Separation, Vol. 2, Plenum Press, New York, 1976, pp.245-278.
11 J. Porath, in R. Epton (Ed.), Chromatography of Synthetic and Biological Polymers, Vol.1, Ellis Horwood Ltd., Chichester, 1978, pp.9-29.
12 M.J. Holroyde, J.M.E. Chester, I.P. Trayer and D.C. Walker, Biochem. J., 153(1976)351-361.
13 V.M. Stepanov, G.I. Lavrenova and M.M. Slavinskaya, Biokhimiya, 39(1974)384-387.
14 V.M. Stepanov, G.I. Lavrenova, K. Adly, M.V. Gonchar, G.N. Balandina, M.M. Slavinskaya and A.Ya. Strongin, Biokhimiya, 41(1976)294-303.
15 P. Štrop and D. Čechová, J. Chromatogr., 207(1981)55-62.
16 B.M. Dunn and I.M. Chaiken, Biochemistry, 14(1975)2343-2349.
17 J. Porath, Biochimie, 55(1973)943-951.
18 G.E. Clement, Progr. Bioorg. Mech., 2(1973)177-201.
19 R. Becker, Y. Schechter and Z. Bohak, FEBS Lett., 36(1973)49-52.
20 D.J. Graves and Y.-T. Wu, Methods Enzymol., 34(1974)140-163.
21 M.C. Sekar, G. Webb and B.D. Roufogalis, Biochim. Biophys. Acta, 613(1980)420-428.
22 A.D. Veličkov and J. Turková, in preparation.
23 J. Porath and T. Kristiansen, in H. Neurath and R.L. Hill (Eds.), The Proteins, Academic Press, New York, 3rd ed., 1975, pp.95-178.
24 S. Ohlson, L. Hansson, P.-O. Larsson and K. Mosbach, FEBS Lett., 93(1978)5-9.

T.C.J. Gribnau, J. Visser and R.J.F. Nivard (Editors),
Affinity Chromatography and Related Techniques

APPLICATION OF IMMOBILIZED PROTEASES TO PEPTIDE SYNTHESIS

H.-D. Jakubke, R. Bullerjahn, M. Hänsler and A. Könnecke

Karl Marx University Leipzig, Department of Biochemistry
DDR-7010 Leipzig, Talstr. 33, German Democratic Republic

ABSTRACT

Several model peptides have been synthesized using chymotrypsin, thermolysin and trypsin covalently attached to silica, enzacryl AA, and enzacryl AH for catalyzing peptide bond formation. The chymotrypsin-mediated coupling reaction provides the best results in biphasic aqueous-organic systems using the silica-bound catalyst and substrates containing a hydrophobic amino acid residue in P_2 position. Furthermore, it could be established that besides immobilized chymotrypsin both silica-bound trypsin as well as thermolysin attached to enzacryl AH are suitable for a series of re-utilization experiments. Preliminary studies with silica-bound chymotrypsin reveal the possibility of peptide coupling reactions without product precipitation which will open up the application to continuous processing.

INTRODUCTION

During the last four years there has been a remarkable interest in application of proteases to peptide synthesis and protein semisynthesis (refs. 1-3). Proteases demand attention chiefly because of the efficiency, stereospecificity, and mildness with which they catalyze peptide bond formation. On the other hand, using the segment condensation approach in the synthesis of biologically active peptides chemical coupling methods are not free from the danger of racemization. As catalysts, proteases do not participate in the reaction stoichiometry and therefore they can be re-utilized for further coupling reactions. In principle, the aqueous-organic two-phase approach of enzyme-catalyzed peptide synthesis (ref. 4) allows the re-utilization after simple phase separation, but the re-use of proteases has been restricted because, being high molecular weight proteins, their stability is limited. Thus, as pointed out in ref. 4, due to the enhanced thermal and chemical stability immobilized proteases should be much more sufficient as recoverable catalysts. To the best of our knowledge, besides a further contribution of our laboratory (ref. 5) controlled peptide syntheses using immobilized proteases have not yet been reported in the literature. Immobilized chymotrypsin was used for esterification of N-protected amino acids (refs. 6-9), but initial attempts to catalyze peptide bond formation in 76 % (v/v) ethanol failed (ref. 10). In addition, the application of immobilized chymotrypsin to plastein reaction has been reported (refs. 11-12).

Generally, immobilized enzymes have potential as recoverable catalysts for synthetic operations, but little has yet been achieved. The attraction of using an immobilized protease in place of a soluble one in enzymic peptide synthesis is two-fold. Firstly, the immobilized protease can be recovered by filtration; secondly, the immobilized enzyme can be re-utilized for further peptide bond forming steps.

In this paper, the preparative scale synthesis of model peptides using immobilized chymotrypsin, trypsin, and thermolysin is described, and the practical value as well as some limitations of this approach are discussed.

RESULTS+

A lot of different methods for the immobilization of proteases have been reported (ref. 13). The methods used include covalent attachment to various inert supports, crosslinking, adsorption, and entrapment within lattices. Avoiding the possibility of desorption of proteolytic activities during the coupling experiments we have exclusively preferred proteases covalently attached to the support material (Table 1).

The data compiled in Table 2 show that all immobilized proteases examined so far successfully catalyze the formation of peptide bonds. Under suitable reaction conditions adapted from coupling

---

+Abbreviations used: IUPAC/IUB rules for peptide are followed, see Eur. J. Biochem., 27(1972)201; Ac = acetyl, Bz = benzoyl, Z = benzyloxycarbonyl, Boc = tert.-butyloxycarbonyl, OMe = methyl ester, OEt = ethyl ester, Glt = 4-carboxybutyryl (glutaryl), Nan = 4-nitroanilide; all amino acids except glycine are of L-configuration. Enzymes: chymotrypsin (EC 3.4.21.1), trypsin (EC 3.4.21.4), thermolysin (EC 3.4.24.4). PEG = polyethylene glycol 600; MeOH = methanol, DMF = dimethylformamide. Further abbreviations used, see Table 1.

experiments employing the same soluble enzymes (refs. 4, 14-15) the yields obtained were satisfying and, as a rule, comparable to

TABLE 1

Immobilized proteases used for coupling reactions

Abbreviation	Enzyme	Support	Amount of bound protein (mg/g)	Relative activity (%)
Si-CTa	Chymotrypsin	Silica gel[a]	30	89[d]
Si-CTb	"	Silica gel[b]	20	-
AA-CT	"	Enzacryl AA[f]	21	95[d]
AH-CT	"	Enzacryl AH[f]	42.5	80[d]
AH-TH	Thermolysin	Enzacryl AH[f]	10	-
Si-TR	Trypsin	Silica gel[c]	30	18[e]

[a]Silica gel 60 (Merck), 35-70 mesh; γ-succinamidopropyl spacer for all silica-supported enzymes.
[b]Macroporous silica, pore diameter 20 nm.
[c]Silica gel 60 (Merck), 70-230 mesh.
[d]Glt-Leu-Phe-Nan as substrate (ref. 16).
[f]Koch & Light (UK). [e]Bz-Arg-OEt as substrate.

those obtained with soluble proteases. However, using immobilized biocatalysts longer reaction times are required. Because the solubility of substrates in aqueous solution is limited, protease-mediated coupling reactions have been usually carried out in buffered homogeneous aqueous-organic mixtures. Furthermore, the addition of an organic cosolvent shifts peptide bond equilibria towards synthesis (ref. 17). On the other hand, it is known that water-miscible organic solvents in high concentrations have a deleterious effect on the catalytic activity of the proteases. Recently we reported on the advantages of using the aqueous-organic two-phase approach in enzymic peptide synthesis (ref. 4). The results in Table 2 indicate that in agreement with soluble chymotrypsin the immobilized catalyst acts more effectively in such solvent systems than in mixtures containing cosolvents miscible with water. In the experiments with immobilized thermolysin the substrates are dissolved completely in buffer, and therefore no organic cosolvent is required at all. For augmenting the solubility of Z-Lys-OMe in trypsin-mediated coupling reaction

10 % (v/v) dimethylformamide was used. Although the synthesis proceeded satisfactorily under the conditions used further experiments with immobilized trypsin employing the aqueous-organic two-phase approach are in progress.

TABLE 2

Synthesis of model peptides catalyzed by immobilized proteases[a]

Peptide[b]	Enzyme	Amount of bound enzyme (mg)	Cosolvent (%,v/v)	Total volume (ml)	Time (h)	Yield (%)
Ac-Leu-Phe+Leu-NH_2	Si-CTb	3	PEG(55)	2.2	1.75	49
"	"	2.5	CH_2Cl_2(60)	2.5	1.5	75
Boc-Leu-Phe+Leu-NH_2	"	2	MeOH(36)	2.2	2	8
"	"	3	PEG(55)	2.2	2.5	24
Z-Ala-Phe+Leu-NH_2	"	3	DMF(35)	2	2	31
"	"	3	CCl_4(56)	2.7	2	70
Z-Phe+Leu-NH_2	AH-TH	0.6	none	2	20	84
Z-Ala-Phe+Leu-NH_2	"	0.5	"	2	20	48
Z-Leu-Met+Leu-NH_2	"	0.4	"	2	20	65
Z-Leu-Met+Ile-Ala-NH_2	"	0.5	"	2	20	69
Z-Lys+Leu-NH_2	Si-TR	6	DMF(10)	2	1.5	71
Bz-Arg+Leu-Ala-NH_2	"	4	none	2	0.25	76

[a]Reaction conditions: i)Si-CTb: 0.2 mmol Z-dipeptide methyl ester, cosolvent and 0.2 M carbonate buffer (pH 10) and Si-CTb were stirred. After filtration the precipitated peptide formed was removed from the Si-CTb with MeOH or DMF, the filtrate was evaporated and the solid product was washed with satd. $NaHCO_3$ soln. and 1 N HCl. The yield of the homogeneous peptide(TLC) was determined by weight (error $\pm$ 3 %). ii) AH-TH: 0.1 mmol Z-Phe-OH, 0.16 mmol nucleophile, AH-TH, and 0.2 M Tris-maleate buffer (pH 7) were stirred and worked up according to i); DMF was used for removal of the peptide. iii) Si-TR: 0.2 mmol Z-Lys-OMe (Bz-Arg-OEt), 0.3 mmol nucleophile, Si-TR, and 1 M (0.5 M) carbonate buffer (pH 10.5) were stirred; after filtration the peptide was removed by MeOH, the filtrate was evaporated, and the product was purified by passing a short Amberlite IRA 400 column after removing inorganic material (water as eluent).

[b] + indicates the peptide bond formed enzymatically.

Table 3 summarizes the results of studies concerning the influence of three different support materials for chymotrypsin on the yield in the syntheses of several Z-tripeptide amides under standard conditions. Slightly different yields were obtained for each peptide using the three immobilized chymotrypsin preparations as a catalyst. Apart from two exceptions, the silica-supported enzyme provided the highest yield, and the largest differences in the yields were found for substrates of high affinity to chymotrypsin, e.g. substrates containing a hydrophobic amino acid residue in P_2 position (Z-Leu-Phe-OMe or Z-Val-Phe-OMe). In some cases the

TABLE 3
Influence of the support materials on chymotrypsin-mediated coupling reactions[a]

Peptide (Cosolvent, %(v/v))	Enzyme preparation Si-CTa	AH-CT	AA-CT
	Yield(%)[b]		
Z-Gly-Phe+Leu-NH_2 (CCl_4, 20)	58	63	51
Z-Ala-Phe+Leu-NH_2 (CCl_4, 40)	60	-	52
Z-Val-Phe+Leu-NH_2 ($CHCl_3$, 20)	52	18	36
Z-Pro-Phe+Leu-NH_2 (CCl_4, 20)	55	-	60
Z-Leu-Phe+Leu-NH_2 (CCl_4, 12)	88	75	62
Z-Leu-Tyr+Leu-NH_2 (CCl_4, 20)	59	-	46

[a]Standard reaction conditions: 0.2 mmol Z-dipeptide methyl ester, 0.4 mmol Leu-NH_2 · HCl, 0.2 M carbonate buffer (pH 10) plus cosolvent (total volume 2.5 ml), amount of immobilized chymotrypsin corresponding to 3 mg (0.12 μmol) bound enzyme, time 1.5 h, DMF for removal of the product; actual pH in the buffer phase 8.5-8.7.

[b]Average yield of two runs.

coupling reaction was incomplete even after prolonged reaction time, and the isolated product was contaminated by the starting ester substrate. It could be removed by either washing with CCl_4 (Boc-Leu-Phe-OMe, Table 2) or by mild base-catalyzed hydrolysis and subsequent washing with $NaHCO_3$ solution (Z-Val-PheOMe).

Attempted coupling of Z-Phe-Phe-OMe with Leu-NH_2 (Si-CTa, 20 %, v/v, $CHCl_3$ as cosolvent) unexpectedly produced Z-Phe-Leu-NH_2 as the main product besides minor amounts of Z-Phe-Phe-Leu-NH_2 due to the cleavage of the substrate at the Phe-Phe bond.

One of the most important features of immobilized enzymes is the intrinsic possibility of re-using the catalyst and the application to industrial processes. Because the peptide products precipita[te] under the conditions used in our experiments, this approach is unsuitable for continuous processing. Nevertheless, we have investigated the discontinuous re-use of the immobilized protease for further coupling reactions. All the immobilized enzyme preparations employed could be re-used successfully for several runs, details are listed in Table 4. In the case of chymotrypsin the

TABLE 4
Re-utilization of immobilized proteases for peptide bond formation[a]

Peptide	Enzyme	Cosolvent	Solvent for removing the product	Yield (%) 1st	2nd	3rd	4th u[se]
Ac-Leu-Phe+Leu-NH_2	Si-CTb	PEG	MeOH	49	52	48	-
"	"	CH_2Cl_2	"	75	71	72	-
"		"	DMF	75	65	-	-
Z-Leu-Phe+Leu-NH_2	Si-CTa	CCl_4	DMF	88	76	71	-
"	AA-CT	"	"	63	57	44	-
Z-Gly-Phe+Leu-NH_2	AH-CT	"	"	64	60	57	-
Z-Phe+Leu-NH_2	AH-TH	none	"	84	78	81	70
Z-Lys+Leu-NH_2	Si-TR	DMF	MeOH	71	73	75	72
Bz-Arg+Leu-Ala-NH_2	"	none	"	76	70	70	-

[a]Reaction conditions were the same as described in Table 2 and 3. The re-utilization of the chymotrypsin preparations was performed after storage of the immobilized catalysts for 4-6 weeks in doubly destilled water; AH-TH and Si-TR were re-used immediately.

solvent used for removing the peptide from the catalyst evidently plays an important role in the re-utilization experiments. With MeOH the yields remained nearly constant, whereas the use of DMF caused a significant decrease of the yield already in the second re-utilization of the immobilized catalyst. In contrast, immobilized thermolysin is not so sensitive to the treatment with DMF, and after the fourth run the yield was still satisfying. Due to the immobilization of trypsin, an enzyme known to be very sensitive to autodigestion, its stability is considerably enhanced. This is convincingly documented by the fact, that immobilized trypsin is a very useful catalyst for re-utilization in peptide bond formation experiments. After the fifth use (not listed in Table 4) the yield of purified Z-Lys-Leu-NH_2 amounted to 70 % which is nearly the same yield obtained in the first coupling experiment.

CONCLUSIONS

It has been demonstrated that immobilized chymotrypsin, trypsin, and thermolysin possess the same efficiency as the soluble enzymes in catalyzing peptide bond formation. Although the reaction time required is often longer, the amount of enzyme needed is mostly lower due to the enhanced stability of the immobilized biocatalyst.

Besides the inherent advantages of enzyme-catalyzed coupling reactions, i. e. stereoselectivity and mild reaction conditions in addition to the minimal side chain protection required, the application of immobilized proteases to peptide synthesis offers additional advantages: i) the immobilized enzyme is readily removable from the reaction mixture and can be re-used for further coupling reactions, ii) the peptide products are obtained nearly pure without any contamination by either proteolytic activities or denatured enzyme protein often observed using soluble proteases for synthetic operations (ref. 18), simplifying the purification of the final products.

As demonstrated, e. g. in the semisyntheses of human insulin (ref. 19) and in carboxypeptidase-Y catalyzed synthesis of model peptides (ref. 20), it is possible to shift the equilibrium kinetically controlled towards synthesis without precipitation of the product. This type of reactions seems to be the most important perspective for the application of immobilized proteases to peptide synthesis and protein semisynthesis. Apart from columns, fluidized-bed, packed-bed, and stirred-tank reactors can be used for this purpose, and future developments will undoubtedly have a considerable impact on the preparative use of immobilized proteases. In preliminary experiments employing immobilized chymotrypsin for the synthesis of Ac-Phe-Ala-NH_2, soluble in the reaction medium (refs. 21-22), from Ac-Phe-OMe and Ala-NH_2 (ratio 1:1.5) we isolated the product in yields of 62 and 56 % after reaction times of 12 and 30 minutes, respectively. This is the first example for protease-catalyzed directed peptide synthesis in homogeneous solution using an immobilized enzyme. Further experiments along this line are in progress.

REFERENCES

1 Y. Isowa, Yuki Gosei Kagaku Kyokaishi, 36(1978)195-205; CA.,89(1978)18 456.
2 F. Brtnik and K. Jost, Chem. Listy, 74(1980)951-964.
3 H.-D. Jakubke and P. Kuhl, Pharmazie,"submitted for publication".
4 P. Kuhl, A. Könnecke, G. Döring, H. Däumer and H.-D. Jakubke, Tetrahedron Lett., 21(1980)893-896.
5 A. Könnecke, R. Bullerjahn and H.-D. Jakubke, Mh. Chem., 112 (1981)469-481.
6 A. M. Klibanov, G. P. Samokhin, K. Martinek and I. V. Berezin, Biotechnol. Bioeng., 19(1977)1351-1361.
7 A. Kapune and V. Kasche, Biochem. Biophys. Res. Commun., 80 (1978)955-962.
8 Y. Nakamoto, I. Karube, I. Kobayashi, M. Nishida and S. Suzuki, Arch. Biochem. Biophys., 193(1979)117-121.
9 R. G. Ingalls, R. G. Squires and L. G. Butler, Biotechnol. Bioeng., 17(1975)1627-1637.
10 W. P. Vann and H. H. Weetall, J. Solid Phase Biochem., 1(1976) 297-306.
11 Z. Varanini, C. Pallavicini, G. Fincati, A. Dal Belin Peruffo, Industrie Alimentari, 18(1979)735-740.
12 I. Karube, Y. Yugeta and S. Suzuki, J. Mol. Catal., 9(1980) 445-451.
13 O. R. Zaborsky, Immobilized Enzymes, Chemical Rubber Co. Press, Cleveland, 1973.
14 T. Oka and K. Morihara, J. Biochem. (Tokyo), 88(1980)807-813.
15 T. Oka and K. Morihara, J. Biochem. (Tokyo), 82(1977)1055-1062.
16 H.-D. Jakubke, H. Däumer, A. Könnecke, P. Kuhl and J. Fischer, Experientia, 36(1980)1039-1040.
17 G. A. Homandberg, J. A. Mattis and M. Laskowski, Jr., Biochemistry, 17(1978)5220-5227.
18 "Unpublished results from our laboratory".
19 K. Inouye, K. Watanabe, K. Morihara, Y. Tochino, T. Kanaya, I. Emura and S. Sakakibara, J. Am. Chem. Soc., 101(1979)751-752; K. Morihara, T. Oka and H. Tsuzuki, Nature, 280(1979)412-413; K. Morihara, T. Oka, H. Tsuzuki, Y. Tochino and T. Kanaya, Biochem. Biophys. Res. Commun., 92(1980)396-402.
20 F. Widmer and J. T. Johansen, Carlsberg Res. Commun., 44(1979) 37-46.
21 J. Fastrez and A. R. Fersht, Biochemistry, 12(1973)2025-2034.
22 K. Morihara and T. Oka, in T. Shiba (Ed.), Peptide Chemistry 1977, Proc. 15th Symp. on Peptide Chemistry, Protein Research Foundation, Osaka, 1978,pp. 79-83; Biochem. J., 163(1977) 531-542.

POSTERS

A

THEORETICAL ASPECTS
POLYMERIC MATRICES AND LIGAND IMMOBILIZATION
APPLICATIONS: ISOLATION AND PURIFICATION RELATED TECHNIQUES

B

APPLICATIONS: DIAGNOSTIC
APPLICATIONS: BIOMEDICAL
APPLICATIONS: ORGANIC DYES - DYE-LIGANDS RELATED TECHNIQUES

4th
INTERNATIONAL SYMPOSIUM
ON AFFINITY CHROMATOGRAPHY
AND RELATED TECHNIQUES

THEORETICAL ASPECTS

A-1 QUANTITATIVE AFFINITY CHROMATOGRAPHIC CHARACTERIZATION OF NEUROPHYSIN-NEUROPEPTIDE INTERACTIONS.

S. Angal and I.M. Chaiken
Lab. Chem. Biol, NIAMDD, NIH, Bethesda, MD 20205, USA

A-2 MATRIX-AROMATIC MOLECULES INTERACTIONS IN GEL CHROMATOGRAPHY

M.G. Cacace, M. Bergami[+] and A. Sada
Institute of Molecular Embryology, CNR, 80072 Arco Felice (NA)
+ Chair of Molecular Biology, University of Rome, Rome, Italy

A-3 THE USE OF AFFINITY CHROMATOGRAPHY IN THE IDENTIFICATION OF CELL SURFACE COLLAGEN RECEPTOR IN PROCARYOTE ACHROMOBACTER IOPHAGUS

I. Emöd, P. Soubigou, N.-T. Tong, L.K. Bagilet and V. Keil-Dlouha
Unité de Chimie des Protéines, Département de Biochimie et Génétique Moléculaire, Institut Pasteur, 28, rue du Docteur Roux, 75724 Paris Cedex 15, France

A-4 USE OF AFFINITY DIFFERENCES BETWEEN COMMERCIALLY AVAILABLE IMMOBILIZED CON-A FOR DEMONSTRATION OF GLYCOPROTEIN MOLECULAR VARIANTS

L. Faye[1] and J.P. Salier[2]
1 Laboratoire de Photobiologie, UER des Sciences, 76130 Mont-Saint-Aignan, France; 2 INSERM Unité-78, 76230 Bois-Guillaume, France

A-5 COMBINED AFFINITY ADSORPTION AND ISOELECTRIC FOCUSING AS A TOOL TO STUDY PHYLOGENETIC RELATIONSHIPS OF CLOSELY RELATED SPECIES OF CHRYSOCARABUS BEETLES

S. Vaje[1], D. Mossakowski[1] and D. Gabel[2]
Departments of [1]Biology and [2]Chemistry, University of Bremen, Box 330 440 D-2800 Bremen 33

A-6 INTERFACIAL PHENOMENA IN AFFINITY CHROMATOGRAPHY OF GLOBULAR PROTEINS

P. Mohr[a)], F. Scheller[a)] and K. Pommerening[b)]
Department of Applied Enzymology[a)] and Department of Heme Catalysis[b)], Central Institute of Molecular Biology of the Academy of Sciences of GDR, GDR-1115 Berlin-Buch

A-7 AFFINITY CHROMATOGRAPHY OF RNA ANALOGUES THROUGH BOVINE PANCREATIC RIBONUCLEASE COVALENTLY BOUND TO SEPHAROSE 4B

M.V. Nogués and C.M. Cuchillo
Dept. Bioquímica, Fac. Ciències and Institut de Biologia Fonamental. Univ. Autònoma de Barcelona. Bellaterra, Spain

A-8 AFFINITY CHROMATOGRAPHY OF RAT IgG SUBCLASSES ON LECTINS COVALENTLY LINKED TO SEPHAROSE 4B

J. Rousseaux, H. Debray, R. Rousseaux-Prevost and G. Biserte
Institut de Recherches sur le Cancer et Unité 124 de l'INSERM, B.P. 311, 59020 Lille Cédex, France

POLYMERIC MATRICES AND LIGAND IMMOBILIZATION

A-14 STUDIES ON THE SPECIFICITY OF THE SO-CALLED NON SPECIFIC ADSORPTION ON INSOLUBLE POLYMERS WITH SPACER ARMS

A. Faure and M. Caron+
Centre National de Transfusion Sanguine BP 100-91400, Orsay, France and + U.E.R. Biomedicale, 93000 Bobigny, France

A-15 ON THE PROBLEM OF NON-SPECIFIC BINDING TO POLYSACCHARIDE MATRICES USED IN AFFINITY CHROMATOGRAPHY. COMPARISON OF DIFFERENTLY DERIVATIZED SHEEP ANTI Fc ANTIBODY POLYSACCHARIDE GELS (CNBR ACTIVATED SEPHAROSE 4B, AFFIGEL 10 (BIO RAD) AND PERIODATE OXIDIZED SEPHADEX G 75)

E.A. Fischer
Dept. Diagnostica, F. Hoffmann-La Roche & Co. AG, Postfach, CH-4002 Basle

A-16 MACROPOROUS OXIRANE ACRYLIC BEADS FOR PROTEIN IMMOBILISATION: EUPERGIT C VERSUS LARGER PORE SIZE TYPE

D.M. Krämer, H. Pennewiss, H. Plainer, R. Schnee and M. Stickler
Röhm GmbH, Research Laboratories, D-6100 Darmstadt, F.R. Germany

A-17 α-AMYLASE PURIFICATION AND SEPARATION FROM GLUCOAMYLASE BY AFFINITY CHROMATOGRAPHY ON CROSS-LINKED AMYLOSE (CL-AMYLOSE)

H.D. Schell, M.A. Mateescu, T. Bentia and A. Petrescu
Institute of Biological Sciences, Spl. Independentei 296, 77748-Bucharest 17, Romania

A-18 SULFONYL CHLORIDE DERIVATIVES FOR IMMOBILIZATION OF PROTEINS AND AFFINITY LIGANDS

K. Nilsson and K. Mosbach
Pure and Applied Biochemistry, Chemical Center, University of Lund, P.O. Box 740, S-220 07 Lund 7, Sweden

A-19 DIFFERENCE OF IgG-BINDING PATTERN OF PROTEIN A ATTACHED TO VARIOUS SOLID PHASES

R. Nilsson and H.O. Sjögren
The Wallenberg Laboratory, University of Lund, Box 7031, 220 07 Lund, Sweden

A-20 NEW APPROACH FOR TARGETED IMMOBILIZATION OF AMINO ACIDS, NEUROTRANSMITTERS AND PEPTIDES ON AGAROSE

B. Penke, J.R. Varga, G.K. Tóth, M. Zarándy and K. Kovács
Institute of Medical Chemistry, Medical School of Szeged, H-6720 Szeged, Hungary

APPLICATIONS: ISOLATION - PURIFICATION

A-26 USE OF A SIMPLE IMMUNOAFFINITY CHROMATOGRAPHY METHOD FOR THE PURIFICATION OF HUMAN IgG ANTI-HEPATITIS B SURFACE ANTIGEN

A. Boniolo and L. Callegaro
Laboratory of Enzymology, Sorin Biomedica S.p.A., Saluggia (Vercelli), Italy

A-27 GENTLE DESORPTION PROCEDURES FOR PURIFYING ENZYMES BY IMMUNOAFFINITY CHROMATOGRAPHY. EXAMPLE WITH BARLEY β AMYLASE

D. Bureau and J. Daussant

Physiologie des Organes Végétaux, C.N.R.S., 4 ter route des Gardes, 92190 Meudon, France

A-28 BIOTINYL STEROID AND IMMOBILIZED AVIDIN AS A TOOL FOR STEROID HORMONE RECEPTOR PURIFICATION

P. Formstecher, P. Lustenberger and M. Dautrevaux

Lab. Biochimie, Faculté de Médecine, 59045 Lille-Cédex, France

A-29 AN OPTIMIZED PROCEDURE FOR THE AFFINITY CHROMATOGRAPHIC ISOLATION OF HETEROGENEOUS AND MONOCLONAL ANTIBODIES

T. Gribnau and A. van Sommeren

Organon Scientific Development Group, P.O.Box 20, 5340 BH Oss, The Netherlands

A-30 ACTIVITY OF ANTISERA DETERMINED BY MEANS OF EIA: EVALUATION OF EXPERIMENTAL DATA USING NON-LINEAR REGRESSION

J.A. van Gorp, T.C.J. Gribnau and A.P.G. van Sommeren

Organon Scientific Development Group, P.O.Box 20, 5340 BH Oss, The Netherlands

A-31 GLYCOPHORIN: USE IN AFFINITY BINDING AND AFFINITY CHROMATOGRAPHY

I. Kahane

Dept. of Membrane and Ultrastructure Research, The Hebrew University-Hadassah Medical School, P.O. Box 1172, Jerusalem 91000, Israel

A-32 ROD PLASMA MEMBRANE ISOLATION BY MEANS OF CON A-COATED POLYSTYRENE BEADS VIA AFFINITY CHROMATOGRAPHY AND DENSITY SEPARATION

K.M.P. Kamps and W.J. De Grip

Dept. of Biochemistry, University of Nijmegen, Geert Grooteplein Noord 21, P.O. Box 6500 HB Nijmegen

A-33 POLYFUNCTIONAL ENKEPHALIN ANALOGUES FOR AFFINITY SEPARATION

A. Koman and L. Terenius
Dept. of Pharmacology, University of Uppsala, Biomedicum, Box 573, 751 23 Uppsala, Sweden

A-34 LARGE-SCALE PURIFICATION OF HEPATITIS B SURFACE ANTIGEN (HBsAg) BY MEANS OF BATCHWISE IMMUNE ADSORPTION

L. Kuijpers, G. Wolters and A. Schuurs
Organon SDG, P.O. Box 20, 5340 BH Oss, The Netherlands

A-35 SEPARATION OF SOLUBLE AMINOPEPTIDASES OF THE HUMAN PLACENTA USING AFFINITY CHROMATOGRAPHY AS THE FINAL STEP

S. Lampelo, K. Lalu and T. Vanha-Perttula
Dept. of Anatomy, University of Kuopio, P.O.Box 138, SF-70101 Kuopio, Finland

A-36 PURIFICATION OF HUMAN PLACENTAL PARTICLE-BOUND AMINOPEPTIDASES (OXYTOCINASE) USING CON A SEPHAROSE 4B AS A FINAL STEP

S. Lampelo, K. Lalu and T. Vanha-Perttula
Dept. of Anatomy, University of Kuopio, P.O.Box 138, SF-70101 Kuopio, Finland

A-37 AFFINITY CHROMATOGRAPHY PURIFICATION OF MEMBRANE 5' NUCLEOTIDASE

J. Dornand, J.C. Bonnafous and J.C. Mani
Laboratoire de Biochimie des Membranes, ER CNRS 228, 8 rue de l'Ecole Normale, F-34075 Montpellier, France

A-38 CONCANAVALIN A AND IMMUNOADSORBANT CHROMATOGRAPHIES DURING THE PURIFICATION OF AMINOPEPTIDASE N AND A FROM INTESTINAL BRUSH-BORDER MEMBRANE

H. Feracci, A. Benajiba and S. Maroux
CBM-CNRS, 31, Chemin Joseph-Aiguier, 13274 Marseille Cédex 2, France

A-39 PROPERTIES OF THE INSULIN RECEPTOR PROTEIN FROM PORCINE LIVER MEMBRANES PURIFIED BY AFFINITY CHROMATOGRAPHY

H. Meyer, H.-J. Bubenzer, L. Herbertz and H. Reinauer
Biochemical department, Diabetes Forschungsinstitut an der Universität Düsseldorf, D-4000 Düsseldorf, FRG

A-40 THE APPLICATION OF AFFINITY CHROMATOGRAPHY FOR PURIFICATION OF PROTEIN KINASES FROM MAIZE SEEDLINGS

G. Muszyńska
Institute of Biochemistry and Biophysics, Polish Academy of Sciences, 36 Rakowiecka St., 02-532 Warszawa, Poland

A-41 PURIFICATION OF INTERACTING MACROMOLECULES USING AFFINITY CHROMATOGRAPHY ON IMMOBILIZED CELLS

M. Ramstorp and B. Mattiasson
Department of Pure and Applied Biochemistry, Chemical Center, University of Lund, P.O.Box 740, S-220 07 Lund, Sweden

A-42 IMMUNOADSORPTION OF STEROIDS AND STEROID-PROTEIN-COMPLEXES

L. Reum, D. Haustein+ and J. Koolman
Physiol.-Chem. Inst. and Inst. Exp. Immunology+ der Universität Deutschhausstr. 1 - 2, D-3550 Marburg, W.-Germany

A-43 ISOLATION OF RAT IgG SUBCLASSES BY AFFINITY CHROMATOGRAPHY ON PROTEIN A SEPHAROSE

J. Rousseaux, R. Rousseaux-Prevost, H. Bazin+ and G. Biserte
Institut de Recherches sur le Cancer et Unité 124 de l'INSERM, B.P. 311, 59020 Lille Cédex, France and + Experimental Immunology Unit, Faculty of Medicine, University of Louvain, 30-1200 Brussels, Belgium

A-44 INTER-α-TRYPSIN-INHIBITOR (ITI): USE OF IMMUNOAFFINITY CHROMATOGRAPHY FOR AN EFFICIENT PREPARATION OF ANTI-ITI ANTISERUM, ITI-FREE SERUM AND PURE ITI

J.P. Salier, R. Sesboüé, D. Vercaigne and J.P. Martin
INSERM Unité-78, 76230 Bois-Guillaume, France

A-45 COMPARISON OF AFFINITY CHROMATOGRAPHY PROCEDURES FOR THE ISOLATION OF PURE HUMAN THYROXINE-BINDING GLOBULIN

G. Sand, P. Petit and D. Glinoer
Laboratory of Radioisotopes, University Hospital Saint-Pierre, 322 rue Haute, B-1000 Brussels, Belgium

A-46 ISOLATION OF SIALOGLYCOCONJUGATES ON COLUMNS OF IMMOBILIZED 5-HYDROXYTRYPTAMINE

D.L. Deane+ and R.J. Sturgeon
Dept. of Brewing and Biological Sciences, Heriot-Watt University, Edinburgh, G.B.; + M.R.C. Clinical and Population Cytogenetics Unit, Western General Hospital, Edinburgh, G.B.

A-47 THE PURIFICATION OF PROTEINS AND ENZYMES SPECIFIC FOR MAMMALIAN MALE GERM CELLS BY AFFINITY-RELATED TECHNIQUES

J. Svasti, P. Toowicharanont and S. Anguravirutt
Dept. of Biochemistry, Faculty of Science, Mahidol University, Rama VI Rd., Bangkok, Thailand

A-48 PURIFICATION OF β-D-GALACTOSIDASE (E.COLI) VIA AFFINITY CHROMATOGRAPHY SYNTHESIS OF A NEW INHIBITOR

M. Varalli, C. Germinario , G. Chieregatti, E. Gnemmi and E. Murador
Richerche e Sviluppo Diagnostici, Farmitalia Carlo Erba S.p.A. - 20090 Rodano (Milano) - Italy

A-49 LARGE SCALE PURIFICATIONS OF RICIN A-CHAIN WITH MINIMUM TOXICITY TO CELLS AND ANIMALS

H. Vidal, P. Gros, J.R. Hennequin, F.K. Jansen and F. Paolucci
Centre de Recherches CLIN-MIDY, Rue du Professeur Joseph-Blayac, 34082 Montpellier Cédex, France

RELATED TECHNIQUES

A-60 APPLICATION OF CHARGE TRANSFER CHROMATOGRAPHY IN ACRIFLAVIN-SEPHADEX AND AFFINITY CHROMATOGRAPHY IN SEPHAROSE 4B-RNASE TO THE PURIFICATION AND GROUP SEPARATION OF NUCLEOTIDES

C. Arús, M.V. Nogués and C.M. Cuchillo
Dept. Bioquímica, Fac. Ciències and Institut de Biologia Fonamental, Univ. Autònoma de Barcelona, Bellaterra (Barcelona), Spain

A-61 AFFINITY CHROMATOGRAPHY BEHAVIOUR OF NOCARDIA INDUCED INTERFERON

R. Barot-Ciorbaru, J.P. Secheresse+ and E. Boschetti+
Institut de Biochimie, Université de Paris Sud, Orsay, France and
+ Réactifs IBF-Pharmindustrie, 35 avenue Jean-Jaurès, Villeneuve-la-Garenne, France

A-62 USE OF PHENYL BORONATE AGAROSE IN ENZYME FRACTIONATION

S. Fulton, E. Carlson, J. Kalinoski
Amicon Corporation, 182 Conant Street, Danvers, Massachusetts 01923, USA

A-63 FRACTIONATION OF CHEMICALLY MODIFIED HEPARINS BY LIGAND EXCHANGE CHROMATOGRAPHY

J. Jozefonvicz+, C. Fougnot+, M. Jozefowicz+ and R.D. Rosenberg*
+ Lab. de Recherches sur les Macromolécules, Université Paris-Nord, Av. J.B. Clément, 93430 Villetaneuse, France and * Harvard Medical School, Sidney Farber Cancer Institute, 44 binney street, Massachusetts 02115, USA

A-64 NEW LIGANDS FOR "MILD" HYDROPHOBIC CHROMATOGRAPHY

T.G.I. Ling and B. Mattiasson
Pure & Applied Biochemistry, Chemical Center, University of Lund, P.O. Box 740, S-220 07 Lund, Sweden

A-65 PREPARATION OF MERSALYL-ULTROGEL FOR -SH GROUP-SPECIFIC COVALENT AFFINITY CHROMATOGRAPHY

J.C. Bonnafous, J. Dornand, J.F. Anberrée, E. Boschetti+ and J.C. Mani
Laboratoire de Biochimie des Membranes, ER CNRS 228, 8 rue de l'Ecole Normale, F-34075 Montpellier, and + Réactifs IBF - Pharmindustrie, 35 avenue Jean-Jaures, F-92390 Villeneuve-la-Garenne

A-66 HYDROPHOBIC INTERACTION CHROMATOGRAPHY AS A TOOL IN THE PURIFICATION OF THE SOLUBLE GLUTATHIONE S-TRANSFERASES FROM RAT LIVER

J.A.T.P. Meuwissen and M. Zeegers
Laboratory of Hepatology, Katholieke Universiteit Leuven, Campus Gasthuisberg, B-3000 Leuven, Belgium

A-67 LIGAND MEDIATED CHROMATOGRAPHY OF YEAST α-GLUCOSIDASE AND HEXOKINASE ON IMMOBILISED PHENYLBORONIC ACID

T.A. Myöhänen, V. Bouriotis and P.D.G. Dean
Dept. of Biochemistry, University of Liverpool, P.O.Box 147, Liverpool L69 3BX, U.K.

A-68 CONFORMATIONAL CHANGES INDUCED ON F_1-ATPase FROM MICROCOCCUS LYSODEIKTICUS BY HYDROPHOBIC INTERACTION CHROMATOGRAPHY

J.P. Pivel and A. Marquet
Unidad de Biomembranas, Instituto de Immunología y Biología Microbiana (C.S.I.C.), Velázquez, 144, Madrid-6, Spain

A-69 PURIFICATION OF PEPTIDE HORMONES (VASOPRESSIN AND OXYTOCIN) FROM GLAND EXTRACTS ON METAL CHELATE ADSORBENTS

M.A. Vijayalakshmi, L. Fanou-Ayi, D. Picque and E. Segard
Institut de Technologie des Surfaces Actives, BP. 233, 60206 Compiègne, France

APPLICATIONS: DIAGNOSTICS

B-1 IMMUNOHETEROGENEITY OF CARCINOEMBRYONIC ANTIGEN (CEA) OR CEA-LIKE SUBSTANCES DETECTED BY IMMUNOAFFINITY CHROMATOGRAPHY

R. Accinni, M.A. Bailo, R. Ferrara, V. Cavalca, C. Biancardi and A. Bartorelli
Dept. of Biochemistry, Istituto Richerche Cardiovascolari, Università di Milano, Via F. Sforza 35, 20122 Milano, Italy

B-2 COLUMN AFFINITY CHROMATOGRAPHY FOR BOUND/FREE SEPARATION IN LIGAND ASSAYS: RADIOIMMUNOASSAY OF HUMAN PLACENTAL LACTOGEN

P. Cornale, M. Bonazzi, C. Multinu, P. Romelli, L. Vancheri and F. Pennisi
Radioimmunoassay Department, Gruppo Lepetit, Via M. Gorki, 30, Cinisello Balsamo (Milano), Italy

B-3 PREPARATION AND AFFINITY CHROMATOGRAPHY PURIFICATION OF UMBELLIFERONE-LABELLED POLYGLUTAMIC ACID-OESTRADIOL-17 CONJUGATE FOR FLUOROIMMUNOASSAY

G.I. Ekeke

School of Chemical Sciences, University of Port Harcourt P.M.B. 5323, Port Harcourt, Nigeria

B-4 AFFINITY CHROMATOGRAPHY AND PROPERTIES OF LECTIN CONTAINING CONJUGATES OF AFFINITINS

H. Franz, J. Mohr and P. Ziska

State Institute of Immunopreparations and Nutrient Media, Klement-Gottwald-Allee 317 - 321, 1120 Berlin, GDR

B-5 IMMUNOCAPILLARYMIGRATION - QUANTITATIVE DETERMINATION OF PLASMA PROTEINS BY AFFINITY CHROMATOGRAPHY IN PAPER STRIPS

C. Glad

Department of Pure and Applied Biochemistry, Chemical Center, University of Lund, P.O.Box 740, S-220 07 Lund, Sweden

B-6 AQUEOUS TWO-PHASE SYSTEMS APPLIED IN THE SEPARATION OF FREE AND BOUND ANTIGEN IN AN IMMUNOASSAY OF β_2-MICRO-GLOBULIN

T.G.I. Ling and B. Mattiasson

Pure and Applied Biochemistry, Chemical Center, University of Lund, P.O.Box 740, S-220 07 Lund, Sweden

B-7 PARTITION AFFINITY LIGAND ASSAY (PALA) IS BASED ON A NEW SEPARATION METHOD IN BINDING ASSAYS

T.G.I. Ling and B. Mattiasson

Pure and Applied Biochemistry, Chemical Center, University of Lund, P.O.Box 740, S-220 07 Lund, Sweden

B-8 IMMUNOADSORBENT DETECTION OF CHAGAS DISEASE SPECIFIC SOLUBLE IMMUNOCOMPLEXES

A. Marcipar, S. Barnes and E. Lentwojt

Institut de Technologie des Surfaces Actives, BP 233, 60206 Compiègne, France

B-9 PARTITION AFFINITY LIGAND ASSAY (PALA) USED IN RAPID QUANTITATION OF BACTERIA

M. Ramstorp, T.G.I. Ling and B. Mattiasson
Pure and Applied Biochemistry, Chemical Center, University of Lund, P.O.Box 740, S-220 07 Lund, Sweden

B-10 USE OF IMMUNOSPECIFIC AFFINITY CHROMATOGRAPHY IN THE PREPARATION OF KIDNEY MEMBRANE ANTIGENS FROM URINE, RELEVANT FOR DIAGNOSTIC PURPOSES

J.E. Scherberich, F. Falkenberg, H.J. Sachse, C. Gauhl, W. Mondorf and W. Schoeppe
Dept. of Nephrology, University of Frankfurt/Main, Dept. of Microbiol. and Immunology, Univers. Bochum, Dept. of Microbiol. and Virology, Univ. Frankfurt/Main, W.-Germany

B-11 DEVELOPMENT OF AN IMMUNORADIOMETRIC ASSAY FOR SERUM hCG WHICH APPLIES ANTIBODIES ISOLATED BY AFFINITY CHROMATOGRAPHY

M.F.G. Segers and C.M.G. Thomas
Dept. of Obstetrics and Gynaecology, Division of Clinical Chemistry, St. Radboud Hospital, Catholic University, Nijmegen, The Netherlands

B-12 APPLICATION OF AFFINITY CHROMATOGRAPHY IN IMMUNOASSAY

B. Terouanne and J.C. Nicolas
INSERM, U. 58 - 60, Rue de Navacelles, 34100 Montpellier, France

APPLICATIONS: BIOMEDICAL

B-18 IN VITRO STUDY OF THE EXTRACORPOREAL REMOVAL OF PLASMA LIPOPROTEINS BY AFFINITY ADSORPTION IN VIEW OF PREVENTION OF CARDIOVASCULAR COMPLICATIONS

E. Boschetti[+], M.D. Akonom-Duquesne, Y. Moschetto, M. Koffigan, C. Cachera and J.C. Fruchart
Laboratoire de Biochimie Chimique et CTB INSERM, Faculté de Pharmacie, 3 rue du Pr. Laguesse Lille, France
+ IBF 92390 Villeneuve-La-Garrene

B-19 DETOXIFICATION BY HAEMOPERFUSION ON IMMOBILIZED ANTIPHENOBARBITAL ANTIBODIES

L. Callegaro and U. Barbieri
Laboratory of Enzymology, Sorin Biomedica S.p.A., Saluggia (Vercelli), Italy

B-20 AGAROSE MACROBEADS (DIAMETER 1-5 MM) AS A SUPPORT PHASE FOR IMMOBILISED PROTEINS OR ENCAPSULATED ADSORBENTS FOR USE IN HAEMOPERFUSION

C.J. Holloway
Institute of Clinical Biochemistry, Medical School, Karl-Wiechert-Allee 9, D-3000 Hannover 61, W.-Germany

B-21 THE ACTIVITY OF SPECIFIC ANTIBODIES COMPLEXING WITH ANTIGEN IN A SERUM/PLASMA INCREASES DURING ADSORPTION WITH A SMALL AMOUNT OF SOLID PHASE PROTEIN A OF STAPHYLOCOCCUS AUREUS

S. Jonsson
Blood Centre, University Hospital, S-214 01 Malmö, Sweden

B-22 REPEATED HEMOPERFUSION OVER AGAROSE BEADS FOR AFFINITY CHROMATOGRAPHY OF PLASMA. BIOCOMPATIBILITY, TOXICOLOGY, AND ANTICOAGULANT REQUIREMENTS AS STUDIED IN VITRO, IN ANIMAL EXPERIMENTS, AND IN A CLINICAL TRIAL WITH IMMUNOGLOBULIN BINDING PROTEIN A-SEPHAROSE^{R}

S. Jonsson, L. Håkansson and U. Nylén
Blood Centre, Univ. Hospital, S-214 01 Malmö, Department of Oncology, Univ. Hospital, S-581 85 Linköping, and Gambro AB, S-220 10 Lund, Sweden

B-23 EXTRACORPOREAL AFFINITY ADSORPTION OF SPECIFIC IMMUNOGLOBULINS (G, M, or A) BY CONTINUOUS FLOW PLASMA PERFUSION OVER IMMUNOADSORBENTS -- USE OF THESE TECHNIQUES IN PATHOLOGIC DISORDERS

P.K. Ray, S. Raychaudhuri, S. Saha, D. McLaughlin, A. Idiculla, J.G. Bassett and D.R. Cooper
Dept. of Surgery, Alma Dea Morani Laboratory of Surgical Immunobiology, The Medical College of Pennsylvania and Hospital, Philadelphia, Pennsylvania 19129, USA

B-24 COVALENT ATTACHMENT OF DRUGS TO CELLULOSE DERIVATIVES. APPLICATION TO THE SYNTHESIS OF POLYMERIC FORMULATIONS FOR PROGRAMMED RELEASE OF DRUGS

F. Lapicque and E. Dellacherie
Laboratoire de Chimie Physique Macromoléculaire - E.R.A. 23, E.N.S.I.C., 1, rue Grandville, 54042 Nancy Cedex, France

APPLICATIONS: ORGANIC DYES - DYE-LIGANDS

B-30 THE USE OF REMAZOL YELLOW GGL-SEPHAROSE FOR THE EXTRACTION OF HUMAN SERUM PROTEINS

P.G.H. Byfield, S. Copping and M.R.A. Lalloz
Endocrinology Research Group, Clinical Research Centre, Harrow, Middlesex HA1 3UJ, UK

B-31 THE VALUE OF SCREENING IN DYE-LIGAND CHROMATOGRAPHY

Y. Hey, S. Qadri and P.D.G. Dean
Dept. of Biochemistry, University of Liverpool, P.O.Box 147, Liverpool L69 3BX, UK

B-32 METAL ION INTERACTIONS FOR PROTEIN PURIFICATION ON TRIAZINE DYE-MATRICES

P. Hughes, C. Lowe and R. Sherwood
P.H.L.S. Centre for Applied Microbiology & Research, Diagnostic Reagents Laboratory, Porton Down, Salisbury, England

B-33 DYE LIGAND CHROMATOGRAPHY IN THE PURIFICATION OF INHIBIN

E.H.J.M. Jansen, J. Steenbergen, F.H. de Jong and H.J. van der Molen
Department of Biochemistry II (division of Chemical Endocrinology), Erasmus University, P.O.Box 1738, Rotterdam, The Netherlands

B-34 PURIFICATION OF NADH-OXIDASE FROM STREPTOCOCCUS FAECALIS BY AFFINITY CHROMATOGRAPHY

J. Danzer, P.T. Kirch, K.-H. Kullmann and B. Limbach
Lehrstuhl für Allgemeine Chemie und Biochemie, Technische Universität München, D-8050 Freising-Weihenstephan, W.-Germany

B-35 THE INTERACTION OF TRIAZINE DYES WITH NUCLEOTIDE-DEPENDENT ENZYMES

C.R. Lowe, Y.D. Clonis and M. Goldfinch

Department of Biochemistry, University of Southampton, Southampton, SO9 3TU, England

B-36 HIGH PERFORMANCE LIQUID AFFINITY CHROMATOGRAPHY (HPLAC) OF PROTEINS ON CIBACRON BLUE F3G-A BONDED MICROPARTICULATE SILICA

C.R. Lowe[+], M. Glad[*], P.O. Larsson[*], S. Ohlsson[*], D.A.P. Small[△], A. Atkinson[△] and K. Mosbach[*]

+ Department of Biochemistry, University of Southampton, Southampton, SO9 3TU, England; * Pure and Applied Biochemistry, Chemical Centre, University of Lund, P.O.Box 740, S-220 07 Lund 7, Sweden and △ PHLS Centre for Applied Microbiology and Research, Porton Down, Salisbury, Wiltshire, England

B-37 THE AFFINITY ELECTRODE: SOME OBSERVATIONS ON THE BINDING OF PROTEINS TO TITANIUM ELECTRODES MODIFIED WITH TRIAZINE DYES

C.R. Lowe and M. Goldfinch

Department of Biochemistry, University of Southampton, Southampton, SO9 3TU, England

B-38 THE IMPORTANCE OF LIGAND PRESATURATION IN DYE LIGAND CHROMATOGRAPHY DEMONSTRATED BY DYE AFFINITY ELECTROPHORESIS

E.C. Metcalf[+], S. Qadri, B. Crow and P.D.G. Dean

Dept. of Biochemistry, University of Liverpool, P.O.Box 147, Liverpool L69 3BX, UK and + Powell & Scholefield Biochemicals, 38 Queensland St., Liverpool, UK

B-39 BLUE ULTROGEL AFFINITY CHROMATOGRAPHY USED AS A PURIFICATION STEP FOR THE EYE DERIVED GROWTH FACTOR (EDGF)

J. Plouet, D. Barritault and Y. Courtois

U 118 INSERM - ERA 842 C.N.R.S., 29 rue Wilhem, 75016 Paris, France

B-40 AFFINITY CHROMATOGRAPHY OF CALMODULIN ON IMMOBILIZED PHENOTHIAZINE ANALOGUES: PURIFICATION AND NATURE OF THE INTERACTIONS

C. Rochette-Egly[+], E. Boschetti[*] and J.M. Egly[Δ]

+ INSERM, U-61, 67300 Strasbourg, France; * Réactifs IBF-Pharmindustrie, 35, avenue Jean-Jaurès, 92390 Villeneuve-La-Garenne, France and Δ INSERM, U-184, Laboratoire de Génétique Moléculaire des Eucaryotes du CNRS, 11, rue Humann, 67085 Strasbourg Cedex, France

B-41 A THERMOSTABLE GLYCEROL DEHYDROGENASE FROM BACILLUS STEAROTHERMOPHILUS: BINDING TO PROCION DYES

M.D. Scawen, K.J. Bown and T. Atkinson

P.H.L.S. Centre for Applied Microbiology & Research, Diagnostic Reagents Laboratory, Porton Down, Salisbury, England

B-42 THE INTERACTION OF HEXOKINASE FROM BACILLUS STEAROTHERMOPHILUS WITH TRIAZINE DYES: A SIMPLE PURIFICATION PROCEDURE

M.D. Scawen, D.R. Hartwell, J. McArdell and T. Atkinson

P.H.L.S. Centre for Applied Microbiology & Research, Diagnostic Reagents Laboratory, Porton Down, Salisbury, England

B-43 TRIAZINE DYES AS AFFINITY LIGANDS IN HIGH PERFORMANCE LIQUID AFFINITY CHROMATOGRAPHY

D.A.P. Small[+], C.R. Lowe[*] and T. Atkinson[+]

+ P.H.L.S. Centre for Applied Microbiology & Research, Diagnostic Reagents Laboratory, Porton Down, Salisbury, England and * Department of Biochemistry, University of Southampton, Southampton, England

B-44 A NOVEL ACTIVE SITE LIGAND FOR HUMAN INTESTINAL ALKALINE PHOSPHATASE

D.G. Williams[+] and P.G.H. Byfield[*]

+ Division of Clinical Chemistry and * Endocrinology Research Group, Clinical Research Centre, Harrow, Middlesex HA1 3UJ, UK

B-45 THE POTENTIAL OF TRIAZINE DYES AS ADSORBANTS FOR THE PURIFICATION OF HUMAN LYMPHOBLASTOID INTERFERON

P.J. Neame, I. Parikh+ and R.T. Acton
Department of Microbiology, University of Alabama in Birmingham, Birmingham, Alabama 35294 USA and + Department of Molecular Biology, Burroughs Wellcome Research Laboratories, Research Triangle Park, North Carolina 27709, USA

RELATED TECHNIQUES

B-55 HIGH PERFORMANCE HYDROPHOBIC (REVERSE PHASE) LIQUID CHROMATOGRAPHY OF PROTEINS: FACTS AND ARTEFACTS EXEMPLIFIED WITH HUMAN FIBROBLAST INTERFERON AND OTHER PROTEINS

H.-J. Friesen [1,3], S. Stein[2] and S. Pestka[2]
1. Chemical Research Department, Hoffmann-La Roche Inc.; 2. Roche Institute of Molecular Biology, Nutley, N.J.; 3. Present address: Research Laboratories, Behringwerke AG, D-3550 Marburg, W.-Germany

B-56 HIGH PERFORMANCE LIQUID AFFINITY CHROMATOGRAPHY (HPLAC) WITH BORONIC ACID SILICA

M. Glad, S. Ohlson+, P.O. Larsson and K. Mosbach
Pure and Applied Biochemistry, Chemical Center, University of Lund, P.O.Box 740, S-220 07 Lund, Sweden
+ Present address: Gambro AB, P.O.Box 10101 Lund, Sweden

B-57 HIGH PERFORMANCE LIQUID AFFINITY CHROMATOGRAPHY (HPLAC) AS A TOOL FOR THE SEPARATION OF BIOMOLECULES

S. Ohlson+, M. Glad, P.O. Larsson and K. Mosbach
Dept. of Pure and Applied Biochemistry, Chemical Center, University of Lund, P.O.Box 740, S-220 07 Lund 7, Sweden
+ Present address: Gambro AB, Box 10101, S-220 10 Lund, Sweden

B-58 ASSESSMENT AND OPTIMIZATION OF SYSTEM PARAMETERS IN THE SIZE EXCLUSION SEPARATION OF PROTEINS AND ENZYMES ON DIOL-MODIFIED SILICAS BY MEANS OF HPLC

P. Roumeliotis, J. Kinkel, K. Unger, G. Taschank+ and G. Brunner+
Institut für Anorganische Chemie und Analytische Chemie, Johannes

Gutenberg-Universität, 6500 Mainz, GFR and + Institut für Immunologie, Johannes Gutenberg-Universität, 6500 Mainz, GFR

B-59 LECTIN INTERACTIONS WITH BRUSH-BORDER HYDROLASES OF HUMAN INTESTINE

N. Bertrand-Triadou and E. Audran
Unité de Recherche de Génétique Médicale Hôpital des Enfants-Malades 149, rue de Sèvres, 75730 Paris Cedex 15, France

B-60 AFFINITY PARTITION BASED ON THE USE OF HEAVILY MODIFIED "SEPARATOR LIGANDS"

C. Glad, M. Ramstorp, T.G.I. Ling and B. Mattiasson
Department of Pure and Applied Biochemistry, Chemical Center, University of Lund, P.O.Box 740, S-220 07 Lund, Sweden

B-61 IMMOBILIZED MODIFIERS FOR PROTEINS ("IMPs")

W.H. Scouten, C. Lewis, J. Treas, R. Haller and W. Iobst
Department of Chemistry, Bucknell University, Lewisburg, PA 17837, USA

B-62 APPLICATION OF IMMOBILIZED PROTEASES TO PEPTIDE SYNTHESIS

H.-D. Jakubke, R. Bullerjahn, M. Hänsler and A. Könnecke
Department of Biochemistry, Karl Marx University Leipzig, DDR-7010 Leipzig, Talstr. 33, German Democratic Republic

PARTICIPANTS

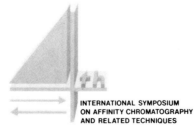

INTERNATIONAL SYMPOSIUM
ON AFFINITY CHROMATOGRAPHY
AND RELATED TECHNIQUES

Accinni, R., Dep. of Biochemistry, Istituto Richerche Cardiovascolari, Università di Milano, Via F. Sforza 35, 20122 Milano, Italy

Albert, R., Janssen Pharmaceutica, Turnhoutseweg 30, 2340 Beerse, Belgium

Allenmark, S.G., Clinical Research Centre, Linköping University, S-58185 Linköping, Sweden

Ariëns, E.J., Department of Pharmacology, Faculty of Medicine, Katholieke Universiteit, Nijmegen, The Netherlands

Armitage, G.D., Sittingbourne Research Centre, Sittingbourne, ME9 8AG Kent, United Kingdom

Arús, C., Dto. Bioquimica, Univ. Autònoma Barcelona, Bellaterra, Spain

Atkinson, A., P.H.L.S. Centre for Applied Microbiology and Research, Diagnostic Reagents Laboratory, Porton Down, Salisbury, SP4 0JG Wilts, United Kingdom

Ayers, J.S., Massey University, Chemical/Biochemical Department, Palmerston North, New Zealand

Batz, H.-G., Boehringer Mannheim GmbH, Bahnhofstrasse 9 - 15, 8132 Tutzing, West Germany

Bayol, A.J.M., Sanofi Recherche, 195, Route d'Espagne, 31024 Toulouse, France

Beliaev, S., Institute of Biorganic Chemistry, Ul. Vavilova 32, Moscow, USSR

Bellon, A., O.L. Vrouw Ziekenhuis, Moorselbaan, 9300 Aalst, Belgium

Biezen, J. v.d., Organon Scientific Development Group, P.O.Box 20, 5340 BH Oss, The Netherlands

Birke, F., Boehringer Ingelheim AG, D-6507 Ingelheim/Rhein, West Germany

Blazey, N.D., N.Z. Pharmaceuticals, Box 1869, Palmerston North, New Zealand

Bloemendal, H., Department of Biochemistry, Faculty of Science, Katholieke Universiteit, Nijmegen, The Netherlands

Boniolo, A., Sorin Biomedica SPA, 13040 Saluggia (VC), Italy

Bonnafous, J.C., Laboratoire de Biochimie des Membranes, ER CNRS 228, 8, Rue de l'Ecole Normale, F-34075 Montpellier, France

Bonnet, M.C., Sanofi Recherche, 195, Route d'Espagne, 31024 Toulouse, France

Borg, H., KabiVitrum AB, 112 87 Stockholm, Sweden

Boschetti, E., Réactifs IBF-Pharmindustrie, 35, Avenue Jean-Jaurès, 92390 Villeneuve-la-Garenne, France

Box, M.E., Limburgs Universitair Centrum, Universitaire Campus, B-3610 Diepenbeek, Belgium

Brackel, A.L.G., Laboratory of Organic Chemistry, Toernooiveld, 6525 ED Nijmegen, The Netherlands

Broeksteeg, A.P.J., Akzo Pharma, Wethouder van Eschstraat 1, 5340 BH Oss, The Netherlands

Brümmer, W., Biochem. Forschung, Postfach 4119, D-6100 Darmstadt 1, West Germany

Bureau, D., Centre Nationale de la Recherche Scientifique, Laboratoire de Physiologi 4 ter, Route des Gardes, 92190 Meudon, France

Byfield, P.G.H., Clinical Research Centre, Watford Road, Harrow, Middlesex, HA1 3UJ, United Kingdom

Byoux, K.E., Institut für Lebensmittelchemie, Stockholmer Strasse 4, 5300 Bonn 1, West Germany

Callegaro, L., Sorin Biomedica SPA, 13040 Saluggia (VC), Italy

Carlsen, S., Nordisk Insulinlab., Sautesvej 11B, DK-2820 Gentofte, Denmark

Caron, M., Université Paris-Nord, 74, Rue Marcel Cachin, 93000 Bobigny, France

Casati, P., Galibia SPA, Via Cilicia 53, 00179 Roma, Italy

Chang, T.M.S., McGill University, 3655 Drummond Street, Montreal, PQ Canada H 1Y6, Canada

Clifford, K.H., Shell Research Ltd., Sittingbourne, Kent United Kingdom

Cornale, P., Gruppo Lepitit SPA, Via Lepetit 8, Milano, Italy

Cueille, G., Rhône-Poulenc Industries, 13, Quai Jules Guesde, 94400 Vitry-sur-Seine, France

Čoupek, J., Laboratory Instruments Works, 162 03 Prague 6, Na Okraji 335 ČSSR, Czechoslovakia

Dam van, J.E.G., Rijksuniversiteit Utrecht, Croesestraat 79, 3522 AD Utrecht, The Netherlands

Dams, G.M., Baroniestraat 7, Boxtel, The Netherlands

Dann, J.G., Wellcome Research Ltd., Langley Court Beckenham, BR3 3BS, Kent, United Kingdom

Dautrevaux, M., Faculté de Médecine, Place du Verdun, 59045 Lille Cédex, France

Dean, P.D.G., Liverpool University, P.O.Box 147, L69 3BX Liverpool, United Kingdom

De Lisi, C., National Institutes of Health, Public Health Service, Bethesda, Md 20034, USA

Dellacherie, E., Ensic Era no 23, 1, Rue Grandville, 54042 Nancy Cédex, France

Dorey, C., Université de Paris 6, 15, Rue de l'Ecole de Médecine, 75006 Paris, France

Duimel, W.J., Centraal Laboratorium van de Bloedtransfusiedienst, P.O.Box 9190, 1006 AD Amsterdam, The Netherlands

Dürholt, M., Pure & Applied Biochemistry, Chemical Center, P.O.Box 740, S-220 07 Lund, Sweden

Egly, J.M., INSERM U-184, Laboratoire de Génétique Moléculaire des Eucaryotes du C.N.R.S., 11, Rue Humann, 67085 Strassbourg, France

Ekeke, G.I., School of Chemical Sciences, University of Port Harcourt, PMB 5323, Port Harcourt, Nigeria

Eketorp, R., Department of Biochemistry, KabiVitrum AB, S-11287 Stockholm, Sweden

Emöd, I., Institut Pasteur, Département de Biochimie et Génétique Microbienne, Unité de Chimie des Proteines, 28, Rue du Docteur Roux, 75724 Paris Cédex 15, France

Eveleigh, J.W., E.I. du Pont de Nemours and Company, E.334 Experimental Station, Wilmington, Delaware 19898, USA

Ezban, M., Nordisk Insulinlab., Sautesvej 11B, 2820 Gentofte, Denmark

Faber, L.H., Miles Nederland, P.O.Box 217, Weesp, The Netherlands

Fargeaud, D., IFFA Mérieux, 254, Rue M. Mérieux 69007, BP n° 9-69342 Lyon Cédex 2, France

Faure, A., C.N.T.S., BP 100, 91400 Orsay-Courtaboeuf, France

Faye, L., Faculté des Sciences, Place Emile Blondel, 76130 Mont Saint Aignan, France

Fischer, E.A., Hoffmann-La Roche & Co., Dept. Diagnostics, Bau 619, Postfach, CH-4002 Basel, Switzerland

Flora, R.M., Millipore Corporation, 80, Ashby Road, Bedford Ma. 01730, USA

Folkesson, R., Institute of Pharmacology, Box 573 BMC, 75123 Uppsala, Sweden

Formstecher, P., Faculté de Médecine, 1, Place de Verdun, 59045 Lille Cédex, France

Fourcart, J., Réactifs IBF, 35, Avenue Jean-Jaurès, 92390 Villeneuve-la-Garenne, France

Fraaye, J.G.E.M., a/d Rijn 12, 6701 PB Wageningen, The Netherlands

Franz, H., S.I.I.N.M., 1120, Berlin-Weissensee, Klem. Gottw. Allee 317/321, DD

Friesen, H.J. Behringwerke AG, Postfach 1140, D-3350 Marburg, West Germany

Fromant, E., Réactifs IBF, 35, Avenue Jean-Jaurès, 92390 Villeneuve-la-Garenne, France

Fulton, S.P., Amicon Corporation, 182, Conant Street, Danvers, Ma 01923, USA

Gabel, D., University of Bremen, Postfach 330440, D-2800 Bremen 33, West Germar

Galunsky, B., University of Bremen, Postfach 330440, D-2800 Bremen 33, West Germany

Geer van der, J., Elsevier Scientific Publishing Company, P.O.Box 330, 1000 AH Amsterdam, The Netherlands

Gelsema, W.J., Analytisch Chemisch Laboratorium, Rijksuniversiteit Utrecht, Croesestraat 77A, 3522 AD Utrecht , The Netherlands

Germinario, G., European Patent Office, Patentlaan 2, Rijswijk (ZH), The Netherlar

Geyer, E., Immuno AG, Industriestrasse 131, A-1220 Wien, Austria

Giessen, K.M. v.d., Applied Science Europe B.V., Röntgenstraat 18, Oud Beijerlar The Netherlands

Glad, J.M., University of Lund, P.O.Box 740, S-220 07 Lund 7, Sweden

Glad, C., Pure & Applied Biochemistry, Chemical Centre, P.O.Box 740, 220 07 Lund, Sweden

Goldberg, E.P., MAE 217, University of Florida, Gainesville, Florida 32611, USA

Goldstein, L., Tel Aviv University, Department of Biochemistry, Ramataviv 69978, Israel

Gorp, J.A. v., Organon International B.V., Kloosterstraat 6, 5340 BH Oss, The Netherlands

Grau, U., Hoechst AG, Pharmaforschung Biochemie, D-6230 Frankfurt 80, West Germany

Gribnau, T.C.J., Organon Scientific Development Group, P.O.Box 20, 5340 BH Oss, The Netherlands

Grund, E., Pharmacia Fine Chemicals AB, Box 175, S-75104 Uppsala 1, Sweden

Guzzi, U., Lepetit SPA, Via Lepetit 8, Milano, Italy

Haas, R., Waters Associates B.V., P.O.Box 166, 4870 AD Etten-Leur, The Netherlands

Hager, H.J., Hoffmann-La Roche, Nutley, N.J. 07110, USA

Haglund, R., Pharmacia Fine Chemicals AB, Box 175, S-75104 Uppsala 1, Sweden

Haug, E., Biotest-Serum-Institut GmbH, Flughafenstrasse 4, D-6000 Frankfurt/ Main 73, West Germany

Henon, M.P., Centre Transfusion Sanguine, 21, Rue Camille Guèrin, 59000 Lille, France

Hetzl, E., Immuno AG, Industriestrasse 131, A-1220 Wien, Austria

Hey, Y., Liverpool University, P.O.Box 147, L69 3BX Liverpool, United Kingdom

Hinkkanen, A., Biochemisches Institut, Albert Ludwigs Universität, Hermann Herderstrasse 7, D-7800 Freiburg, West Germany

Hjertén, S., Institute of Biochemistry, Biomedical Center, Box 576, S-751 23 Uppsala, Sweden

Hochuli, E., Hoffmann-La Roche AG, Department ZFE 62/206, CH-4002 Basel, Switzerland

Holloway, C.J., Institut für Klinische Biochemie, OE 4342, Karl Wiechert Allee 9, D-3000 Hannover 61, West Germany

Houdenhoven van, F.E.A., Diosynth BV, P.O.Box 20, 5340 BH Oss, The Netherlands

Hughes, P., PHLS Diagnostic Reagents Laboratory, Porton Down, Salisbury, Wilts SP4 OJG, United Kingdom

Inglis, R., Pharmacia Fine Chemicals AB, Box 175, S-75104 Uppsala 1, Sweden

Inman, J.K., National Institutes of Health, Building 10, Room 11N252, Bethesda, Md 20205, USA

Jack, G.W., PHLS Centre for Applied Microbiology & Research, Porton Down, Salisbury, Wilts SP4 OJG, United Kingdom

Jakubke, H.-D., Sektion Biowissenschaften der Karl Marx Universität, Bereich Biochemie, Talstrasse 33, DDR-7010 Leipzig, DDR

Jandl, J., Veterinary Research Institute, Hudcova 70, 621 32 Brno, Czechoslovakia

Jansen, E.H.J.M., Erasmus Universiteit, P.O.Box 1738, 3000 DR Rotterdam, The Netherlands

Janson, J.C., Pharmacia Fine Chemicals AB, P.O.Box 175, S-75104 Uppsala, Sweden

Jennissen, H.P., Institut für Physiologische Chemie der Ruhr-Universität Bochum, Postfach 10 21 48, D-4630 Bochum, West Germany

Jonsson, U.R.S., Blood Center, University Hospital, S-21401 Malmö , Sweden

Jozefonvicz, J., Université Paris-Nord, Centre Scientifique et Polytechnique, Laboratoire de Recherche sur les Macromolecules, C.S.P. Avenue J.B. Clément, 93430 Villetaneuse, France

Jørgensen, P.N., Novo Industri A/S, Novo Allé, DK-2880 Bagsvaerd, Denmark

Kahane, I., The Hebrew University Israel, Department of Membrane and Ultrastructural Research, Hadassah Medical School, P.O.Box 1172, Israel

Kálal, J., Institute of Macromolecular Chemistry, Czechoslovak Academy of Science Prague 6 - Petriny, Czechoslovakia

Kamps, K.M.P., Department of Biochemistry, Geert Grooteplein Noord 21, 6500 HB Nijmegen, The Netherlands

Kasche, V., University of Bremen, Physikalische Biologie, Postfach 330440, D-2800 Bremen 33, West Germany

Kelemen, A., Chemical Works G. Richter, P.O.Box 27, H-1475, Budapest 10, Hungary

Kester, H.C.M., Landbouwhogeschool Wageningen, Generaal Foulkesweg 53, 6703 BM Wageningen, The Netherlands

Kirch, P.T., E. Merck Biochem. Res., Postfach 4119, D-6100 Darmstadt, West Germany

Kofod, B., Dakopatts A/S, 22 Guldborgvej, Denmark

Koman, A., Department of Pharmacology Biomedicum, University of Uppsala, Box 573, S-75123 Uppsala, Sweden

Krämer, D., Research Laboratories/Röhm GmbH, Postfach 4242, D-6100 Darmstadt, West Germany

Kremers, J.H.W., LKB-Produkten BV, P.O.Box 216, 2700 AE Zoetermeer, The Netherlands

Krisam, G., Gambro Dialysatoren KG, P.O.Box 1323, D-7450 Hechingen, West Germany

Kruse, V., Novo Research Institute, Novo Allé, DK-2880 Bagsvaerd, Denmark

Kuijpers, L.P.C., Organon International BV, P.O.Box 20, 5340 BH Oss, The Netherlands

Kula, M.R., Gesellschaft für Biotechnologische Forschung mbH, Mascheroderweg 1, D-3300 Braunschweig, West Germany

Kågedal, L., Pharmacia Fine Chemicals AB, Box 175, S-75104 Uppsala, Sweden

Lalu, K., Department of Anatomy, University of Kuopio, P.O.Box 138, SF-70101 Kuopio 10, Finland

Lampelo, S., Department of Anatomy, University of Kuopio, P.O.Box 138, SF-70101 Kuopio 10, Finland

Lange, N.E.K., Swiss Ferment Ltd., Mülhauserstrasse 70, CH-4056 Basel, Switzerland

Lapicque, F., Ensic Era no. 23, 1. Rue Grandville, 54042 Nancy Cédex, France

Lasch, J., Physiol.-Chem. Institut, Hollystrasse 1, PSF 184, DDR-4020 Halle/ Salle, DDR

Laszlovsky, I., Chemical Works G. Richter Ltd., P.O. Box 27, H-1475 Budapest 10, Hungary

Lehmann, Gambro Dialysatoren KG, P.O. Box 1323, D-7450 Hechingen, West Germany

Lehmann, M., Kernforschungszentrum, Weberstrasse 5, D-7500 Karlsruhe 1, West Germany

Linde van der, P.C.G., Botanical Laboratory, Nonnesteeg 3, Leiden, The Netherlands

Ling, T.G.I., Pure & Applied Biochemistry, Chemical Center, P.O.Box 740, S-220 07 Lund, Sweden

Liveyns, R., Smith Kline-RIT, 89, Rue de l'Institut, B-1330 Rixensart, Belgium

Lowe, C.R., Department of Biochemistry, Bassett Crescent East, S09 3TU Southampton, United Kingdom

Lustenberger, P., Laboratoire de Biochimie, Faculté de Médicine, 1, Rue G. Veil, 44035 Nantes, France

Lyklema, J., Laboratorium voor Fysische en Kolloidchemie, de Dreyen 6, 6703 BC Wageningen, The Netherlands

Madden, J.K., Unilever Research, Colworth House, Sharnbrook, MK44 1LQ Bedford, United Kingdom

Mallia, A., Pierce Chemical Co., P.O.Box 117, Rockford, Illinois 61105, USA

Mandelli, C., F.LLI Lamberti SPA, Via Piave 18, 21041 Albizzate (VA), Italy

Manecke, G., Institut für Organische Chemie, F.U.B. Takustrasse 3, D-1000 Berlin 33, West Germany

Marchand, J., Centre de Recherches, Rue du Professeur Blayac, 13009 Marseille, France

Margaritella, P., Immuno AG, Industriestrasse 72, 1220 Wien, Austria

Maroux, S., CBM-CNRS, 31, Chemin Joseph Aiguier, 13274 Marseille Cédex, France

Marquet, A., Instituto de Immunologia, Velázques 144, Madrid-6, Spain

Metcalf, E.C., Powell & Scholefield Biochemicals Ltd., 38, Queensland Street, L7 3JG Liverpool, United Kingdom

Meuwissen, J.A.T.P., Laboratory of Hepatology, Campus Gasthuisberg, Building Research and Teaching, Herestraat 49, B-3000 Leuven, Belgium

Meyer, H.E., Diabetes-Forschungsinstitut, Auf'm Hennekamp 65, D-4000 Düsseldor West Germany

Migliaccid, M., Institute of Biochemistry, Via Saffi 2, 61029 Urbino, Italy

Miguel, E., C.E.P.S.A., Centro de Investigacion, C/Picos de Europa, 7, S. Fern. de Henares, Madrid, Spain

Mohr, P., Zentralinstitut für Molekularbiologie, Lindenbergerweg 70, DDR-1115 Berlin-Buch, DDR

Mol, C.R., Waters Associates BV, Postbus 166, 4870 AD Etten-Leur, The Netherlan

Mooi, G.J., Boehringer Mannheim BV, Postbus 828, 1000 AV Amsterdam, The Netherlands

Morales, T., Kernforschungszentrum, Weberstrasse 5, D-7500 Karlsruhe 1, West Germany

Mosbach, K., Pure and Applied Biochemistry, Chemical Center, P.O.Box 740, S-22 07 Lund 7, Sweden

Müller, W., University of Bielefeld, P.O.Box 8640, D-4800 Bielefeld, West German

Muszyńska, G., Institute of Biochemistry and Biophysics, 36, Rakowiecka St., 02-532 Warszawa, Poland

Myöhänen, T.A., Liverpool University, P.O.Box 147, L69 3BX Liverpool, United Kingdom

Månsson , M., Pure & Applied Biochemistry, Chemical Center, P.O.Box 740, S-220 07 Lund, Sweden

Neame, P.J., 814 Diabetes Hospital UAB, University Station, Birmingham A1 35294 USA

Nelboeck, M., Boehringer Mannheim, 21, Bareiselweg, D-8132 Tutzing, West Germany

Nilsson, R., The Wallenberg Laboratory, Box 7031, 220 07 Lund, Sweden

Nilsson, K., Pure and Applied Biochemistry, Chemical Center, University of Lund, P.O.Box 740, 220 07 Lund, Sweden

Nishikawa, A.H., Hoffmann-La Roche Inc., Biopolymer Res B-102, Nutley, N.J. 07110, USA

Niss, U., Gambro AB, Box 10101, S-220 10 Lund, Sweden

Nivard, R.J.F., Department of Organic Chemistry, Katholieke Universiteit, Toernooiveld, 6525 ED Nijmegen, The Netherlands

Nogués, M.V., Dept. Bioquímica, Univ. Autònoma de Barcelona, Bellaterra, Barcelona, Spain

Noordeloos, P.J., LKB-Produkten BV, P.O.Box 216, 2700 AE Zoetermeer, The Netherlands

Normier, G., Laboratoire Pierre Fabre, 17, Avenue Jean Moulin, 81106 Castres, France

Nortier, B., Amicon BV, Mechelaarstraat 11, 4903 RE Oosterhout (NB), The Netherlands

Nylen, U., Gambro AB, Box 10101, 220 10 Lund, Sweden

Ohlsen, S., Gambro AB, Box 10101, 220 10 Lund, Sweden

Onkelinx, E., Limburgs Universitair Centrum, Universitaire Campus, B-3610 Diepenbeek, Belgium

Oss van, C.J., State University of New York at Buffalo, 207 Sherman Hall, Buffalo, NY 14214, USA

Ouderaa van der, F.J.G., Unilever Research Laboratorium, P.O.Box 114, 3130 AC Vlaardingen, The Netherlands

Over, J., Centraal Laboratorium van het Nederlandse Rode Kruis, Plesmanlaan 125, 1006 AD Amsterdam, The Netherlands

Parikh, I., The Wellcome Research Laboratories, 3030 Cornwallis Road, Research Triangle Park, NC 27709, USA

Patchornik, A., Weizmann Institute of Science, Department of Organic Chemistry, Rehovot, Israel

Pels, J., Landbouwhogeschool Wageningen, Generaal Foulkesweg 53, 6703 BM Wageningen, The Netherlands

Penke, B., University of Szeged, Dom Tèr 8, H-6720 Szeged, Hungary

Pestel, J., Institut Pasteur, 15, Rue Camille Guérin, 59019 Lille Cédex, France

Peyrouset, A., Elf Aquitaine, Centre de Recherche de LACQ, 64170 Artix, France

Pierce, R.J., Institut Pasteur, 15, Rue Camille Guérin, 59019 Lille Cédex, France

Pieterson, W.A., Diosynth BV, P.O.Box 20, 5340 BH Oss, The Netherlands

Plum, I.G., Novo Industri A/S, Novo Allé, DK-2880 Bagsvaerd, Denmark

Pluzek, K.J., DAKO-Immunoglobulins A/S, Guldborgvej 22, DK-2000 Copenhagen F, Denmark

Porath, Institute of Biochemistry, Uppsala University Biomedical Center, Box 576, Uppsala, Sweden

Ramstorp, M.I., Pure & Applied Biochemistry, Chemical Center, University of Lund, P.O.Box 740, S-220 07 Lund, Sweden

Räsänen, V., OY Medix AD, P.O.Box 819, SF-00101 Helsinki 10, Finland

Rauenbusch, E., Bayer AG, Department of Biochemistry, Friedrich-Ebert-Strasse 217 - 319, D-5600 Wuppertal 1, West Germany

Ravestein, P., Plataanweg 18, 3053 LP Rotterdam, The Netherlands

Reum, L., Physiol.-Chem. Institut I, Deutschhausstrasse 1-2, D-3550 Marburg/L, West Germany

Riikola, L.H., OY Medix AD, P.O.Box 819, SF-00101 Helsinki 10, Finland

Robertson, B.W., Shell Research Ltd., Sittingbourne Research Center, Sittingbourne Kent, United Kingdom

Röder, A.H., Boehringer Mannheim, Bahnhofstrasse 5, D-8132 Tutzing, West Germar

Rombouts, F.M., Landbouwhogeschool Wageningen, Department of Food Science, De Dreyen 12, 6703 BC Wageningen

Rosa, M., Unilever Research, Den Haag, The Netherlands

Roumeliotis, P., Institut für Anorganische Chemie, Johannes Gutenberg Universität, D-6500 Mainz, West Germany

Rousseauz, R., U-124 INSERM, BP 311, 59020 Lille Cédex, France

Rouseaux, J., U-124 INSERM, BP 311, 59020 Lille Cédex, France

Rouze, P., INRA, Station de Virologie & Immunologie, 78850 Thiverval-Grignon, France

Saint-Blancard, J., C.T.S.A. "Jean Julliard", 1, Rue Raôul Batany, 92141 Clamart, France

Salier, J.P., U-78 INSERM , 543, Chemin de la Bretagne, 76230 Bois-Guilaume, France

Sand, G., Hôpital Saint Pierre, 322, Rue Haute, B-1000 Bruxelles, Belgium

Scawen, M.D., PHLS Centre for Applied Microbiology and Research, Porton Down, SP4 0JG Salisbury, United Kingdom

Scherberich, J.E., Department of Nephrology, University of Frankfurt, Theodor Stern Kai 7, D-6000 Frankfurt am Main, West Germany

Schmidt, K., Institut für Pharmakodynamik, Universitätsplatz 2, A 8010 Graz, Austria

Schnee, R., Research Laboratories Röhm GmbH, Postfach 4242, 61 Darmstadt, West Germany

Schneider, P., Novo Industri A/S, Novo Allé, DK-2880 Bagsvaerd, Denmark

Schroeder, D.D., Cutter Laboratories Inc., 4th and Parker Street, Berkely, Cal. 94710, USA

Schroeder, P., Fa. Dr. Rentschler, P.O.Box 320, 7958 Laupheim, West Germany

Schutt, H., Bayer AG, VE Biochemie, Friedrich-Ebertstrasse 217 - 319, D-5600 Wuppertal 1, West Germany

Schutyser, J.A.J., Akzo Corporate Research, Velperweg 76, 6800 AB Arnhem, The Netherlands

Schuurs, A.H.W.M., Organon International BV, P.O.Box 20, 5340 BH Oss, The Netherlands

Scopes, R.K., Department of Biochemistry, La Trobe University, Bundoora Victoria 3083, Australia

Scouten, W.H., Bucknell University, Department of Chemistry, Lewisburg, Pa 17837, USA

Segers, M.F.G., Laboratorium voor Obstetrie en Gynaecologie, Geert Grooteplein Zuid, 6500 HB Nijmegen, The Netherlands

Seris, J.L., Elf Aquitaine, 64170 Artix, France

Shami, Y., Inter-Yeda Ltd., Kiryat Weizmann, Rehovot 76110, Israel

Shayn, H., Pharmacia Fine Chemicals AB, Box 175, S-75104 Uppsala 1, Sweden

Sherrington, D.C. Dept. of Pure & Applied Chemistry, University of Strathclyde, Glasgow G1 1XL, United Kingdom

Sherwood, R., PHLS Diagnostic Reagents Laboratory, Porton Down, Salisbury, SP4 0JG Wilts, United Kingdom

Siegbahn, N., DPAB University of Lund, Chemical Center, P.O.Box 740, 220 07 Lund, Sweden

Sijpesteyn, T., Elsevier Scientific Publishing Company, P.O.Box 330, 1000 AH Amsterdam, The Netherlands

Sitrin, R., Smith Kline Laboratories F-32, 1500 Spring Garden St., P.O.Box 7929, Philadelphia, Pa 19101, USA

Small, D.A.P., PHLS CAMR, Porton Down, Salisbury, SP4 0JG Wilts, United Kingdom

Sommeren van, A.P.G., Organon Scientific Development Group, P.O.Box 20, 5340 BH Oss, The Netherlands

Sparrman, M., Institute of Biochemistry, University of Uppsala, Box 576, S-75123 Uppsala, Sweden

Štamberg, J., UMCH CSAV, Heyrovský sq. 1888, 16206 Praha 6, Czechoslovakia

Stocchi, V., Univ. Inst. Biochemistry, Via Saffi 2, 61029 Urbino, Italy

Sturgeon, R.J., Heriot-Watt University, Department of Brewing and Biological Sciences, EH1 1HY Edinburgh, United Kingdom

Sulkowski, E., Rosswell Park Memorial Institute, 666, Elm Street, Buffalo, N.Y. 14263, USA

Tardy, M., Institut Mérieux, Marcy l'Etoile, 69260 Charb. Les Bains, France

Terouanne, B., INSERM U-58, 60, Rue de Navacelles, 34100 Montpellier, France

Thom, D., Unilever Research, Colworth House, Sharnbrook MK44 1LQ Bedford, United Kingdom

Tocilescu, F., Paul Ehrlich Institut, Paul Ehrlichstrasse 40 - 42, D-6000 Frankfurt/ Main, West Germany

Toll, P.J.M., Faculty of Science, Toernooiveld, 6525 ED Nijmegen, The Netherlands

Törnblad, B.K., Pharmacia Diagnostics AB, Box 17, S-751 03 Uppsala 1, Sweden

Traas, D.W., Gaubius Instituut TNO, Herenstraat 5d, 2313 AD Leiden, The Netherlanc

Turková, J., Institute of Organic Chemistry and Biochemistry, Flemingovo Namesti 2, 166 10 Praha 6, Czechoslovakia

Uitzetter, J.H.A.A., Landbouwhogeschool Wageningen, Generaal Foulkesweg 53, 6703 BM Wageningen, The Netherlands

Unger, K.K., Institut für Anorganische and Analytische Chemie, Joh. Gutenberg Universität, D-6500 Mainz, West Germany

Varalli, M., Farmitalia Carlo Erba, RS Diagnostici, 20090 Rodano-Milano, Italy

Varró, R., Human Institut, Szállás-u. 5, H-1107 Budapest, Hungary

Verhofstad, A.A.J., Department of Anatomy and Embryology, P.O. Box 9101, 6500 HB Nijmegen, The Netherlands

Vidal, H., Centre de Recherche CLIN-MIDY, Rue du Professeur Blayac, 34082 Montpellier Cédex, France

Vijayalakshmi, Institut de Technologie des Surfaces Actives, B.P. 233, 60206 Compiègne, France

Vikelsoe, J., Nordisk Insulin Laboratory, Niels Steensenvej 1, 2820 Gentofte, Denmark

Visser, J., Agricultural University, Generaal Foulkesweg 53, 6703 BM Wageningen, The Netherlands

Vliet van, Th.B., Botanical Laboratory, Nonnensteeg 3, Leiden, The Netherlands

Vretblad, P., Pharmacia Fine Chemicals AB, Box 175, S-75104 Uppsala, Sweden

Waart van der, M., Organon International B.V., Kloosterstraat 6, 5340 BH Oss, The Netherlands

Wezel van, A.L., Rijksinstituut voor de Volksgezondheid, Postbus 1, 3720 BA Bilthoven, The Netherlands

White, D.D., Organon Laboratories Ltd., Newhouse Industrial Estate, Lanarkshire, Scotland, United Kingdom

Wiedner, H., Boehringer Mannheim GmbH, Werk Penzberg Abt. N-CSM, 8122 Penzberg, West Germany

Wijnendaele van, F.V.W., Smith Kline-Rit, Rue de l'Institut, 1330 Rixenart, Belgium

Wilchek, M., The Weizmann Institute of Science, Department of Biophysics, Rehovot, Israel

Wulff, G., Universität Düsseldorf, Universitätsstrasse 1, D-4000 Düsseldorf, West Germany

Ysewijn-van Brussel, K.A.R.N., Academisch Ziekenhuis, De Pintelaan 135, 9000 Gent, Belgium

Zoest van, R., Pharmacia BV, Koperwerf 16, 2544 EN Den Haag, The Netherlands

SUBJECT INDEX

The numbers listed in this index refer to the first page of the relevant articles

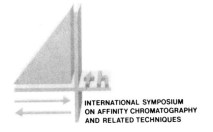

INTERNATIONAL SYMPOSIUM
ON AFFINITY CHROMATOGRAPHY
AND RELATED TECHNIQUES

A

B

H

I

M

N

O

P

Q

R

S